MEMBRANES AND
MEMBRANE PROCESSES

MEMBRANES AND MEMBRANE PROCESSES

Edited by

E. Drioli

University of Calabria
Cosenza, Italy

and

M. Nakagaki

Kyoto University
Kyoto, Japan

PLENUM PRESS • NEW YORK AND LONDON

7346-7583

CHEMISTRY

Library of Congress Cataloging in Publication Data

Europe–Japan Congress on Membranes and Membrane Processes (1984: Stresa, Italy)
  Membranes and membrane processes.

  "Proceedings of the Europe–Japan Congress on Membranes and Membrane Processes, held June 18–22, 1984, in Stresa, Italy"–T.p. verso.
  Includes bibliographies and index.
  1. Membranes–Congresses. I. Drioli, E. II. Nakagaki, Masayuki, 1923–   . III. Title.
TP159.M4E57 1984   660.2′842   86-5076
ISBN 0-306-42270-0

Proceedings of the Europe–Japan Congress on Membranes and
Membrane Processes, held June 18–22, 1984, in Stresa, Italy

© 1986 Plenum Press, New York
A Division of Plenum Publishing Corporation
233 Spring Street, New York, N.Y. 10013

Printed in the United States of America

PREFACE

During the past two decades Membrane Science and Technology has made tremendous progress and has changed from a simple laboratory tool to large scale processes with numerous applications in Medicine and Industry.

In this volume are collected papers presented at the First Europe-Japan Congress on Membrane and Membrane processes, held in Stresa in June 1984. Other contributions to the Conference will be published in a special issue of the Journal of Membrane Science.

This Conference was organized by the European Society of Membrane Science and Technology and the Membrane Society of Japan, to bring together European Scientists and Engineers face to face with their colleagues from Japan; in both countries membrane processes will play a strategic role in many industrial areas in the 1990s, as predicted by the Japanese project for Next Generation Industries and by the EEC Project on Basic Technological Research (BRITE).

The large number of participants, of about four hundred from twenty-six countries including USA, Australia, China and Brazil, the quality of the Plenary Lectures and Scientific Communications made the Conference a significant international success.

We are deeply grateful to the following organizations that contributed financial support for the Conference:

Ajinomoto Co., Japan
Alitalia SpA, Italy
Asahi Chemical Industry Co. Ltd., Japan
Asahi Glass Co. Ltd., Japan
Bellco Spa, Italy
Culligan Int., USA
Daicel Chemical Industries Ltd., Japan
Daiichi Seiyaku Co. Ltd., Japan
De Martini Spa, Italy
O.de Nora Imp. Elettrochimici Spa, Italy
Diemme Filtri SpA, Italy

Eisai Co. Ltd., Japan
ENEA, Italy
ENKA AG, West Germany
Farmitalia Spa, Italy
Fuji Standard Research Inc., Japan
Gelman Sciences Japan, Ltd., Japan
IRCHA, France
Istituto Bancario San Paolo di Torino, Italy
Japan Gore-Tex Inc., Japan
Japan Organo Co. Ltd., Japan
Kao Corporation, Japan
Kuraray Co. Ltd., Japan
Kurita Water Industries Ltd., Japan

Mitsubishi Rayon Co. Ltd., Japan
Montedison Spa, Italy
Nomura Microscience Co. Ltd., Japan
Nitto Electric Industrial Co. Ltd., Japan
Pierrel Spa, Italy
Shionogi & Co. Ltd., Japan
Snia Fibre Spa, Italy

Snow Brand Milk Products Co. Ltd., Japan
Solvay & Cie, Belgium
Suntory Co., Japan
Taisho Pharmaceutical Co. Ltd., Japan
Tokuyama Soda Co. Ltd., Japan
Toray Industries Inc., Japan
Toyobo Co. Ltd., Japan

We acknowledge also in particular the Italian Oriented Program on Fine Chemistry (CNR) and its Director Prof. L. Caglioti for the continuous support to this event.

All the papers published in this volume have been reviewed by members of the Scientific Committee or international experts in the field. We are particularly grateful for their cooperation to: G. Alberti, P. Aptel, T. Fane, C. Fell, A. Gliozzi, T. Hoshi, G. Jonsson, N. Kamo, S. Katoh, O. Kedem, H. Kimizuka, S. Kimura, Y. Kobatake, T. Kondo, P. Meares, T. Nakagawa, S. Nakao, J. Neel, A. Nidola, Y. Nozawa, H. Ohya, D. Paterson, M. Pegoraro, R. Rautenbach, G. Sarti, G. Semenza, M. Seno, H. Strathmann.

Enrico Drioli
Presidente European Society of
Membrane Science and Technology

Masayuki Nakagaki
President Membrane Society
of Japan

CONTENTS

Transport Mechanisms of Organic Ions in Rat Renal
    Brush Border and Basolateral Membrane Vesicles   1
    R. Hori, K.-I. Inui, M. Takano and T. Okano

Reconstitution of Ionic Channels into Planar Lipid
    Membranes by Fusion with Proteoliposomes   9
    G. Menestrina and F. Pasquali

Photochemically Activated Electron Transport Across
    Unilamellar Liposomes Using Coenzyme Q and
    Vitamin K_1 as Carriers   17
    D. Coutts and R. Paterson

Disorder of Coupled Transport in Renal Brush Border
    Membrane Vesicles from Rabbits with Experimental
    Fanconi Syndrome Induced by Anhydro-4-Epitetracycline
    as a Model of Membrane Disease   23
    Y. Orita, Y. Fukuhara, T. Nakanishi, M. Horio and H. Abe

Advance and Perspective in Medical Application of Liposomes   31
    Y. Nozawa

Measurement and Interpretation of Friction Coefficients of
    Water and Ions in Membrane Polymers   39
    M. A. Chaudry and P. Meares

Tentative Classification of Transport-reaction Systems by
    SeCDAR-Analysis (Uphill, facilitated and active processes)   55
    E. Sélégny

The Response of Membrane Systems when Exposed to Oscillating
    Concentration Waves   69
    P. Doran and R. Paterson

Facilitated Transport of Ions in Asymmetric Membranes   77
    M. Nakagaki and R. Takagi

Ion and Solvent Transports Through Amphoteric Ion Exchange
    Membrane   85
    H. Kimizuka, Y. Nagata and W. Yang

A Selective Parameter for Ionic Membrane Transport
    Selectivity and Porosity of Several Passive Membranes   93
    A. Hernandez, J. A. Ibañez and A. F. Tejerina

Ionic Permeability of the S18 Saft Carboxylic Membrane   101
    A. Lindheimer, D. Cros, B. Brun and C. Gavach

Preparation of Microporous Membranes by Phase Inversion
    Processes   115
    H. Strathmann

Chitosan Hollow Fibers: Preparation and Properties   137
    F. Pittalis and F. Bartoli

Research on the Preparation of a New Type of
    Polyarylsulfone Membrane for Ultrafiltration   143
    S.-J. Han, K.-F. Wu, G.-X. Wu and Y.-B. Wang

Outer Skinned Hollow-fibers-spinning and Properties   151
    J. M. Espenan and P. Aptel

Factors Affecting the Structure and Properties of
    Asymmetric Polymeric Membranes   163
    A. Bottino, G. Capannelli and S. Munari

Separation Characteristics of Ultrafiltration Membranes   179
    G. Jonsson and P. M. Christensen

Separation of Aminoacids by Charged Ultrafiltration
    Membranes   191
    S. Kimura and A. Tamano

Predicting Flux-time Behavior of Flocculating Colloids
    in Unstirred Ultrafiltration   199
    R. M. McDonogh, C. J. D. Fell and A. G. Fane

Separation Characterization of Ultrafiltration Membranes   209
    G. Trägårdh and K. Ölund

Influence of Surfactants on the Transport Behavior of
    Electrolytes Through Synthetic Membranes   215
    D. Laslop and E. Staude

Effects of Antifoams on Cross-flow Filtration of Microbial
    Suspension   223
    K. H. Kroner, W. Hummel, J. Völkel and M.-R. Kula

Yeast Cell Entrapment in Polymeric Hydrogels: a Kinetic
    Study in Membrane Reactors   233
    M. Cantarella, V. Scardi, A. Gallifuoco, M. G. Tafuri
    and F. Afani

Performance of Whole Cells Possessing Cellobiase Activity
    Immobilized into Hollow Fiber Membrane Reactors   241
    A. Adami, C. Fabiani, M. Leonardi and M. Pizzichini

Recent Developments of Ultrafiltration in Dairy Industries   255
    J. L. Maubois

Electrodialysis of Dilute Strontium Cations in Sodium
    Nitrate Concentrated Solutions   263
    C. Fabiani and M. De Francesco

Ion-selective Membranes for Redox-flow Battery   273
    H. Ohya, K. Emori, T. Ohto and Y. Negishi

Zero Gap Membrane Cell and Spe Cell Technologies vs.
    Current Density Scale Up   281
  A. Nidola

Electrodialysis in the Separation of Chemicals   299
  W. A. McRae

Carrier Facilitated Transport and Extraction Through
    Ion-exchange Membranes: Illustrated with Ammonia,
    Acetic and Boric Acids   309
  D. Langevin, M. Metayer, M. Labbe, M. Hankaoui
  and B. Pollet

A Thin Porous Polyantimonic Acid Based Membrane as a
    Separator in Alkaline Water Electrolysis   319
  R. Leysen, W. Doyen, R. Proost and H. Vandenborre

Porous Membranes in Gas Separation Technology   327
  U. Eickmann and U. Werner

Separation of Hydrocarbon Mixtures by Pervaporation
    Through Rubbers   335
  J.-P. Brun, C. Larchet and B. Auclair

An Energy-efficient Membrane Distillation Process   343
  T. J. van Gassel and K. Schneider

Use of Hydrophobic Membranes in Thermal Separation of
    Liquid Mixtures: Theory and Experiments   349
  G. C. Sarti and C. Gostoli

The Water Structure in Membrane Models Studied by
    Nuclear Magnetic Resonance and Infrared
    Spectroscopies   361
  C. A. Boicelli, M. Giomini and A. M. Giuliani

Preparation and in vitro Degradation Properties of
    Polylactide Microcapsules   371
  K. Makino, M. Arakawa and T. Kondo

Electromicroscopic Study of Ultrathin Solute Barrier
    Layer of Composite Membranes and their Solute
    Transport Phenomena by the Addition of Alkali
    Metal Salts   379
  T. Uemura and T. Inoue

Active Transport in Artificial Membrane Systems   387
  Y. Kobatake, N. Kamo and T. Shinbo

Electrophysiological Aspects of Na^+-coupled Cotransport
    of Organic Solutes and Cl^- in Epithelial Cell
    Membranes   405
  T. Hoshi

Disinfection of Escherichia coli by Using Water
    Dissociation Effect on Ion-exchange Membrane   421
  T. Tanaka, T. Sato and T. Suzuki

The Characterization of Polymer Membranes as Produced by
    Phase Inversion   429
  C. T. Badenhop and A. L. Bourgignon

Transport Equations and Coefficients of Reverse Osmosis
      and Ultrafiltration Membranes   447
   S. Kimura

Transfer of Solutes Through Composite Membranes
      Containing Phospholipids   455
   E. Sada, S. Katoh and M. Terashima

New Inorganic and Inorganic-organic Ion-Exchange Membranes   461
   G. Alberti, M. Casciola, U. Costantino and D. Fabiani

The Influence of the Surfactants on the Permeation of
      Hydrocarbons Through Liquid Membranes   475
   P. Pluciński

Separation of Isomeric Xylenes by Membranes Containing
      Clathrate-forming Metal Complexes   483
   S. Yamada and T. Nakagawa

Mathematical Modeling of Colloidal Ultrafiltration:
      a Method for Estimation Ultrafiltration Flux Using
      Experimental Data and Computer Simulations   497
   R. Lundqvist and R. von Schalien

Permeability and Structure of PMMA Stereocomplex Hollow
      Fiber Membrane for Hemodialysis   507
   T. Kobayashi, M. Todoki, M. Kimura, Y. Fujii,
   T. Takeyama and H. Tanzawa

An Improved Model for Capillary Membrane Fixed Enzyme
      Reactors   515
   J. E. Prenosil and T. Hediger

Ethanol Production by Coupled Enzyme Fermentation and
      Continuous Saccharification of Cellulose, Using
      Membrane Cell-recycling Systems   533
   H. Hoffmann, M. Grabosch and K. Schügerl

Enzymatic Determination of Cholic Acids in Human Bile
      by Amperometric Oxygen Electrode as Detector   543
   L. Campanella, F. Bartoli, R. Morabito and
   M. Tomassetti

Synthesis of Polyurethane Membranes for the Pervaporation
      of Ethanol-water Mixtures   549
   L. T. G. Pessoa, R. Nobrega and A. C. Habert

Separation of Benzene-methanol and Benzene-cyclohexane
      Mixtures by Pervaporation Process   563
   E. Nagy, J. Stelmaszek and A. Ujhidy

Industrial Separation of Azeotropic Mixtures by
      Pervaporation   573
   E. Mokhtari-Nejad and W. Schneider

Economics of Industrial Pervaporation Processes   581
   H. E. A. Brüschke and G. F. Tusel

Nonisothermal Mass Transport of Organic Aqueous Solution
      in Hydrophobic Porous Membrane   587
   Z. Honda, H. Komada, K. Okamoto and M. Kai

Observations on the Performances of Pervaporation Under
    Varied Conditions   595
  R. Rautenbach and R. Albrecht

Chemical Engineering Aspects of Membrane Processes and
    Adaptation of the Equipment to the Application   609
  J. Wagner

Control of Gas Transfer in Polymeric Membranes   613
  H. Ohno and K. Suzuoki

Development of Hydrogen Separation Membranes for
    "C_1 Chemistry" in Japan   621
  K. Haraya, Y. Sindo, N. Ito, K. Obata, T. Hakuta
  and H. Yoshitome

Application of Gas-liquid Permporometry to
    Characterization of Inorganic Ultrafilters   629
  C. Eyraud

An Experimental Study and Mathematical Simulation of
    Membrane Gas Separation Process   635
  E. Drioli, G. Donsi', M. El-Sawi, U. Fedele,
  M. Federico and F. Intrieri

Index   647

TRANSPORT MECHANISMS OF ORGANIC IONS IN RAT RENAL

BRUSH BORDER AND BASOLATERAL MEMBRANE VESICLES

Ryohei Hori, Ken-ichi Inui, Mikihisa Takano and
Tomonobu Okano

Department of Pharmacy, Kyoto University Hospital
Faculty of Medicine, Kyoto University
Sakyo-Ku, 606, Japan

Renal handling of organic ions is a complex phenomenon involving glomerular filtration, tubular secretion and tubular reabsorption[1,2]. The primary function of the renal organic ion secretory system is the elimination of foreign compounds, and organic anions and cations are actively secreted by the proximal renal tubules. The sequence of movement of organic ions is transport across basolateral membranes, accumulation in the cells, followed by efflux from the cell across brush border membranes into tubular fluid. However, it has been difficult to characterize the specific membrane events underlying the transepithelial transport of organic ions, because of its complex structure, being composed of two distinct membranes, luminal brush border and contraluminal basolateral membranes. The two membranes differ in the enzyme composition and in the transport mechanisms for solutes.

In recent years a methodology has been developed to use vesicles of the isolated brush border and basolateral membranes for the analysis of tubular transport. In particular, many studies have been presented on the mechanisms for D-glucose and neutral amino acid transport in brush border membranes[3]. However, there have been only a few reports concerning the plasma membrane transport of organic anions and cations[4]. Thus we were prompted to assess the transport of various organic ions by brush border and basolateral membrane vesicles: p-aminohippurate (anion)[5], tetraethylammonium (cation)[6], and cephalexin (amphoteric compound)[7,8].

ISOLATION OF BRUSH BORDER AND BASOLATERAL MEMBRANES

Brush border membrane vesicles were isolated from the renal cortex of rats according to the calcium precipitation of Evers et al.[9]. On the other hand, the isolation of basolateral membranes has been difficult, because the densities of brush border and basolateral membranes are so close. We have recently developed a simple method for the isolation of basolateral membranes using Percoll density gradient centrifugation[10]. As shown in Figure 1, $(Na^{+}+K^{+})$-ATPase, the marker enzyme for basolateral membranes, was enriched 22-fold in the basolateral membrane preparation compared with that found in the homogenate. Alkaline phosphatase and aminopeptidase, the marker enzymes for brush border membranes, were enriched 10-fold in brush border membrane preparation. In both prepar-

1

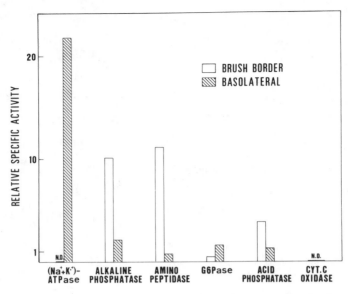

Fig. 1. Relative specific activities of marker enzymes in brush border and basolateral membranes.

ations, the contamination by mutual membranes, mitochondria, lysosomes, endoplasmic reticulum and cytosol was small.

The uptake of substrates was measured by a rapid filtration technique. In the regular assay, the reaction was initiated rapidly by adding 20 µl of substrate mixture to 20 µl of membrane vesicle suspension at 25°C. At the stated time points, the incubation was stopped by diluting a reaction sample with 1 ml of ice-cold stop solution. The tube contents were immediately poured onto Millipore filters (0.45 µm). The radioactivity of p-amino[³H]hippurate or [³H]tetraethylammonium trapped on the filters was determined by liquid scintillation counting.

p-AMINOHIPPURATE TRANSPORT

The effect of probenecid and 4,4'-diisothiocyano-2,2'-disulfonic stilbene (DIDS) on p-aminohippurate uptake by brush border and basolateral membrane vesicles was studied. DIDS, a specific inhibitor for anion transport into red blood cells, inhibited more strongly the uptake of p-aminohippurate than probenecid. The degree of the inhibition in brush border and basolateral membrane vesicles was similar. In the concentration dependence of p-aminohippurate uptake, the relationship between concentration and rate of uptake was nonlinear in basolateral membranes, providing evidence for saturability. However there was no evidence for saturation of the uptake by brush border membrane vesicles.

In order to confirm the differences in p-aminohippurate transport by brush border and basolateral membrane vesicles, we have studied the effect of countertransport on p-aminohippurate uptake. As is evident from Figure 2, vesicles preloaded with a high concentration of unlabeled p-amino-hippurate showed enhancement of p-amino[³H]hippurate accumulation by countertransport only in basolateral membranes, while no change of the uptake was observed in brush border membranes[5].

Furthermore, the temperature dependence of p-aminohippurate uptake by brush border and basolateral membrane vesicles was studied. The Arrhenius

2

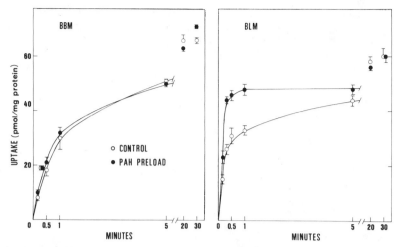

Fig. 2. Countertransport effect on p-aminohippurate uptake by brush
         border (BBM) and basolateral membrane (BLM) vesicles. [From
         Reference 5].

plot for the uptake by brush border membrane vesicles was linear over the
temperature range used (5–37°C). In contrast, the Arrhenius plot for the
uptake by basolateral membrane vesicles was biphasic. As Smedt and
Kinne[11] also reported a biphasic Arrhenius plot with respect to Na^+-
dependent D-glucose transport by brush border membrane vesicles, the
present data suggest the contribution of a carrier-mediated transport
system for p-aminohippurate in basolateral membranes.

The role of membrane potential as a driving force for p-aminohippurate
uptake by brush border and basolateral membrane vesicles was studied by
applying different anion gradients with sodium directed into the vesicles.
The more permeant lipophilic anion, SCN^-, thought to facilitate a more
rapid development of interior negative membrane potential, was compared
with less permeant anions, such as Cl^- and SO_4^{2-}. Anion permeability to
biological membrane generally follows in the order of $SCN^- > Cl^- > SO_4^{2-}$.
p-Aminohippurate uptake by brush border membrane vesicles was higher when
chloride was replaced by sulfate, and lower when chloride was replaced by
thiocyanate. On the other hand, this effect for the uptake induced by
anion gradients was small in extent in basolateral membrane vesicles.
These results suggest that a decrease of the inside negative membrane
potential increases p-aminohippurate uptake, and this effect is more
evident in brush border membranes compared with basolateral membranes.

Based on the above results, the uptake of p-aminohippurate by basolateral membrane vesicles satisfies some of the criteria for carrier-
mediated process; namely, the process is saturable, temperature dependent,
inhibited by anion transport inhibitors, and undergoes a countertransport
effect. In contrast, brush border membrane vesicles failed to display the
capacity to accelerate the exchange of p-aminohippurate, saturability of
the uptake, and biphasic Arrhenius plot, although probenecid and DIDS
reduced p-aminohippurate transport. Okamoto et al.[12] discussed that if
an ion is conducted via a carrier, the transport is affected by temperature, which controls the fluidity of lipids in the membrane, and that if
it is conducted via a channel, the effect of fluidity is rather small.
Therefore, it may be reasonable to assume that p-aminohippurate is transported across brush border membranes by a gated channel, which responds to
anionic charge, rather than by a simple diffusion. Furthermore, p-amino-

hippurate uptake by brush border membrane vesicles was influenced more
sensitively by the alteration of the membrane potential compared with that
by basolateral membrane vesicles, and it was significantly stimulated by
the membrane potential induced with various anion gradients, which renders
the intravesicular space more positive. This finding is compatible with
the secretion of p-aminohippurate at the luminal side in vivo, because the
intracellular compartment has more negative electrical potential than the
luminal fluid compartment (Figure 5).

TETRAETHYLAMMONIUM TRANSPORT

It is well-known that the organic cation transport system is clearly
separable from the anion transport system. In the concentration dependence
of [^{3}H]tetraethylammonium uptake by brush border and basolateral membrane
vesicles, the relationship between concentration and uptake rate was
nonlinear in both membranes, providing evidence for saturability. After
the correction for the nonsaturable component, the values of Km and Vmax
were 0.8 mM and 7.4 nmol/mg protein per min in brush border membranes, and
2.5 mM and 5.6 nmol/mg protein per min in basolateral membranes, respectively. Vesicles preloaded with high concentration of unlabeled tetraethylammonium showed enhancement of [^{3}H]tetraethylammonium accumulation by
countertransport in brush border and basolateral membranes. These data
suggest the contribution of a carrier-mediated transport system for tetraethylammonium in both membranes. Furthermore, the temperature dependence
of tetraethylammonium uptake by brush border and basolateral membrane
vesicles was studied. The Arrhenius plot for the uptake was biphasic in
both membranes.

Based on the above results, the uptake of tetraethylammonium by brush
border and basolateral membrane vesicles satisfied some of the criteria for
carrier-mediated process; namely, the process is saturable, temperature
dependent, inhibited by other organic cations, and undergoes a countertransport effect.

In order to get further information about the driving force for the
transport, we have studied the effect of various ionic conditions on the
uptake of tetraethylammonium. As is evident from Figure 3, there was no

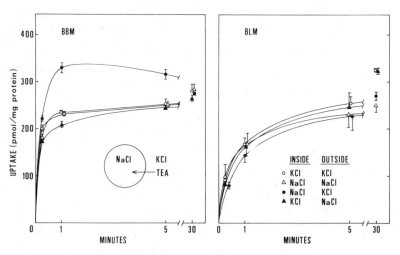

Fig. 3. Effect of various ionic conditions on tetraethylammonium uptake by
brush border (BBM) and basolateral membrane (BLM) vesicles. [From
Reference 6).

4

effect of ionic conditions on tetraethylammonium uptake by basolateral membranes[6]. However, there was the stimulation of tetraethylammonium uptake by brush border membrane vesicles in the presence of Na^+ gradient: sodium chloride is inside of the vesicles and potassium chloride is outside. Thus, this ionic condition can be a driving force for tetraethylammonium transport.

As intravesicular acidification of brush border membrane vesicles was confirmed in the presence of $[Na^+]i$ and $[K^+]o$ using Acridine orange, the effect of H^+ gradient on tetraethylammonium uptake by brush border and basolateral membrane vesicles was studied (Figure 4)[6]. In brush border membranes, the presence of an H^+ gradient (inside pH 6.0, outside pH 7.5) induced a transient uphill transport of tetraethylammonium (overshoot phenomenon). When the gradient was reversed (lower pH in outside), uphill transport was not observed. The final levels of the uptake were all identical, indicating that the pH gradient did not effect the vesicle size. In contrast, an H^+ gradient was ineffective in basolateral membrane transport of tetraethylammonium. The concentrative uptake of tetraethylammonium driven by H^+ gradient was completely inhibited by mercuric chloride.

Furthermore, it is important to clarify the role of membrane potential as a driving force for tetraethylammonium uptake by brush border and basolateral membrane vesicles. Valinomycin in the presence of K^+ gradient (inside>outside) was employed to produce an inside-negative membrane potential. Tetraethylammonium uptake by brush border membrane vesicles was unaffected by valinomycin, suggesting electroneutral antiport of H^+ and tetraethylammonium. In contrast, a valinomycin-induced inside-negative membrane potential stimulated significantly the initial rate of tetraethylammonium uptake by basolateral membrane vesicles.

Based on the above results, tetraethylammonium is transported from blood to cell across basolateral membrane via a carrier-mediated system and this process is stimulated by the intracellular negative potential. Tetraethylammonium transport from cell to urine across brush border membranes is driven by an H^+ gradient via an electroneutral H^+-tetraethylammonium antiport system. This H^+ gradient can be created by an Na^+-H^+ antiport system and/or an ATP-driven H^+ pump in brush border membranes (Figure 5).

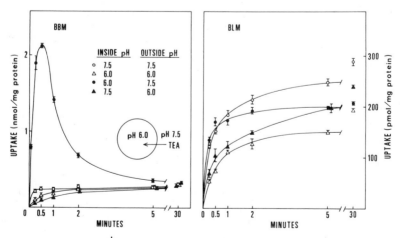

Fig. 4. Effect of H^+ gradient on tetraethylammonium uptake by brush border (BBM) and basolateral membrane (BLM) vesicles. [From Reference 6].

5

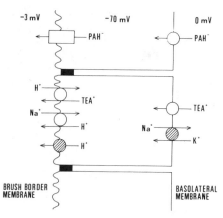

Fig. 5. Model for transepithelial transport of p-aminohippurate (PAH)
and tetraethylammonium (TEA) in proximal tubular cells.

From the studies described above, it is evident that renal epithelial
cells possess a striking polarity with respect to the transport properties
of organic ions across brush border and basolateral membranes (Figure 5).
These results can represent useful information to further the study of
tubular transport mechanisms of organic anions and cations.

REFERENCES

1. J. M. Irish, III and J. J. Grantham, Renal handling of organic anions
    and cations, in: "The Kidney," B. M. Brenner and F. C. Rector, Jr.,
    eds., W. B. Saunders Co., Philadelphia (1981).
2. A. Kamiya, K. Okumura, and R. Hori, Quantitative investigation on
    renal handling of drugs in rabbits, dogs and humans, J.Pharm.Sci.,
    72:440 (1983).
3. H. Murer and R. Kinne, The use of isolated membrane vesicles to study
    epithelial transport processes, J.Membrane Biol., 55:81 (1980).
4. C. R. Ross and P. D. Holohan, Transport of organic anions and cations
    in isolated renal plasma membranes, Ann.Rev.Pharmacol.Toxicol.,
    23:65 (1983).
5. R. Hori, M. Takano, T. Okano, S. Kitazawa, and K. Inui, Mechanisms of
    p-aminohippurate transport by brush border and basolateral membrane
    vesicles isolated from rat kidney cortex, Biochim.Biophys.Acta,
    692:97 (1982).
6. M. Takano, K. Inui, T. Okano, H. Saito, and R. Hori, Carrier-mediated
    transport systems of tetraethylammonium in rat renal brush border
    and basolateral membrane vesicles, Biochim.Biophys.Acta, in the
    press.
7. K. Inui, T. Okano, M. Takano, S. Kitazawa, and R. Hori, Carrier-
    mediated transport of amino-cephalosporins by brush border membrane
    vesicles isolated from rat kidney cortex, Biochem.Pharmacol.,
    32:621 (1983).
8. K. Inui, T. Okano, M. Takano, H. Saito, and R. Hori, Carrier-mediated
    transport of cephalexin via the dipeptide transport system in rat
    renal brush border membrane vesicles, Biochim.Biophys.Acta, 769:449
    (1984).
9. C. Evers, W. Haase, H. Murer, and R. Kinne, Properties of brush border
    vesicles isolated from rat kidney cortex by calcium precipitation,
    Membrane Biochemistry, 1:203 (1978).
10. K. Inui, T. Okano, M. Takano, S. Kitazawa, and R. Hori, A simple
    method for the isolation of basolateral plasma membrane vesicles

from rat kidney cortex: Enzyme activities and some properties of glucose transport, Biochim.Biophys.Acta, 647:150 (1981).

11. H. D. Smedt and R. Kinne, Temperature dependence of solute transport and enzyme activities in hog renal brush border membrane vesicles, Biochim.Biophys.Acta, 648:247 (1981).

12. H. Okamoto, N. Sone, H. Hirata, M. Yoshida, and Y. Kagawa, Purified proton conductor in proton translocating adenosine triphosphatase of a thermophilic bacterium, J.Biol.Chem., 252:6125 (1977).

RECONSTITUTION OF IONIC CHANNELS INTO PLANAR

LIPID MEMBRANES BY FUSION WITH PROTEOLIPOSOMES

G. Menestrina and F. Pasquali

Dipartimento di Fisica
I-38050 Povo (TN)
Italy

INTRODUCTION

In the past few years a new approach to the study of membrane transport has been developed, called "membrane reconstitution". It relies upon
the idea that the researcher can gain a higher experimental control by
transferring a transporting protein out of its complicated native membrane
into a simpler artificial one. The method requires: homogenization of the
native tissue; solubilization of fragments by nondenaturing detergents
which extracts intrinsic proteins; fractionation of the lipid, protein,
detergent micellar solution; removal of the detergent from the appropriate
fractions. As a result small lipid vesicles containing only the transporting protein with a correct orientation can be obtained[1]. Unfortunately
small lipid vesicles are unsuitable for studying the transport properties
of proteins which have a very high turnover like ionic channels of excit-
able cells. Only very recently the reconstitution technique has been
successfully applied to natural ionic channels, and this was accomplished
by adding a new step to the method: the purified vesicles were made to
fuse with a planar bilayer, thus transferring their protein content into
the host membrane. Planar bilayer membranes are electrically accessible,
hence transport rates can be measured with high precision as ionic cur-
rents[2].

We describe here the application of this new technique to the reconstitution of two proteic channels into a planar lipid bilayer. The first
is the ionic channel formed by hemocyanin, the oxygen carrier in molluscs,
which is a well known pore-former and can be used as a probe to optimize
the conditions under which the reconstitution procedure works[3,4]. The
second is the channel formed by complement. Studies on the mechanism of
immune cytolysis have shown that complement dependent membrane damage is
mediated by the assembly of the five complement components C5-C9 on the
target-membrane[5]. These serum proteins aggregate to form a macromolecular complex called C5b-9 which appears at the electron microscope
as a hollow cylinder protruding from the membrane[6]. Current evidences
indicate that a hydrophilic trans-membrane channel is thus created and that
this may represent the mechanism of cytolysis by complement[7].

MATERIALS AND METHODS

Reagents and Buffers

The reagents used were all the best grade available. KCl, NaCl, HCl, KOH, glucose and sucrose were purchased by Carlo Erba; $NaNO_3$, EDTA, DOC and trypsin by Merck; Tris was Riedel de Haën and HEPES was from Sigma. Buffer solutions used were: buffer A, 100 mM KCl, 1 mM EDTA, 5 mM Hepes pH adjusted by KOH as specified in the text; buffer B, 50 mM NaCl, 5 mM $NaNO_3$, 10 mM Tris, pH 8.0 adjusted with HCl; buffer C, 25 mM NaCl, 75 mM KCl, 5 mM $NaNO_3$, 5 mM Tris, pH 7.0. The lipids used were: phosphatidylcholine, PC, phosphatidylethanolamine, PE, and phosphatidylserine, PS, purchased respectively by Lipid Products, Fidia Res. Lab. and Calbiochem, if not otherwise specified.

Preparation of Vesicles Containing Hemocyanin

Liposomes were prepared as follows: 30 mg of phospholipids (PC/PE/PS in a molar ratio 2:5:3) dissolved in chloroform were dried first under a stream of nitrogen in a rotary evaporator (Büchi) and then under rotative vacuum for several hours at room temperature. Two ml of buffer A, pH 8.0, containing DOC were then added to the dry film, to a final lipid/detergent molar ratio 1:1, and the solution was gently stirred to clarity (6-12 h); DOC was then removed from this solution of mixed liquid/detergent micelles by 18 h flow-through dialysis ca.600 vol of buffer A, pH 8.0, in a LIPOPREP dialyzer, Dianorm, equipped with 10,000 dalton cut off membranes; this procedure leads to the formation of unilamellar lipid vesicles with an average ratius of 30÷50 nm[8]. Megathura, crenulata hemocyanin (Calbiochem) dialyzed overnight against buffer A, pH 7.0, and centrifuged at 10.000 rpm for 20 min was then added to the liposomes to a final lipid/protein (W/W) ratio of 10:1. After several hours of incubation the protein that did not incorporate into the vesicles was separated from the proteoliposomes by centrifugation on a discontinuous sucrose density gradiente made of three layers of buffer A, pH 8.0, containing 5%, 40%, 50% sucrose from top to bottom with a volume ratio of 2:1:1 in the same order. Centrifugation was run to equilibrium at 42.000 rpm in a Sorvall Ti-865 rotor for 14 h at 4°C. Five fractions, approximately all of the same volume, were collected puncturing the tubes, numbered from the bottom and dialyzed each against ca. 400 vol. of buffer A, pH 7.0, for 24 h at 4°C. Protein and vesicles content of each fraction was evaluated by absorption spectra in the range 200÷400 mm. Hemocyanin eluted in the bottom fractions 1,2 as assayed by the height of the absorption peak at 278 nm, whereas vesicles eluted in the top fraction 4,5 as assayed by light-scattering at 330 nm.

Preparation of Vesicles Containing Complement

Complement was reconstituted into lipid vesicles by a procedure similar to that described above except for the following changes. Phospholipids used were PC/PE/PS, all purchased by Calbiochem, in a 3:4:3 molar ratio. Guinea pig complement, containing fluid phase SC5b-9 complex, was purchased liophilized by Miles Lab., diluted in buffer B and subjected to mild proteolysis by trypsin in presence of DOC (protein/detergent, W/W, ratio 1:10) for 12 h at 35°C. During this step the S protein is detached from the complex and substituted by detergent molecules[9]. It was then added to the lipid, DOC solution (to a final protein/lipid/detergent ratio of 1:8:18 in weight) and dialyzed as above against ca. 800 vol of buffer B. As reported by Bahkdi et al.[9], this procedure leads to incorporation of the complement into the vesicles in a form which is closely similar to the terminal membrane complex C5b-9 (m) in vivo. Proteoliposomes were then separated from free protein by density gradient centrifugation as above except that the vesicles were layered in the second of four layers of equal volume and of sucrose concentration 5%, 10%, 35%, 45%, from top to bottom. Ten fractions were collected and numbered from the bottom of the tubes, dialyzed extensively against buffer B and assayed for protein and vesicles contents as above.

Planar Bilayer Experiments

Black lipid membranes (BLM) were prepared by the usual technique with a Pasteur pipette on a circular hole of 0.5 mm diameter between two equal, 4 ml, compartments, machined in a Teflon cup, filled with buffer. The lipid solution used was PC/PS, W/W ratio 2:1, dissolved at 50 mg/ml in n-decane. Amounts from the different fractions of the preparations described above were added to one compartment, cis side, after complete blackening of the membrane. Divalent cations and sugars were then added as specified in Results to achieve fusion of proteoliposomes into the planar bilayer. Current passing through Ag-AgCl electrodes was converted into voltage by a virtual grounded operational amplifier (AD 515 K) with a 100 M Ω feed-back resistor paralleled by a 20 pF capacitor. Voltages, applied by a battery operated generator, are referred to the the cis compartment. Experiments were run at room temperature.

RESULTS AND DISCUSSION

Fusion of Hemocyanin Containing Vesicles with BLM

M. crenulata hemocyanin is a well known channel former; when added in small amounts (μg/ml) to a BLM bathing solution it increases the conductance of the membrane in discrete steps by up to three orders of magnitude. Such increase is largely independent both of the lipid composition of the membrane and of the ionic composition of the electrolytic solution, and follows within a few minutes the addition of the protein to the bath. When added to only one side of the bilayer, cis side, hemocyanin confers to the BLM a strongly non linear instantaneous current-voltage curve and an even more strongly rectifying steady-state current-voltage characteristic. This has been explained assuming that all channels are oriented in parallel into the bilayer and that they can experience conformational transitions induced by the applied electric field[10,11]. Hemocyanin channels are known to be cation selective at neutral pH, i.e. cations can pass through the channel with almost the same mobility as in the aqueous solution, whereas anions are excluded; this selectivity pattern is most probably due to the effects of the net negative charge fixed on this protein at pH 7.0[11]. When proteoliposomes which had been incubated with hemocyanin and then purified by ultracentrifugation on a discontinuous sucrose gradient are added to a bilayer, an incorporation pattern is observed that is substantially different from that outlined above for the pure protein. As shown in Figure 1 proteovesicles from fraction 4 of the preparation described in Methods, are not able to augment the permeability of the membrane by themselves; a strong conductance increase is observed only when divalent cations are present and the osmotic pressure of the cis compartment is suddenly increased. The complete set of conditions necessary to observe such increase are: the presence of negatively charged lipids, as PS, in both the vesicles and the planar bilayer; the presence of low concentrations, few mM, of divalent cations; an increase in the osmolarity at least of the cis compartment, achieved by addition of glucose or sucrose[12]. A symmetrical increase of the osmolarity of the two compartments is as effective as an increase on the cis side only, whereas an increase in the trans compartment osmolarity only, is completely ineffective as is also an increase in the cis compartment when divalent cations are not present[12]. The special set of conditions necessary to have an increase in the planar membrane permeability strongly indicates that it is

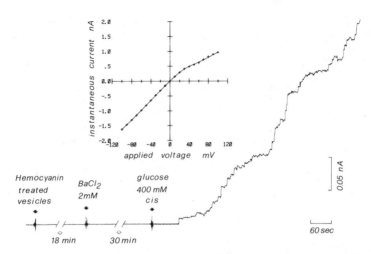

Fig. 1. Time course of membrane current after the addition of hemocyanin
doped vesicles to a BLM cis bathing solution, clamping voltage –
40 mV. The conductance increases only under fusogenic conditions,
i.e. after an osmotic jump in the cis compartment in the presence
of divalent cations. Arrows indicate the various additions to the
solution. Pieces of flat trace have been cut whose duration is
indicated below the interruptions. The final conductance in this
trace corresponds to the incorporation of more than one hundred
hemocyanin channels into the planar bilayer. In the inset the
instantaneous current-voltage curve of the same membrane at a
later moment is reported. The non-linear characteristic is fitted
using a two-state voltage-gating model, as described[11]. Solid
line is a least square adaptation which gives a gating charge, the
charge which moving under the applied electric field promotes the
transition of the channel between the two states, of 1.7 elementary charges and an equilibrium potential, the voltage at which the
two states are equally populated, of + 35 mV. Buffer A, pH 7.0.

induced by fusion of the proteoliposomes with the BLM. Actually negative
lipids and divalent cations are necessary, since they promote the vesicles-
planar membrane adhesion[13], but not sufficient. The installation of an
osmotic gradient across the vesicles walls, which causes them to swell, is
also required, in good agreement with the more recent experimental findings
on membrane fusion[14,15]. As shown in the inset of Figure 1 the instantaneous current-voltage curve of the modified bilayer is virtually indistinguishable from that obtained adding pure hemocyanin in the solution[10].
This result indicates first that the current increase is due to the transfer of hemocyanin channels from the vesicles to the planar bilayer; second
that all channels are oriented in parallel into the bilayer i.e. that
during fusion the proteins are transferred from the vesicles to the BLM
without loss of their orientation.

Fusion of Complement Containing Vesicles with BLM

The fluid-phase complex of complement SC5b-9 can be transformed by
mild proteolysis into an amphiphilic molecule with the same neoantigens
characteristic of the C5b-9 membrane attack complex of the complement[9].
This proteolytically altered SC5b-9 complex incorporates into lipid
vesicles prepared by detergent removal yielding a membrane bound structure
resembling the C5b-9 complement lesion[7,9]. Such vesicles can be separated from unbound complement by density gradient centrifugation. Figure 2
shows the elution pattern from the centrifugation tube of such a prepar-

ation. Vesicles floated on the top of the gradient in one major peak, whereas the protein divided up into two peaks: one with higher density and one which cofloated with the lipid vesicles, representing the fraction of complement bound to the membranes. When vesicles from fractions 7 and 8 are incubated on one side of a BLM a strong conductance increase can be observed when fusogenic conditions are applied (Figure 3). As in the case

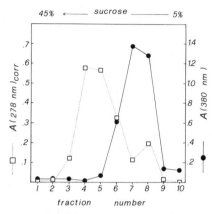

Fig. 2. Elution profile from the test tube after discontinuous density gradient sucrose centrifugation for purification of vesicles which interacted with complement. Protein concentration was measured by the absorbance at 278 nm and corrected for light-scattering, assuming a linear contribution between 254 and 302 nm. Vesicles concentration was measured by the absorbance at 380 nm arising from light-scattering. Absorptions are referred to an optical path length of 1 cm. The direction of the sucrose gradient is given.

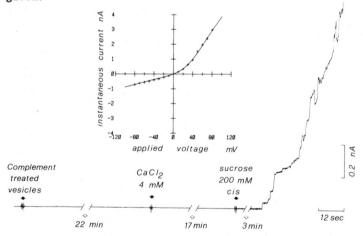

Fig. 3. Time course of membrane current after addition of complement treated liposomes to the cis compartment in a BLM experiment, holding voltage + 40 mV. Vesicles from fraction 7 of Figure 2 have been used. Also in this case membrane conductance grows quickly after fusogenic conditions have been applied. Additions and interruptions are indicated as in Figure 1. The instantaneous current-voltage curve, shown in the inset, is strongly rectifying with a voltage dependence opposite to that of hemocyanin doped bilayers. Least square parameters are: gating charge-1.2 elementary charges, equilibrium potential + 38 mV. Buffer C, pH 7.0.

discussed above for hemocyanin doped vesicles, negative lipids, divalent cations and an osmotic gradient across the vesicles are required together to change the permeability of the membrane. We think that also in this case fusion of vesicles containing the complement channel with subsequent transfer of the pores into the host membrane is at the basis of the permeability changes observed. Control experiments indicated that neither the native complement nor the proteolyzed fractions can increase the BLM conductance in any condition. This is expected since native SC5b-9 is hydrophilic, whereas the trypsinized complex tends to aggregate in absence of excess detergent[9]. The typical conductance of this complement mediated channel is about 1 nS, as can be seen also in Figure 3, a very large value which suggest that the channel should be rather unselective and which is consistent with experiments run on erythrocyte ghosts which indicated that pore diameter could range from 1 to larger than 3 nm[16]. The instantaneous current-voltage curve of these membranes is shown in the inset of Figure 3. As for the case of hemocyanin the characteristic is strongly non linear, but negative voltages are required to depress the conductance in this case. We think that this fact is interesting but that more work is necessary in order to establish if it can have a physiological role in the immune defense.

Acknowledgements

We wish to thank Ms. Maria Teresa Giunta for her skilful work in typing this manuscript. F.P. is the recipient of a grant of Istituto di Ricerca Scientifica e Tecnologica (TN).

REFERENCES

1. C. Miller and E. Racker, Reconstitution of membrane transport functions, in: "The Receptors, a Comprehensive Treatise," Vol.I, R. D. O'Brien, ed., Plenum Press, New York (1979).
2. A. Darszon, Strategies in the reassembly of membrane proteins into lipid bilayer systems and their functional assay, J.Bioenerg. Biomemb., 15:321 (1983).
3. R. Latorre and O. Alvarez, Voltage-dependent channels in planar lipid bilayer membranes, Physiol.Rev., 61:77 (1981).
4. T. J. McIntosh, J. D. Robertson, H. P. Ting-Beall, A. Walter, and G. Zampighi, On the structure of the hemocyanin channel in lipid bilayers, Biochim.Biophys.Acta, 601:289 (1980).
5. P. J. Lachmann and R. A. Thompson, Reactive lysis: the complement mediated lysis of unsensitized cells. II. The characterization of activated reactor as C5b and the participation of C8 and C9, J.exp.Med., 131:644 (1970).
6. R. R. Dourmashkin, The structural events associated with the attachment of complement components to cell membranes in reactive lysis, Immunology, 35:205 (1978).
7. S. Bhakdi and J. Tranum-Jensen, Molecular nature of the complement lesion, Proc.Natl.Acad.Sci.(U.S.A) 75:5655 (1978).
8. O. Zumbuehl and H. G. Weder, Liposomes of controllable size in the range of 40 to 180 nm by defined dialysis of lipid/detergent mixed micelles, Biochim.Biophys.Acta, 640:252 (1981).
9. S. Bhakdi, B. Bhakdi-Lehnen, and J. Tranum-Jensen, Proteolytic trans-formation of SC5b-9 into an amphiphilic macromolecule resembling the C5b-9 membrane attack complex of complement, Immunology, 37:901 (1979).
10. G. Menestrina, D. Maniacco, and R. Antolini, A kinetic study of the opening and closing properties of the hemocyanin channel in artificial lipid bilayer membranes, J.Membrane Biol., 71:173 (1983).

11. G. Menestrina and R. Antolini, The dependence of the conductance of the hemocyanin channel on applied potential and ionic concentration with mono and divalent cations, Biochim.Biophys.Acta, 688:673 (1982).

12. F. Pasquali, G. Menestrina, and R. Antolini, Fusion of large unilamellar liposomes containing hemocyanin with planar bilayer membranes, Z.Naturforsch., 39c:147 (1984).

13. C. Miller, P. Arvan, J. N. Telford, and E. Racker, Ca^{++}-induced fusion of proteoliposomes. Dependence on transmembrane osmotic gradient, J.Membrane Biol., 30:271 (1976).

14. F. S. Cohen, M. H. Akabas, and A. Finkelstein, Osmotic swelling of phospholipids vesicles causes them to fuse with a planar phospholipid bilayer membrane, Science, 217:458 (1982).

15. L. R. Fisher and N. S. Parker, The mechanism of osmotic control of lipid bilayer fusion, Biophys.J., 45:169a (1984).

16. L. E. Ramm, M. B. Whitlow, and M. M. Mayer, Size distribution and stability of the trans-membrane channel formed by complement complex C5b-9, Molec.Immun., 20:155 (1983).

PHOTOCHEMICALLY ACTIVATED ELECTRON TRANSPORT ACROSS UNILAMELLAR LIPOSOMES

USING COENZYME Q AND VITAMIN K_1 AS CARRIERS

Duncan Coutts and Russell Paterson

Department of Chemistry
University of Glasgow
Glasgow, Scotland

It is well-known that coenzyme Q and vitamin K_1 are electron transporting agents, which, with plastoquinone, play essential roles in both
oxidative phosphorylation and photosynthesis. These are complex processes,
which involve many coordinated steps in order to convert metabolic or light
energy into electrochemical potential gradients which are, in turn, coupled
to the mechanism for ATP synthesis.

Coenzyme Q and vitamin K_1 are lipid quinones. Having shown earlier
that these quinones could transport electrons (and protons) through bulk
liquid membranes[1], it was of interest to determine whether this process
could be made to occur across single bilayer lipid membranes (BLMs), and,
if so, the degree to which the other lipids of the BLM would affect the
repetitive oxidation-reduction cycles by which these processes must occur,
Figure 1.

Preliminary tests showed that supported planar BLMs could be formed
containing these quinones[2], but could not be used, because of their
limited life and the slow kinetics of bulk reduction across such small
membrane areas.

It was therefore decided to use large unilamellar vesicles[3] (LUVs)
which would combine greater physical stability with a greatly increased
surface area. Continuous transport of electrons across membranes, requires
that the reducing agent is added (or created) on one side of the transporting membrane. An obvious method would be to encapsulate the substrate to
be reduced and then to add the reducing agent when required, to the bulk
solution. One disadvantage of this is that the substrate is present as a
very minor component of the system. A more interesting prospect was to use
liposomes (LUVs) as complex reducing agents, which might be added to a
substrate solution, reduce it and be removed subsequently. Liposomes
cannot be prepared and handled containing highly reactive reducing agents.
It was decided therefore that they should be prepared in an inactive form
and 'activated' when required.

The method used in this study was to encapsulate a photoreduction
system within the liposomes and to activate them once they were dispersed
in the substrate by means of a suitable light source. Flavinmononucleotide
(FMN), a prosthetic group of the flavoprotein enzymes, including Complex 1
of the mitochondrial respiratory chain, seemed an excellent choice for the

primary reductant, and Radda and Calvin[4] showed that it could be incorporated in a photosolution and reduced by visible light.

EXPERIMENTAL

A typical photo-reduction solution used in these studies had composition: FMN (sodium salt), 0.01M, EDTA, 0.12M; Tris (buffer) 0.01M and sodium chloride, 0.112M. It was adjusted to pH 7.4 by a very small addition of 1M sodium hydroxide. Tris buffers were prepared at this pH and made isotonic with addition of sodium chloride. The liposomes were prepared by the method of Deamer and Bangham[3] except that all preparations and processes were conducted at minimal light levels in a darkened laboratory. In this preparation photosolution was placed in the thermostatted (60°C) Leibig condenser and an ethereal solution of lipids injected slowly by a motor-driven micrometer syringe. The final preparations were cooled and sonicated (at a very low power) for a short time immediately prior to filtration through a 5u Millipore and separation on a Sephadex G-50 column. (The apparently irrational sonication proved a practical method of removing (clumps of liposomes) which otherwise would block filters or spoil the gel filtration step). This separation was achieved by washing the dispersion through the column with Tris buffer which was isotonic (or slightly hypertonic) compared with the suspension. A clean separation was achieved and no photosolution was detectable in the liposome dispersions collected from the columns. These dispersions were 'standardized' by adjusting their optical densities to an arbitrary value of 1.5 at 445 nm (at this wavelength large scattering occurred). These standard dispersions were then mixed with equal volumes of cytochrome c solution (2×10^{-5}M), in isotonic buffer. There was no detectable leak of photosolution in this mixing process. Half of this dispersion was placed in two 1 cm UV absorption cells fitted with small magnetic stirrers. One of these (the test dispersion) was fitted with a Subaseal rubber cap containing two hypodermic needles, wrapped in foil, and degassed, by bubbling, with purified argon.

Both cells were placed in a UV-Visible spectrophotometer in turn and their absorbances, measured at 550 nm and recorded over the range 600-450 nm. These measurements were repeated with the degassed dispersion as test and the other as reference, providing a base-line comparison before illumination of the test sample was begun. At five or ten minute intervals the test cell was removed placed on a magnetic stirred and illuminated by a Wotan halogen lamp (150W) house in a 35 mm slide projector. The light was first filtered through a concentrated solution of 2,6-dichlorophenol indophenol (DCIP) with path length 1.2 cm. (With this filter transmission between 500 and 700 nm and below 360 nm was less than 1%. Maximum transmission (28%) was at 430 nm in the range of absorption of the FMN photosystem; 270-450 nm). Spectra were recorded after each period of illumination, as before. These samples were apparently stable and no reduction was observed to continue out of the beam. At the end of the experiment a few crystals of sodium dithionite were added to complete the reduction and the spectrum of the reference suspension checked to ensure that no drift had occurred. The degree of reduction of the cytochrome was obtained from the absorbances at 550 nm.

The remaining half of the standard dispersion in cytochrome c solution was now used as a control. It had been in the same laboratory conditions for the same length of time, but without illumination. (Previous studies showed that the filtered light caused no leakage of photosolution from the liposomes). There was however the problem of leakage due to the natural processes of decay of the liposome population. To test this the control dispersion was centrifuged at 9000 rpm for 30 min to remove liposomes and further filtered through a 0.22u Millipore to ensure the removal was complete. The resulting samples contained the original external cytochrome solution and any photosystem components which might have leaked from the liposomes since the preparation of the standard dispersions.

This solution was divided into two, one half degassed, and the cycle of illuminations and recordings repeated, as with the original dispersions.

RESULTS AND DISCUSSION

Preliminary results showed that there was a significant reduction of cytochrome c by liposomes containing vitamin K_1 even if there was no photo system present. This effect was traced direct effects of the light source on vitamin K_1 and was solved by using a solution of DCIP as a filter. The effect may be similar to that observed by Bayer[5] who observed that isolated rat mitochondria partially lost their ability to carry out oxidative phosphorylation after irradiation with near UV light and that normal activity could be restored on addition of cytochrome c and vitamin K_1. Control experiments were based upon illumination of the filtered reference suspension, after the experiment. In such controls it was found that some reduction of the substrate occurred in all cases. Careful tests showed that there was no residual photosystem in the initial dispersion of liposomes in cytochrome c solution and that there was no rupture of liposomes caused by stirring or exposure to (filtered) light (compared to controls) and that cytochrome c substrate solution was itself unaffected by this light on twice the time scale of standard experiments. The amount of leakage, estimated as equivalent to an addition of 0.1 ul of photosolution directly to 2ml of substrate solution: a free FMN concentration of 10^{-6}M. It is obvious therefore that the homogeneous solution reaction of reduced FMN with cytochrome c, directly is orders of magnitude more efficient than electron transport across the vesicle bilayer.

All liposomes preparations contained the same photosolution. Standard suspensions were obtained by dilution with buffer to an arbitrary optical density of 1.5 at 445 nm and in all cases these were dispersed in the same cytochrome solution. As far as was possible therefore the procedures were standardised so that the observation would reflect the electron transport properties of the liposomal membranes.

Typical results are shown in Figure 2 for a membrane with 95% egg lethicin (labelled as PC subsequently) and 5% vitamin K_1. The difference between the dispersion and its own control is marked and shows that the addition of vitamin K_1 causes electron transport across bilayer membrane of the liposome. Similar results were obtained when dioleolylphosphatidyl choline (DOPC) replaced PC and also when vitamin K_1 was replaced by co-enzyme Q. At this point it was clear that both these natural quinones were essential to electron transport in our model system and that the process involved reduction processes through the liposomal BLM. The results of duplicate experiments were always qualitatively the same but quantitative reproduction of results was not possible due to the sensitivity of the system to the minor effects. Comparisons with the control meant however that the criteria for assigning phenomena to lipid composition were self consistent within any experiment.

The accepted mechanism for this process, Figure 1, would be for the quinone carrier to be reduced at the inner membrane surface and in this (hydroquinone) form to flip from one side of the bilayer to the other, where, it might reduce the substrate cytochrome c. In the oxidized state once more it would, at some later time, flip back to start the cycle once more. These processes require that the membrane phase to be fluid. A practical test of this in our system was to make liposomes with dimyristylphosphatidylcholine (DMPC) which has a liquid crystal transition temperature of 23°C. At the temperature of preparation such bilayers would be fluid but below the transition temperature they would be in the gel state, effectively preventing cross-membrane migration of large molecules. In these experiments the temperature was kept well below the transition temperature and the results showed that there was no electron transport, due to the quinone carrier, Figure 3.

As a further test of the mechanism a membrane was prepared with 20% deoxyhydrocholesterol, 75% PC and 5% vitamin K_1. Thi_ too gave no electron transport. Since cholesterol is well-known for reducing the fluidity of lipid hydrocarbon of the BLM, this provides additional evidence for the flip-flop mechanism of both vitamin K_1 and coenzyme Q in bilayer membranes.

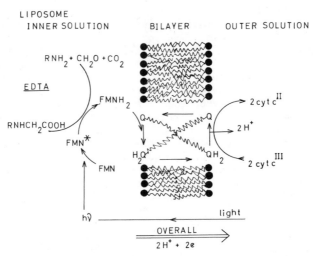

Fig. 1. Schematic representation of the photochemically activated electron transport mechanism.

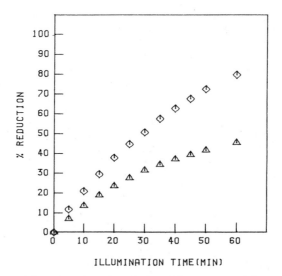

ILLUMINATION TIME (MIN)

Fig. 2. A typical reduction involving liposomes with membrane composition; 95% PC (egg lethicin) and 5% Vitamin K_1. (atom %).

20

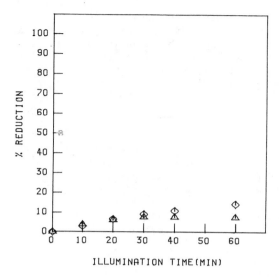

Fig. 3. A demonstration of the loss of carrier function using synthetic liposomes with saturated fatty acid chains, membrane composition; 95% DMPC and 5% Vitamin K_1.

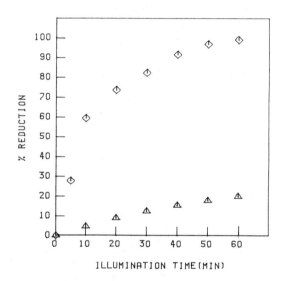

Fig. 4. Test of liposomes with composition similar to the inner mito-chondrial membrane, composition; 38% PC, 38% PE, 19% CL and 5% Coenzyme Q.

The inner mitochondrial membrane is known to contain high level of cardiolipin (negatively charged) which binds positively charged cytochrome, aiding electron transport. Preliminary studies with liposome membranes containing phosphatidyl serine (PS), which is also anionic at these pHs, showed that leakage was high when these were dispersed in cytochrome c solution, due (it appeared) to strong electrostatic binding, and partial incorporation of the resulting complex in the membrane, which is well-documented elsewhere[6]. The problem of leakage was removed in this work by using solutions of high ionic strength which screened the electrical charges and reduced the initial electrostatic binding. As a result lipo-somes containing PS behaved normally and allowed electron transport, as

Figure 2. A particularly rapid reduction was achieved with a liposomal membrane with approximately the same composition as the inner mitochondrial membrane, 38% PC, 38% PE (phosphatidylethanolamine), 19% CL and 5% coenzyme Q. This shows that the lipid composition of the inner membrane is (even in transposition to such model environments) a very suitable medium for electron transport by lipid quinones.

REFERENCES

1. S. S. Anderson, I. G. Lyle, and R. Paterson, Nature, 259:147 (1976).
2. I. G. Lyle, PhD Thesis, Glasgow University (1978).
3. D. W. Deamer and A. D. Bangham, Biochim.Biophys.Acta, 443:629 (1976).
4. G. K. Radda and M. Calvin, Biochemistry, 3:384 (1964).
5. R. E. Bayer, J.Biol.Chem., 234:688 (1959).
6. D. Papahadjopoulos, M. Cowden, and H. Kimelberg, Biochim.Biophys.Acta, 330:8 (1973).

DISORDER OF COUPLED TRANSPORT IN RENAL BRUSH BORDER MEMBRANE
VESICLES FROM RABBITS WITH EXPERIMENTAL FANCONI SYNDROME INDUCED
BY ANHYDRO-4-EPITETRACYCLINE AS A MODEL OF MEMBRANE DISEASE

Yoshimasa Orita, Yoshifumi Fukuhara, Takeshi Nakanishi
Masaru Horio and Hiroshi Abe

The First Department of Medicine
Osaka University Medical School
Fukushima, Osaka, Japan

INTRODUCTION

The acquired Fanconi Syndrome induced by toxic substances, such as
outdated tetracycline and maleic acid, is characterized by a reversible
disorder in renal proximal tubule transport affecting glucose, amino acids
and other substances[1]. It is of interest that each of the molecular or
ionic species whose urine excretion is increased in this condition is
coupled to Na^+ influx across the brush border membrane of the proximal
tubule[2]. In the present study we used isolated brush border membrane
vesicles (BBMV) from the kidneys of anhydro-4-epitetracycline(epi-TC)-
administered rabbits and carried out a series of experiments to investigate
an inhibition of Na^+-dependent D-glucose transport.

MATERIALS AND METHODS

We injected 100 mg/kg body weight of epi-TC intravenously to male
albino rabbits to induce Fanconi syndrome. Glucosuria was detected in all
rabbits one hour after injection, then the rabbits were sacrificed by
decapitation. The kidneys were perfused in situ with Dulbecco's PBS until
blood-free. Cortices were dissected and BBMV were prepared by the Ca^{2+}
precipitation method[3].

Uptake Measurements (Rapid Filtration Technique)

Unless otherwise noted, the procedure for uptake and exchange measurements was as follows. A 20 or 50 μl aliquot of vesicles (1.5-3.0 mg/ml)
was placed in a glass test tube and at time zero a 50 or 100 μl aliquot of
incubation medium containing radioactively labelled ligands and other
constituents as required was added. After an appropriate time the reaction
was terminated by the addition of a 10-fold dilution of ice-cold stop
solution. After addition of the stop solution the vesicles were applied to
a Millipore filter (HAWP 0.45 μm) under light suction. The filter was then
washed by a further 4.5 ml of the stop solution. The filter, which retained BBMV, was dissolved in scintillation fluid and counted along with
samples of the incubation medium and appropriate standards.

The detailed compositions of the various media used in each experiment are given in the figure legends. In general 10 mM Tris-HEPES (pH 7.4) containing 100 mM mannitol (Buffer A) plus 100 mM KSCN (Buffer AK) was used as the basis for all media. In this way 100 mM KSCN was present in equilibrium across the vesicle membrane at all times. When 12.5 µg/mg protein of valinomycin was added, the transmembrane potential difference was sufficiently short-circuited[3]. The stop solution was 10 mM Tris-HEPES with 300 mM NaCl, 1 mM phlorizin and sufficient mannitol to compensate for intravesicular osmolarity.

Phospholipid Analysis of BBMV

The lipid of BBMV (1 mg protein/ml) was extracted by the method of Bligh and Dyer[4]. The extract was dried under nitrogen gas and dissolved with chloroform/methanol (1:2) solution. The samples and the authentic phospholipid standards were applied to high-performance thin-layer chromatography plate (HPTLC; Merck)[5]. The separation was performed with the solvent methyl acetate:n-propanol:chloroform:methanol:0.25% aqueous potassium chloride (25:25:25:10:9). Phospholipids were detected by using iodine. Phosphatidylcholine (PC), phosphatidylethanolamine (PE) and phosphatidylserine (PS) were well separated and their amounts of control BBMV and BBMV from epi-TC-administered rabbits were judged by visual inspection.

The content of phospholipids in vesicle membranes was determined colorimetrically using thiocyanatoiron reagent, which forms hydrophobic complexes with phospholipids[6]. Briefly a 0.2 ml aliquot of BBMV (5mg/ml) was mixed with the same volume of ethanol, then the mixture was heated for 5 min at 60°C. Added with 0.5 ml of thiocyanatoiron reagent and 0.3 ml of 0.17 N hydrochloric acid, the mixture was further incubated for 5 min at 35°C. Resultant thiocyanatoiron-phospholipids complexes were extracted with 1.5 ml of dichloroethane, shaking on the thermomixer and centrifuged at 1,500 rpm for each 2 min. The absorbance was observed at 470 nm.

Chemicals

Epi-TC was prepared by the method of McCormick et al.[7] and the purity (above 90%) was checked by high-performance liquid chromatography. D-[^{14}C]-glucose, D-[^{3}H]-glucose and L-[^{3}H]-glucose were obtained from New England Nuclear Corp. (U.S.A). Phlorizin was purchased from Sigma Chemical Co. (U.S.A.). Other chemicals were of highest purity available from commercial sources. All solutions were filtered (Sartorius 0.2 µm) before use.

All experiments were carried out at least in triplicate at 25°C. The errors shown in the figures (provided they are large enough to illustrate) are standard errors. Results of representative experiments are given and evaluated statistically by the independent t-test.

RESULTS

Enzymatic Characterization of Vesicle Preparations (Table 1)

The enrichment of brush border membrane marker enzyme activities was approximately 10-fold relative to the starting tissue homogenate in both vesicle preparations. Ratios of maltase activity before and after digestion by 1% Triton X-100 were 97.5±2.3% in control BBMV and 93.4±2.2% in BBMV from epi-TC-administered rabbits, indicating that both vesicle preparations are right-side out.

24

Distribution Space of D-glucose (Figure 1)

The plots of uptake vs. inverse osmolarity in both vesicle preparations were linear and superimposable, indicating that D-glucose is equilibrating with the same osmotically active space.

Table 1. Effect of EPI-TC Administration on Enzyme Activities of Renal Homogenate and Brush Border Membrane

enzyme	homogenate		brush border membrane	
	control	epi-TC	control	epi-TC
alkaline phosphatase	0.198±0.011	0.184±0.010	1.692±0.119(8.6)	1.500±0.134(8.2)
trehalase	0.126±0.010	0.143±0.015	1.105±0.123(8.8)	1.255±0.169(8.8)
γ-glutamyl transpeptidase	0.798±0.037	0.711±0.047	7.098±0.647(9.9)	7.012±0.694(9.9)
5'-nucleotidase	0.033±0.003	0.031±0.004	0.209±0.029(6.3)	0.169±0.015(5.5)
Na^+-K^+ ATPase	0.056±0.005	0.062±0.006	0.009±0.003(0.2)	0.011±0.005(0.2)
glucose-6-phosphatase	0.098±0.005	0.084±0.006	0.060±0.004(0.6)	0.052±0.008(0.6)
succinate dehydrogenase	0.020±0.002	0.022±0.001	0.005±0.001(0.3)	0.006±0.001(0.3)

Data are expressed as μmoles/min./mg protein.   mean±S.E.

All results were from ten determinations.

Parenthesis represents enrichment, brush border membrane/homogenate.

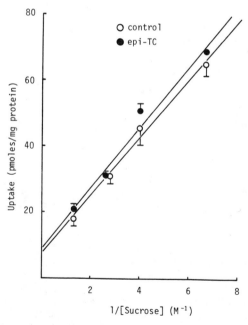

Fig. 1. Effect of extravesicular osmolarity on the equilibrium uptake of D-glucose. Vesicles (○, control BBMV; ●, BBMV from epi-TC-administered rabbits) were prepared in Buffer AK containing 100 mM NaCl plus valinomycin. The incubation medium was the same buffer containing 0.1 mM D-[^{3}H]-glucose and 150-750 mM sucrose (final concentration). Uptake was measured after 5 min of incubation.

Time Course of D-glucose Uptake into BBMV (Figure 2)

Figure 2 shows the time course of D-glucose uptake into BBMV. The overshoot uptake was biggest in the presence of NaSCN and smallest in the presence of Na_2SO_4 under zero trans condition, indicating that the magnitude of the overshoot is due to the chaotropic action or the membrane permeability of anion of each sodium salt. When NaCl or Na_2SO_4 was used, the uptake of D-glucose from 10 to 30 s into BBMV isolated from epi-TC-administered rabbits (dashed line) was significantly lower than that of control (solid line).

Time Course of D-glucose Uptake into BBMV Under Voltage-clamped Condition

The uptake of D-glucose into BBMV isolated from epi-TC-administered rabbits was significantly less than that from control rabbits from 5 to 30 s (Figure 3). The intravesicular volume, which was estimated by the uptake at 60 min, was not affected by epi-TC administration.

Permeability of Brush Border Membrane (Figure 4)

The permeability of BBMV from epi-TC-administered rabbits was the same as that of control BBMV as reflected in the rate of L-glucose efflux. These data indicate that there is no significant change in vesicle membrane permeability properties with epi-TC administration in vivo.

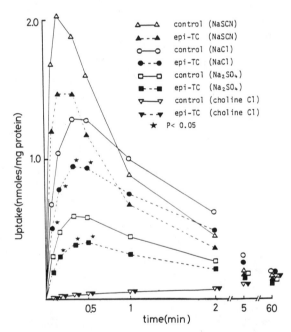

Fig. 2. Time course of D-glucose uptake into BBMV. Vesicles were pre-loaded with Buffer A (pH 7.4) and the incubation medium was the same buffer containing 0.1 mM D-[^{14}C]-glucose and 100 mM NaCl (○,●), 100 mM NaSCN (△,▲), 50 mM Na_2SO_4 (□,■) or 100 mM choline chloride (▽,▼) (final concentration). Open symbols represent uptake into vesicles from normal control rabbits and closed symbols represent uptake into vesicles from epi-TC-administered rabbits. Asterisks indicate significant difference between uptake into control vesicles and that into vesicles from epi-TC-administered rabbits (p<0.05). In this figure the error bars were removed for the clarity of illustration.

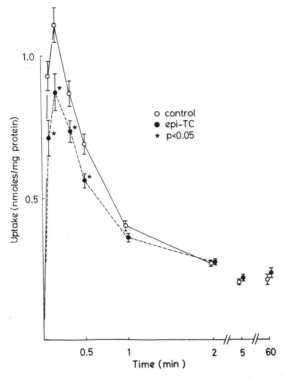

Fig. 3. Time course of D-glucose uptake into BBMV under voltage-clamped
condition. Vesicles were preloaded with Buffer AK (pH 7.4)
containing 200 mM mannitol plus 12.5 µg valinomycin/mg protein.
The incubation medium was the same buffer containing 0.1 mM
D-[^{14}C]-glucose and 100 mM NaCl (final concentration). Open and
closed circles represent uptake into vesicles from normal control
rabbits (o) and those from epi-TC-administered rabbits (●),
respectively. Asterisks indicate significant difference between
uptake into control vesicles and that into vesicles from epi-TC-
administered rabbits ($p < 0.05$).

Kinetics of D-glucose Transport (Figure 5)

    D-Glucose fluxes were measured into vesicles as a function of
D-glucose concentration under equilibrium exchange condition[8,9]. Sodium-
dependent component of D-glucose flux was calculated from the total flux by
subtracting the sodium-independent flux determined with choline replacing
sodium. Under equilibrium exchange condition we can remove the possibility
of dissipation of Na$^+$-gradient by epi-TC. Significant decrease in
D-glucose flux into BBMV from epi-TC-administered rabbits was observed.
Figure 5 shows the Eadie-Hofstee plots of the experiments. Least square
fits of these plots yield K_m = 27.4±1.8 vs.   22.3±0.9 mM and V_{max} = 141.3±9.9
vs. 79.3±7.6 nmoles/min/mg protein (control vs. BBMV from epi-TC-
administered rabbits, respectively).

Phospholipids Of BBMV

    HPTLC of the lipid extracts from control BBMV and epi-TC-administered
BBMV reveal PC, PS, PE and an unknown band between PS and PE, respectively
(data not shown). We cannot find any difference in the migration of each
band from the origin, or in the density of each band between control BBMV
and BBMV from epi-TC-administered rabbits.

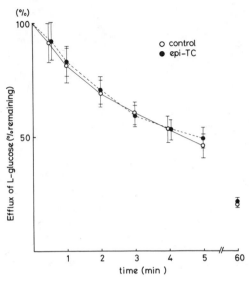

Fig. 4. The time dependence of L-glucose efflux from BBMV. Vesicles (o, control BBMV; ●, BBMV from epi-TC-administered rabbits) were preloaded with Buffer A containing 2.4 μM L-[³H]-glucose, and then diluted (1:6) into the same medium without L-glucose. L-Glucose retained in these vesicles was measured as a function of time. Volumes of control and epi-TC-administered BBMV at time zero (equilibrium volume) were not significantly different from each other (1.56±0.12 and 1.73±0.04 μl/mg protein, respectively).

Colorimetric determination of total phospholipids in BBMV gave the values of 0.15 mg lipid/mg protein in control BBMV and 0.14 mg lipid/mg protein in BBMV from epi-TC-administered rabbits. In this method, the dithiocyanatoiron reagent reacts with PC most sensitively, while the sensitivities to other phospholipids (PE, PS, phosphatidylinositol and phosphatidic acid) are 82.9, 37.8, 23.2 and 1.2%, respectively, relative to PC at 100%[6].

DISCUSSION

In this paper we present the results of a series of experiments which examine the properties of Na[+]-dependent D-glucose transporter in BBMV from epi-TC-administered rabbits.

As shown in Table 1, the purity of isolated membrane vesicles, as judged by enzyme activity, is not different between control BBMV and BBMV from epi-TC-administered rabbits. Both vesicle preparations are oriented right-side-out, evidenced by the latency of maltase activity upon digestion of the membranes with Triton X-100. We find that D-glucose is taken up into osmotically active space in BBMV from epi-TC-administered rabbits and exhibits the same distribution volume and the same degree of nonspecific binding and trapping as in control BBMV (Figure 1). There is no difference in the passive permeability properties between control BBMV and BBMV from epi-TC-administered rabbits, measured by the time dependence of the L-glucose efflux from BBMV (Figure 4). Moreover, we demonstrate that the contents of phospholipids in both vesicle preparations are the same and that HPTLC's are identical. Thus, we cannot find any alteration in the integrity of the membrane property of BBMV from epi-TC-administered rabbits. Therefore, the decreased Na[+]-dependent D-glucose transport into

28

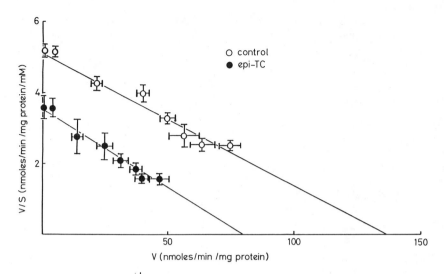

Fig. 5.   Kinetics of Na$^+$-dependent D-glucose transport. Vesicles were
preequilibrated with Buffer AK (pH 7.4) containing D-glucose and
100 mM NaCl plus 12.5 μg valinomycin/mg protein for at least 1 h
before. The incubation medium was the same buffer containing
trace amounts of D-[^{3}H]-glucose. Flux at each concentration of
D-glucose was determined by the slope of the regression line from
the plots of t vs. -ln(1-a/A). Flux observed in the presence of
choline were subtracted from that found in the presence of sodium
to obtain the sodium-dependent component of D-glucose flux.
Eadie-Hofstee plots of sodium-dependent components of D-glucose
flux were shown. Open and closed circles were from each single
membrane preparation (○; from normal control rabbits; ●; from
epi-TC-administered rabbits, respectively). Least squares fits
to these plots yield K_m = 27.4±1.8 vs. 22.3±0.9 mM, and V_{max} =
141.3±9.9 vs. 79.3±7.6 nmoles/min/mg protein, respectively.

BBMV from epi-TC-administered rabbits (Figure 2 and Figure 3) is not due to
any change of vesicle size, of tracer binding or trapping by the membrane,
or of the leakage of uptaken glucose.

      Kinetic analyses of Na$^+$ dependent components of D-glucose flux in
control BBMV and BBMV from epi-TC-administered rabbits indicate transport
sites obeying Michaelis-Menten kinetics with K_m=27.4±1.8 vs. 22.3±0.9 mM,
and V_{max}=141.3±9.9 vs. 79.3±7.6 nmol/min/mg protein, respectively (Figure
5). Thus we demonstrate that V_{max} in BBMV from epi-TC-administered rabbits
is significantly smaller than that in control BBMV (p<0.01), while the
apparent K_m values are not different from each other. One simple explanation for these results is a decrease in the number or in the translocation
of D-glucose transporters which are affected by epi-TC or its metabolites
in vivo.

SUMMARY

      We used isolated brush border membrane vesicles (BBMV) and investi-
gated how epi-TC, known to induce experimental Fanconi syndrome, inhibited
Na$^+$-dependent D-glucose transport.

      The purity, the sidedness, the lipids, the distribution space of
D-glucose, or the passive permeability properties between control vesicles

and those from epi-TC-administered rabbits were not different from each other. The overshoot uptake of D-glucose into BBMV from epi-TC-adminis-tered rabbits was significantly smaller than that of control with or without voltage-clamping. We measured D-glucose flux under the equilibrium exchange condition to obtain kinetic parameters, and V_{max} in BBMV from epi-TC-administered rabbits (79.3 ± 7.6 nmoles/min/mg protein) was signifi-cantly smaller than in control BBMV (141.3 ± 9.9 nmoles/min/mg protein), while K_m values were not different from each other (22.3 ± 0.9 and 27.4 ± 1.8 mM, respectively). These results suggest that sodium-dependent D-glucose carriers per se are affected by epi-TC in vivo.

Acknowledgement

This work was supported in part by a Grant-in Aid for Scientific Research (A) 56440039 from the Ministry of Education, Science and Culture, Japan.

REFERENCES

1. J. H. Stein and R. T. J. Kunau, Miscellaneous disorders of tubular function, in: "Diseases of the Kidney," 3rd ed., L. E. Eearly and C. W. Gottschalk, eds., Little, Brown and Co., Boston (1979).
2. M. B. Burg, Renal handling of sodium, chloride, water, amino acids, and glucose, in: "The Kidney," 2nd ed., B. M. Brenner and F. C. Rector, eds., W. B. Sauuders, Philadelphia (1981).
3. R. J. Turner and A. Moran, Heterogeneity of sodium-dependent D-glucose transport sites along the proximal tubule: evidence from vesicle studies, Am.J.Physiol., 242:F406 (1982).
4. E. G. Bligh and W. J. Dyer, A rapid method of total lipid extraction and purification, Can.J.Biochem.Physiol., 37:911 (1959).
5. F. Vitiello and J. P. Zanetta, Thin-layer chromatography of phospholipids, J.Chromatogr., 166:637 (1978).
6. Y. Yoshida, E. Furuya and K. Tagawa, A direct colorimetric method for the determination of phospholipids with dithiocyanatoiron reagent, J.Biochem., 88:463 (1980).
7. J. R. D. McCormick, S. M. Fox, L. L. Smith, B. A. Bitler, J. Reichenthal, N. E. Origoni, W. H. Muller, R. Winterbottom, and A. P. Doerschuk, Studies of reversible epimerization occurring in the tetracycline family. The preparation, properties and proof of structure of some 4-epi-tetracyclines, J.Am.Chem.Soc., 79:2849 (1957).
8. U. Hopfer, Kinetics of Na^+-dependent D-glucose transport, J.Supramol. Struct., 7:1 (1977).
9. H. de Smedt and R. Kinne, Temperature dependence of solute transport and enzyme activities in hog renal brush border membrane vesicles, Biochem.Biophys.Acta, 648:274 (1981).

ADVANCE AND PERSPECTIVE IN MEDICAL

APPLICATION OF LIPOSOMES

Yoshinori Nozawa

Department of Biochemistry
Gifu University School of Medicine
Tsukasamachi-40, Gifu, Japan

INTRODUCTION

Since the concept of liposomes emerged in 1965, a considerable amount of information has been accumulating regarding the physicochemical properties of liposomes. A vast knowledge on the artificial lipid vesicles has advanced the prospect of their practical use as carriers of enzymes and drugs in therapeutic as well as preventive medicine. The first trial of the enzyme replacement therapy was performed in 1976 for a patient with a lysosomal storage disease, Type II glycogenosis (Pompe's disease)[1]. Although the stability of liposomes in blood circulation was assessed, several difficult problems remained to be circumvented prior to the extended applications to other cases of enzyme-deficient diseases. On the other hand, liposomes have received considerable attention as highly useful vehicles for various drugs such as antibiotics, hormones and antitumor agents. In this chapter I will review the general aspects of liposome's medical applications and then discuss a few topics including the targeting.

CRITERIA OF LIPOSOMES FOR MEDICAL APPLICATIONS

For the effective use of liposomes for pharmacological or clinical purpose, in addition to the fundamental requirements of non-toxicity and bio-degradability, liposomes should satisfy the following criteria.

1. High encapsulation efficiency
2. High stability and rapid clearance
3. Selective direction to target tissue (targeting)
4. Effective intracellular delivery

There are several methods of incorporating materials into liposomes; aqueous phase entrapment of water-soluble substances, hydrophobic interaction with lipid bilayers, electrostatic binding with charged lipids and chemical linkage with liposomal surface. And the rate of entrapment may vary depending on the type of the incorporating methods. Furthermore, the critical factors which control the encapsulation efficiency are size (small, large), architecture (unilamellar, multilamellar), and surface charge (neutral, negative, positive). A high degree of liposomal stability is a great advantage as drug carriers and a number of different factors influence the liposomal integrity after administration into blood circulation. Two major factors are lipolytic activity in plasma and exchange or transfer of liposomal lipids to plasma lipoprotein. The most important criterion is the selective transfer of entrapped liposomes to target tissues or cells, and extensive efforts have been made to achieve this purpose by tailoring the liposomal characteristics. Once liposomes arrive at the target sites, the entrapped contents should be delivered with an efficient rate into the cells.

STABILITY IN AND CLEARANCE FROM BLOOD CIRCULATION

The liposomal integrity after administration into the blood circul-ation can be disturbed by lipid hydrolysis due to lipases present in plasma. It was also shown that plasma is able to degrade liposomes by an active transfer of lipid molecules to the high density lipoprotein (HDL)[2]. Thus the entrapped contents are leaked before they are trans-ferred to the sites where they are needed. Our results demonstrate that the release of liposome-entrapped calcein can be repressed by incorporating cholesterol into liposomes. For example, calcein is released very rapidly after liposomes composed of egg lecithin (PC) without cholesterol are incubated at 37°C in human plasma (unpublished data). However, incorpor-ation of equimolar cholesterol allows the liposomes to retain the entrapped compound with no leakage. A plausible explanation for the great stability of cholesterol-rich liposomes is that the presence of cholesterol inhibits the phospholipid loss to plasma lipoproteins. The recent study using lipoprotein-deficient mice has explored the role of HDL in destabilizing action on liposomes, providing direct evidence that among various lipo-protein fractions only HDL decreases liposomal stability[3].

The stability of liposomes in the circulation shows an inversed relationship with the rate of clearance; increase of stability produces decrease of clearance. Therefore, the manipulation of lipid composition can change permeability and clearance of liposomes.

TARGETING

A major barrier to the therapeutic use of liposomes is the difficulty of directing them to the specific site of action. Numerous studies have shown that untargeted liposomes in the blood circulation are preferentially removed by reticuloendothelial cells, e.g. Kupffer cells of the liver and macrophages of the spleen[1,4,5]. This natural homing is of limited advantage for the medical application of liposomes.

The distribution of liposomes in vivo can be modulated to some extent by their size, charge and fluidity. For example, the large liposomes are more readily taken up than the small liposomes by the cells of the retic-uloendothelial system. The uptake of phosphatidylcholine (PC)/phospa-tidylserine (PS) (4:1) liposomes into the liver is higher than that of distearoyl-PC/cholesterol (Chol) (2:1), and the opposite situation is observed with the spleen. This may imply that liposomes, which have the reduced fluidity due to the presence of cholesterol and phospholipid with high transition temperature, are stable in the circulation and transferred at the reduced rate into the liver[6]. Furthermore, the uptake of these small solid liposomes is rapid without a time-lag in non-parenchymal cells and slow with a lag phase in parenchymal cells.

A slower transport in parenchymal cells than in Kupffer cells has been also observed with small, positively charged liposomes consisting of dipalmitoyl-PC/Chol/stearylamine. In general, for the effective direction of liposomes to parenchymal cells, their size should be smaller than 0.1 μm in diameter. An affinity of neutral and positively charged liposomes for lymphoid tissue is indicated by enhanced uptake of the liposomes by the spleen. The same trend is observed with lymph nodes when liposomes are injected in the footpad of rats[7].

Since such a rather unspecific method to manipulate the carrier's lipid composition does not satisfy fully the criteria as an ideal pharmacological vehicle, a more specific approach is needed. This can be achieved by binding the recognition molecules to the liposome surfaces and some examples are listed below:

1. Glycoproteins
a. asialofetuin (AF), asialoorosomucoid (AO)
b. pronase-treated asialofetuin (AFSC)

2. Glycolipids
a. cerebrosides (galacto-, lacto-)
b. ganglioside (GM_1, asialoGM_1)
c. sulfatide

3. Sugar-linked PE and cholesterol
a. galactose-PE
b. lactose-PE
c. aminomannose(galactose)-cholesterol

4. Antibody
a. polyclonal
b. monoclonal

Of these various marker molecules the carbohydrate moiety linked to proteins or lipids plays a crucial role in influencing the targeting of liposomes. Extensive studies have been done on the selective direction to the liver, since the plasma membrane of liver cells has a receptor capable of recognizing the terminal β -galactose residues. Two representative compounds, asialofetuin (AF) and asialoorosomucoid (AO) can be bound to liposomes by using a spacer, maleimidophenylbutyrate (MPB) which interacts with PE at the one end and asialoprotein at the other end, so that the galactose residue is oriented to be exposed on the liposomal surface[8]. To reduce the antibody producing potency, the asialofetuin (AF) was subjected to pronase digestion to remove the large peptide portion, leaving asialosuger moiety intact. When the pronase treated asialofetuin (AFSC)-containing liposomes (PC/Chol/PS, 4:5:1) labelled with ^{14}C-tripalmitin were injected intravenously in the rat, a rapid increase of the radioactivity was observed in the liver with a concomitant decrease in the blood. And the results of cell fractionation showed that a substantial proportion of radioactivity accumulates in the lysosomal fraction shortly after intravenous administration (Table 1).

Table 1. Subcellular Distribution of Pronase-Treated Asialofetuin Liposomes Injected in Rat Liver

Fractions	Relative percent of total radioactivity	
	3 min	120 min
	(%)	
Nuclei	8.1	14.3
Mitochondria-lysosomes	56.2	48.7
Microsomes	11.9	15.6
Postmicrosomal supernatant	9.5	2.1

PC/Chol/PS (4:5:1)-AFSC (^{14}C-tripalmitin)

In addition to these macromolecule markers, glycolipids with low molecular weight were introduced as an effective homing device on the surface of liposomes. GM_1 ganglioside is found to mediate the preferential direction of the liposomes to hepatic parenchymal cells[9]. Subsequently, lactosylceramide (LacCer) was also employed as a recognition glycolipid marker[10]. Incorporation of LacCer into small unilamellar liposomes (DMPC)/Chol/PS, 4:5:1) decreased their half-life in the blood and instead increased the rate of uptake in the liver. This greatly enhanced uptake was mostly due to the uptake in the parenchymal cells. A small increase in the uptake by the non-parenchymal cells was observed but its mechanism was unknown. Recently, a successful in vivo transfer of exogenous gene to rat liver cells was accomplished by using the LacCer-containing liposomes[11]. The results of the intrahepatic cellular distribution of the exogenous gene DNA showed a surprising finding that a considerable amount of DNA was present in the endothelial cells which have no galactose-recognizing receptor.

We have compared the uptake rate in the liver cells of liposomes targeted with AFSC and LacCer. As depicted in Figure 1, there is no difference between the untargeted control liposome and the AFSC-containing liposome in the non-parenchymal cells, whereas in parenchymal cells both targeted liposomes are taken up to a much larger extent than the control liposome; the uptake rate is higher in the LacCer-liposome than in the AFSC-liposome. The effect of LacCer was abolished by the injected AF, indicating the implication of a galactose-specific receptor in endocytosis of LacCer-liposomes. When the invertase-encapsulated liposomes with LacCer were injected in the rat, the recovered enzyme activity in the parenchymal cells was higher as compared with the case of the untargeted liposomes. No significant changes in the activity were observed regardless of the presence or absence of the glycolipid (Table 2).

Taken together with the liposomal size, the effective targeting to the parenchymal and Kupffer cells can be achieved by using small uni-lamellar and large multilamellar liposomes with recognition markers, respectively[12].

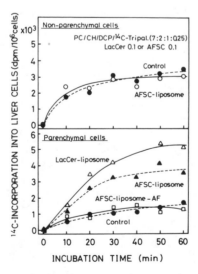

Fig. 1. Effects of targeting molecules upon incorporation of liposomes into parenchymal and non-parenchymal cells of rat liver. LacCer, lactocerebroside; AF, asialofetuin; AFSC, pronase-treated asialofetuin.

Table 2. Distribution of Invertase Encapsulated in Targeted Liposomes in Hepatocytes

	Control liposomes	Lactosylceramide liposomes
Parenchymal cells		
Activity / cells	$0.23U/10^7$ cells	$0.58U/10^7$ cells
Total activity	20.6 ± 0.6 %	49.0 ± 4.0 %
Non-parenchymal cells		
Activity / cells	$2.09U/10^7$ cells	$1.44U/10^7$ cells
Total activity	39.4 ± 4.0 %	26.7 ± 1.1 %

PC/Chol/DCP (7:2:1)-LacCer (0.1)

The targeting to the brain is one of the most difficult obstacles to overcome, and without this achievement the enzyme replacement therapy of various inherited enzyme deficiencies is far from being successful. Although great efforts have been made to accomplish the passage of the enzyme-loaded liposomes through the blood-brain barrier, no satisfactory approaches have been obtained as yet. But recently evidence was presented which indicates that incorporation of human brain sulfatide in the conventional liposomes (PC/Chol, 7:2) enables the vesicles to pass through the barrier[13]. The enzyme entrapped in the liposomes (PC/Chol/sulfatide, 7:2:1) was administered intravenously to a rat and 5% of the total enzyme activity was recovered in the brain. This finding was confirmed by electron microscopic examination of horseradish peroxidase- or ferritin-encapsulated liposomes. The rate of the liposomal transport across the blood-brain barrier appears to vary depending on the source of sulfatide.

The liposomal surface can be also modified by incorporating into liposomes of aminosugar derivatives of cholesterol[14]. Addition of a 6-aminomannose derivative of cholesterol produces initial accumulation of liposomes in the lung after intravenous injection followed by retention in the liver and spleen. The subcutaneous administration of the liposomes bearing the aminomannosyl cholesterol confers substantial specificity for polymorphonuclear leukocytes.

The liposomal targeting with antibody is considered to be a promising clue for efficient delivery of their contents to specific cells or tissues. Liposomes bearing heat-aggregated IgM molecules and noncovalently associated IgM molecules have been observed to interact selectively with target cells. However, these procedures produce a limited enhancement of liposome - cell interaction. This disadvantage was circumvented by a novel method for the covalent binding of Fab' antibody fragments to the liposomal surface[15], which involves a disulfide interchange reaction between the thiol group on the Fab' fragment and a pyridyldithio derivative of PE molecule in liposomes.

A novel approach of targeting without any recognition molecules on the liposomal surface is local hyperthermia and its mechanism is based on the rationale that liposomes become leaky and release their entrapped contents near the lipid-crystalline transition temperature[16]. The potency to kill E. coli cells was enhanced by heating the neomycin-encapsulated liposomes (DPPC) with phase transition temperature at 41°C. This procedure was applied in the tumor target by using the methotrexate-bearing DPPC-liposomes, resulting in a greatly enhanced accumulation of drug in the hyperthermic area[17].

An alternative physical approach of targeting is incorporation of a pH-sensitive molecule, palmitoyl homocysteine into liposomes. Liposomes release their entrapped contents at the regions where pH is lower than the normal value, such as tumors and sites of inflammation[18].

APPLICATIONS IN MEDICINE

Extensive studies in the past decade in animals and limited trials in patients have greatly advanced the concept that liposomes can be used as carriers of therapeutic or diagnostic materials. Various approaches in medical applications have been made as outlined below.

1. Enzyme replacement therapy (lysosomal storage disease)
2. Antimicrobial therapy (protozoa, fungus)
3. Cancer chemotherapy
4. Metal storage disease (chelation therapy)
5. Intra-articular administration (arthritis, cancer)
6. Diabetes (insulin carrier)
7. Immunopotentiation (adjuvant)
8. Radioimmunodetection (cancer)
9. Gene carrier

The entrapment of enzymes provides a valuable strategy for both enzyme delivery and protection against degradation and immunologic complications. Furthermore, since enzymes transported via liposomes were found to be largely localized in lysosomes, the initial suggestion of enzyme replacement therapy was for the inherited lysosomal storage disease. Actually, enzyme-loaded liposomes were used for two patients; the liposome containing glucoamylase from Aspergillus niger for Pompe's disease[1] and the liposome containing β-glucosidase for Gaucher's disease[19]. The practical use of enzyme-entrapped liposomes is still hopeful but at the present time it has to await an achievement of the effective targeting to desired tissues including the brain.

The liposomal drugs are useful for treatment of tropical diseases, leishmaniasis, malaria and trypanosomiasis. Especially, in leishmaniasis in which the causative agent resides in the reticuloendothelial system, the intravenous administration of liposome-entrapped antimonial drugs results in successful therapy[20]. Liposomes (DPPC/Chol/DCP, 1:0.75:0.11) containing neutral glycolipids (gluco-, galacto-, lacto-cerebroside) suppress the sporozoite-induced malaria in mice[21]. Recently, liposomes have been used as carriers of amphotericin B in the treatment of fungal diseases, histoplasmosis, crypto-coccosis, and candidiasis in animals[22].

For cancer therapy chemotherapy, numerous drugs have been entrapped in liposomes, because most of the anti-cancer agents are highly toxic to normal tissues such as bone marrow, kidney, and heart. Entrapment of drugs is expected to reduce toxicity and excretion rate from blood and to fulfill an effective drug delivery to the lesions. However, no approach has achieved clinical acceptance. A promising clue for cancer chemotherapy is local hyperthermia combined with the liposomal administration of drugs[16]. An alternative contribution of liposomes to cancer therapy can be made by activation of resting macrophages with lymphokine encapsulated in vesicles, which leads to tumoricidal action.

Metal storage disease also is one of the reticuloendothelial system diseases and the intravenously given liposomes containing a chelator, DTPA have been observed to exert therapeutic effects for animals loaded with plutonium, mercury or iron[23].

Liposomes possess the appropriate properties as efficient adjuvant for use in humans. This merit of liposomes has been assessed using various antigens, such as hepatitis B antigen, cholera toxic, and diphtheria toxoid etc.[24]. For example, the injection of hepatitis B antigen-loaded liposomes produces a higher antibody response in guinea pigs immunized with the viral antigen.

Radioimmunodetection attracts much attention as a very useful technique of cancer diagnosis. Patients with cancer metastases were given liposomes containing secondary antibody after injection of ^{131}I-labelled primary antibody. The examination by external scintigraphy showed a greatly improved discrimination of tumors by using the secondary antibody-laden liposomes[25].

The use of liposomes to transfer gene into culture cells permits incorporation of exogenous DNA into the cells and also expression of the introduced gene. A recombinant plasmid encoding rat preproinsulin I was entrapped in large liposomes (PC/PS/Chol, 8:2:10) and given intravenously to rats. The results suggest that insulin is synthesized in liver and spleen for a short time and also that this gene expression correlates with the blood insulin level and with glycemia[26].

Acknowledgements

This work was supported by grants from the Ministry of Education, Science and Culture and from the Ministry of Health and Welfare of Japan.

REFERENCES

1. D. A. Tyrrell, B. E. Ryman, B. R. Keeton, and V. Dubowitz, Use of liposomes in treating Type II glycogenosis, Br.Med.J., ii:89 (1976).
2. G. Gregoriadis and A. C. Allison, "Liposomes in Biological Systems," John Wiley & Sons, New York (1980).
3. J. Senior, G. Gregoriadis, and K. A. Mitropoulos, Stability and clearance of small unilamellar liposomes: Studies with normal and lipoprotein-deficient mice, Biochim.Biophys.Acta, 760:111 (1983).
4. G. Gregoriadis, Targeting of drugs, Nature, 265:407 (1977).
5. G. Gregoriadis, Targeting of drugs with molecules, cells and liposomes, TIPS, July:304 (1983).
6. C. F. Gotfredsen, S. Fokjer, E. L. Hjorth, K. D. Jorgensen, and M. -C. Debroux-Guisset, Disposition of intact liposomes of different compositions and of liposomal degradation products, Biochem.Pharmacol., 32:3381 (1983).
7. B. F. Ryman, R. F. Jewkes, K. Jeyasingh, M. P. Osborne, H. M. Patel, V. J. Richardson, M. H. N. Tattersall, and D. A. Tyrrell, Potential applications of liposomes to therapy, Annu.N.Y.Acad.Sci., 308:281 (1978).
8. Y. Banno, K. Ohki, and Y. Nozawa, Targeting of asialofetuin sugar chain-bearing liposomes to liver lysosomes, Biochem.Int., 7:455 (1983).
9. A. Surolia and B. K. Bachhawat, Monosidoganglioside liposome-entrapped enzyme uptake by hepatic cells, Biochim.Biophys.Acta, 497:760 (1977).
10. H. H. Spanjer and G. L. Scherphof, Targeting of lactosylceramide-containing liposomes to hepatocytes in vivo, Biochim.Biophys.Acta, 734:40 (1983).
11. P. Soriano, J. Dijkstra, A. Legrand, H. Spanjer, D. Londos-Gagliardi, F. Roerdink, G. Scherphof, and C. Nicolau, Targeted and nontargeted liposomes for in vivo transfer to rat liver cells of a plasmid containing the preproinsulin I gene, Proc.Natl.Acad.Sci.USA, 80:7128 (1983).

12. Y. E. Rahman, E. H. Cerny, K. R. Patel, E. H. Lau, and B. J. Wright, Differential uptake of liposomes varying in size and lipid composition by parenchymal and Kupffer cells of mouse liver, Life Sci., 31:2061 (1982).

13. K. Yagi, M. Naoi, H. Sakai, H. Abe, H. Konishi, and S. Arichi, Incorporation of enzyme into the brain by means of liposomes of novel composition, J.Appl.Biochem., 4:121 (1982).

14. M. R. Mauk, R. D. Gamble, and J. D. Baldescwieler, Vesicle targeting: Timed release and specificity for leukocytes in mice by subcutaneous injection, Science, 207:309 (1980).

15. F. J. Martin, W. L. Hubbell, and D. Papahadjopoulos, Immunospecific targeting of liposomes to cells: A novel and efficient method for covalent attachment of Fab' fragments via disulfide bonds, Biochemistry, 20:4229 (1981).

16. M. B. Yatvin, J. N. Weinstein, W. H. Dennis, and R. Blumenthal, Design of liposomes for enhanced local release of drugs by hyperthermia, Science, 202:1290 (1978).

17. J. N. Weinstein, R. L. Magin, M. B. Yatvin, and D. S. Zaharko, Liposome and local hyperthermia: Selective delivery of methotrexate to heated tumors, Science, 204:188 (1979).

18. M. B. Yatvin, W. Krentz, B. A. Horwitz, and M. Shinitzky, pH-sensitive liposomes: Possible clinical applications, Science, 210:1253 (1980).

19. P. E. Belchetz, I. P. Braidman, J. C. W. Grawley, and G. Gregoriadis, Treatment of Gauchers disease with liposome-entrapped glucocere-broside: β-Glucocidase, Lancet, ii:116 (1977).

20. R. R. C. New, M. L. Chance, S. C. Thomas, and W. Peters, Antileishmaniasis activity of antimonials entrapped in liposomes, Nature, 272:55 (1978).

21. C. R. Alving, I. Schneider, G. M. Swartz, and E. A. Steck, Sporozoite-induced malaria: Therapeutic effects of glycolipids in liposomes, Science, 205:1142 (1979).

22. G. Lopez-Berestein, R. Mehta, R. H. Hopfer, K. Mills, L. Kasai, K. Mehta, V. Fainstein, M. Luna, E. M. Hersh, and R. Juliano, Treatment and Prophylaxis of disseminated infection due to Candida albicans in mice liposome-encapsulated amphotericin B, J.Infect. Dis., 147:939 (1983).

23. Y. -E. Rahman, M. A. Rosenthal, and E. A. Cerny, Intracellular plutonium: Removal by liposome-encapsulated chelating agent, Science, 180:300 (1973).

24. G. Gregoriadis, Liposomes: A role in vaccines, Clin.Immunol.News Lett., 2:33 (1981).

25. G. M. Barratt, B. E. Ryman, R. H. J. Begent, P. A. Keep, F. Searle, J. A. Boden, and K. D. Bagshawe, Improved radioimmunodetection of tumors using liposome-entrapped antibody, Biochim.Biophys.Acta, 762:154 (1983).

26. C. Nicolau, A. L. Pape, P. Soriano, F. Fargette, and M. -F. Juhel, In vivo expression of rat insulin after intravenous administration of the liposome-entrapped gene for rat insulin I, Proc.Natl.Acad. Sci.USA, 80:1068 (1983).

MEASUREMENT AND INTERPRETATION OF FRICTION COEFFICIENTS

OF WATER AND IONS IN MEMBRANE POLYMERS

M. A. Chaudry and P. Meares

Chemistry Department
University of Aberdeen
Scotland

INTRODUCTION

The potential of hyperfiltration as a large scale desalination technique was revealed when Loeb and Sourirajan[1] prepared asymmetric membranes that combined high permeability to water with good salt rejection. These membranes were prepared from cellulose acetate. Previously, many readily available polymers had been screened, in the form of thin dense films, for water permeability and salt rejection. Cellulose acetate was identified from that work as being the most promising material[2]. By the middle 1960's vigorous research had begun to prepare membrane polymers superior to cellulose acetate. In particular, the mechanical strength and resistance to hydrolysis and degradation in hostile environments of cellulose acetate were seen to be inadequate for many applications. Furthermore, although the best cellulose acetate membranes showed excellent salt rejection, they were difficult to prepare repeatably and were sensitive to minor variations in the casting and annealing routines.

The search for new membrane materials has followed two divergent lines. In the first, cellulose acetate has been modified in various ways e.g. by using mixed cellulose esters or cellulose ethers and by the attachment of graft chains along the backbone. In the second, attempts have been made to invent and prepare polymers, chemically unrelated to cellulose acetate, that might have desirable properties as hyperfiltration membranes. The outcome of twenty years research has been the discovery of five or six materials with permeabilities and rejection properties approaching, but rarely surpassing, those of cellulose acetate. However, these synthetic materials have somewhat superior mechanical, thermal and chemical stabilities[3].

Evidently membrane scientists have been unable to specify the requirements of an ideal membrane material with sufficient precision to enable polymer chemists to devise suitable molecules systematically. The objective of the program described here was to add to the understanding of how efficient desalination by hyperfiltration takes place at the molecular level.

Although various, and sometimes bizarre, mechanisms have been proposed to explain desalination by hyperfiltration, it is now widely accepted that in polymeric membranes the so-called "solution-diffusion" theory is essentially correct[4]. According to this theory the active layer of the membrane, whether it be asymmetric or composite, acts as a homogeneous phase in which water and salt dissolve as uniformly distributed solutes. A definite equilibrium distribution is set up at the membrane/solution interfaces with respect to each of the components: water and salt. This distribution is a function of temperature and concentration; it may also be a function of pressure although little attention has been paid to this possibility to date. The ratio of the concentrations of salt and water in the membrane is lower than in the solutions and this is an important factor in determining the salt rejection.

Water and salt in the membrane are transported across it, driven by the gradients of chemical potential. Their separate fluxes are influenced also by their mobilities in the membrane. The mobilities are functions of many factors including the physico-chemical states of water and salt e.g. whether the latter exists as ions or as ion pairs and the extent of their hydration in the membrane. Finally, the original solution – diffusion theory assumed zero coupling between salt and water fluxes in the membrane.

Although the solution – diffusion model is essentially sound, certain details are open to question; in particular it is not obvious that salt and water in the membrane will be uniformly distributed at the molecular level. If the water formed hydrogen bonded molecular clusters and the ions retained specific hydration to some extent then one could not readily dismiss the possibility of some degree of coupling between their fluxes.

The research to be described attempts to characterize in finer detail than hitherto the elementary processes that occur when aqueous salt solutions are forced through cellulose acetate under pressure. It is hoped that such enhanced understanding will help guide steps towards improved membrane materials.

SPECIFIC PROPERTIES OF MEMBRANE POLYMERS

In order to confine attention to the intrinsic properties of the membrane polymer, unclouded as far as possible by the technique of membrane preparation, it was decided to use homogeneous solvent-cast films annealed in water in the way commonly applied to asymmetric membranes. The thickness of such films can easily be measured accurately and the parameters of the polymer, such as permeability coefficient, specific conductance etc. can then be determined. It was recognized that the measurement of relatively small fluxes would be necessary.

To gain a picture of the ways in which water and ions are distributed in and move through the membrane, and how they interact with it and with one another, it is necessary to measure a set of transport and equilibrium properties. The choice of properties was governed by two considerations: convenience of measurement and their correlation through linear non-equilibrium thermodynamics so as to yield binary molecular friction coefficients.

Several presentations of non-equilibrium thermodynamics adapted to hyperfiltration have been published. Spiegler and Kedem[5] recognized that the steep concentration gradient in the membrane and its dependence on flow rate militated against the use of a simple discontinuous treatment. They introduced instead a system of differential equations for the two fluxes: salt and water. This aspect of the problem has been further explored by Vink, first by expanding the phenomenological transport coefficients in a power series of concentration[6,7] and, more recently by introducing a frictional coefficient formalism[8]. A treatment directed towards high flow rates has been developed by Pusch[9]. All these treatments have regarded water and salt as the independent fluxes.

Relatively little experimental work has been published suitable for testing these theories and all of it is on asymmetric membranes of uncertain effective thickness. Furthermore, several studies have confirmed that the ionic charge of cellulose acetate cannot be ignored[10,11]. Consequently a formalism which takes account of the anions and cations as separate species is needed to describe fully the interactive processes important during hyperfiltration.

A suitable treatment was published several years ago[12] based on the definition of differential discontinuous phenomenological coefficients as proposed by Michaeli and Kedem[13]. This theory was tested and then used to characterize transport in an ion-exchange membrane[14] and the data were transformed into binary friction coefficients[15]. The same approach can be applied, essentially unaltered, to the study of transport in a membrane under hyperfiltration conditions. Only a brief outline of the theory is given here because the details can be found in the original publications.

THEORETICAL BACKGROUND

The linear non-equilibrium thermodynamic development of membrane transport can take many forms. Different sets of independent forces or fluxes can be chosen and their conjugates identified through the invariance of the entropy production. A convenient set of fluxes for the present purpose is i the electric current density, ϕ_1 the molar flux density of cations, ϕ_v the volume flux density defined by

$$\phi_v = V_s \phi_1 / \nu_1 + V_w \phi_w \qquad (1)$$

where ϕ_w is the molar flux density of water, V_w and V_s are the molar volumes of salt and water and ν_1 the number of cations per "molecule" of neutral salt. The forces conjugate to these fluxes were derived first by Kedem and Katchalsky[16].

In the limit that these forces are infinitessimally small, the linear equations between fluxes and forces contain the desired differential conductance coefficients $\mathcal{L}_{\alpha\beta}$. The flux equations take the form

$$\partial\phi_1 = \mathcal{L}_\pi (\partial\pi/\nu_1 c_s) + \mathcal{L}_{\pi p} \partial(p-\pi) + \mathcal{L}_{\pi E} \partial E \qquad (2)$$

$$\partial\phi_v = \mathcal{L}_{p\pi} (\partial\pi/\nu_1 c_s) + \mathcal{L}_p (p-\pi) + \mathcal{L}_{pE} \partial E \qquad (3)$$

$$\partial i = \mathcal{L}_{E\pi} (\partial\pi/\nu_1 c_s) + \mathcal{L}_{Ep} (p-\pi) + \mathcal{L}_E \partial E \qquad (4)$$

Here p, π and E are, respectively, the pressure difference, osmotic pressure difference and electric potential difference, measured with electrodes reversible to the anions and immersed in the bulk solutions, between opposite sides of the membrane. c_s is the concentration of salt in the external solutions which, in the limit of infinitessimal forces, have the same concentration on both sides. Onsager reciprocity applies to the conductance coefficients viz.

$$\mathcal{L}_{\alpha\beta} = \mathcal{L}_{\beta\alpha} \qquad (5)$$

$\mathcal{L}_E$ is given by the membrane conductance per unit area, κ, and $\mathcal{L}_{pE}$ by the product of κ and the electro-osmotic volume flow per unit current, W. $\mathcal{L}_{\pi E}$ may be obtained from the transport number of cations in the membrane,

t_1, by using the relation

$$\mathcal{L}_{\pi E} = t_1 \kappa / z_1 F \tag{6}$$

The remaining three coefficients have to be extracted from data obtained under conditions where a concentration gradient exists across the membrane. The relevant expressions are[14]

$$\mathcal{L}_p = \frac{\mathcal{L}_{pE}^2}{\mathcal{L}_E} + \left[\left(\frac{\partial \phi_v}{\partial \pi}\right)^{i=0} - \left(\frac{\partial \phi_v}{\partial \pi}\right)^{i=0}_{p=0}\right]\left(\frac{\partial \pi}{\partial p}\right)^{i=0} \tag{7}$$

$$\mathcal{L}_\pi = \frac{\mathcal{L}_{\pi E}^2}{\mathcal{L}_E} + \nu_1 c_s \left(\frac{\partial \phi_1}{\partial p}\right)^{i=0} \Bigg/ \left(\frac{\partial \pi}{\partial p}\right)^{i=0}$$

$$+ \nu_1 c_s^2 \left[\left(\frac{\partial \phi_v}{\partial p}\right)^{i=0} - \left(\frac{\partial \phi_v}{\partial \pi}\right)^{i=0}_{p=0}\left(\frac{\partial \pi}{\partial p} - 1\right)^{i=0}\right]\left(1 - \frac{\partial p}{\partial \pi}\right)^{i=0} \tag{8}$$

$$\mathcal{L}_{p\pi} = \frac{\mathcal{L}_{pE}\mathcal{L}_{\pi E}}{\mathcal{L}_E} + \nu_1 c_s \left[\left(\frac{\partial \phi_v}{\partial p}\right)^{i=0} - \left(\frac{\partial \phi_v}{\partial \pi}\right)^{i=0}_{p=0}\left(\frac{\partial \pi}{\partial p} - 1\right)^{i=0}\right] \tag{9}$$

It is seen that to evaluate these conductance coefficients, four differential coefficients are needed. $(\partial \phi_v / \partial \pi)^{i=0}_{p=0}$ is the variation of osmotic volume flow with osmotic pressure difference across the membrane. $(\partial \phi_v / \partial p)^{i=0}$, $(\partial \phi_1 \partial p)^{i=0}$ and $(\partial \pi / \partial p)^{i=0}$ are obtained from hyperfiltration data. The determination of the $\mathcal{L}_{\alpha\beta}$ coefficients is described in the experimental section.

These conductance coefficients are complex and not suitable for the interpretation of molecular phenomena. The particle flux densities ϕ_1, ϕ_2 and ϕ_w, where subscripts 1, 2 and w refer to cations, anions and water respectively, can be evaluated from ϕ_1, ϕ_v and i. They conjugate with the electro-chemical potential differences of the molecular species as driving forces.

The coefficients relating these fluxes and forces can be designated $\mathcal{L}_{ik}$ where i and k are 1, 2 or w. Their relation to the $\mathcal{L}_{\alpha\beta}$ coefficients is given by[12]

$$(\mathcal{L}_{ik}) = \Gamma^{-1}(\mathcal{L}_{\alpha\beta})\,\Gamma^{-1T} \tag{10}$$

where Γ is the matrix

$$\Gamma = \begin{pmatrix} 1 & 0 & 0 \\ V_s/\nu_1 & 0 & V_w \\ z_1 F & z_2 F & 0 \end{pmatrix} \tag{11}$$

42

Coefficients such as $\mathcal{L}_{ik}$ indicate how far the flux of i is modified by the thermodynamic force on k. The mechanism of this coupling is indirect and a further transformation of the coefficients permits a set characterizing binary molecular events to be derived. Before making this transformation it is convenient to note that, because the $\mathcal{L}_{ik}$ refer to conditions of infinitessimal forces in a homogeneous membrane, it is legitimate to transfer from the discontinuous model of the membrane system to a continuous description of the membrane phase itself in which the driving forces are the negatives of the gradients of electro-chemical potentials $\partial\mu_i/\partial x$. This involves a dimensional change from the discontinuous forces $\Delta\bar{\mu}_i$. If the corresponding set of continuous coefficients is denoted by (ℓ_{ik}^i) their relation to the $(\mathcal{L}_{ik})$ is

$$(\ell_{ik}) = \delta(\mathcal{L}_{ik}) \tag{12}$$

where δ is the thickness of the membrane to which the $(\mathcal{L}_{ik})$ refer.

Instead of expressing the fluxes as linear functions of every force, a set of equations can be written representing the forces as linear functions of the fluxes; this is

$$\begin{aligned}
\partial\mu_1/\partial x &= r_{11}\phi_1 + r_{12}\phi_2 + r_{1w}\phi_w \\
\partial\mu_2/\partial x &= r_{21}\phi_1 + r_{22}\phi_2 + r_{2w}\phi_w \\
\partial\mu_w/\partial x &= r_{w1}\phi_1 + r_{w2}\phi_2 + r_{ww}\phi_w
\end{aligned} \right\} \tag{13}$$

The coefficients r_{ik} can be evaluated from ℓ_{ik} by making use of

$$r_{ik} = A_{ik}/|\ell| \tag{14}$$

where A_{ik} is the co-factor of ℓ_{ik} and $|\ell|$ is the determinant of the matrix of (ℓ_{ik}).

The coefficients r_{ik} are called resistance coefficients. Their physical meaning is clear and can be deduced in the following way. If two of the fluxes in equations (13), ϕ_2 and ϕ_w say, are set at zero and the other, ϕ_1, is set at unity one has

$$\partial\mu_1/\partial x = r_{11} \tag{15}$$

$$\left. \begin{aligned}
\partial\mu_2/\partial x &= r_{21} \\
\partial\mu_w/\partial x &= r_{w1}
\end{aligned} \right\} \tag{16}$$

Evidently r_{11} equals the force per mole on component 1 needed to drive it at unit flux while all other components are held at rest. r_{21} and r_{w1} are equal to the forces per mole which have to be applied to components 2 and w respectively to hold them at rest while 1 is driven through the membrane at unit flux. These forces are expected to be opposite in direction from the flux of 1 and to measure the drag on 1 exerted by the other mobile components, 2 and w. Consequently, the resistance to the motion of component 1 exerted by the membrane matrix itself, r_{m1}, is given by

$$r_{11} = -(r_{21} + r_{w1} + r_{m1}) \tag{17}$$

There has been a long held desire to represent the resistance to flow of molecules as arising from a frictional interaction between the species moving at different average velocities. An excellent discussion and formulation has been given by Spiegler[17]. If the components are uniformly distributed at concentrations $\bar{c}_i$, the mean velocities, v_i, in the flux direction are given by

$$v_i = \phi_i/\bar{c}_i \qquad (18)$$

On the frictional analogy, the force F_{ik} on i per mole due to the relative motion of k can be expressed as

$$F_{ik} = f_{ik}(\phi_i/\bar{c}_i - \phi_k/\bar{c}_k) \qquad (19)$$

i.e. f_{ik} is the force per mole on i caused by the amount of k normally in its environment at unit relative velocity of i and k. The coefficients f_{ik} are called molar friction coefficients; they obey the relations

$$\bar{c}_i f_{ik} = \bar{c}_k f_{ki} \qquad (20)$$

Equations (13) and (19) are mutually consistent provided

$$r_{ii} = \sum_{k \neq i}^{m} f_{ik}/\bar{c}_i \qquad (21)$$

and

$$r_{ik}_{i \neq k} = f_{ik}/\bar{c}_k \qquad (22)$$

An examination of the various f_{ik} and their variations with concentration and temperature can throw a good deal of light on the distributions and interactions of the mobile components in and with the membrane[18]. The remainder of this paper describes such an examination for the transport of potassium chloride and water in cellulose acetate.

EXPERIMENTAL AND DATA HANDLING

Membranes

The measurements were made on homogeneous membranes of Eastman Kodak 398-3 cellulose acetate because there have been probably more hyper-filtration data published on this than on any other single material. The membranes were cast from 2% w/v solution in pure dry acetone by pouring on to clean, levelled glass plates with an appropriate number of tape strips affixed around the edges to produce a film of the desired thickness. Complete evaporation of solvent was allowed to take place very slowly in a controlled atmosphere chamber; after about 20 hr the sheet was transferred to a vacuum oven for 3 hr at 40°C to remove all traces of acetone.

The membrane was removed from the plate by soaking in distilled water for at least 24 hr and then annealed in water at 80°C for 30 min. Membranes were stored in water. Wet thicknesses were measured by using a transducer gauge. Hyperfiltration and osmosis experiments were carried out mainly on membranes 10 μm thick but the range 4 μm – 40 μm was examined to demonstrate that flux was inversely proportional to thickness. Electrical measurements were made on membranes about 40 μm thick.

Electrical Measurements

An account has already been given[11] of the measurement of membrane conductance and membrane potential in cellulose acetate. The electro-osmotic permeability was studied by using the cell and methods developed for use with ion-exchange membranes[19].

No electro-osmotic flow was detected in these membranes when in contact with chloride solutions in the range 0.025 - 0.20M at 25°, 35° and 45°C. It was concluded therefore that W was zero. Hence the analysis of the kinetic and equilibrium conductance data and the membrane concentration potential data used previously[11] was valid and the Hittorf transport number of the cations could be evaluated from the membrane potentials without correction for solvent transference.

This method of analysis gave not only $\mathcal{L}_E$, $\mathcal{L}_{pE}$ and $\mathcal{L}_{\pi E}$ over the concentration range covered by the experiments, it gave also more detailed information about the nature of the membranes and their equilibrium with the electrolyte solutions. In particular, the concentration of fixed negative charges on the cellulose acetate and the ratio of the activity coefficients of salt in bulk solution and absorbed by the membrane could be evaluated. The latter enables the Donnan principle to be used to relate the concentrations of salt in the solution and membrane at equilibrium. The data also enabled the diffusion coefficients or mobilities of cations and anions in the membrane to be evaluated. These quantities are important in the determination of the frictional coefficients, as will become apparent.

Hyperfiltration Experiments

Hyperfiltration measurements were made by using a modified version of the cell described already[14]. The Tufnol high pressure reservoir was encased in a steel jacket to permit the cell to be used safely at pressures up to 30 atm. Some redesigning of the membrane holder was required to enable very thin and hard membranes to be mounted in a leak-free way.

Fluxes were small, 0.1 - 0.2 g/hr was typical, and 2-3 days were required to establish steady flow rates and product concentration because of the time required scavenge the low pressure side of the cell with product solution. The volume flow rate was found by weighing the effluent collected in a measured time and its concentration was found by conductance determination. The density of the solution was read from a calibration curve once its conductance was known.

Appropriate calibration experiments were carried out to demonstrate that fluxes were inversely proportional to membrane thickness and that pure water flux was directly proportional to applied pressure. A full account of these hyperfiltration experiments will be published elsewhere.

Although measurements have been made on four salts data on only potassium chloride will be reported here. The pressure range covered was 1.25 - 2.75 MPa. At the higher pressures the rejection of KCl exceeded 97% at all temperatures and concentrations studied. Figures 1 and 2 show, respectively, the volume and salt fluxes as functions of pressure at three temperatures for 0.1M feed solution.

Osmotic Flow Rates

The precise measurement of small osmotic flow rates is rather difficult. A special cell was designed for the purpose but space does not permit a full description here. It consisted of glass chambers about 70 ml in volume so that concentration changes due to salt and water flux during experiments were entirely negligible. The membrane was held against a thin perforated metal foil by a small hydrostatic pressure. Flow was measured from water into the various salt solutions. Measurement was based on observing the movement of a calibrated plunger, driven by a micrometer screw, required to maintain stationary the meniscus in a capillary con-

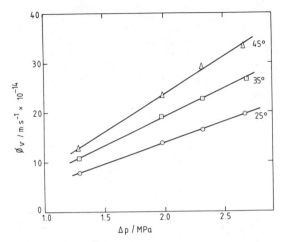

Fig. 1.   Volume flux density ϕ_v from 0.1M KCl as a function of pressure
at three temperatures.   The data are normalized to a standard
membrane 1 m thick.

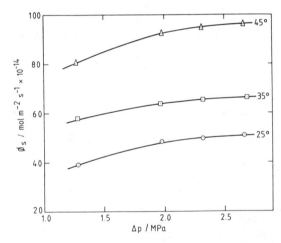

Fig. 2.   Salt flux density ϕ_1 from 0.1M KCl as a function of pressure
at three temperatures.   The data are normalized to a standard
membrane 1 m thick.

nected to the water-filled chamber.   Steady flow was attained after 5-6 hr,
it was inversely proportional to membrane thickness.

The ratio of the osmotic flow at unit osmotic pressure and the volume
flow at unit hydrostatic pressure from any given solution lay in the range
0.96 - 0.98; the value being highest at the highest temperature.   This
value may be identified with the reflection coefficient of the membrane.

Calculation of Differential Conductance Coefficients, $\mathcal{L}_{\alpha\beta}$

The three coefficients $\mathcal{L}_E$, $\mathcal{L}_{pE}$ and $\mathcal{L}_{\pi E}$ are found from the electrical
measurement in which there is no concentration difference across the
membrane.   The concentration dependence of the coefficients therefore
introduces no problems in their evaluation.

$\mathcal{L}_p$, $\mathcal{L}_\pi$ and $\mathcal{L}_{p\pi}$ are found by using equations (7), (8) and (9). Four differential coefficients have to be extracted from the hyperfiltration and osmotic data. The somewhat complex procedure for doing this has been described in detail in a previous paper[14]. Essentially the same procedure was used here and the work was simplified because $(\partial\phi_v/\partial p)^{i=0}$ and $(\partial c_e^{-1}/\partial p)^{i=0}$ were found to be constants at constant temperature. This enabled the relation

$$\frac{\partial\phi_1}{\partial p} = c_e \left[\frac{\partial\phi_v}{\partial p} - \phi_v c_e \frac{\partial(1/c_e)}{\partial p} \right] \tag{23}$$

to be used with advantage. Here c_e is the effluent or product concentration. Representative data are shown in Figures 2 and 3.

The theoretical development[14] makes it clear that the differentials have all to be evaluated with a constant concentration on the high concentration side (0.1M in the work described here) and at values of p and π that correspond with a particular effluent or low concentration c_e. The values of $\mathcal{L}_{\alpha\beta}$ so evaluated relate to the membrane in contact with solutions of concentration c_e. It was convenient to choose three values of c_e, namely 3, 4 and 5 mol m^{-3}.

The presence of fixed charges in cellulose acetate made kinetic and equilibrium conductance measurements sensitive to pH at very low concentrations and also t_1 was too close to unity for a reliable estimation of $(1-t_1)$. These difficulties were overcome by using the data on electrical properties at higher concentrations already published[11] to evaluate the concentrations of cations and anions that would be present in the membrane at 3, 4 and 5 mol m^{-3}. These concentrations were used with the diffusion coefficients of the ions in the membrane to calculate κ and t_1 at the concentrations and temperatures of interest. In other words, the theoretical treatment of the electrical properties[11] was used to extrapolate the data to the required low concentrations. Since W was found to be zero the calculation of $\mathcal{L}_E$, $\mathcal{L}_{pE}$ and $\mathcal{L}_{\pi E}$ was then straightforward.

$\mathcal{L}_p$, $\mathcal{L}_{p\pi}$ and $\mathcal{L}_{p\pi}$ were then determined from the data already described. The six $\mathcal{L}_{\alpha\beta}$ coefficients are given in Table 1 for KCl at three concentrations and temperatures. It should be noted that the values are expressed in S.I. units and hence refer to a notional membrane 1 m thick.

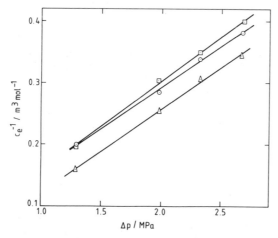

Fig. 3. Reciprocal of the product concentration c_e plotted against pressure for 0.1M KCl feed at three temperatures.

Table 1. $\mathcal{L}_{\alpha\beta}$ Coefficients for KCl in Cellulose Acetate (398-3)

c_e mol m^{-3}	$10^8 \times \mathcal{L}_E$ ohm^{-1} m^{-2}	$10^{20} \times \mathcal{L}_p$ m^3 N^{-1} s^{-1}	$10^{17} \times \mathcal{L}_\pi$ mol^2 m^{-3} N^{-1} s^{-1}	$10^{13} \times \mathcal{L}_{\pi E}$ mol V^{-1} m^{-2} s^{-1}	$\mathcal{L}_{pE}$ m V^{-1} s^{-1}	$10^{20} \times \mathcal{L}_{p\pi}$ mol N^{-1} s^{-1}
			$T = 25^\circ C$			
3	7.75	8.93	2.41	7.88	0	1.79
4	7.79	8.97	4.95	7.91	0	2.39
5	8.12	9.02	5.58	7.99	0	3.07
			$T = 35^\circ C$			
3	9.78	12.30	2.92	9.95	0	1.01
4	9.99	12.35	4.54	10.03	0	1.34
5	10.26	12.64	5.11	10.13	0	1.68
			$T = 45^\circ C$			
3	11.60	15.46	2.80	11.78	0	0.41
4	11.88	15.52	4.82	11.88	0	0.54
5	12.24	15.66	5.51	12.02	0	0.68

By using equations (10), (11) and (12) the $\mathcal{L}_{\alpha\beta}$ coefficients have been converted to the corresponding ℓ_{ik} coefficients. Equation (14) was then used to invert them and so obtain the r_{ik} resistance coefficients.

Equations (21) and (22) show how the binary friction coefficients f_{ik} can be obtained from the r_{ik} but their calculation requires a knowledge of the molar concentration of the ions and water in the membrane. The membranes used here correspond with the heat treated type IIa of Meares, Craig and Webster[20]. Their value of the water content at 25°C was taken because the deswelling in solutions of only 3, 4 and 5 mM salt concentration must be negligible. The water contents at 35° and 45°C were estimated from this value by using the enthalpy of sorption, -5.62 kJ mol^{-1}, given by Merten[21] together with the Clapeyron equation.

To avoid relying on the Donnan equation to estimate the ion concentrations at such low external solution concentrations, use was made of the relation

$$\bar{c}_i = RT/D_i r_{ii}, \quad i=1 \text{ or } 2 \tag{24}$$

to estimate $\bar{c}_1$ and $\bar{c}_2$ from the resistance coefficients r_{11} and r_{22} and the ion diffusion coefficients D_1 and D_2 determined previously[11].

The summations in Equation (21) include terms for the ion of water-polymer matrix interactions e.g.

$$\bar{c}_1 r_{11} = f_{12} + f_{1W} + f_{1m} \tag{25}$$

To evaluate coefficients of the type f_{mi} from

$$\bar{c}_i f_{im} = \bar{c}_m f_{mi} \tag{26}$$

a value is needed for the matrix concentration $\bar{c}_m$ i.e. the molar concentration of fixed charges. This is given by the concentration of fixed charges per kg of sorbed water[11] multiplied by the number of kg of water per m^3 of swollen membrane[20].

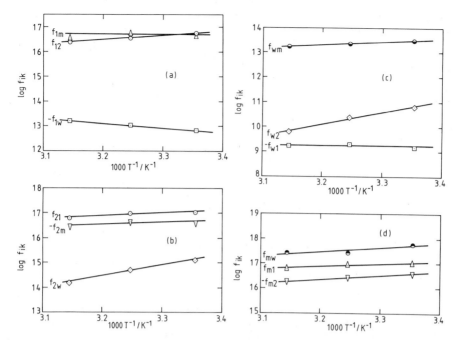

Fig. 4.   Decadic logarithms of binary molar friction coefficients, f_{ik}, expressed in $J \, sm^{-2} \, mol^{-1}$, plotted against the reciprocal of the absolute temperature, T/K. $1=K^+$, $2=Cl^-$, w=water, m=matrix.   (a) frictional interactions of the cations (b) frictional interactions of the anions (c) frictional interactions of water (d) frictional interactions of the matrix.

The decadic logarithms of the f_{ik}, averaged over the three values of c_e, are plotted in Figure 4 against 1000/T.

COUPLING OF FLUXES AND FRICTIONAL INTERACTION OF PARTICLES

The resistance coefficients r_{ik} obey the Onsager reciprocity principle and indicate the extent to which a flux of one component exerts a dragging force on another.   It is a requirement of thermodynamics that

$$r_{ii} \, r_{kk} \geqslant r_{ik}^{2} \qquad (27)$$

and Caplan has suggested[22] that the efficiency of coupling between components can be measured by the coupling coefficients q_{ik} defined by

$$q_{ik} = -r_{ik}/(r_{ii} \, r_{kk})^{\frac{1}{2}} \qquad (28)$$

q_{ik} can take negative and positive values provided $q_{ik} \leqslant 1$.   Positive values indicate that a flow of i contributes a force on k in the direction of the flow of i.   The value $q_{ik} = 1$ implies a completely tight coupling between the molar flows of i and k.   Negative values of q_{ik} indicate that a flow of i in one direction causes a force on k in the other direction. Negative values are often found where i and k are ionic species of like charge.

Table 2 gives the values of q_{ik} for potassium chloride and water in cellulose acetate.   It is immediately obvious from q_{12} that the flows of potassium and chloride ions experience very strong positive coupling,

Table 2. Coupling Coefficients q_{ik} for KCl in Cellulose Acetate (398-3) (1=K⁺, 2=Cl⁻, w=water)

$c_e/mol\ m^{-3}$	q_{12}	$10^4 \times q_{1w}$	$10^2 \times q_{2w}$
T = 25°C			
3	0.81	-0.49	0.86
4	0.91	-1.47	0.49
5	0.92	+0.50	0.55
T = 35°C			
3	0.81	-1.61	0.38
4	0.87	-1.51	0.33
5	0.87	-0.69	0.36
T = 45°C			
3	0.75	-1.96	0.17
4	0.85	-1.98	0.12
5	0.87	-1.33	0.11

perhaps decreasing slightly with increasing temperature and with decreasing concentration. This observation is far more significant than that, in the absence of electric current, the fluxes of potassium and chloride ions must be equal. In a well hydrated cation-exchange membrane q_{12} has values less than 0.15 [15]. The conclusion must be that in cellulose acetate, where the degree of hydration and the dielectric constant are much lower than in a swollen ion exchanger, the cations and anions exist and diffuse overwhelmingly as associated ion pairs.

It will be noted that, in Table 2, q_{1w} is four orders of magnitude and q_{2w} two orders of magnitude lower than q_{12}. Evidently the coupling between the ion and water fluxes is only relatively weak. This has been thought, for a long time, to be the case and, indeed, the elementary solution-diffusion model of hyperfiltration entirely neglects such coupling. Its effect on the leak of salt is however not negligible because in the ion flux equations the relatively small coefficients r_{1w} and r_{2w} are multiplied by the large flux of water.

Possibly no great significance can be attached to the negative sign of q_{1w}. Within the experimental uncertainty it is probably zero. Because most of the potassium ions in the membrane are counterions to the fixed negative charges on the polymer and, in the low dielectric constant of the medium, they spend most of their time closely associated with the fixed charges, it is easily understood that the average effect of the water flux per potassium ion is small and lower than would be anticipated if the potassium ions were uniformly distributed in the membrane.

The chloride ions have entered the membrane against the Donnan exclusion effect of the fixed charges. Electrostatic repulsion will cause them, at the molecular level, to avoid the fixed charges and so to occupy preferentially regions of the polymer that have opened up to admit the absorbed water. Thus, although there is no specific interaction between the anions and water, their coupling is larger and more significant than that of the cations and water. Hence q_{2w} is positive and much larger than q_{1w}.

The balances between the individual frictional interactions experienced by each species can be seen more clearly in Figure 4. Figure 4a shows that the cations undergo about the same degree of interaction with the anions (f_{12}) and the polymeric matrix (f_{1m}). Their interaction with the water (f_{1w}) is, as noted above, lower by three orders of magnitude and negative in sign. Figure 4b shows similarly that the anions undergo interactions of comparable magnitude also with the cations (f_{21}) and the matrix (f_{2m}) but it is important to note that f_{2m} is negative. This is consistent with the observations which have been made on frictional coefficients in ion-exchange membranes and in aqueous electrolyte solutions. The "frictional interaction" between species of like charge is negative. Of course, friction cannot be negative and this observation indicates that the frictional model is inappropriate to describe the interaction between mutually repelling species.

Implicit in the frictional model is the assumption that all species are, on average, uniformly distributed through the total volume because the mean net molecular velocities are calculated from the quotient of the flux density and the concentration per unit volume. In fact long range electrostatic repulsions and attractions lead to a non-uniform distribution. The repelled mobile particles are pushed into closer interaction with other species, in this case with cations and water. Thus one finds the anion-water interaction (f_{2w}) still relatively small but at least an order of magnitude larger than the cation-water interaction.

It can be deduced that the low values of f_{1w} and f_{2w} arise in these results largely from the observation of zero electro-osmotic transport coefficient W. The observations were made in the concentration range 25-200 mM but the calculations refer to concentrations 3-5 mM. Other workers[23,24] have observed that when the solution concentration greatly exceeds the fixed charge concentration, about 6 millimolal in cellulose acetate[11], W falls to zero but it is non-zero at sufficiently low external concentrations. Some doubt must exist therefore as to whether, at the values of c_e studied here W is truly zero.

In Figure 4c it may be seen that the frictional interaction experienced by the water is almost wholly with the membrane matrix f_{wm}. Interactions with the ions are three orders of magnitude lower and play no significant role in determining the drag on, and hence the flux of, water.

The frictional coefficients of the matrix, seen in Figure 4d, are all significant. Not only is f_{mw} clearly the largest but, because the relative velocity of water and matrix is larger than the relative velocity of salt and matrix, the net force experienced by the polymeric chains during hyperfiltration is dominated by their interaction with the water flux. Except for the small and uncertain coefficient f_{1w}, the friction coefficients are seen to decrease as temperature increases but the temperature range studied is too small to permit activation energies to be deduced with confidence.

CONCLUSIONS

This study should be regarded as the first, and to some extent preliminary, installment in a program to compare the particle interactions occurring in various polymer membranes of potential value in hyperfiltration. Some measurements have already been made on an aromatic polyamide and on a polysulphone sulphonic acid. It will be desirable also to confirm the present findings by trying to extend the electrical measurements to solutions of lower concentration.

The clearest findings that have emerged from the results reported here are that the fixed charge concentration of cellulose acetate, although low, exerts a significant influence on the thermodynamic and kinetic behavior of the ions since it helps to restrict the salt uptake and exerts a strong retardation on the movement of the cations. Interaction between ions and water is remarkably small, in contrast with the findings in normal ion-exchange membranes. As a result, only very weak coupling exists between the flows of water and salt. This fact is very important in enabling cellulose acetate membranes to show excellent rejection properties in desalination.

The ion-matrix frictional coefficients are of order 10^{16} while the water-matrix friction is of order only 10^{13}. This permits the water molecules to enjoy a much greater mobility than ions in the membrane.

REFERENCES

1. S. Loeb and S. Sourirajan, Sea water demineralization by means of an osmotic membrane, Adv.Chem.Ser., 38:117 (1962).
2. S. Loeb, in: "Desalination by Reverse Osmosis," U. Merten, ed., MIT Press, Cambridge, Mass. (1966).
3. H. K. Lonsdale, The growth of membrane technology, J.Membrane Sci., 10:81 (1982).
4. H. K. Lonsdale, U. Merten and R. L. Riley, Transport properties of cellulose acetate osmotic membranes, J.Appl.Polym.Sci., 9:1341 (1965).
5. K. S. Spiegler and O. Kedem, Thermodynamics of hyperfiltration (reverse osmosis) criteria for efficient membranes, Desalination, 1:311 (1966).
6. H. Vink, Theory of ultrafiltration, Acta Chem.Scand., 20:2245 (1966).
7. H. Vink, Thermodynamic treatment of membrane processes, Z.Phys.Chem. N.F., 71:51 (1970).
8. H. Vink, Material Transport in Hyperfiltration, J.Chem.Soc.Faraday Trans., I, 71:51 (1970).
9. W. Pusch, Determination of transport parameters of synthetic membranes by hyperfiltration experiments I, Ber.Busenges, Phys.Chem., 81:269 (1977).
10. H. -U. Demisch and W. Pusch, Ion exchange capacity of cellulose acetate membranes, J.Electrochem.Soc., 123:370 (1976).
11. M. A. Chaudry and P. Meares, An electrical study of ion transport in cellulose acetate, in: "Synthetic Membranes Volume 1," A. F. Turbak, ed., ACS Symp.Ser., 153:101 (1981).
12. H. Kramer and P. Meares, Correlation of electrical and permeability properties of ion-selective membranes, Biophys.J., 9:1006 (1969).
13. I. Michaeli and O. Kedem, Description of the transport of solvent and ions through membranes in terms of differential coefficients, Trans.Faraday Soc., 57:1185 (1961).
14. T. Foley, J. Klinowski and P. Meares, Differential conductance coefficients in a cation-exchange membrane, Proc.Royal Soc., A 336:327 (1974).
15. P. Meares, Some uses for membrane transport coefficients, in: "Charged and Reactive Polymers, Vol. 3," E. Selegny, ed., Reidel, Dordrecht (1976).
16. O. Kedem and A. Katchalsky, Permeability of composite membranes Part 1, Trans.Faraday Soc., 59:1918 (1963).
17. K. S. Spiegler, Transport processes in ionic membranes, Trans.Faraday Soc., 54:1408 (1958).
18. P. Meares, Coupling of ion and water fluxes in synthetic membranes, J. Membrane Sci., 8:295 (1981).
19. W. J. McHardy, P. Meares, A. H. Sutton and J. F. Thain, Electrical transport phenomena in a cation-exchange membrane Part II, J.Colloid Sci., 29:116 (1969).

20. P. Meares, J. B. Craig and J. Webster, Diffusion and flow of water in homogeneous cellulose acetate membranes, in: "Diffusion Processes, Vol. 1," J. Sherwood et al., eds., Gordon and Breach, London (1970).

21. U. Merten, "Desalination by Reverse Osmosis," M.I.T. Press, Cambridge, Mass. 169 (1966).

22. S. R. Caplan, The degree of coupling and efficiency of fuel cells and membranes, J.Phys.Chem., 69:3801 (1965).

23. M. Tasaka, S. Tamura, N. Takemura and K. Morimoto, Concentration dependence of electro-osmosis and streaming potential across charged membranes, J.Membrane Sci., 12:169 (1982).

24. G. Dickel and G. Backhaus, Invariance of the point of inversion in binary isotonic osmosis, J.Chem.Soc.Faraday Trans., II 74:124 (1978).

TENTATIVE CLASSIFICATION OF TRANSPORT-REACTION SYSTEMS BY SeCDAR-ANALYSIS

(UPHILL, FACILITATED AND ACTIVE PROCESSES)

Eric Sélégny

University of Rouen, Faculty of Sciences
Membrane Group, UA 500
76130 Mont Saint Aignan, France

ABSTRACT

By considering the involved species: permeant (S), electron (e), carrying species (C), driving species (D), activators (A) and the nature and location of the reactions (R) in logical, symmetrical or dissymmetrical arrangements, the more complex systems are derived from simpler ones: facilitated extractions or transports use equilibrium S+C reactions; D drives S uphill by cotransport (co), symport (s) or antiport (a). In Secdar-1 pumps D is generated (source) or combined (sink) while A reacts irreversibly. In SeCDAR-2 pumps either C or D or S or e sink and source reactions drive the transport process using A_1 activator and A_2 deactivator. The retaining and valve effects (V), energy source (E) and balances involved (B) help in the identification of the motive and pumping couplings. "SeCDAR VEB co usa" (u=uniport) summarizes the elements of this analysis. Representation of the different systems and experimental examples taken from the literature are given in tables. The advantages of this rational analysis are briefly discussed.

In the last two decades a great variety of abiotic, biomimetic, biophysical or biotic transport reaction systems was considered, analyzed and experimented. The thermodynamics of linear or non-linear couplings and the understanding of asymmetry conditions have made fundamental progress[1-6]. Mitchell[7] has proposed a "chemi-osmotic" classification of biotic systems. The first artificial "secondary" active transport pump is already 17 years old[8] and we published the first kinetically established "primary" pump 15 years ago. We have also previously analyzed immobilized enzyme systems[4]. Progressively the synthetic systems, with greater freedom in the selection of the components, became nearly as varied as their biotic equivalents. Meanwhile, the basic principles have been tested and verified. Some clarifications are perhaps needed now in order to distinguish easily the different reaction-facilitated, uphill or reaction-driven, i.e. "active" systems. We want to show here briefly, within the space alloted, how the more complex systems can be derived logically from simpler ones by systematic addition and repetition of the basic components and elements summarized by our classification scheme "SeCDAR VEB co usa". The symbols for this scheme are detailed in Table 1.

A pumped permeant species $\vec{S}$ is accumulated ($S\uparrow$) on the upper side in a form identical to that on the lower side of the membrane. This needs the

Table 1. Notations and Conventions (SeCDAR) (VEB)

S		the <u>permeant</u> studied: moves from the left to the right
	$\bar{S}$	is retained: by charge coupling or cannot link to the carrier
	S*	is a derivative of S: complexed, protonated, oxidized, reduced..
	$\vec{S}$	moves
	S↑	accumulates behind a valve (is pumped).
e		<u>electron</u>.
C		<u>carrying species</u>; links S reversibly: its presence or reactivity are confined to membrane or to membrane and one compartment only. In membrane: <u>mobile</u> carrier (shuttle), <u>fixed</u> carrier (hopping of S), <u>tunnel</u> (ionophores). Simplified notation for SC is S*.
	C*	carrier <u>activated</u> by chemical transformation using A/A' reaction;
	$\bar{C}$	inactivated carrier: cannot link S.
D		<u>driving species</u> is able to drive S uphill.
	$\overset{\leftarrow}{D}$	by antiport: ion exchange (S and D same sign of charge) or competition for C (SC + $\overset{\leftarrow}{D}$ ⇌ C$\overset{\leftarrow}{D}$) or both.
	$\vec{D}$	by symport: charge compensation (S and D or different sign of charge) or reacting on different sites with C (S + $\vec{D}$ + C ⇌ S$\vec{C}$D) or both.
	$\overset{\rightarrow}{DS}$	by cotransport ($\overset{\rightarrow}{DS}$ + C ⇌ $\vec{D}$SC)
A		<u>activating species</u> gives reaction product(s) A' on <u>one</u> side; (A/A').
	A'	represents A → A' transformation; (A/AD) means A + D → AD;
	(A/A')	(A/A')/$\vec{S}$ means A → A' + S where only S moves; (A$_1$/A$_1'$) and (A$_2$/A$_2'$) are different couples on different sides (acid/base, ox$_1$/red$_1$...).
R		<u>reactions</u>: in (locally) irreversible (i.e. dissymmetrical) <u>sink</u> reactions S or D or C reacts with A (S or D or C sink); in <u>source</u> reactions A reacts producing S or D or C*. The (symmetrical) equilibrium S+C or D+C reactions may be <u>displaced</u>. Reactions to be distinguished: a) without modification of the charge (e.g. complexing); b) charge splitting (e.g. hydrolysis); c) charge fusion (e.g. neutralization); d) charge transfer (redox).
V		<u>valves</u> or valve-effects prevent backward transport: a) the barrier is not permeable to the <u>uphill</u> form S or S* (i.e. charge, size, solubility "<u>screenings</u>"); b) the uphill form is retained by charge-coupling (charge balance) or cannot link to C (i.e. $\bar{S}$); c) the carrier sites are occupied by a competing species (D) (Lechatelier effect); d) carrier is inactivated on the upper side ($\bar{C}$); e) refluxing permeant recycled in a reaction sink is repumped. <u>Retention forces</u> from a) to d) are the same as above but <u>downhill</u>.
E		<u>energy</u> coupled onto the pump: a) physical (Lechatelier factors); b) added species (D) (i.e. "osmotic" or "concentration"); c) chemical; d) electrochemical; e) photochemical.
B		<u>balances</u> are responsible for couplings i.e. retentions and motive transport-reaction forces; a) mass balance for each species; b) charge balance in each section (membrane or compartment); c) balance of oxidation numbers; d) conservation of momentum (not considered here volume flows being neglected).
u, s, a, co		uniport, symport, antiport (see D) and schemes 1.0 and 1.3. cotransport.
$\vec{d}$, $\overset{\leftarrow}{d}$,		are sym or antiport "partners" of S that, in opposition with D, <u>do not</u> drive the process but ensure electroneutrality.

cancelling of any retaining forces downhill and a valve effect (V) uphill to prevent back-diffusion. The energy source (E) ("osmotic", chemical, electro, photo-chemical...) coupled onto motility characterizes the process; pumping needs irreversibility and directional transport or reaction site asymmetry[3-9]. Couplings in these systems result from transport and reaction balances (B): of mass, of charge and of oxidation numbers (that reflect electron (e) transfers); it is important to identify the one that is responsible for the driving of the motion.

With the definition adopted for the "carrying species" (C) which is "able to complex reversibly S (in S*) but unable to have access to the totality of the system", all the systems controlled by the law of mass action and the corresponding factors are regrouped, in a self consistent manner, in the class of facilitated and reaction-mediated processes. Furthermore, the developed considerations become independent from the nature (liquid, ion-exchange, lipidic...) of the membrane as long as S and C (that is S and membrane) "match" together. The "driving species" $\vec{D}$, $\vec{D}$ or $\vec{DS}$ link to C reversibly and cross the membrane in antiport (a), symport (s) or cotransport (co) couplings with S. These qualities enable them to drive uphill S pumps and transduce an "osmotic" into another "osmotic" energy.

The point is, that all the "active" (i.e. reaction driven) pumps derive from the preceding processes by adjunction of an activator or deactivator (A) or of both activator and deactivator (A_1, A_2). An irreversible transformation, noted (A/A'), (A_1/A_1') or (A_2/A_2'), is chemically coupled to the activation or generation of one of the species S, e, C, D in "source reactions" and to their consumption or inactivation in "sink reactions". These denominations[3-6,9] recall the second principle and the mathematical equivalence of diffusion-reactions and heat pumps. Irreversible sinks and sources, characterized by the "asymmetry of the chemical pathway" (of the global chemical scheme) furnish energy. However, vectorial transport also needs spatial asymmetry (by the Curie-Prigogine principle); this is obtained either by asymmetrical distribution of a sink, of a source or of their association, or alternatively by asymmetry of boundary transport properties. Sink, source, sink and source, their distribution in space and time and the species concerned differentiate the various active pumps. The reactions (R) involved (complexing, charge splitting or fusion or transfer) give complementary means for classification.

In Table 2 the systems are numbered and listed in order of increasing complexity. They are represented using the symbols of Table 1, in a compact notation where the membrane separating two compartments is reduced to the space bounded by two hyphens. The denominations and the chemical equations corresponding to each system are also given. Typical experimental examples are listed with quotations in Table 3; understandable preference was given to systems studied in our group or to those able to illustrate the independence of the classification from the material of the membrane. However, with the exception of any unwanted omission the list of active pumps is supposed to be comprehensive.

The Systems and their Notation

(Note that in Table 2 and 3 S travels (is pumped) from the left to the right. In the text numbers in brackets will refer to the enumeration of the systems in the Tables).

Table 2. Representation of Transport-reaction Systems

<u>Notation</u>: species in compartment 1 - in membrane - in compartment 2 S
moves <u>from the left to the right</u>. <u>Mixture</u>: S,D <u>reaction</u>: S.A.

1. <u>Passive Transports</u> (no reaction)

(1.0) S-S-S   (1.1) $\vec{S}$,d-$\vec{S}$,$\overleftarrow{d}$-S,$\overleftarrow{d}$ (1.2) $\vec{S}$,$\vec{d}$-$\vec{S}$,$\vec{d}$-S,d   (1.3) Sd-Sd-Sd

  S uniport $\vec{S}$,$\overleftarrow{d}$ antiport $\vec{S}$,$\vec{d}$ symport   Sd cotransport

2. <u>S + C $\rightleftharpoons$ S* Systems</u> (S* only exists where C can penetrate; C is not
   represented)

(2.1) S-S-S*   (2.2) S-S*-S* "facilitated" extractions
(2.3) S*-S-S   (2.4) S*-S*-S "facilitated" desorptions
(3.0) S-S*-S "carrier facilitated" shuttle, absorption-desorption
(4.0) S*-S-S* "extramembrane carrier" (desorption-absorption)
(5.0) S°-S°-S* from carrier ° to carrier * transfer
(5.1) S°-S$^{\bullet}$-S° facilitated electron transfer (ion symport)

<u>S, C, D Uphill Transport</u> pumps

(6.1) D,$\vec{S}$-$\vec{S}$,$\overleftarrow{D}$-$\overleftarrow{D}$,S↑ antiport S+C$\overleftarrow{D}$ $\rightleftharpoons$ SC+$\overleftarrow{D}$ exchange
(6.2) $\vec{S}$,$\vec{D}$-$\vec{S}$,$\vec{D}$-D,S↑ symport S+D+C$\rightleftharpoons$ SDC double carrier

(6.3) $\vec{DS}$-$\vec{DS}$-D,S cotransport S+D $\longrightarrow$$\vec{DS}$ (+C$\rightleftharpoons$$\vec{DSC}$)

<u>One Sink or one Source (SeCDAR-1) Systems</u>

(7.1) S.(A/A')/$\vec{S}$*-$\vec{S}$*-S*   S sink, <u>left</u>   A+S $\rightarrow$ A'+$\vec{S}$*

(7.2) S$\overset{*}{}$(A/A')/$\vec{S}$ -$\vec{S}$ -S   reactive desorption   A+S^{+} $\rightarrow$ A'+$\vec{S}$

(7.3) S-S-(A/A')/S*↑ $\Big\{$ Reactive valves by   A+S $\rightarrow$ A'+S*↑

(7.4) S*-S*-(A/A')/S↑   reaction on the <u>right</u> A+S^{+} $\rightarrow$ A'+S↑

<u>D Sinks</u> (D-$\overline{D}$)

(8.1) $\vec{S}$,(A/AD)-$\vec{S}$,$\overleftarrow{D}$-$\overleftarrow{D}$,S↑ $\overleftarrow{D}$ sink   A+D $\rightarrow$ AD

(8.2) $\vec{S}$,$\vec{D}$-$\vec{S}$,$\vec{D}$-(A/AD),S↑ $\vec{D}$ sink   "

(8.3) $\vec{DS}$ - $\vec{DS}$-(A/AD),S↑ $\vec{DS}$ sink   "

<u>D Sources</u> (D^{+})

(9.1) D,$\vec{S}$-$\vec{S}$,$\overleftarrow{D}$-S↑, (A/A')/$\overleftarrow{D}$ $\overleftarrow{D}$ source   A $\rightarrow$ A'+$\overleftarrow{D}$

(9.2) S,(A/A')/$\vec{D}$-$\vec{S}$,$\vec{D}$-D,S↑ $\vec{D}$ source   A $\rightarrow$ A'+$\vec{D}$

(9.3) S.(A/A')/$\vec{DS}$-$\vec{DS}$-DS $\vec{DS}$ source   A+S $\rightarrow$ A'+$\vec{DS}$

(9.4) -(A/A')/$\overleftarrow{D}$,$\vec{D}$- $\overleftarrow{D}$+$\vec{D}$ source (secondary active transport)
                                                    (continued)

Table 2. Continued

Sink and Source Asymmetry, (SeCDAR-2) Systems

Carrier source and sink (C*-$\bar{\text{C}}$)

(10.1) $\vec{\text{S}}$, (A_1/A'_1)-C*S...$\bar{\text{C}}$,$\bar{\text{S}}$-(A_2/A'_2),S↑ $S+A_1 \rightarrow C*S+A'_1$; $C*S+A_2 \rightarrow \bar{\text{C}}+S{\uparrow}+A'_2$

(10.2) $\vec{\text{S}}$,$\vec{\text{D}}$-C*S...$\bar{\text{C}}$,$\bar{\text{S}}$-(A_2/A_2D),S↑ $S+D+\bar{\text{C}} \rightarrow C*S$ "

D source and sink (D*-$\bar{\text{D}}$)

(11.1) $\vec{\text{S}}$,(A_1/A_1D)-$\vec{\text{S}}$,$\overleftarrow{\text{D}}$-S↑,(A_2/A'_2)/$\overleftarrow{\text{D}}$

(11.2) $\vec{\text{S}}$,(A_1/A'_1)/$\vec{\text{D}}$-$\vec{\text{S}}$,$\vec{\text{D}}$-(A_2/A_2D),S↑

(11.3) S.(A_1/A'_1)/$\vec{\text{DS}}$-$\vec{\text{DS}}$-(A_2/A_2D),S↑

$A_1 \rightarrow A'_1+D*$

$D*+A_2 \rightarrow \bar{\text{D}}$

$A_1+A_2 \rightarrow A'_1+\bar{\text{D}}$

S sources and sinks

(12.1) S.(A_1/A'_1)/$\vec{\text{S}}$*-$\vec{\text{S}}$*-$(A_2/A\frac{*}{2})$/S↑   S and * cotransport pump

(12.2) S*(A_1/A'_1)/$\vec{\text{S}}$-$\vec{\text{S}}$-(A_2/A'_2)/S*↑   S* pump

(12.3) $\bar{\text{S}}$.$(A_1/\bar{\text{A}}_1)$/$\vec{\text{S}}$-$\vec{\text{S}}$-(A_2/A'_2)/$\bar{\text{S}}$↑   S replacement and capture

Oscillating-reaction Pump: sink-source oscillations

(13) $\vec{\text{S}}$-(A_1/A'_1)/C*S(time 1)...(A_2/A'_2)/$\bar{\text{C}}$.$\bar{\text{S}}$(time 2)-S↑
                    (dissymmetrical barriers)

Passive Systems (no reaction)

The diffusing species are represented unchanged in the membrane and the compartments (between and on both sides of the hyphens). This reflects the fact that by "continuity" spontaneous transport is driven by the species (S) that "matches the membrane and the system". This means that the species is not retained in a complex or by charge coupling and that it can penetrate through the membrane. Neutral species can travel by uniport (1.0); if S bears a charge, the coupled transport of a partner d is needed to equilibrate the charge balance by exchange-antiport (1.1) or symport ($\vec{\text{d}}$) (1.2) with S. Cotransport is shown in (1.3). (In opposition with D, d does not drive the process (see Table 1)). We neglect volume flows. Physical forces (pressure P, temperature T, electrical field ΔE) that can change the limits and the course of the process are reflected in the changes of the electrochemical potential of the permeating species. These effects are summed up in the equations of irreversible thermodynamics.

Facilitations by S+C$\rightleftharpoons$S* Equilibria (S, C systems)[4,10-16]

The particularity of these systems is that C cannot penetrate or react in the totality of the system. By permutation we get dissymmetrical and symmetrical schemes if C is confined to one of the compartments (2.2, 2.4), or to one compartment and the membrane (2.1, 2.3), or only to the membrane (3.0), or inversely, is excluded from the membrane (4.0). An interesting simplification results from the fact that S* can only exist where C can penetrate; for this reason C was omitted in the schemes. Considering the direction of S-transport, (2.1) and (2.2) are "facilitated extraction" systems, (2.3) and (2.4) are "facilitated desorption" systems. However, due to chemical reversibility, (chemical symmetry), (2.1) and (2.3) on the one hand and (2.2) and (2.4) on the other hand are equivalent two by two. (4.0) represents desorption-absorption systems with "extra-membrane carriers"[4]. (3.0) includes the classical "carrier facilitated transport" liquid membrane systems and also ionexchange membranes or ionophore mediated transports. The schemes written for uniport could be easily extended to sym, anti and cotransport. In all these systems the process is at equilibrium when the electro-chemical potential of free S is equal on both membrane boundaries.

Equilibrium displacements. All the factors that by mass action or following Lechatelier-Van t'Hoff-Ostwald principles can displace the S, C, S* equilibria, are susceptible to change the limits and the direction of S transport. They concern physical factors (P, T, ΔE), concentration levels and C concentration relative to S. Most of these effects have been explored in detail; they can promote efficient, high rate, non polluting extractions or desorptions[12]. The transport reaction rate is enhanced by catalysts (inorganic or enzymes)[4], and by high C concentration found for example in ion-exchange membranes that are now intensively studied in our group; (see also the Métayer et al., paper presented at this meeting).

Remarks. a) Mobile carriers have been especially developed and added to liquid membranes[14]. We should observe that the S-C couples include all the reversible associations: Brønsted acids and bases, metallic, macrocyclic complexes, chelating or charge associations and fixed groups. In most systems the respective roles of S and C can be reversed by changing the membrane and the boundary conditions. b) In some cases (5.0) S sits already on a "carrier" (as H^+ in H_3O^+ or 0 in H_2O_2) and the true process is a carrier to carrier transfer[15]. This is always the case with "facilitated electron-transports" (5.1)[14].

From now on we shall consider only membrane crossings with fixed or mobile carrier mechanisms even if for the sake of simplicity C is omitted in the schemes of Table 2.

Uphill Transports with Driving Species (S, C, D Systems)[5,14,16-18]

These pumpings concern multicomponent systems in which S and D cross the membrane by coupled transport. As shown in schemes and equations (6.1) to (6.3) antiported $\overrightarrow{S}$ and $\overleftarrow{D}$ compete for the same site of C, symported $\overrightarrow{D}$ links to a different site and with $\overrightarrow{DS}$ we get the cotransport pump. Charge effects can reinforce chemical couplings. Many different species have been accumulated with such pumps. The process is driven by the (excess) boundary (electro-)chemical potential asymmetry of added D. In the compartments the concentration levels and reactivities of D and antiported S must be kept in the limits that allow absorption and desorption of both D and S on opposite sites in order to avoid the blocking of carrier sites (see 6.1.3, Table 3).

Table 3. Examples for Systems 2 to 12

Notations: compartment - membrane - compartment; [ ] = reference

(2.1; 2.3) CO_2gas-silicone-HCO_3^- [10]; NH_3gas-silicone-NH_4^+ [5]

(2.2) CO_2gas-polyvinylalc.-HCO_3^- [11]; NH_3gas-R.SO_3H^+-NH_4^+ [12]

(3.0) CO_2gas-HCO_3^-, carbonic anhydrase-CO_2gas [13]

(continued)

Table 3. Continued

(3.1) K^+ picrate-dibenzo 18 crown 6-K^+ picrate [14]

(4.0) HCO_3^--silicone, CO_2-HCO_3^- [10]; blood-silicone-HCO_3^- [11]

(5.0) H_2O_2-hydrophillic membrane, catalase-hemoglobine, O_2 [15]

(5.1) M^+, (Red_1/Ox_1) $-C_{ox}/C_{red}$ organic phase- (Ox_1/Red_1), M^+ [14]

(6.1) K^+,Bz.phenylalanine-diaza crown ether- K^+, Bz.phenylalanine↑[14]

  H^+,Na^+ -polyelectrolyte in octanol- $\overleftarrow{H^+,Na^+}$↑ [16]

  $\bar{N}a^+$,Ca^{++}-polyelectrolyte in octanol-$\overleftarrow{Ca^{++}},\bar{N}a^+$ [16]

  (inhibition of C)

(6.3) $NH_3 \cdot \bar{H}$ - RSO_3^- - K^+,NH_4^+↑ [5]

(7.1) $Cu^{++} \cdot (Red/Ox)/\overrightarrow{Cu^+},X^-$-Bathocuproine, toluene-$Cu^+$,$X^-$ [19]

(7.2) $NH_4^+ \cdot (OH^-/H_2O)/\overrightarrow{NH_3}$ -silicone, $\overrightarrow{NH_3}$- NH_3↑ [5]

  (glucosephosphate=$\overleftarrow{G_6P \cdot H_2O/P^-}$)/glucose($\bar{G}$)-agarose,p.acr.ac-G↑ [6]

(7.3) $\overrightarrow{glucose}$ - agarose, p.acr.ac.- (ATP/ADP)/G_6P^- [6]

(8.1) $\overrightarrow{CH_3CO_2}$, (H/H_2O) -Anion exchanger, $\overleftarrow{OH^-}$ - $\overleftarrow{OH^-}$, CH_3CO_2↑ [12]

  $\overrightarrow{Na^+}$, (OH^-/H_2O) -Monensin, hexanol $\overrightarrow{Na^+}$, $\overleftarrow{H^+}$- H^+, Na^+↑ [18]

(8.2) $\overrightarrow{H^+,Br^-}$ -weak anion exch (chitosan), $\overrightarrow{HBr}$ - (OH^-/H_2O), Br^-↑ [20]

(8.3) $\overrightarrow{NH_3H^+}$ -cation exchanger K^+- (OH^-/H_2O), NH_3↑ [5]

(9.1) NH_4^+, $\overrightarrow{Na^+}$ -cation exch.- (urea, urease)/$\overleftarrow{NH_4^+}$,Na^+↑ [21]

  $H^+,\overrightarrow{Na^+}$ -cation exch.- (glucose.ATP/ADP)/$\overleftarrow{H^+}$, Na^+↑ [6]

(9.4) Anion exchanger -Enzyme, Amino ac.ester- Cation exchanger [3]

(10.1)

Pumped	Energy	Activation	|SC*|	|$\bar{C}$|	Desactivation	By product	Ref
glucose (G)	ATP	hexokinase	G_6P^-	P^-	H_2O,Pase	ADP	[8]
glucose (G)	ATP	hexokinase (H^+)	G_6P^-	P^-	H_2O,Pase (OH^-)	ADP	[6]
$NO_3^-(-e)$	ΔE extern	-e	$FeNO_2^+$	Fe^{3+}	+e	external	[22]
$2H^+$,2e	hν redox	Red_1+2e	hydro-quinone	quinone Vit.K3	-2e,Ox_2	Ox_1+Red_2	[14]
picrate(-e)	redox	Ox_1,-e	complex	Ox Red	$^+$e,Red_2	Red_1+Ox_2	[23]
ClO_4^-							
$O_2(-4e)$	G/O_2	G oxidase	H_2O_2	H_2O	peroxidase	gluconic ac	[24]

(10.2)

Cl^-	ΔH^+	tautomer	[amino,HCl]	amide	OH^-	H_2O	[25]
Na^+,Cl^-	light	isomer	spiropirane derivative		dark		[26]

(11.1) $\overrightarrow{Na^+}$,(OH^-/H_2O) -Cation exchanger- (G.ATP/ADP)/$\overleftarrow{H^+}$ [27]

(11.3) $NH_3 \cdot (H_2/H^+)/\overrightarrow{NH_4}$ -Cation exchanger, $\overrightarrow{e}$-$(OH^-/H_2O)/NH_3$↑ [5]

  $Cu^{++} \cdot AcH.(OH^-/H_2O)/\overrightarrow{CuAc_2}$ -$\overrightarrow{CuAc_2}$,$\overrightarrow{AcH}$- $(CuAc_2.H^+/\overrightarrow{AcH})/Cu^{++}$ [28]

(12.1) $Cu^{++} \cdot (hν Red_1/Ox_1)/[Cu^{++},\overrightarrow{e},X^-]$-Bathocuprine-(dark,air)/$[Cu^{++},X^-]$↑ [19]

(12.3) (H_2/H^+),$\overrightarrow{Na^+}$- $\left\{ \begin{array}{c} \overrightarrow{e} \text{ by wire} \\ \text{Ion Exch, } \overrightarrow{Na} \end{array} \right\}$ -(O_2/OH^-),Na^+↑ [29]

Also: e carrier + M^+ carrier $\Longrightarrow \overrightarrow{M^+}$, $\overrightarrow{e}$ symport [14]

One could say that a D activates the S-C link on the one side and
deactivates it on the other side of the membrane. It was also observed
early[18] that the coupled loss and accumulation in the compartments of
diffusing D decreases from both sides the concentration gradient and hence
the efficiency of the pumping. Jumping over historical chronology and
other approaches, we can find here logical justifications for chemical
generation and capture of D or activation and deactivation of C and S.

One Source or one Sink: SeCDAR-1 systems[3,5,6,12,18-21]

A priori we are concerned with sink or source reactions involving A
and S, C or D. In fact a careful analysis shows that any C produced or
consumed would play the role of $\vec{D}S$, consequently C is discarded.

An S sink or a source gives reactive desorptions or extractions and no
pumps as revealed by schemes (7.1) to (7.4). Reactions on the left are
suitable for the chemical cancellation of retaining forces (source) or for
the conversion of S into a chemical form that can react with C (sink); they
are proper to promote the passage through a membrane. Reactions on the
right have the opposite function, transform species that have already
passed through the membrane and create chemical valves. The chemical form
of S being different on the two sides consecutive to the reaction, these
systems do not accomplish pumping[5,6,19].

D source pumps. The simplest ones correspond to the generation of a
D on the side from which it diffuses: $\vec{D}$ on the right, $\vec{D}$ and $\vec{D}S$ on the left.
In schemes (9.1) to (9.3) the reactions are symbolized by $(A/A')/\vec{D}$ or
$(A/A')/\vec{D}$ or $S.(A/A')/\vec{D}S$. (For illustrations see Table 3). It is worth
remembering that a generated D can still accumulate in the compartment
into which it diffuses. Scheme (9.4) deserves a special mention as it
corresponds to the hydrolytic enzyme reaction bounded by two different
membranes that constituted the first example of an artificial "secondary"
pump[3].

D-sink pumps. With a sink placed on the side where low D-concen-
trations are desired, D is trapped and D-accumulations are avoided. Sink
reactions are represented by (A/AD) in schemes (8.1) to (8.3). In the
important case of Arrhenius acid-base pumps the system is basically composed of an acid and a base separated by a membrane; there are however
six possible combinations. Depending on C (i.e. the membrane) and of
interactions with the pumped partner, either H^+ or OH^- playing the role
of $\vec{D}$, D or $\vec{D}S$, crosses the membrane and AD water is formed (with charge
fusion) in the sink. On the uphill side of the pump a salt is obtained
or a neutral S is liberated from its $\vec{D}S$ complex. The popularity of these
systems is shown by various examples of the literature using liquid or
charged membranes. Weak acids or bases can be efficiently pumped. (See
e.g. Métayer et al., at this meeting)[12].

Complexing, redox or other (locally) irreversible D-sink reactions that
belong to this class of systems have been also experimented.

In conclusion with D-sources the back-diffusion of D remains possible,
D-sinks form D-valves. It is worth mentioning that these pumps are usually
quoted as "secondary pumps". The cotransport system (8.3) may escape this
qualification.

Associated Sink and Source Reactions: SeCDAR-2 Systems

In these systems S or C or D activated on the one side is deactivated
on the other side (or part) of the membrane in chemical couplings with the
(A_1/A_1') and the (A_2/A_2') transformations respectively; they may or may not

involve electrons. At least one of these transformations furnishes the
energy; a stoichiometrical amount of byproduct formed for each S pumped
(or repumped if there is a reflux), shows the irreversibility of the
process. In an enlarged definition the "byproduct" can represent degraded light or heat.

C^* and D^* designating the activated species and $\bar{C}$ and $\bar{D}$ those
unable to carry or drive S, we get the $C^*-\bar{C}$ (10.1)(10.2) and the $D^*-\bar{D}$
(11.1) to (11.3) systems. With S we must consider also "retained" $\bar{S}$
and take in account the permeability of the membrane to S or to its
derivative S^*; this leads to the $\bar{S}-S-\bar{S}$, $S-S^*-S$ and S^*-S-S^* (12.1) to
(12.3) pumps.

C^*-C Pumps (10.1) and (10.2)[5,6,8,14,22,23]

These "primary" pumps are characterized[6,9] by the "spacewave"
concentration profile of the pumped S. We can subdivide these systems
in three subgroups depending on the $C^*/\bar{C}$ couple.

$\bar{C}$ is formed by the hydrolysis of C^*, electron transfers are not
involved. With the first "primary" system, experimented in our group
shortly after identification of the space-wave, glucose was pumped[9].
Using ATP, glucose was phosphorylated in the glucose-sink and hydrolysis
of the phosphate liberated glucose in the source. Taking advantage of
enzyme catalyzes with splitting reactions the two reaction sites were
localized asymmetrically inside the membrane. It was demonstrated that
such a reactive membrane asymmetry could communicate concentration asymmetry to the otherwise symmetrical system. This was the issue at stake
as we obtain there the reciprocal of passive systems. Other enzymatic
glucose[6] or synthetic NH_3[5] "molecule pumps" reinforced these results:
the reactions are directly coupled by S.

$C^*/\bar{C}$ form a RedOx couple, C is an electron carrier: in these systems
electron flows through the membrane in "electroneutrality" with the charge
flow due to the species pumped between the compartments. In order to
ensure asymmetry a different reducing, oxidizing or photo-oxidizing agent
is needed in each compartment. However, as shown in the first of these
systems[22], with a neutral S (i.e. NO) electrons furnished from outside
and circulating through wire and carrier in series, can drive the pump with
external energy. It seems correct to say that "primary" electron flow
drives all these pumps.

$\bar{C}$ is formed by the splitting of (oxidized) SC^*: the formation of H_2O_2
during enzyme catalyzed oxidation of glucose by air and further splitting
of H_2O_2 constitute a system able to pump O_2[24].

C^* and $\bar{C}$ are tautomers or convertible isomers (10.2): light and dark,
pH difference can promote the existence of two forms of the carrier on
different sides, one of them being only able to link S. These cases are
clearly distinct from simple sym or cotransport. Other forms of energy may
be envisaged[25,26].

$D^*-\bar{D}$ Systems (11.1) to (11.3)[5,27,28]

$\overset{\leftarrow}{D}$ or $\vec{D}$ or $\overset{\rightarrow}{DS}$ being chemically generated and inactivated by any of the
reaction couples cited already for $C^*/\bar{C}$, (acid-base, redox, photochemical
etc.), is driven through the membrane by the effect of the "primary"
chemical D-mass balance. In sym and antiport[27] systems S does not
participate in the reactions but is pumped by "uphill transport" mechanism,
driven by the $\vec{D}$ or $\vec{D}$ movements. We get "secondary S pumps" driven by the
primary D-pump.

In two-reaction driven cotransports (11.3) the chemical complex $\overrightarrow{DS}^*$ is generated in the sink-reaction and is destroyed in the source-reaction. As S enters the general chemical mass balance of both reactions, these systems belong to the class of "primary S pumps". An interesting example is that of acid-base couples generated in the compartments by redox reactions, $(H_2+O_2 \rightarrow H^+ +OH^-)$, where the electrons needed can travel by wire or by carrier. Any species S complexed by one of the ions ($DS = NH_4^+$ or $DS = HCO_3^-$) can be transported through a membrane permeable to the complex; water ($D = H_2O$) is formed in the receiving compartment and a D is consumed for each S transported. It was, however, demonstrated early[28] that a weak acid complexing agent D can be recycled by a $\overrightarrow{DS}^* = \overline{D}$ shuttle through the membrane ($\overrightarrow{D}^* = Ac_2Cu$; $\overline{D} = AcH$, Table 3 (11.3)) while water is a byproduct. The analogy between $C^*/\overline{C}$ intramembrane-carrier driven and transmembrane $\overrightarrow{DS}^*/\overline{D}$ driven pumps is striking; (a very detailed analysis would show even the existence of an intermediary situation with a reaction inside and the other one outside the membrane).

S Sink and Source Pumps[14,29]

(S-S*-S) (12.1) is the most straightforward of these systems. The membrane is impermeable to S but allows the passage of (oxidized or reduced) S* formed in the downhill S-sink; the reverse reaction traps S uphill. It is seen from the example of Table 3 that e is cotransported (or antiported) with S. The analogy with cotransport (9.3) is so striking that by writing D=e the two systems (9.3) and (12.1) could be fused in a unique class.

(S*-S-S*) (12.2) is a S* pump reciprocal of (12.1), with a membrane permeable to S but not to S*: a downhill S-source and an uphill-sink can solve this problem. Deprotonation, dehydroxylation, decomplexation and other splittings of S* with the help of A_1 and restitution with the help of A_2 enter this category. ("Negative carrier" could be a good denomination).

($\overline{S}$-S-$\overline{S}$) (12.3) is a charge-replacement system. With a membrane permeable to S the charge of S is replaced downhill and compensated uphill by the reaction product. As shown by the examples this may be accomplished by electron transfer/transport between A_1 and A_2 either by wire or with an electron carrier circulating in the membrane. These are secondary S movements driven by an electron pump. The membrane selectivity uncouples the chemical mass balance of the reactions but not their charge balance. Hydrolytic equivalence of these redox systems could be quoted.

In all the three S sink-source systems the selectivity of the membrane is a major factor. C constitutes a link in the reaction-transport chain but is not a functionally necessary component of the pump and can be absent.

Time Oscillation Pumps[29]

The systems considered until now possess a steady state regime. We have also shown the theoretical possibility of pumping with open entertained oscillatory reaction systems bounded by different membranes, where the sink and the source reaction alternate as a function of time. It was found by computation and by stability analysis that when reactions are quicker than transport the oscillations are stabilized by the transport step, and give rise to a sort of a "clock pump". Experimental approaches are under way with hydrolytic or electron carrier systems. These pumps are susceptible of developments similar to that of the "space wave" type systems.

Multimembrane Composite Systems

The situations resulting from the various associations of the enumerated basic systems deserve a separate examination elsewhere, however we should make here three remarks:

1) in composite systems the role of a given permeant can change from one membrane to another membrane. For instance, S pumped into a compartment through one boundary can show a D-type behavior on the other boundary and drive in (or out) a different permeant. This means that pumps are susceptible to drive pumps.
2) the pumping effect is transmitted to adjacent compartments through S permeable membranes[5,28]. Such membranes may be considered as links between the pump and the other parts of the system. For instance, in a four compartment system, where the sink and the source compartments of a two-reaction pump were inserted between S-permeable membranes, the net pumping proceded from the non-reactive compartment number 1 to non-reactive compartment number 4[5]. Such experiments were considered necessary for theoretical reasons in order to demonstrate symmetry breaking pumping between identical bounding solutions.
3) by inverting the sink and the source reactions and the membranes, an S pump may be converted into an S* (i.e. SC*, SD*, Se) pump: the same reaction couple may be used to pump S or S*.

DISCUSSION AND CONCLUSION

The first objective of this analysis was to show, or to recall that the all membrane systems may be analyzed and classified in a logical sequence with our SeCDAR scheme. It is also the occasion to make a few comments. Once this SeCDAR construction is made, the classification becomes a tool: we can distinguish the similarities and the differences between systems. Lacking cases and effects may be identified; with explicit energy coupling mechanisms thermodynamic and kinetic analyses are facilitated and the factors governing the process may be easily recognized. It is also easy to establish, as it was sometimes suggested, similarities between the different transport systems and classical chemical cells or concentration, acid-base and redox electrochemical cells. We may observe that "primary active transport pumps" comprise a system in which the driving force is generated by "reaction asymmetry" and where S is a part of the reaction mass balance. We have found three types of systems that correspond to that definition involving either S and C or S and DS or S and e in the reactions. The one-reaction C-sink or DS-sink systems are intermediary. All the other systems constitute "secondary pumps", namely all the one-or two-reaction systems where $\vec{D}$ or $\overset{\leftarrow}{D}$ is involved. One-reaction systems necessitate transport asymmetry (one membrane or two different membranes) in order to pump. Uphill transport pumps get there (osmotic) energy from D concentration asymmetry. The asymmetrical displacement of carrier-permeant complexing equilibria can drive permeant fluxes. Carrier facilitated transports and extractions are distinguishable by the symmetrical or dissymmetrical accessibilities of C to membrane and compartments. We see that such qualifications as "pump" or "active transport" or "primary active transport" or "secondary" are useful but not very accurate denominations; the identification of the SeCDAR class to which the system belongs contains more helpful information both on the conceptual level and in respect to applications.

Another aspect of this message is that in the present state of the art we have sufficient knowledge for generally adopting a classification in the form of our SeCDAR scheme. This would allow everyone to use the same terms for the same systems in the literature. This would often be less confusing.

SeCDAR is naturally not frozen in this state; it can and should be extended to multicarriers[12], reaction-chain links between sink and source, multitransports by a pump, couplings of identical or different systems and pumps with feed-back control and inductions[27], combined bio-synthetic pumps for biomeasuring[31] and other non-considered or new effects.

A parallel system, now under way, is centered on the monitoring of chemical reactions[23-32], pumps and oscillations by forces mediated by the transport; the two classifications could meet.

To conclude we should recall that the SeCDAR classification is dependent neither on the material of the membrane nor on the exact nature or phase of the permeant (gas, liquid, dissolution) or of the reactants. We see that the phenomena are related to such properties and interrelations as permeabilities, reversible or irreversible reactions and their locations and sometimes on their timing. This is what renders the SeCDAR classification so general and ready for kinetic or energetic (irreversible thermodynamics) quantifications.

REFERENCES

1. P. Glansdorf and I. Prigogine, "Structure, stabilité, fluctuations", Masson, Paris (1971).
2. O. Kedem and A. Katchalsky, Trans.Faraday Soc., 57:1185 (1961); P. Mears, paper presented at this meeting.
3. J. A. De Simon and S. R. Caplan, J.Theor.Biol., 39:235 (1973); Chem. Eng.Symp., Series 68:43 (1970).
4. E. Sélégny, Some systems coupling enzyme reactions and other phenomena. Energy conversions, in: "Polyelectrolytes," E. Sélégny, ed., Reidel, Dordrecht, 418-82 (1974).
5. E. Sélégny and D. Langevin, J.Chem.Res., (S):278 (1978); (M):3466 (1978); C.R. Ac.Sci.Paris, 278D:431 (1977).
6. E. Sélégny and J. C. Vincent, J.Biophys.Chem., 12:93 (1980); 12:107 (1980); J. C. Vincent, E. Sélégny and M. Métayer, J.Biophys.Chem., 14:159 (1981).
7. P. Mitchell, "Chemiosmotic coupling and energy transduction," Glynn Res. Ltd., Bodmin (1968).
8. R. Blumenthal, S. R. Caplan and O. Kedem, Biophys.J., 7:735 (1967).
9. E. Sélégny, G. Broun and D. Thomas, C.R. Ac.Sci.Paris, 269C:1377 (1969); 271D:1423 (1970).
10. G. Broun, E. Sélégny, T. M. Canh and D. Thomas, FEBS Lett., 7:223 (1970).
11. T. M. Canh "Transfer et generation des gaz par des Membranes," Dr. Sc. Thesis, Rouen (1971).
12. M. Métayer, D. Langevin, M. Labbé and E. Sélégny in: "Physical Chemistry of Transmembrane Ion Motions," G. Spach, ed., Elsevier, 115 (1983); J.Chim.Phys., 79:661 (1982) and M. Métayer, D. Langevin, M. Labbé and M. Hankaoui, paper presented at this meeting.
13. S. R. Suchdes and J. S. Schulz, Chem.Eng.Sci., 28:13 (1974); N. C. Otto and J. A. Quinn, Chem.Eng.Sci., 26:949 (1971).
14. J. M. Lehn, in: "Physical Chemistry of Transmembrane Ion Motions," G. Spach, ed., Elsevier, 181:206 (1983).
15. G. Broun, T. M. Canh, D. Thomas, D. Domurado and E. Sélégny, Trans. Amer.Soc.Artif.Intern.Organs., 17:341 (1971).
16. R. Varoqui and E. Pfeffercorn, in: "Charged Gels and Membranes II," E. Sélégny, ed., Reidel, Dordrecht (1976).
17. H. Tsukube, J.Chem.Soc.Perkin Trans., I:2359 (1982).
18. E. L. Cussler, AIchE J., 17(6):1300 (1971).
19. A. Ohki, S. Matsuno, T. Tukeda, M. Tukagi and K. Ueno, Sep.Sci.&

Tech., 17(10):1237 (1982); Chem.Lett., 1529 (1982).

20. T. Uragami, F. Yoshida and M. Sugihara, J.Appl.Pol.Sci., 28:1361 (1983).

21. A. M. Talbot-Chobert, Thesis 3°C, Rouen (1983).

22. J. W. Ward, Nature, 227:162 (1970).

23. T. Shinbo, K. Kurihara, Y. Kobatake and N. Kamo, Nature, 270:277 (1977); T. Shinbo, M. Suginra, N. Kamo and Y. Kobatake, J.Membr. Sci., 9:1 (1981).

24. E. Sélégny, Unpublished.

25. N. Ogata, K. Sagui and H. Figimura, J.Pol.sci.Polym.Lett., 17:753 (1979).

26. T. Shimidzu and M. Yoshikawa, J.Membr.Sci., 13:1 (1983).

27. J. C. Vincent, Dr. Sc. Thesis, Rouen (1980).

28. R. Paterson, Nature, 217:545 (1968).

29. C. Eyraud and C. Daneyrolle, C.R. Ac.Sci.Paris, 284C:9 (1977).

30. E. Sélégny, and J. C. Vincent, J.Chim.Phys., 77:1083 (1980); J. C. Vincent and E. Sélégny, C.R. Ac.Sci.Paris, 292III:173 (1981); J.Non-equil.Thermodyn., 7:259 (1982).

31. G. A. Junter, E. Sélégny and J. F. Lemeland, Bioelectrochem.Bioenerg., 9:679 (1982); 9:699 (1982).

32. E. Sélégny, J. M. Valleton, J. C. Vincent, Biophys.Chem., 15:235 (1982); Bioelectrochem.Bioenerg., 10:133 (1983).

THE RESPONSE OF MEMBRANE SYSTEMS WHEN EXPOSED

TO OSCILLATING CONCENTRATION WAVES

Paul Doran and Russell Paterson

Department of Chemistry
University of Glasgow
Glasgow, Scotland

In network thermodynamics, membranes are described in terms of generalized resistors and capacitors[1]. These reflect the ability of a membrane both to dissipate power and to store chemical energy internally in each local volume during transport processes. For quantitative modelling, the membrane is subdivided conceptually and mathematically into homogeneous 'lumps' or slices, each characterized by its own resistance and capacitance. The accuracy of quantitative modelling of diffusion processes depends upon the degree of reticulation of the lumped model; since the model approaches evermore closely to a true continuum of states as the number of lumps (n) is increased. The error is approximately proportional to n^{-2} [2].

For a system which obeys Fick's law of diffusion (analogous to Ohm's law) in which effort, e, (the analog of voltage) is identified as concentration[3,4], it is easily shown that the chemical resistance, R, for each lump of an n-lump model is given by Equation (2), and the corresponding capacitance, defined from q = C.e, by Equation (3).

$$R = n^{-1} \ell/(D \; A\alpha) \tag{2}$$

$$C = \alpha A \; \ell/n \tag{3}$$

In these equations it is assumed that the membrane is regular, of area, A; thickness, ℓ and has diffusion coefficient, D, defined by Fick's first law. (The distribution coefficient, α, was introduced[4] to define the concentration of the equilibrium solution as a common effort in all phases).

As an example, to model a membrane which was exposed on one side to a source of effort (SE) and connected to a closed (homogeneous) volume of solution (or gas) on the other (Ct) we may use a bond graph, Figure 1a or an equivalent circuit, Figure 1b. These alternative representations show a 3-lump model of the membrane. (The bond graph notation although less familiar is the more powerful for modelling complex, coupled, transport phenomena[5]).

These networks may be used algorithmically, to compute the system dynamics of a membrane assembly for any experiment or membrane, in which bond graph parameters and initial conditions are defined[3,4,6].

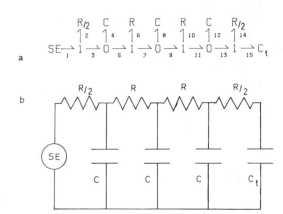

Fig. 1. (a) A bond graph representation of a 3-lump model of a membrane exposed to a source of effort, SE, (either constant or oscillating) on one side and connected to a reservoir or collecting volume on the other, represented here by the terminal capacitor, Ct. (b) The equivalent circuit representation of the same membrane system.

The experimentalist presented with an unknown membrane is faced with the inverse problem, that of extracting the transport parameters of a membrane by observing the progress of experiments of his or her own design. With the circuit and bond graph representations clearly in mind it is obvious that the membranologist would be likely to use a much narrower range of experiments in such investigations than the electronic engineer, faced with an analogous black-box problem. In particular, the membranologist would lack the powerful AC methods used routinely to investigate electrical circuitry.

Using mathematical models and bond graph simulations, we set out to examine the system dynamics of the membrane assembly represented in Figure 1 when exposed to a regular source of concentration waves with fixed frequency and amplitude. From the properties of the analogous electrical circuit, Figure 1b, we would expect the membrane to act as an AC filter, reducing the amplitude and causing phase shifts in the emergent waves (detected in the terminal capacitor/collecting volume). In this paper we examine the possibilities of using these effects for membrane characterization. By modelling and simulation of experiments a variety of wave forms, frequencies and amplitudes were tested and, from the results of these calculations, equipment and procedures were devised to perform test experiments in the laboratory.

During the course of this research we discovered some interesting precedents for the use of forced oscillations, in studies on thermal diffusion, by Ångstrom[7], Thomson (Lord Kelvin)[8] and more recently in gas diffusion[9].

MATHEMATICAL MODELS

The membrane is considered to be exposed on one side to forced oscillations in the concentration of the contact solution and to be connected on the other to a closed reservoir or collecting volume, Vt, whose initial concentration, co, is the mean concentration of the ingoing wave. It is assumed also that the membrane is in equilibrium with its contacting solutions, at all times, and that (if the membrane is in the form of a clamped sheet) 'edge effects' can be neglected[10,2]. Fick's laws are assumed with constant diffusion coefficient, D. A further simplification is the assumption that the collecting volume, Vt, is effectively infinite. In practice this is not a severe limitation, as shown below.

Three wave forms were considered, cosine, square, and triangular. Only the first two will be discussed here because no further information was obtained by the use of triangular waves. For our system the phase shift altered by only 4° on changing from square to triangular waves, but for triangular waves there was, in addition, a significant loss of amplitude. The fundamental responses to a cosine wave were considered initially. The wave form and boundary conditions for a solution of Fick's laws were, Equations (4) and (5).

$$SE = c = co(1 + \cos(\omega t)) \qquad \text{at } x = 0 \qquad (4)$$

and

$$ct = co \qquad\qquad \text{at } x = \ell \qquad (5)$$

The time variation of concentration in the collecting volume, after initial transients have decayed is given by ct, Equation (6) which was calculated by dividing the infinite volume expression for the moles of permeant through the membrane, qt, by the actual volume of the collecting vessel.

$$ct = B(\omega) \cos(\omega t + \phi(\omega) - \pi/2) + co \qquad (6)$$

where $B(\omega)$ is the amplitude of the emerging wave, Equation (7),

$$B(\omega) = 2^{3/2} \frac{AK\alpha co \; D \; (2(\cosh(2K1) - \cos(2K1)))^{-\frac{1}{2}}}{\omega Vt} \qquad (7)$$

and $\phi(\omega) - \pi/2$ is the phase change, Figure 2, where $\phi(\omega)$ is defined by Equation (8).

$$\phi(\omega) = \tan^{-1} \left\{ - \frac{\tan(K1)}{\tan(K1)} \right\} + \pi/4 \qquad (8)$$

It is clear that the diffusion coefficient can be obtained from the phase change, Equation (8), where $K = \sqrt{\omega/(2D)}$, while the amplitude of the emerging wave is a function of permeability, $P = D \; \alpha/\ell$. For a typical membrane, Figure 2, the output amplitude is very small, approximately 0.2% of the source. Since many membrane parameters are strongly concentration dependent it is not practical to use large concentration oscillations, accordingly the most useful application of these methods appeared to be direct and repetitive measurements of diffusion coefficients from the phase shift. For typical membrane conditions large phase shifts were predicted for oscillations with periods in the range 10-1000s which was sufficiently slow to encourage us to consider generating concentration waves using microcomputer controlled equipment.

Although performing sinusoidal oscillations in concentration by ingenious use of controlled syringes is feasible, by far the most convenient wave form is the square wave, which, may be obtained using sprays, as described below.

A solution for source oscillations in the form of a square wave were similar to those given above. The boundary conditions were as before except that the source, SE, was now a square wave given by the Fourier series, Equation (9). The corresponding steady state oscillations (also as a summation) are given by Equation (10).

$$SE = co\left(1 + \frac{4}{\pi}\left(\sum_{n=1,3,5,..}^{\infty} \frac{(-1)^{\frac{n-1}{2}}}{n} \cos(n\omega t)\right)\right) \qquad (9)$$

$$ct = \frac{4}{\pi}\left(\sum_{n=1,3,5,.}^{\infty} \frac{B(n\omega)\cos(n\omega t + \phi(n\omega) - \pi/2)(-1)^{\frac{(n-1)}{2}}}{n}\right) + co \qquad (10)$$

The symbols have the same significance as before and now $\phi(n\omega)$ are the values of $\phi(\omega)$ for each of the contributory waves of Equation (9), and similarly for $B(n\omega)$. The results are shown in Figure 3. Once more

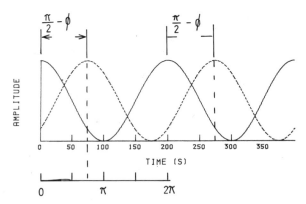

Fig. 2.   Representation of the phase shift with a source, SE = co $(1+\cos(\omega t))$ and the collecting volume is large. Membrane parameters: D = 2.751 x 10^{-6}, $cm^2 s^{-1}$, ℓ = 0.205 mm, A = 0.771 cm^2, α = 0.702, co = 0.065M. Arbitrary scales have been used and the input and output amplitudes have been made equal for ease of comparison.

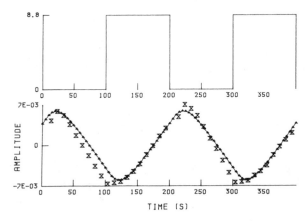

Fig. 3.   Concentration waves in the collecting volume, generated by a square wave source. ⌇ , experimental; - , from the mathematical model, Equations (9) and (10); and ▲ , using a 20-lump bond graph, as Figure 1a. Membrane parameters as Figure 2. Amplitudes are expressed as experimental conductances (mmho).

infinite collecting volume, Vt, was assumed. Parallel derivations for limited collecting volumes were made, but are not presented here, since it was shown that the two solutions effectively coincide when $\sqrt{2}$ DAαK/(ωVt)<<1. In these studies thisfunction 0.002 and the change in ϕ, even for a collecting volume of 2 ml, as used in this work, was 0.09°.

These predictions were verified by bond graph simulations. The calculations were performed using the methods described earlier for the SE-C model[3,4], but now using the cosine, square, or other wave expressions for SE, as above. Lumped models were constructed according to the membrane and cell parameters of the model to be predicted and resistances and capacitances defined by Equations (2, 3). The state space equations defining the flows into each of the capacitors (defined in these papers) were integrated numerically to give flows into and concentrations in the collecting volume. It is worth noting that the whole experiment was modelled including the initial period in which transients exist and additional information as to the effects of variable sources or concentration profiles in the membrane are obtained easily, and as a matter of course.

EXPERIMENTAL

The diffusion cell was made from a rectangular block of Perspex (Pexiglass) approximately 3x4x8 cm. This was drilled horizontally and vertically to provide two intersecting cylinders, Figure 4. The vertical one was sealed at the bottom to hold a 'Spin fin' stirrer (Bel Art) which was driven magnetically by from outside the cell. The upper portion of the vertical tube was threaded and sealed with a Teflon screw top which held a pair of platinized platinum wires which served as conductance electrodes. The bottom of this screw top was a hemispherical hollow to improve mixing in the cell. It also contained a capillary to aid the removal of entrapped bubbles. The horizontal hole passed through the block. On one side, Figure 4, it defined the exposed membrane area, on the other it too was threaded and sealed with a Teflon screw, which held a thermistor temperature probe.

The cell, which had a volume of 2 ml, was bolted to a large perspex box: the membrane aperture was defined the coincident holes in these two

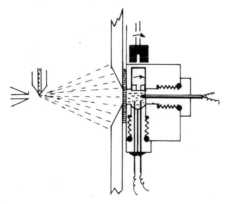

Fig. 4. A diagrammatic representation of the experimental cell, used for generating square concentration waves as input, and detecting transmitted concentration waves in the collecting vessel. For clarity, only one of the two spray guns is shown.

parts. The perspex box contained two sprays (Badger Air Brush Coy), attached to thermostatted reservoirs of the two solutions to be sprayed. In this work these were usually water and dilute potassium chloride. The cell was filled initially with water or salt at the mean concentration of the two solutions. The square wave was obtained by turning the air brushes on and off alternately at regular intervals, using solenoids attached to the nitrogen gas lines (20 psi) which operated them. The switching was performed by a small computer and the concentration waves in the cell (collecting) volume were detected by a Wayne Kerr conductance bridge (precision 0.01%) and displayed. Temperature effects on conductivity could be corrected independently by taking account of the thermal coefficient of conductance of the salt.

Membranes were prepared from Visking dialysis tubing and pre-equilibrated with water and solution before use.

RESULTS AND DISCUSSION

The results of experiments using Visking dialysis tubing are summarized in Table 1. Square waves were produced with amplitudes of around 0.05M by spraying alternately with water and potassium chloride solutions of 0.1M. The concentration of salt in the collecting volume was obtained from the conductance of the solution, using standard calibration procedures. For the very dilute salt solutions detected in this work the conductance and concentrations were virtually proportional (after correction for the conductivity of the original water). Very careful thermostatting of the solution was required. This was achieved largely by the thermal conductivity of the membrane itself transmitting heat from the sprays (alternative thermostatting by water circulation to the half cell may also be used as necessary). Conductivities were recorded continuously during the experiment. Suitable periods of oscillation for Visking membranes were calculated to be close to 200s. Most experiments were made under these conditions, but to test the frequency response, the range of oscillations was varied between 60 and 400s. From independent time-lag determinations of diffusion coefficients[11], observed phase shifts could be compared with those calculated Equation (10). The agreement was usually within one or two degrees, and, on conversion of ϕ to diffusion coefficient, the forced square wave and time lag methods agree to within 5-6%. This may be

Table 1. Observed and Predicted Wave Forms in Collecting Solutions

disc	thick-ness (mm)	period (s)	phase shift obs.	calc.* $-\phi$ (degrees)	diffusion coefficient obs**	calc.* ($cm^2 s^{-1}$ x 10^6)	percent
a	0.213	60	125.1	113.0	-	-	-
b	0.210	200	38.7	37.4	2.84	2.94	3.5
c	0.204	200	34.6	36.7	3.00	2.83	5.8
d	0.196	200	52.7	49.9	1.81	1.91	5.4
e	0.205	200	36.2	38.2	2.90	2.75	5.3
f	0.205	400	17.9	20.3	-	-	-
g	0.205	400	23.8	19.0	-	-	-

Square waves (0 - 0.13M) with small variations from one experiment to another, were used in all cases. Membrane area: 0.771 cm. Volume of collecting vessel, Vt: 2.06 cm^3.
* from eqn(10). ** by independent measurements.

regarded as substantive proof of the validity of the method, since the agreement was within the experimental error of each method and could well be reduced with further refinements.

The basic system of oscillating waves has therefore been shown to provide an extremely rapid and repetitive method for determining diffusion coefficients which may be easily modified to follow the course of diffusion of almost any diffusant. The accuracy of determination is as good as all but the most rigorous double isotope methods. It has only one disadvantage and that is the requirement that the detection methods must be very sensitive due to the low ratio of the amplitude of output to source. It is also to be noted that the output is further attenuated the thicker the membrane, Equation (10). Although the amplitude to permeability relationship was not pursued, here, it is to be noted that the simulation, Figure 3, reproduces the observed output amplitude and frequency from independently measured diffusion coefficients and distribution coefficients.

As a final word on the wave form, it is to be noted that, should the distribution coefficient be a function of concentration (as for charged membranes obeying Donnan conditions), all wave forms, except the square wave, will be distorted on transposition from the solution into the membrane phase. Although we have shown that phase shift (but, importantly, not amplitude) is virtually unaffected by the wave form, nevertheless, this must be an additional factor in favor of the square wave and our present experimental system.

REFERENCES

1. G. Oster, A. S. Perelson, and A. Katchalsky, Network Thermodynamics: Dynamic modelling of biophysical systems, Q. Review Biophysics, 6:1 (1973).
2. J. Crank, "The Mathematics of Diffusion," Oxford U. Press (1975).
3. R. Paterson and Lutfullah, Simulation of Membrane Processes Using Network Thermodynamics, Proc. Society of Chemical Industry (SCI) Conference, IEX'84, Cambridge, England, "Ion Exchange Technology," D. Naden and M. Streat, eds., Ellis Horwood, Chichester, pp.242-256 (1984).
4. R. Paterson and Lutfullah, Simulation of Transport Processes Using Bond Graph Methods, Part I, Gas Diffusion Through Planar Membranes and Systems Obeying Fick's Laws, J. Membrane Science, 23:59 (1985).
5. R. Paterson, Network Thermodynamics, in: "Membrane Structure and Function," E. E. Bittar, ed., John Wiley, N.Y., 2:1 (1980).
6. R. Paterson, Lutfullah and P. Doran, in preparation.
7. A. J. Ångstrom, Neue Methode das Warmeleitungsvermogen der Korper zu bestimmen, Ann.Physic.Lpz., 114:513 (1861).
8. W. W. Thomson (Lord Kelvin) On the reduction of observations of underground temperature; with application to Professor Forbes' Edinburgh Observations and the Calton Hill Series, Trans.Roy.Soc. Edinburgh, 22:405 (1861).
9. K. Evnochides and E. J. Henley, Simultaneous measurement of vapour, diffusion and solubility coefficients in polymers by frequency response techniques, J.Polymer Science, Part A-2 8:1987 (1970).
10. R. M. Barrer, J. A. Barrie and M. G. Rogers, Permeation through a membrane with mixed boundary conditions, Trans.Faraday Soc., 58:2473 (1962).
11. P. Doran and R. Paterson, unpublished results.

FACILITATED TRANSPORT OF IONS IN ASYMMETRIC MEMBRANES

M. Nakagaki and R. Takagi

Faculty of Pharmaceutical Sciences
Kyoto University, Sakyo-ku
Kyoto 606, Japan

INTRODUCTION

It is well known that the asymmetry of membrane structure is closely related to its functions. If the membrane is asymmetric with respect to a partition coefficient, the ion concentration gradient within the membrane will be different from that expected from the bulk concentration ratio. When the concentration gradient within the membrane is larger than that within a symmetric membrane, the facilitated transport will take place in a passive transport. When the direction of concentration gradient within the asymmetric membrane is opposite to that expected from the bulk concentration ratio, the reverse transport will take place.

If the membrane is asymmetric with respect to the membrane charge distribution, the concentration difference of cation between two membrane surfaces is not equal to that of anion. Therefore, the facilitated or the reverse transport may also take place. Thus, if the asymmetric membrane in which the facilitated or the reverse transport of ions takes place can be designed, the membrane separation is expected to be performed with high efficiency in a passive transport.

In this paper, the ion transport through an asymmetric membrane is theoretically discussed under the condition of steady state where no electric current flows across the membrane. The two kinds of asymmetry are treated: one is the asymmetric distribution of membrane charge and the other is the asymmetry with respect to the partition coefficient of ion at the membrane/bulk solution interface.

GENERAL EQUATION

Neglecting the volume flow across the membrane, the flux of k-th ion, $[J_k]^*$ (m mol·cm^{-2}·sec^{-1}), is given by Equation (1), where k indicates any of all ionic species including both cation and anion.

$$[J_k]^* = - f \frac{[k]^* B_k^*}{N_A} \left(\frac{\partial \tilde{\mu}_k^*}{\partial x}\right) ,$$

(1)

where

$$\tilde{\mu}_k^* = \mu_k^{o*} + RT \ln \gamma_k^* [k]^* + z_k F_A E^* . \tag{2}$$

Here, f is the membrane coefficient[1], R the gas constant, T the absolute temperature, F_A the Faraday constant and N_A Avogadro's number. B_k^* is the mobility of k-th ion and E_* the potential within the membrane. z_k^* is the valence(algebraic) and $[k]^*$ is the molar concentration of the k-th ion, where * denotes the quantity in the membrane. Neglecting the concentration dependence of activity coefficient[2], Equation (1) is rewritten as follows:

$$\frac{[J_k]^*}{f} = - \frac{RT}{N_A} B_k^* (\frac{\partial [k]^*}{\partial x}) - z_k \frac{F_A}{N_A} B_k^* [k]^* (\frac{\partial E^*}{\partial x}) . \tag{3}$$

The membrane system is assumed to be in a steady state in which no electric current flows across the membrane and $[J_k]^*$ is constant throughout the system. When an external electric field does not exist, the electric current I, is equal to zero as is given by Equation (4).

$$I = F_A \sum_k z_k [J_k]^* = 0 . \tag{4}$$

From Equations (3) and (4), the potential gradient within the membrane is given by Equation (5).

$$\frac{\partial E^*}{\partial x} = - \frac{RT}{F_A \sum_k z_k^2 B_k^* [k]^*} \sum_k z_k B_k^* (\frac{\partial [k]^*}{\partial x}) . \tag{5}$$

The distribution of the ionic concentration within the membrane is assumed to be given by Equation (6)[3] according to Henderson[4].

$$[k]^* = (1-\delta(x))[k]^{I*} + \delta(x)[k]^{II*} . \tag{6}$$

Here, $\delta(x)$ is common to all kinds of ionic species, and $\delta(0)=0$ and $\delta(L)=1$, where L is the thickness of the membrane. I and II denote the surface of the membrane at $x = 0$ and $x = L$, where the membrane surface is in contact with solutions I and II, respectively.

Equation (7) is obtained by integrating Equation (3) over the membrane thickness L, utilizing the relations of Equations (5) and (6).

$$\frac{[J_k]^* N_A L}{fRT} = - B_k^* \Delta [k]^* + \frac{b}{d} + \frac{ad-bc}{d^2} \ln(\frac{d}{c} + 1) , \tag{7}$$

where

$$\Delta [k]^* = [k]^{II*} - [k]^{I*} , \tag{8}$$

$$a = (z_k B_k^* [k]^{I*})(\sum_k z_k B_k^* \Delta [k]^*) , $$

$$b = (z_k B_k^* \Delta [k]^*)(\sum_k z_k B_k^* \Delta [k]^*) , \tag{10}$$

78

$$c = \sum_k z_k^2 B_k^*[k]^{I*} \tag{11}$$

and

$$d = \sum_k z_k^2 B_k^* \Delta[k]^* \tag{12}$$

The ion flux is rewritten by Equation (13)[5].

$$\frac{[J_k]^*}{f} = -\frac{\Delta[k]^*}{L} D_k^* \rho_k \,, \tag{13}$$

where

$$D_k^* = \frac{RTB_k^*}{N_A} \frac{\sum_j z_j (z_j - z_k) B_j^* \Delta[j]^*}{\sum_j z_j^2 B_j^* \Delta[j]^*} \tag{14}$$

and

$$\rho_k = 1 - \frac{z_k \sum_j z_j^2 B_j^* ([k]^{I*}[j]^{II*} - [j]^{I*}[k]^{II*})}{\Delta[k]^* \sum_j z_j (z_j - z_k) B_j^* \Delta[j]^*}$$

$$x \frac{\sum_j z_j B_j^* \Delta[j]^*}{\sum_j z_j^2 B_j^* \Delta[j]^*} \ln\left(\frac{\sum_j z_j^2 B_j^* [j]^{II*}}{\sum_j z_j^2 B_j^* [j]^{I*}}\right) \tag{15}$$

DISCUSSION

Uni-univalent Electrolyte System

For simplicity, hereafter, we treat the uni-univalent electrolyte system (i^+, a^-). It is assumed that the generalized Donnan equilibrium condition holds at the membrane/bulk solution interface. In the Donnan equilibrium condition, however, the partition coefficients of an ion are different between two interfaces. The membrane is also assumed to have an asymmetric distribution in membrane charge.

The generalized Donnan equilibrium condition is given by Equation (16), where S is I or II, indicating the respective solutions in contact with the membrane surfaces.

$$\frac{[k]^{S*}}{[k]^S} = \frac{1}{g_k^S} \exp(-z_k F_A \Delta E^S/RT) \,, \tag{16}$$

where g_k^S is the partition coefficient of the k-th ion at the membrane/bulk solution interface and ΔE^S the surface potential. By eliminating the surface potential, Equation (16) yields.

$$\left(g_i^S \frac{[i]^{S*}}{[i]^S}\right) = \left(g_a^S \frac{[a]^{S*}}{[a]^S}\right)^{-1} \,. \tag{17}$$

The electroneutrality condition is given by Equation (18).

$$[i]^{S*} - [a]^{S*} + \theta^S = 0 \tag{18}$$

where θ^S is the membrane charge density.

Putting $[i]^I = [a]^I = rC$, $[i]^{II} = [a]^{II} = C$, $g^I = \sqrt{g_i^I g_a^I} = r_g g$,

$g^{II} = \sqrt{g_i^{II} g_a^{II}} = g$, $\theta^I = r_\theta \theta$ and $\theta^{II} = \theta$, the ionic concentrations at the

membrane surfaces are obtained from Equation (17) and (18) as follows[6]:

$$[i]^{I*} = \frac{C}{g} r_\theta \left(\sqrt{\nu^2 + R_{g\theta}^2} - \nu\right), \quad [i]^{II*} = \frac{C}{g} \left(\sqrt{\nu^2 + 1} - \nu\right) \tag{19}$$

$$[a]^{I*} = \frac{C}{g} r_\theta \left(\sqrt{\nu^2 + R_{g\theta}^2} + \nu\right), \quad [a]^{II*} = \frac{C}{g} \left(\sqrt{\nu^2 + 1} + \nu\right) \tag{20}$$

where $R_{g\theta} = r/r_g r_\theta$ and $\nu = \theta g/2C$.

Substituting Equations (19) and (20) into Equations (13), we have

$$[J_k]^* = f \frac{D_{ia}^{o*}}{gL} C \frac{W^2 - X^2}{W - \tau^* X} \rho \tag{21}$$

where

$$D_{ia}^{o*} = \frac{2RT}{N_A} \frac{B_i^* B_a^*}{B_i^* + B_a^*}, \tag{22}$$

$$\tau^* = (B_i^* - B_a^*)/(B_i^* + B_a^*), \tag{23}$$

$$W = r_\theta \sqrt{\nu^2 + R_{g\theta}^2} - \sqrt{\nu^2 + 1}, \quad W' = \sqrt{\nu^2 + R_{g\theta}^2} - \sqrt{\nu^2 + 1}, \tag{24}$$

$$X = (r_\theta - 1)\nu \tag{25}$$

and

$$\rho = 1 - \frac{\nu r_\theta (\tau^* W - X) W'}{(W - \tau^* X)(W^2 - X^2)} \ln \frac{\sqrt{\nu^2 + 1} - \tau^* \nu}{r_\theta(\sqrt{\nu^2 + R_{g\theta}^2} - \tau^* \nu)}. \tag{26}$$

Being put $R_{g\theta} = r$ (i.e. $r_g = 1$, $r_\theta = 1$), Equation (21) reduces to the equation for a symmetric membrane.

80

Asymmetry with Respect to the Partition Coefficient

In this section, we discuss the membrane system in which $r_g \neq 1$ and $r_\theta = 1 (\theta^I = \theta^{II} = \theta)$. The asymmetry discussed here may be realized by chemically modifying one surface of the symmetric membrane with a substance such as carbohydrate. If the solvent of bulk solution I is different from that of bulk solution II, the same situation as the asymmetric membrane will also be realized even if the membrane itself is a symmetric membrane.

In this system, Equation (21) can be rewritten as follows:

$$\frac{[J_k]^* N_A L \, g}{f R T (B_i^* + B_a^*) C} = \frac{1 - \tau^{*2}}{2} (\sqrt{\nu^2 + R_{g1}^2} - \sqrt{\nu^2 + 1})$$

$$- \tau^* \nu \ln \frac{\sqrt{\nu^2 + 1} - \tau^* \nu}{\sqrt{\nu^2 + R_{g1}^2} - \tau^* \nu}) , \tag{27}$$

where $R_{g1} = r/r_g$. From Equation (27), it can be seen that the parameter r_g does not appear in this equation explicitly, but implicitly in $R_{g1} (= r/r_g)$. This indicates that the asymmetry of the partition coefficient gives the membrane system the same effect as a change of bulk concentration ratio, r. Therefore, the facilitated transport will take place under the condition of $r_g < 1$. The reverse transport will also take place under the condition of $r_g > r$.

Figure 1a shows the flux as a function of $1/r_g$ under the condition of $r = 2$, $r_\theta = 1$ and $\tau^* = -0.7$. From Figure 1a, it is clear that the facilitated transport takes place at $r_g < 1$ and the reverse transport at $r_g > r$. If $r = 2$, the flux is zero at $1/r_g = 0.5$, and if the curves in Figure 1a can be assumed to be straight lines, the flux at e.g., $1/r_g = 2$ is about three times larger than the flux in the symmetric membrane $(r_g = 1)$, since $(2 - 0.5)/(1 - 0.5) = 3$. It is also found that the flux at $\tau^* \nu > 0$ (curves a and b in Figure 1a) is larger than that at $\tau^* \nu < 0$ (curves e and d, respectively) for constant $|\nu|$. This indicates that the flux at $B^* > B^*$ is larger than the flux at $B^* < B^*$, where B_c^* and B^* are the mobilities of co-ion and gegen-ion, respectively, in the membrane.

Asymmetry with Respect to the Membrane Charge Density

In this section, the membrane system with an asymmetric membrane charge distribution is discussed, in which $r_\theta \neq 1$ and $r_g = 1$. This membrane system may be realized by chemically modifying the membrane with a charged substance. The ion flux is given by Equation (21) by replacing $R_{g\theta}$ with $R_{1\theta} (= r/r_\theta)$.

Figure 1b shows the ion flux as a function of r_θ under the condition of $r = 2$, $r_g = 1$ and $\tau^* = -0.7$. For the calculation, the average membrane charge density (θ_{av}) defined by $(\theta^I + \theta^{II})/2$ is kept constant. In Figure 1b, ν_{av} is $\theta_{av} g / 2C$.

In Figure 1b, it is shown that the flux is almost independent of r_θ if the membrane charge density $|\nu_{av}|$ is small (curves c, d and e), but the flux increases with r_θ if $|\nu_{av}|$ is large (curves a, b, f and g).

In lines a, b, f and g, the facilitated transport takes place at $r_\theta \gg 1$ compared with the flux at $r_\theta = 1$ (symmetric membrane). The reverse

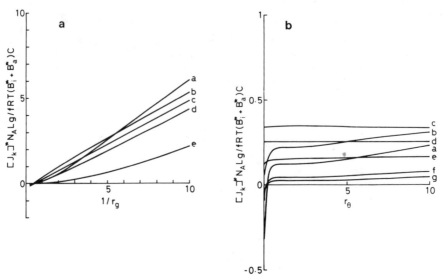

Fig. 1. Flux in the asymmetric membranes. (a) asymmetry with respect to
the partition coefficient $\tau* = -0.7$, $r = 2$, $r_\theta = 1$. $\nu = -10(a)$,
$-1(b)$, $0(c)$, $1(d)$, $10(e)$. (b) asymmetry with respect to the
membrane charge density $\tau* = -0.7$, $r = 2$, $r_g = 1$. $\nu_{av} = -10(a)$,
$-5(b)$, $-1(c)$, $0(d)$, $1(e)$, $5(f)$, $10(g)$.

transport takes place at $r_\theta \ll 1$ in lines a, f and g. It is found also from
Figure 1b that the value of flux at $B*_c > B*_g$ is larger than that at $B*_c < B*_g$ for
the same $|\nu_{av}|$.

The decrease of the value of r_θ decreases the charge density at the
membrane surface I and increases the charge density at membrane surface II.
When $r_\theta \ll 1$, the concentration of gegen-ion becomes much higher at the
membrane surface II than I, and the concentration gradient of the gegen-ion
is reversed from what is expected from the concentrations of solutions
outside the membrane. Thus, the reverse transport takes place under the
condition of $r_\theta \ll 1$ for large $|\nu_{av}|$.

Simulation of Ionic Flux

In order to discuss the applicability of asymmetric membrane to the
dialysis, we try to simulate the time dependency of each flux in a closed
system where solutions on both sides of the membrane have the same and a
constant volume.

Being put the surface area of membrane (A) as 1 cm^2, the volume of
bulk solutions (V) as 100 ml, $(RT/N_a)(B*_i + B*_a)$ as 1×10^{-5}, g as 1, L as
1×10^{-3} cm, f as 1 and one time unit (tu) as 10000 second, the flux during
one time unit is given by Equation (28) from Equation (21).

$$\frac{[J_k]* A \text{ tu}}{V} = \frac{1 - \tau*^2}{2} \frac{W^2 - X^2}{W - \tau* X} \rho \text{ tu} \quad (\text{mol}/1) \qquad (28)$$

The flux is calculated at 0.5 tu interval by Equation (28). The
concentration of each bulk solution and ν are determined by every 0.5 tu
and substituted into Equation (28).

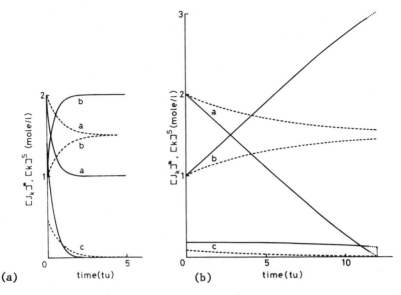

(a)   time(tu)   (b)   time(tu)

Fig. 2.   Simulation of time dependence of flux.   (a) flux in the asymmetric
membrane with respect to the partition coefficient at $\tau^* = -0.18$,
$r_g = 0.5$ and $r_\theta = 1$ initial conditions: $[k]^I = 2$ mole/l, $[k]^{II} =$
1 mole/l, $\nu = -1$.   Dotted line shows the flux in the symmetric
membrane under the same condition as the asymmetric membrane
except $r_g = 1$.   Lines a, b and c show $[k]^I$, $[k]^{II}$ and $[J_k]^*$,
respectively.   (b) flux in the asymmetric membrane with respect to
the membrane charge density at $\tau^* = -0.18$, $r_g = 1$ and $r_\theta = 10$
initial conditions: $[k]^I = 2$ mole/l, $[k]^{II} = 1$ mole/l, $\bar{\nu}_{av} = -10$.
Dotted line shows the flux in the symmetric membrane under the
same condition as the asymmetric membrane except $r_\theta = 1$.   Lines a,
b and c show $[k]^I$, $[k]^{II}$ and $[J_k]^*$, respectively.

Figure 2a shows the time dependence of flux in the asymmetric membrane
under the condition of $\tau^* = -0.18$, $r_g = 0.5$ and $r_\theta = 1$.   The value of -0.18
of τ^* corresponds to the value for NaCl in a cellulose membrane[6].   The
corresponding value for NaCl in an aqueous solution is about -0.21.   A dotted line shows the flux in the symmetric membrane under the same condition
as the asymmetric membrane except $r_g = 1$.   From Figure 2a, it is found that
initial flux in the membrane asymmetry with respect to the partition coefficient is very large compared with that of the symmetric membrane.   Ions
flow from the bulk solution I to II until r becomes equal to r_g.

Figure 2b shows the flux under the condition of $\tau^* = -0.18$, $r_g = 1$
and $r_\theta = 10$.   A dotted line shows the flux in the symmetric membrane under
the same condition as the asymmetric membrane except $r_\theta = 1$.   From Figure
2b, it is found that the initial flux is smaller than the initial flux in
the asymmetric membrane shown in Figure 2a, but larger than that in the
symmetric membrane.

The point especially interesting in the asymmetric membrane system
with respect to the membrane charge density as shown in Figure 2b is that
the flux scarcely varies with time.   The ionic flow from bulk solution I to
II decreases the bulk concentration ratio and decreases the concentration
gradient.   The concentration gradient-dependent part of flux decreases with
time.   Thus, the flux in the symmetric membrane decreases with time.
However, the flux in the asymmetric membrane with respect to the membrane
charge density is kept almost unchanged with time, compared with the
decrease of the flux in the symmetric membrane.

From Equation (3) with Equation (5), the flux of a uni-univalent electrolyte, $[J_{ia}]^*$, is expressed as follows:

$$[J_{ia}]^* = - f \ D_{ia}^{o*} \ \{ (\frac{[c]^* + |\theta|/2}{[c]^* + (1 - t_c^*)|\theta|}) \ (\frac{\partial [c]^*}{\partial x})$$

$$+ \frac{1}{2} (\frac{[c]^*}{[c]^* + (1 - t_c^*)|\theta|}) \frac{\partial |\theta|}{\partial x} \} \ , \tag{29}$$

where c designates the co-ion. t_c^* denotes the transference number of co-ion and θ the membrane charge density. The concentration gradient of the gegen-ion, on the other hand, is dependent very much on the gradient of the membrane charge density $(\partial |\theta|/\partial x)$, which consists a part of the driving force for the flux of the electrolyte. Thus, the asymmetric membrane in the membrane charge distribution has the driving force for the electrolyte transport in the membrane itself, and the transport continues at almost constant rate. The flux suddenly becomes zero at the point where the concentration of bulk solution I becomes zero, since the gegen-ion has been partitioned at the membrane surface I according to the electoneutrality condition until the concentration of bulk solution I becomes zero.

CONCLUSION

Ion transports across the asymmetric membrane have been studied theoretically. It was concluded that the facilitated transport as well as the reverse transport takes place both in the asymmetric membranes with respect to the partition coefficient and in the membrane with asymmetric charge distribution. If the asymmetric membrane as discussed in this paper can actually be prepared, the membrane separation by dialysis may be expected to be performed with high efficiency.

REFERENCES

1. M. Nakagaki, Seitaimaku ni okeru Busshituyuso, in: "Zoku Seitaimaku to Makutoka," M. Nakagaki, ed., Nankodo, Tokyo, Japan (1974).
2. M. Nakagaki and S. Kitagawa, The theory of the reverse diffusion of ions in mixed electrolyte solutions and experimental results on HCl-CaCl$_2$-H$_2$O ternary system, Bull.Chem.Soc.Japan, 49:1748 (1975).
3. M. Nakagaki, Theory of membrane potential and the effect of hydration of ions (2), Maku(Membrane), 1:321 (1976).
4. P. Henderson, Zur Thermodynamik der Flüssigkeitsketten, Z.Physik. Chem., 59:118 (1907).
5. M. Nakagaki, Makuyuso Gensho no Kiso, in: "Makugijutsu no Kiso," Membrane Soc. Japan, ed., Kitamishobo, Tokyo, Japan (1983).
6. M. Nakagaki and R. Takagi, Theoretical study on asymmetric membrane potential, Chem.Pharm.Bull., in press.

ION AND SOLVENT TRANSPORTS THROUGH

AMPHOTERIC ION EXCHANGE MEMBRANE

H. Kimizuka, Y. Nagata and W. Yang

Department of Chemistry
Faculty of Science, Kyushu University
Fukuoka, Japan

SUMMARY

Transport properties of the amphoteric ion exchange membrane-aqueous sodium chloride system have been studied. The six elements of a conductance matrix (Phenomenological coefficients) have been experimentally determined according to the previous theory[1] and the membrane properties were discussed in terms of the conductance elements. It has been shown that the diffusional ion conductances are ca. ten times greater than the electrical ion conductances and the ion-solvent correlation in the transport processes is much less than the interionic correlations. A failure of the Nernst-Einstein relation as well as Despić and Hills' equation for the electroosmotic effect on the membrane conductance was also pointed out.

INTRODUCTION

It has been reported in a previous paper[1] that the steady ion and solvent fluxes through the membrane in the isothermal concentration cell consisting of the two aqueous single electrolyte solutions separated by a membrane are expressed by the phenomenological equation as follows:

$$
\begin{bmatrix} i_\alpha \\ i_\beta \\ F j_w \end{bmatrix} = - \begin{bmatrix} g_{\alpha\alpha} & g_{\alpha\beta} & g_{\alpha w} \\ g_{\beta\alpha} & g_{\beta\beta} & g_{\beta w} \\ g_{w\alpha} & g_{w\beta} & g_{ww} \end{bmatrix} \begin{bmatrix} V - V_\alpha \\ V - V_\beta \\ - V_w \end{bmatrix}
\tag{1}
$$

were the subscripts α, β and w refer to cation, anion and water, respectively; i denotes an ion current; F, Faraday constant; j_w, the water flux; $g_{\gamma\delta}$ ($\gamma, \delta = \alpha, \beta, w$), the elements of conductance matrix; $^w V_i$ (i = α, β), the pseudo equilibrium potential; V_w, the effective driving potential for water or the water potential. V_i and $^w V_w$ are given by

$$
z_i F V_i = RT \ln \alpha_i^I / \alpha_i^{II} - \bar{v}_i \Delta P
\tag{2}
$$

and

$$FV_w = RT \ln \alpha_w^I / \alpha_w^{II} - \bar{v}_w \Delta P = \bar{v}_w (\Delta \pi - \Delta P) \ ,$$

(3)

respectively, where the superscripts I and II refer to phases I and II, respectively; R and T denote gas constant and absolute temperature, respectively; Z the charge; α the activity; $\bar{v}_\gamma$ the partial molar volume of species γ which is assumed constant; $\Delta \pi$ the osmotic pressure difference; ΔP ($\Delta P = P^{II} - P^I$), the external pressure difference. Such an expression as Equation (1) has been shown to be advantageous for treating the electrical properties of the membrane. A number of relations among the membrane properties and the elements of the conductance matrix have been given in the previous paper[1].

The aim of this paper is to determine the elements of the conductance matrix in the amphoteric ion exchange membrane–aqueous sodium chloride system and to discuss the membrane properties in terms of the conductance matrix.

EXPERIMENTAL

Materials

A 1.0-PA-29 amphoteric ion exchange membrane is a thin cross-linked polystyrene layer with sulfonate and quaternary ammonium groups, 1 mequiv. (g-wet membrane)$^{-1}$ in the ion exchange capacity each, and is tight contact with a porous support of polyvinyl chloride sheet with ca. 0.1 mm in thickness and kindly supplied by Central Research Laboratory of Kanegafuchi Chemical Industry Co., Ltd. Before use, the membrane was equilibrated with the concentrated solution of the salt under study, leached in deionized water, and equilibrated with the solution to be used in the measurements.

Sodium chloride and sucrose were of analytical grade and used without further purification. The salt concentration in phase I was varied from 10^{-3} to 10^{-1} mol/dm^3 and that in phase II was kept at 10^{-2} mol/dm^3. The sucrose concentration used for the osmotic flux measurements were 0.02 to 0.12 mol/dm^3.

Procedures of Experiments

The procedures of the experiments were essentially the same as those described in the preceding papers[2,3]. The experiments were carried out under the regulated room temperature at 25±1°C.

RESULTS AND DISCUSSION

The equation for the membrane current I, is obtained from Equation (1) as follows[1]

$$I = i_\alpha + i_\beta = -G_m (V - V_0)$$

$$= -g_\alpha (V - V_\alpha) - g_\beta (V - V_\beta) + g_w V_w$$

(4)

where g_α, g_β and g_w denote the conductances for permeating species α, β and w, respectively

$$g_\gamma = g_{\gamma\alpha} + g_{\gamma\beta} \ ,$$

(5)

G_m, the membrane conductance

$$G_m = g_\alpha + g_\beta ,$$ (6)

and V_o, the membrane potential at zero membrane current

$$V_o = t_\alpha V_\alpha + t_\beta V_\beta + t_w V_w$$ (7)

$$t_\gamma = g_\gamma / G_m ,$$ (8)

t_γ, the transport number of a species γ. Equation (7) may be rewritten as referred to the electrode reversible to anion

$$E_o = V_o - V_\beta = t_\alpha(V_\alpha - V_\beta) + t_w V_w$$ (9)

$$= (t_+)_{app} (V_\alpha - V_\beta) .$$ (10)

Figure 1 shows the membrane potential E_o, the membrane conductance G_m and the apparent transport number of cation $(t_+)_{app}$ as functions of the mean electrolyte activity $a_\pm^I$ in phase I. This figure indicates that the membrane is preferentially selective towards the anion.

The equations for the ion and solvent fluxes in the presence and absence of the membrane current are also obtained from Equation (1) as follows[1]

$$i_\alpha - i_\alpha^0 = - g_\alpha(V - V_0) = t_\alpha I$$

$$F(j_w - j_w^0) = - g_w(V - V_0) = t_w I$$ (11)

where i_α^0 and j_w^0 denote the cation current and the water flux in the absence of the membrane current, respectively. Figure 2 shows that i_α and j_w are linear against V and hence the conductances and transport numbers can be estimated according to Equation (11). The deviation of the apparent transport number from the true one was very small and within the experimental error. This may be ascribed to the minor contribution of $t_w V_w$ to E_o in Equation (9). The electrical ion and water conductances are plotted against the mean electrolyte activity of phase I as shown in Figure 3. It is seen in this figure that the water conductance or the electroosmotic solvent flow conductance g_w is 50 to 100 times greater than the ion conduc-

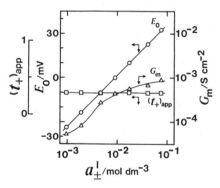

Fig. 1. Membrane potential E_o, membrane conductance G_m, and apparent transport number $(t_+)_{app}$ as functions of mean electrolyte activity of phase I.

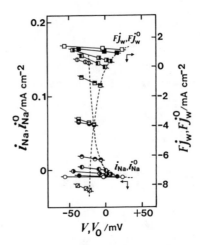

Fig. 2. Cation and water fluxes *vs.* membrane potential. Broken lines
connect the values at zero electric current. Circle and square
marks indicate i_{Na} (i_{Na}^0) and Fj_w (Fj_w^0), respectively. ○, □, 10^{-3};
●, ■, 2×10^{-3}; ◑, ◲, 5×10^{-3}; ◐, ◨, 10^{-2}; ◓, ◩, 2×10^{-2}; ◔, ◪,
5×10^{-2}; ◫, ◙, 10^{-1}. (Numericals indicate the concentrations of
phase I in mol dm^{-3}).

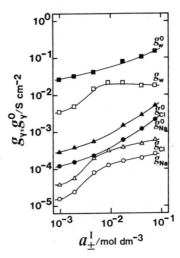

Fig. 3. Electrical ion and water conductances and diffusional ion and
water conductances *vs.* mean electrolyte activity of phase I.

tances g_α and g_β, an indicative of the greater electroosmotic membrane
permeability to water. It will be mentioned later about the diffusional
ion conductances g_γ^0's, plotted together with g_γ's in this figure.

The fluxes, i_α^0 and Fj_w^0, are expressed by the equation[1]

$$
\begin{bmatrix} i_\alpha^0 \\ Fj_w^0 \end{bmatrix} = - \begin{bmatrix} G_{\alpha\alpha}^0 & G_{\alpha w}^0 \\ G_{\alpha w}^0 & G_{ww}^0 \end{bmatrix} \begin{bmatrix} V_\beta - V_\alpha \\ - V_w \end{bmatrix}
\tag{12}
$$

where

$$G^0_{\alpha\alpha} = g_{\alpha\alpha} - G_m t^2_\alpha$$

$$G^0_{\alpha w} = g_{\alpha w} - G_m t_\alpha t_w$$

$$G^0_{ww} = g_{ww} - G_m t^2_w \quad . \tag{13}$$

The change in the water potential, ΔV_w, under the restriction of constant V_α and V_β may cause the changes in the fluxes, Δi^0 and Δj^0_w, which are given as[1]

$$\Delta i^0_\alpha = G^0_{\alpha w} \Delta V_w$$

$$\Delta F j^0_w = G^0_{ww} \Delta V_w \tag{14}$$

provided there is no appreciable change in the activity coefficients of ions as well as the phenomenological coefficients due to the addition of sucrose in one of the two aqueous phases. Figure 4 shows the changes in the ion and water fluxes at zero membrane current as functions of the change in the water potential. The result shown in Figure 4 enables us to determine the three phenomenological coefficients, $G^0_{\alpha\alpha}$ $G^0_{\alpha w}$ and G^0_{ww} according to Equations (12 and (14).

Comparing Equation (13) with Equation (5), we obtain the set of relations for estimating the elements of the conductance matrix as follows

$$g_{\alpha\alpha} = G_m t^2_\alpha + G^0_{\alpha\alpha} \; ; \qquad g_{\alpha w} = G_m t_\alpha t_w + G^0_{\alpha w} \; ;$$

$$g_{\alpha\beta} = G_m t_\alpha t_\beta - G^0_{\alpha\alpha} \; ; \qquad g_{\beta w} = G_m t_\beta t_w - G^0_{\alpha w} \; ;$$

$$g_{\beta\beta} = G_m t^2_\beta + G^0_{\alpha\alpha} \; ; \qquad g_{ww} = G_m t^2_w + G^0_{ww} \quad . \tag{15}$$

Thus, the measurements of G_m, t_α, t_w, $G^0_{\alpha\alpha}$, $G^0_{\alpha w}$ and G^0_{ww} enable us to estimate the six elements of conductance matrix according to Equation (15). Figure 5 shows these elements as functions of the mean electrolyte activity

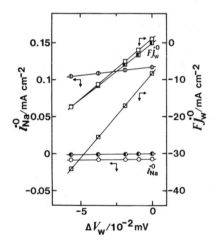

Fig. 4. Effect of water potential difference on ion and water fluxes at zero electric current. Circle and square marks indicate i^0_{Na} and $F j^0_w$, respectively. $\circ$, $\square$, 10^{-3}; $\bullet$, $\blacksquare$, 10^{-2}; $\oplus$, $\boxtimes$, 10^{-1}. (Numericals indicate the concentrations of phase I in mol dm^{-3}).

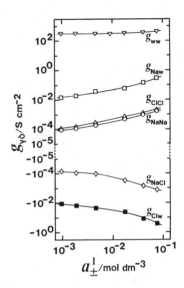

Fig. 5. Elements of conductance matrix $vs.$ mean electrolyte activity of phase I.

of phase I. It may be said that these elements characterize the membrane properties and provide explanations for all the membrane phenomena concerning the system under study.

Figure 5 indicates that $g_{NaNa} \simeq g_{ClCl} \simeq - g_{NaCl}$ and all these coefficients are the same order of magnitude in contrast to the result in the cation exchange membrane – aqueous calcium chloride system[2], $g_{CaCa} > g_{CaCl} > g_{ClCl}$. The same amounts of cations and anions are migrating in the amphoteric ion exchange membrane, while major migrating ions in the ion exchange membrane are the counter ions. This may be reflected to the different order of the interionic conductance elements in these two membranes. Further, it should be pointed out in Figure 5 that the cross coefficient is negative in the amphoteric membrane in contrast to the positive cross coefficient in the ion exchange membrane[2]. The negative cross coefficient may correspond to the positive L_{NaCl} coefficient in aqueous sodium chloride solution[4,5]. Figure 5 also indicates that g_{NaW} is positive and g_{ClW}, negative and their absolute values are greater than those of the interionic conductance elements by a factor of 2. The greatest is the hydrodynamic water conductance g_{ww} which is greater than the interionic conductance by a factor of 5-6.

The diffusional conductance g_{γ}^{0} calculated according to the relation[2,6].

$$g_{\gamma}^{0}/g_{\gamma} = 1 - g_{\alpha\beta}G_{m}/g_{\alpha}g_{\beta} \tag{16}$$

is plotted together with g_{γ} as shown in Figure 3. Since $g_{\alpha\beta}$ is negative, g_{γ}^{0} is greater than g_{γ} as shown in this figure. It was shown in the preceding paper[7] that the mean electrical absolute mobility $\bar{U}_{i}$ and the mean diffusion coefficient $\bar{D}_{i}$ of ion i within the membrane can be related to g_{i} and g_{i}^{0}, respectively, as follows

$$g_{i} = (Z_{i}F)^{2}\bar{U}_{i}\bar{C}_{i}/\Delta x = |Z_{i}| \bar{C}_{i}\lambda_{i}/\Delta x \tag{17}$$

$$g_{i}^{0} = (Z_{i}F)^{2}\bar{D}_{i}\bar{C}_{i}/RT\Delta x = |Z_{i}| \bar{C}_{i}\lambda_{i}^{0}/\Delta x \tag{18}$$

where $\bar{C}_i$, λ_i and λ_i^0 denote the mean concentration, the electrical equivalent ion conductance and the diffusional equivalent ion conductance, in the membrane, respectively; Δx, the membrane thickness. The fact, $g_i < g_i$, indicates the failure of the Nernst-Einstein relation. Combining Equation (16) with Equations (17) and (18), we have

$$\lambda_i - \lambda_i^0 = |Z_i| \, F^2 \, (\bar{U}_i - \bar{D}_i/RT) = g_{\alpha\beta}\Delta x/|Z_i|\bar{C}_i(1-t_i) \,. \tag{19}$$

Equation (19) implies that the difference between the electrical and diffusional equivalent ion conductances should be related to the interionic interaction rather than the electroosmotic effect as proposed by Despić and Hills[8].

The electroosmotic effect on the membrane conductance can be expressed by using the relations in the previous paper[1]

$$\lambda - \lambda' = g_v^2 \, \Delta x/|Z_i| \, |\bar{C}_i|g_p \tag{20}$$

where λ and λ' denote the equivalent conductance with and without the electroosmosis, respectively; g_v the electroosmotic volume flow conductance; g_p the hydrodynamic volume flow conductance. The elements of the conductance matrix in Equation (1) were related to g_v and g_p by the relations[1]

$$g_v = g_w + g_\alpha\bar{v}_s/Z_\alpha v_\alpha\bar{v}_w$$
$$g_p = g_{ww} + (g_{\alpha v} + g_{\alpha w}) \, \bar{v}_s/Z_\alpha v_\alpha\bar{v}_w$$
$$g_{\alpha v} = g_{\alpha w} + g_{\alpha\alpha}\bar{v}_s/Z_\alpha v_\alpha\bar{v}_w \tag{21}$$

where $\bar{v}_s$ is the partial molar volume of salt and v_α, the stoichiometric coefficient of ion α. The electroosmotic contribution to the equivalent conductance may be evaluated by the factor, $g_v^2\Delta x/|Z_i|\bar{C}_i\lambda g_p$, which was found to be a negligible order of 10^{-3}. Equation (20) could be reduced to Despić and Hills' equation under several assumptions[1]. It is seen by comparing Equation (20) with Equation (19) that the replacement of λ' by λ^0 is the misinterpretation in the theory of Despić and Hills.

In the preceding paper[2], the interionic conductance elements were determined from the membrane conductance, the apparent transport number and the apparent diffusional ion conductance under the assumption of no solvent effect. Comparing the present theory with the previous one[2,6], we derived[1]

$$(g_{\alpha\beta})_{app} = \frac{(1+f_\alpha)(1+f_\beta)}{1+t_\alpha f_\alpha+t_\beta f_\beta} \left[g_{\alpha\beta} - \frac{t_\alpha t_\beta(t_\alpha f_\alpha+t_\beta f_\beta)}{t_\alpha t_\alpha + t_\beta f_\beta} \, Gm \right]$$

$$(g_{ii})_{app} = G_m(t_i)_{app} - (g_{\alpha\beta})_{app} \tag{24}$$

where $f_i(i=\alpha,\beta)$ is given by

$$f_i = -[g_{iw} - g_{\alpha\beta}t_w/(1-t_i)]V_w/g_i^0(V_0-V_i) \,. \tag{25}$$

The apparent values estimated according to Equations (24) and (25) were close to the true ones and the deviations were less than 5%, although the electroosmotic conductance elements $g_{\alpha w}$ and $g_{\beta w}$ were much greater than the interionic conductance elements. This may be due to small f_i's, to which the small V_w caused by the small concentration difference may be responsible as deduced from Equation (25).

The membrane properties such as the mechanical filtration coefficient, the hydrodynamic permeability, the solute permeability coefficient, the reflection coefficient and the coupling coefficient could be expressed as functions of the elements of conductance matrix[1] and hence all these membrane properties may be evaluated from the data obtained in the present study.

Acknowledgement

The present work was supported by a Grant-in-Aid for Scientific Research No. 58470007 from the Ministry of Education, Science and Culture, Japan.

REFERENCES

1. H. Kimizuka, Y. Nagata, and K. Kaibara, Nonequilibrium thermodynamics of the ion and solvent transports through ion-exchange membrane, Bull.Chem.Soc.Jpn., 56:2731 (1983).
2. H. Kimizuka, K. Kaibara, E. Kumamoto, and M. Shirozu, Diffusional and conductive membrane permeabilities in the cation exchange membrane-aqueous electrolyte system, J.Membrane Sci., 4:81 (1978).
3. E. Kumamoto and K. Kimizuka, Transport properties of the barium form of a poly(styrenesulfonic acid) cation-exchange membrane, J.Phys. Chem., 85:635 (1981).
4. A. Katchalsky and P. F. Curran, "Nonequilibrium Thermodynamics in Biophysics," Harvard University Press, Cambridge (1965).
5. E. Kumamoto and H. Kimizuka, Nonequilibrium thermodynamics of ionic diffusion coefficients in binary electrolyte solutions, Bull.Chem. Soc.Jpn., 52:2145 (1979).
6. H. Kimizuka and K. Kaibara, Nonequilibrium thermodynamics of ion transport through membranes, J.Colloid & Interface Sci., 52:516 (1975).
7. K. Nomura, A. Matsubara, and H. Kimizuka, Theoretical and experimental studies of the membrane permeabilities in liquid membrane, Bull. Chem.Soc.Jpn., 54:1324 (1981).
8. A. Despić and G. J. Hills, Electro-osmosis in charged membranes. The determination of primary solvation numbers, Dis.Faraday Soc., 21:150 (1956).

A SELECTIVE PARAMETER FOR IONIC MEMBRANE TRANSPORT

SELECTIVITY AND POROSITY OF SEVERAL PASSIVE MEMBRANES

A. Hernandez, J. A. Ibañez and A. F. Tejerina

Dept. of Thermal Sciences, Fac. Sciences
University Valladolid
Spain

SUMMARY

In this paper we consider the case when a homogeneous passive membrane separates two isothermal aqueous solutions of a binary electrolyte at different concentrations c_1 and c_2 ($c_1 < c_2$) with no gradients of applied electric potential and pressure. In such conditions we define an experimental parameter of selectivity and study its dependence on the relative ionic permeability and on the relative saline concentration. We also study the relation between the porosity and the selectivity of several passive membranes. We work with NaCl solutions at a constant temperature of 298.0 ± 0.1 K.

INTRODUCTION

The efficiency with which a passive membrane transports selectively an ionic species i, is characterized by the ionic permeability P_i, or by the transport number t_i, with $i = +, -$. We will show that this efficiency may also be expressed by means of an experimental parameter K, which is related to the transport numbers and the ionic permeabilities; and which will be called selective parameter. This will be done by integrating the Nernst-Planck[1,2,3] equations using the Goldmann's hypothesis[4] which assumes a linear gradient of the membrane potential. The applicability of the Goldmann's constant field assumption or the electroneutral approximation has often been examined[5-11]. It has been shown that either approximation may be appropiate depending upon whether the square of the ratio of the junction thickness to the Debye length corresponding to the mean concentration in the junction is large or small. If it is very small, the assumption of constant field should be a good approximation, while if it is large, electroneutrality is the simplification of choice. But in the steady state, when the diameter of the solvated ions are much smaller than the average diameter of the membrane pores, the thickness of the junction is not important. And, if the concentrations and charge density of the membrane are small, the Goldmann's hypothesis is the right one.

We will study the selectivity and porosity of several passive membranes like Nucleopore filters (thin sheet of polycarbonate perforated by an array of almost parallel and cylindrical pores) M02, M04, M06, M08, M10 and M20, Millipore filters (a mixture of cellulosic acetate and nitrate) VC and Gs and a Pall filter (pure Nylon 66) Ultipore NM which separate two phases consisting of diluted solutions of NaCl. For these membranes and solutions the Goldmann's hypothesis is a good approximation in the steady state. The characteristics of the membranes are given in Table 1.

THEORY

Ionic Permeabilities and Membrane System Permeability

The equations of Nernst-Planck establishes that in the case of dilute solutions[12], the unidimensional flow J_i, of the species i through the membrane is given in module by

$$J_i = - D_i \left(\frac{d\,c_i}{d\,x} + z_i \frac{c_i}{RT} \frac{F}{dx} \frac{d\psi_m}{} \right), \quad i = +, - \tag{1}$$

Here we have assumed the x-axis to be perpendicular to the direction of the membrane surfaces. Besides D_i is the diffusion coefficient of ion i, z_i its valence ($z_+ = 1$ and $z_- = -1$) and c_i and ψ_m the molar concentration and electric potential at x position inside the membrane. The other symbols have their usual meaning.

When the electrolyte is 1:1 type the salt flow J and the ionic flows are related by[12,13,14]

$$J = J_+ = J_- \tag{2}$$

The ionic permeability of the species i is

$$P_i = D_i/d \tag{3}$$

d being the thickness of the membrane system which is formed by the membrane and the diffusion layers adhered to it[12].

The membrane system permeability is

$$P_S = J/(c_2 - c_1) \tag{4}$$

Table 1.   Characteristics of the Membranes Used. The Average Pore Radius has been Obtained from a Variant of the Teorell's "Half Time"[12]. The Pore Density has been Measured from Microscopic Methods.

Membrane	Average Pore Radius r (μm)	Pore Density N $(10^9 por./m^2)$	Membrane Thickness d (μm)
M20	0.800 ± 0.020	22.9	10 ± 1
M10	0.411 ± 0.008	50.7	"
M08	0.226 ± 0.004	300.0	"
M06	0.218 ± 0.004	270.0	"
M04	0.100 ± 0.002	1000.0	"
M02	0.060 ± 0.001	3000.0	"
GS	0.220 ± 0.004	19730.0	90 ± 10
VC	0.100 ± 0.002	95493.0	"
NM	0.100 ± 0.002	94220.0	"

94

where c_1 and c_2 are the saline concentrations of the two solutions separated by the membrane system $(c_2 > c_1)$.

The Goldmann's approximation is

$$\frac{d\,\psi_m}{d\,x} = \frac{\Delta\psi_m}{d} \tag{5}$$

where $\Delta\psi_m$ is the membrane potential.

Integrating the Equation (1) and using (2-5) we obtain

$$n_p = \frac{\xi n_c - 1}{n_c - \xi} \tag{6}$$

with

$$\xi = \exp(F\,\Delta\psi_m/RT) \tag{7}$$

$$n_c = c_2/c_1 \tag{8}$$

$$n_p = P_-/P_+ \tag{9}$$

where n_c is the relative saline concentration and n_p the relative ionic permeability.

For dilute solutions, the membrane potential is given by[15,16,17]

$$\Delta\psi_m = m\,\ln(c_2/c_1) \tag{10}$$

m being an experimental characteristic of each membrane, solute and temperature. If we consider the new parameter K, defined by

$$K = Fm/RT \tag{11}$$

the Equation (7) becomes

$$\xi = n_c^K \tag{12}$$

and the Equation (6)

$$n_p = \frac{n_c^{K+1} - 1}{n_c - n_c^K} \tag{13}$$

In this way we have established the expressions which give us the ionic relative permeability n_p, as a function of the relative concentration n_c, and the experimental values of K.

In the case of passive transport with only a gradient of concentration as a driving force, only the values of K, such as

$$|K| \leq 1 \tag{14}$$

must be considered, since according to Equation (13) values such as $|K| > 1$ would imply ionic permeabilities of opposite signs, that is to say, flows of cations and anions in opposite directions, which is not possible in our conditions.

Transport Numbers

The experimental parameter K, is related to the transport numbers in the way that we shall show below.

The transport numbers t_+ and t_-, are defined as the fraction of electric current transported by cations and anions respectively, and for a 1:1 electrolyte they are given by[12]

$$t_i = u_i/(u_+ + u_-) \tag{15}$$

where u_i is the ionic mobility of the species i. On the other hand, there exists a direct relation between the mobilities and the permeabilities[18]

$$u_i = A_g d \, P_i/A_d RT \tag{16}$$

where A_g is the geometric area of the membrane and A_d is the membrane area available for diffusion. Using Equations (15) and (16) we obtain

$$t_+ = 1/(1+n_p) \tag{17}$$

$$t_- = n_p/(1+n_p) \tag{18}$$

The Equations (13), (17) and (18) give us the transport numbers as functions of n_c and K.

Porosity

It can be expected that the selectivity of a passive membrane will be related to the degree of its structural complexity. To obtain a quantitative measure of this relation it is necessary to express the degree of complexity by a measurable magnitude. The simplest way to do this is by using the porosity θ, which is

$$\theta = A_d/A_g = r^2 N \tag{19}$$

where r is the average pore radius of the membrane and N is its superficial pore density (see Table 1).

RESULTS AND DISCUSSION

The Equation (13) is given by a surface in a space whose axes are K, n_p and n_c in Figure 1. In this figure it can be seen that for a given value of n_c the ionic relative permeability n_p rises uniformly with K.

The transport number t_- if K > 0 (t_+ if K < 0) versus n_c, is given in Figure 2 for $|K|$ 1. In this figure it can be seen that the transport numbers depend upon the relative saline concentration n_c, and t_- (K > 0) and t_+ (K < 0) rise uniformly with K. From Equations (17) and (18) it is evident that $t_+ + t_- = 1$.

In Figure 3 we propose a plot of the selective parameter K, obtained from Equations (10) and (11), versus porosity θ, Equation (19), see Table 2. Figure 3 and Table 2 show that the selectivity increases with porosity and its dependence upon the membrane thickness is not important in our conditions (all the Nucleopore membranes have the same thickness but several selectivities).

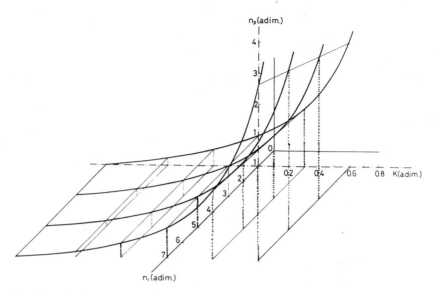

Fig. 1. The ionic relative permeability n_p, as depending upon the relative saline concentration n_c, and the selective parameter K. See Equation (13). The values of K with $|K|<1$ are the only ones we can find in passive transport.

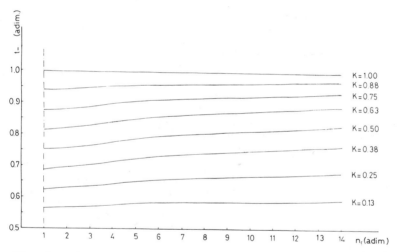

Fig. 2. The transport number t_- (K>0) or t_+ (K<0) versus n_c.

The increase of K with the increase of θ is difficult to understand. In this respect we want to point out that:

a) The experimental parameter K refers to the ionic selectivity at the steady state (n_p = P_-/P_+ at steady state) with well defined saline concentrations c_1 and c_2 (there is not any net separation process of ions), (n_c = c_2/c_1, c_1 = const.)

b) All our membranes have a pore radius of the same order and do not restrict the transport of the Cl⁻ nor Na⁺ ions, because the diameter of the resolvate ions are smaller than the mean pore diameters by a factor which is approximately equal[21] to 10[3].

c) In these conditions an increase of porosity means an increase of the membrane surface which is accessible to the ions.

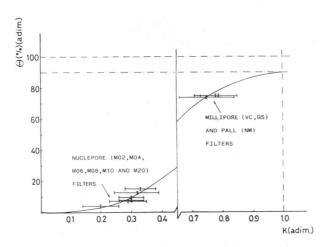

Fig. 3.   Proposed plot of the selective parameter K versus the porosity θ.
          Assuming cylindrical pores the porosity can only be of 90% as a
          maximum without destroying its structure.

Table 2.   Experimental Values of the Selective Parameter K, and Porosities
           θ, Obtained from Equation (19) and Table 1

Membrane	K (adim.)	$\theta(10^3$ adim.)	θ% (adim.)
M10	0.20 ± 0.06	39.82	4.0
M06	0.29 ± 0.07	70.69	7.1
M20	0.30 ± 0.04	71.94	7.2
M02	0.30 ± 0.04	94.20	9.4
M04	0.32 ± 0.07	126.00	12.6
M08	0.33 ± 0.05	150.80	15.8
NM	0.75 ± 0.09	740.00	74.0
GS	0.78 ± 0.07	750.00	75.0
VC	0.79 ± 0.06	750.00	75.0

d) The ionic selectivity given by the value of K comes up from the interaction between ions and fixed charges on the membrane surfaces, both
   the external ones and these of the pores.

So it was foreseeable to find that the bigger the membrane surface,
which is accessible to ions, the bigger the ionic selectivity. That is to
say that an increase of porosity produces an increase of ionic selectivity.

Values for the relative ionic permeabilities have been obtained[19] in
good agreement with those obtained from K experimentally measured. Transport numbers calculated from K are also in good agreement with those obtained by other methods[20].

Acknowledgement

This work has been supported by the National Advising Committee of
Scientic and Technical Research, CAICYT, Project No. 1275 (Spain).

REFERENCES

1. W. Nernst, Die Electromotorische Wirksamkeit der Ionen, Z.Phys.Chem., 4:561 (1889).
2. M. Planck, Ueber die Errengung von Elecktricität und Wärme in Elektrolytem, Ann.Phys.Chem.N.F., 39:161 (1890).
3. M. Planck, Ueber die Potentialdifferenz zwischen zwei verdünnten Lösungen binärer Elektrolyte, Ann.Phys.Chem.N.F., 40:561 (1890).
4. D. E. Goldman, Potential impedance and rectification in membranes, J.Gen.Physiol., 27:37 (1943).
5. R. A. Arndt, J. D. Bond, and L. D. Roper, An exact constant field solution for a simple membrane, Biophys.J., 10:1149 (1970).
6. R. A. Arndt, J. D. Bond, and L. D. Roper, Electroneutral approximate solutions of the steady state electrodiffusion equations for a simple membrane, J.Theoret.Biol., 34:265 (1972).
7. L. Bass, Electric structures of interfaces during steady electrolysis, Trans.Faraday Soc., 60:1656 (1964).
8. F. Conti and G. Eisenman, The steady state properties of ion exchange membranes with fixed sites, Biophys.J., 5:511 (1965).
9. A. D. MacGillivray, Nernst-Planck equations and the electroneutrality and Donnan equilibrium assumptions, J.Chem.Phys., 48:2903 (1968).
10. A. D. MacGillivray and D. Hare, Applicability of Goldmann's constant field assumption, J.Theoret.Biol., 25:113 (1969).
11. R. J. French, Numerical Solutions to the Nernst-Planck Equations. Applications to Biological and Synthetic Membranes, Thesis, Washington State University, U.S.A. (1973).
12. N. Laksminarayanaiah, Integration of Nernst-Planck flux equation, c.3, in: "Transport Phenomena in Membranes," Academic Press, New York (1972).
13. H. Vink, Transport properties of charged membranes in low charge densities, Acta Chem.Scandinavica, A3:547 (1979).
14. E. Kumamoto, Effect of unstirred layers on the membrane potential in an concentration cell, J.Membrane Sci., 9:43 (1981).
15. J. A. Ibáñez, A. F. Tejerina, J. Garrido, and J. Pellicer, Diffusion salt flow throughout membranes and permeability determination from cell potential measurements, J.Non-equilib.Thermodyn., 5:313 (1980).
16. J. Benavente, Contricución al estudio de los fenomenos Electro-cinéticos en Membranas Porosas, Thesis, University of Malaga, Spain (1981).
17. A. Rejou-Michel, Perméabilités et Potentiels Electriques des Systemes Membranaires. Mesures en Non-equilibre imposé, Thesis, University Paris VII, France (1978).
18. A. Gunn and P. E. Curran, Membrane potentials and ion permeability in cation exchange membranes, Biophys.J., 11:559 (1972).
19. J. A. Ibáñez, A. Hernández, and A. F. Tejerina, Determination of ionic permeabilities from passive membrane potentials, J.Non-equilib. Thermodyn., 7:159 (1982).
20. J. A. Ibáñez, A. F. Tejerina, and A. Hernández, Transport numbers of sodium and water in Nuclepore membranes from membrane potential, J.Non-equilib.Thermodyn., 8:293 (1983).
21. J. O'M. Bockris and A. K. N. Reddy, Modern Electrochemistry. An Introduction to an Interdisciplinary Area, Plenum Publishing Co., New York (1980).

IONIC PERMEABILITY OF THE S18 SAFT CARBOXYLIC MEMBRANE

A. Lindheimer, D. Cros, B. Brun and C. Gavach

Lab. Associé au C.N.R.S. No.330, "Physicochimie
des Systèmes Polyphasés," Université des Sciences
et Techniques du Languedoc Place Eugène Bataillon
34060 Montpellier-Cédex, France

SUMMARY

The S18 SAFT membrane is a battery separator composed of a polyethylene film grafted with acrylic acid. The selectivity of the ion transport through this carboxylic membrane is studied by means of two different
techniques: measurements of monoionic membrane potential and flux measurements using radioactive tracers. The most striking result is the value
higher than unit of the apparent transport number of potassium ion when
this membrane is in contact with concentrated KOH solutions. In contact
with NaOH, the transport number is practically equal to unit.

INTRODUCTION

With sulfonic polymeric membranes, a great deal of experimental work
has been achieved with the object of drawing some correlations between the
internal structures of the membranes and their permselectivities or their
ionic transport properties. On the other hand, just a few studies have
been devoted to polymeric cationic membranes bearing carboxylic groups in
spite of the fact that these kinds of membrane are used more and more in
electrochemical cells. Their swelling degree is not fixed by the ionic
hydration[1,2]. Bonamour and Millequant[3] have tried to relate the
swelling degree to the composition of polymeric membranes containing carboxylic and vinyl pyridine ionogenic groups. These authors also give some
information about the conductance and the ionic transport of these membranes in the presence of sodium chloride and potassium chloride. By means
of IR spectroscopy, Hurwitz and his colleagues[4] have specially investigated the hydration and the water molecules interactions in carboxylic
membranes with a "Teflon" matrix. Various carboxylic membranes are used as
battery separators in place of cellophane diaphragms, particularly with
alkaline batteries where the carboxylic membrane is in contact with highly
concentrated potassium hydroxyde solutions. Generally, these films are
grafted with acrylic acid to increase the wettability. So, the main goal
of this work is to draw some quantitative information about the transport
of potassium, sodium, hydroxyl and chloride ions in the S18 carboxylic
membrane which is used as a separator in a Nickel-Cadmium alkaline battery.
The ionic permeability is studied by means of two different techniques:
potentiometry and flux measurement by means of radioactive tracers. This

latter technique enables one to determine the transmembrane transport number without changing the composition of the aqueous media. Under certain well defined conditions, a radioactive labelled ion acts as an ideal tracer for the transmembrane transport[5,6]. Therefore, the technique has been widely used since the pioneer work of Meares and his co-workers[7,8].

EXPERIMENTAL

The S18 SAFT membrane is obtained by an irradiation grafting of acrylic acid on a polyethylene support followed by a special treatment which results in increasing its conductance. The thickness is about 50 μm and the exchange capacity is 4 meg/g. After soaking 1 g of dry membrane has a volume of 0,905 cm^3.

Measurements of the Zero Current Membrane Potential

The measurements of the zero current membrane potential have been carried out in a thermostated teflon cell, using a couple of saturated calomel electrodes and liquid junctions between the reference electrodes and the studied solutions. The membrane potentials are recorded by means of a chart recorder with an imput impedance higher than $10^{12}\Omega$ (Tacussel EPLI + TVED).

Measurements of the Ionic Transmembrane Fluxes

The current generator used is a potentiostat (Tacussel PRT 20). Again we used a thermostated teflon cell, the two compartments of which are initially filled with the same electrolytic solutions. A small amount of labelled aqueous solution is added to the anodic side. The same amount of unlabelled electrolyte is also added to the cathodic side. The variation of the activity appearing in the cathodic compartment is determined by measuring the radioactivity of samples with a Packard Tri-Carb liquid Scintillation Spectrometer. Immediately after each sampling, the same volume of electrolyte at the same concentration is added to the cationic compartment. When the electrolyte concentration was low, the number of labelled ions added with the radioactive solution has been taken into consideration.

The two electrodes are platinum electrodes (area 2 cm^2). In order to prevent the migration of H$^+$ and OH$^-$ resulting from the water electrolysis with solutions at neutral pH, ion exchange membranes are interposed between the electrodes and the aqueous solutions in contact with the studied membrane. At the end of each experiment, we checked that the pH was not changed in each compartment on both sides of the membrane.

In order to determine the value of the fluxes with the rate of the radioactivity change, we have taken into consideration the number of labelled ions crossing the membrane and the dilution effect of labelled ions. The symbols ' and " represent respectively the compartments where the labelled ions are added and where they arrive. The change of radioactivity is measured in both compartments and corrections are made in order to take into account the radioactive decrease when the period is short.

Every five minutes, a volume p is sampled from each compartment and just after the sampling, the same volume of unlabelled solutions are again added. Let us call t_1, t_2 ... t_i ...... t_{i+1} the instants of the sampling; r'_{t_i} and r''_{t_i} are the values of the radioactivity measured in the samples from the compartments ' and " at the instant t_i. Taking into account the effect of the dilution of the tracer due to the addition of unlabelled ion after each sampling, the variation of the radioactivity in the compartment " between the sampling at t_{i-1} and t_i is:

$$R''_{t_i} - R''_{t_{i-1}} = (r''_{t_i} - r''_{t_{i-1}}) \frac{V + r}{p} t_{i-1}$$

V is the volume of the compartments.

The radioactivity of compartment ' is $R' = r' \dfrac{V}{p}$.

Applying the law of ideal tracer, it is possible to deduce the number of K^+ ions having crossed the membrane between the instants t_{i-1} and t_i:

$$n''_{t_i} - n''_{t_{(i-1)}} = \frac{R''_{t_i} - R''_{t_{(i-1)}}}{R'_{t_{(i-1)}}} \, C.V.$$

C is the considered ion concentration in the compartment.

RESULTS

Zero Current Membrane Potential

Diffusion potentials are measured by pairs of aqueous solutions containing the same electrolyte at different concentrations. In almost all cases, the electrical potential of the dilute solution is higher than that

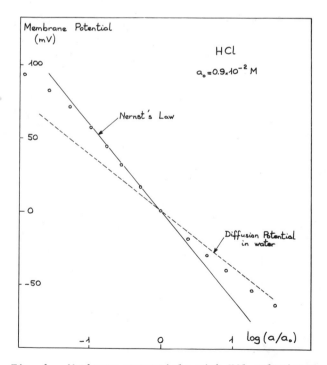

Fig. 1. Membrane potentials with HCl solutions.

of the more concentrated one. This result suggests a cationic character of
the membrane. The electrolyte concentration in the reference compartment
is kept constant (10^{-2}M). The electrolyte concentration in the other side
is varied from 10^{-4}M to 1M. The values of the diffusion membrane potential
are plotted as a function of the logarithm of ratio of the electrolyte
activities. Figure 1 shows the variation of the diffusion membrane potential with aqueous solutions containing only HCl. In this case, the potentiometric response is slightly lower than theoretical nernstian response.
On Figure 2 are plotted the variations of the diffusion membrane potential
with aqueous solutions containing KCl, the value of the pH being 6,5 on
each side. When the activity ratio is between 10 and 10^{-1}, a linear response is obtained but the slope is smaller than the theoretical nernstian
one. Outside this range, the membrane potential keeps a more constant
value.

This figure also shows the variations of the diffusion membrane potential when the electrolyte is KOH. The results are rather similar to those
obtained with KCl but, in this case, we can notice a decrease of membrane
potential when the activity ratio increases. When KOH is changed by NaOH,
the obtained results are the same.

In the measurements of the biionic membrane potentials, the membrane
is put between two aqueous homoanionic electrolyte solutions at the same
concentration, the pH value being 6,5 on both sides. In the case of 1-1
electrolytes, the electrolyte of the reference solution is KCl 10^{-2}M. In
the case of 2-1 electrolytes, $CaCl_2$ is used as reference electrolyte. In
the Figure 3, the values of the biionic potential are plotted as a function
of the crystallographic ratio of the ions.

On Figure 9, are plotted the numbers of KCl and NaCl molecules sorbed
in the membrane as a function of the electrolytic activity of the solution
with which the membrane is in equilibrium.

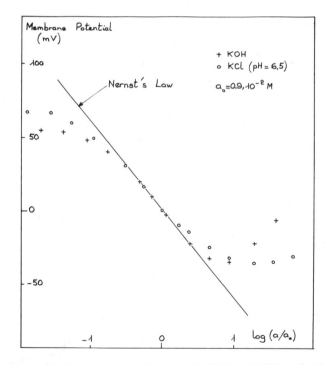

Fig. 2. Membrane potentials with KOH and KCl solutions

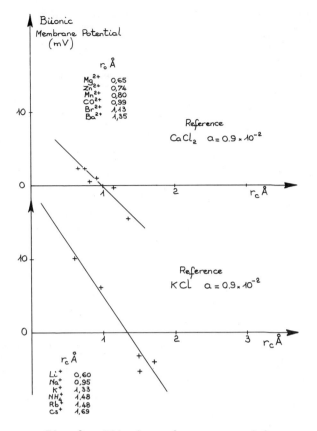

Fig. 3.   Biionic membrane potentials.

Measurements of the Exchange Ionic Fluxes

On Table 1, are gathered the values of the unidirectional fluxes of Na^+, Cl^- and K^+ when NaCl, KOH and NaOH are on both sides of the membrane and when no current is applied.

Measurements of the Transport Numbers

Figure 4 shows the typical variation of the number of K^+ ions transported through the membrane when a constant electrical current is maintained. The values of the fluxes are deduced from the linear part of the curves when a steady state is established. We can define for instance the transport number of K^+ through the membrane according to the classical equation:

$$t_{K^+} = \frac{F(J_+ - J_-)}{I}$$

I being the intensity of the electrical current.

On Figure 5 are plotted the values of t + and t + when KOH and NaOH are on both sides of the membrane. The Figures 6 and 7 show the variations of the cationic and anionic transport numbers with the electrolytic concentrations at pH = 6,5 respectively for NaCl and KCl.

Table 1. Exchange Fluxes: J_o (ion gram.s^{-1}. cm^{-2})

C_{mole} l^{-1}	K^*OH	Na^*OH
0,03	-	$1,4.10^{-8}$
0,05	-	$2,8.10^{-8}$
0,1	6.10^{-8}	$4,6.10^{-8}$
0,25	-	9.10^{-8}
0,50	-	$1,6.10^{-7}$
0,75	-	$2,1.10^{-7}$
1	$4,6.10^{-7}$	-
6,7	$1,2.10^{-6}$	-

C_{mole} l^{-1}	NaCl	
	Na^+	Cl^-
10^{-1}	$4,2.10^{-8}$	5.10^{-9}
5×10^{-1}	$1,6.10^{-7}$	$6,5.10^{-8}$
1	$3,1.10^{-7}$	$1,7.10^{-7}$
1,5	$2,9.10^{-7}$	2.10^{-7}
2	$3,1.10^{-7}$	$2,3.10^{-7}$

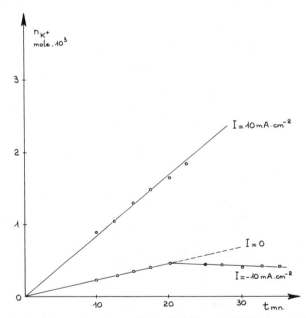

Fig. 4. Typical determination of ionic fluxes K*OH   t = 25°C   C = 1M.

On Figure 8 are presented the variations of K^+ and Cl^- transport numbers with the pH when KCl = 10^{-1}M. The dotted curve shows the variations of $[1 - (t_{K^+} + t_{Cl^-})]$ which can be identified with H^+ transport numbers.

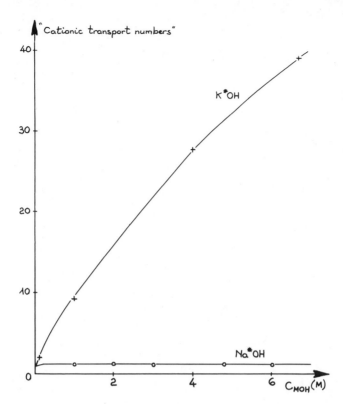

Fig. 5. The cationic transport numbers dependency with electrolytic concentration I = 10 mA.cm^{-2}.

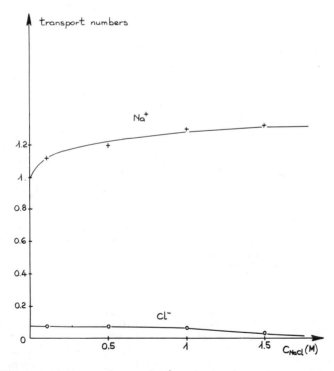

Fig. 6. The transport numbers of K$^+$ and Cl$^-$ dependency with KCl concentration pH = 6.5   I = 10 mA.cm^{-2}.

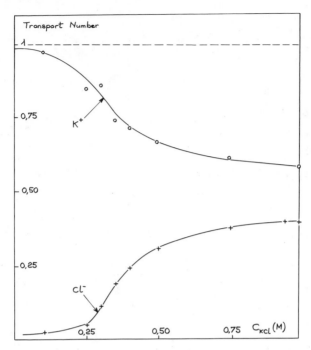

Fig. 7. The transport numbers of Na$^+$ and Cl$^-$ dependency with NaCl concentration pH = 6.5   I = 10 mA.cm^{-2}.

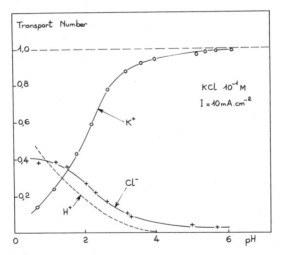

Fig. 8. The transport numbers of K$^+$ and Cl$^-$ dependency with the pH KCl 10^{-1}M   I = 10mA.cm^{-2}.

DISCUSSION

When the aqueous solution contains only HCl, the diffusion membrane potentials vary linearly with the logarithm of the ratio of the mean activity of the electrolyte. In the presence of a strong acid such as HCl, we could expect that the carboxylic groups inside the membrane would be neutralized and therefore that the membrane acts as an uncharged membrane. In this case, the diffusion membrane potential should be nothing but a simple diffusion potential between two HCl solutions separated by a simple dia-

108

phragm. The variations of this potential diffusion with the logarithm of the activity ratio are satisfactorily expressed by Henderson's equation:

$$V_{diff.} = \frac{RT}{F} \frac{u_+ - u_-}{u_+ + u_-} \ln \frac{a'}{a''}$$

The individual limiting conductances of H^+ and Cl^- being respectively 350 and 76,35 $\Omega^{-1}cm^2$ (g equiv)$^{-1}$[9], the slope of the variation of the diffusion potential as a function of the logarithm of the activity ratio is 38 mV per decade. With the carboxylic membrane, the slope is higher than 38 mV per decade and the variations are linear up to a ratio of activity equal to 0,01. This result means that the transport number of protons inside the membrane is higher than in water, suggesting a protonic conduction despite the fact that the carboxylic groups are not ionized.

With KCl, at pH = 6,5 the membrane diffusion potential varies linearly with the logarithm of the activity ratio only inside a limited range where the slope is about 50 mV per decade. This results demonstrates that the carboxylic membranes are cationic; a simple diffusion potential between two KCl solutions separated by a diaphragm would give a zero value because the mobilities of K^+ and Cl^- are practically equal. In the higher concentrations range, the variations of the diffusion membrane potential deviate from the linearity. This discrepancy can be explained by the classical theories developed by Teorell, Meyer and Sievers[10,13] in order to account for the potentiostatic response of ion exchange membranes. The deviations from nernstian responses are due to the penetration of co-ions into the membrane and to the diffusion of the electrolyte inside the membrane. These considerations cannot explain the deviations occurring in the low concentrations range. However, the permselectivity of the membrane towards the proton which is suggested from the results obtained with HCl can explain that, in the low K^+ concentration range, when the ratio of the number of H^+ over the number of K^+ is increasing, the permeability to the protons makes the potentiometric response smaller.

With KOH, the slopes of the potentiometric responses are similar but the deviations are more important. In the higher concentration range, this effect can be easily explained considering that the mobility of OH^- is higher than that of Cl^- and despite the fact that all the carboxylic groups are dissociated, OH^- ions can penetrate into the membrane. This assumption is confirmed by the results of the measurements of the transport numbers using radioactive tracers.

The transport numbers of K^+ have values higher than unit. These values increase with KOH concentrations and decrease when the current density increases, (see Table 2). This observation suggests that inside the membrane, there exist associated forms of KOH with the general formula $[K_n(OH)_p]^{(n-p)}+$ so that when an electrical charge equal to one faraday crosses the membrane, more than 1 gram-equivalent of K^+ is transferred from the anodic compartment to the cathodic one. As in all aggregation processes, the number of associated form increases with the concentration and decreases when the electrical field increases. This ionic association involving OH^- can be explained by the formation inside the membrane of hydrogen bonding induced by the carboxylic groups. Levy, Jenard and Hurtwitz[4] have shown, from IR measurements, that networks of hydrogen bond exist in carboxylic membranes with a Teflon matrix. The water structure inside the membrane is probably ice-like and this structure can explain not only the aggregation of KOH which gives rise to transport numbers higher than unit but also the quasi nernstian response obtained with HCl and which suggests a permeability of the membrane to H^+ as well as to OH^-.

Table 2. Transport Numbers of K^+ Ion in KOH Solutions

$Cmol.l^{-1}$	Current densities $I\ mA\ cm^{-2}$	$t = +25°C$
$10^{-2}M$	5	0,61
	10	0,31
	20	0,15
$10^{-1}M$	5	3,8
	10	2,0
	20	1,0
1 M	5	19
	10	9,3
	20	5,7
6,7 M	5	61
	10	39
	20	16

Regenerated cellulose films are also used as battery separators. In a recent work, Jenkins, Maskell and Tye[14] have shown that, with these membranes, the transport number of K^+ in KOH remains lower than unit and comprised between 0,25 and 0,42 while the limiting transport number of K^+ in KOH solution is 0,27. So the transport properties inside the regenerated cellulose membrane seems to be near to its value in the bulk aqueous medium. On the other hand probably because of the formation of hydrogen bonding, the carboxylic membranes show transport properties very different from the aqueous bulk solutions. It must be noted that the behavior of Na^+ in NaOH solutions is very different from the K^+ one; Figure 5 shows that, for all the studied concentrations, the value of t_{Na^+} is close to one. The results which are presented on Figure 8 lead to the determination of the value of the dissociation degree of carboxylic groups in the membrane. In the higher pH range, where these groups are not dissociated, the membrane acts as an uncharged diaphragm. As the pH increases in the aqueous solutions, the carboxylic groups become deprotonated and, in accordance with Donnan's law, the transport number of K^+ increases while that of Cl^- decreases. At the highest pH, the transport numbers respectively reach the following values: 0,98 for K^+ and 0,02 for Cl^-. Combining these results with those presented on Figure 9, we can deduce the values of θ (dissociation degree of the carboxylic groups in the membrane; its variations with KCl concentrations are plotted on Figure 10.

The results relative to the ion transport numbers at neutral pH, show again an important difference between NaCl and KCl (see Figures 6 and 7). A detailed discussion about these results is proposed elsewhere[15], and leads to the following conclusion: all the results relative to NaCl can be interpreted by the existence inside the membrane of a very strong amount of ion pairs which do not participate in the electrical transport (t_{Cl^-} nearly zero as shown on Figure 2 but number of sorbed NaCl inside the membrane different from zero as shown on Figure 9). Such a model is not convenient for KCl solutions and a higher and unsymmetrical association must be assumed to take into consideration the results relative to this electrolyte.

The results shown on Table 1 can be used to estimate the ionic diffusion coefficients inside the membrane; the diffusional flux can be expressed by the classical equation[16]:

$$\frac{1}{J_o} = \frac{d}{\bar{D}\,\bar{C}} + \frac{2\delta}{D} \cdot \frac{1}{C}$$

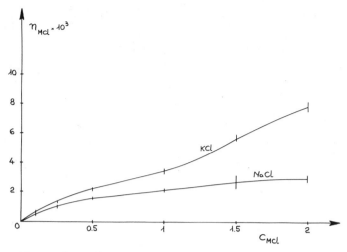

Fig. 9. The number of sorbed molecules for one gram of dry membrane plotted against the electrolytic concentration.

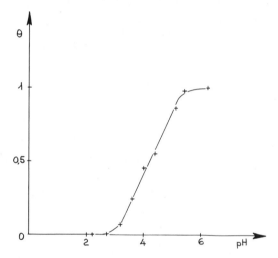

Fig. 10. The dissociation degree of the membrane versus the pH for KCl 10^{-1}M.

The plot of $\frac{1}{J_0}$ versus $\frac{1}{C}$ must be a straight line which allows the determination of δ: film thickness and $\bar{D}$: ionic self diffusion inside the membrane. Figure 11 is an illustration of this equation for Na*Cl, NaCl* and Na*OH. It gives the following parameters: $\delta = 133$ µm, $\bar{D}_{Na+} = 1,1.10^{-6}$ cm^{2}s^{-1} to be compared with the sodium diffusion coefficient in aqueous solution $D_{Na+} = 1,27 \times 10^{-5}$ cm^{2}s^{-1}. It must be noted that, the straight lines relative to Na*Cl and Na*OH are the same. So, the ionic association for these electrolytes inside the membrane seems to be complete.

Figure 12 shows a comparison between Na*OH and K*OH: the straight lines obtained with these two electrolytes give the same value. The cationic self diffusion in KOH inside the membrane is $\bar{D}_{K+} = 6,7 \times 10^{-6}$ cm^{2}s^{-1}.

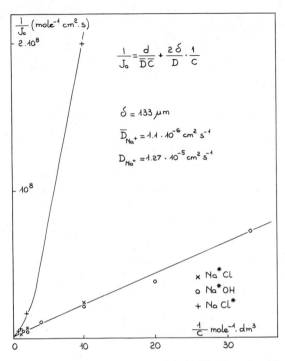

Fig. 11. Determination of the ionic self diffusion inside the membrane.

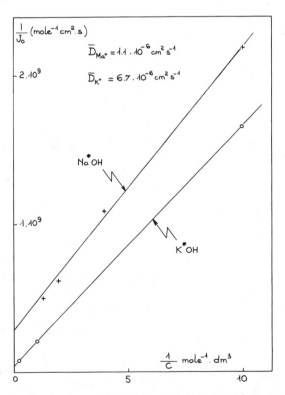

Fig. 12. Comparison of the self-diffusion behavior between NaOH and KOH.

REFERENCES

1. V. I. Soldatov and L. V. Novitskaya, Russ.J.Phys.Chem., 39:1453 (1965).
2. H. P. Gregor, L. B. Luttinger, and E. M. Loebl, J.Phys.Chem., 59:34 (1965).
3. A. M. Jendrychowska-Bonamour and J. Millequant, European Polymer J., 16:25 (1980)
4. L. Y. Levy, A. Jenard, and H. D. Hurwitz, J.Chem.Soc.,Faraday Trans., 78:29 (1982).
5. J. H. Jacquez, in: "Physiol. Memb. Disord," Andreoli, Hoffman, ed., p.147 (1978).
6. O. Kedem and A. Essig, J.Gen.Physiol., 48:1047 (1965).
7. P. Meares and H. H. Ussing, Trans.Faraday Soc., 55:244 (1959).
8. P. Meares and A. H. Sutton, J.Colloid Int.Sci., 28:118 (1968).
9. R. A. Robinson and R. H. Stokes, in: "Electrolyte Solutions," Butterworths Scientic Publications, London, p.452 (1955).
10. T. Teorell, Proc.Soc.Exptl.Biol.Med., 282:33 (1935); Proc.Natl.Acad. Sci.,U.S., 21:152 (1935).
11. T. Teorell, Z.Elektrochem., 55:460 (1951).
12. T. Teorell, Progr.Biophys.Chem., 3:305 (1953).
13. K. H. Meyer and J. G. Sievers, Helv.Chim.Acta, 19:649 (1936).
14. A. A. Jenkins, W. C. Maskell, and F. L. Tye, J.Memb.Sci., 11:231 (1982).
15. C. Gavach, A. Lindheimer, D. Cros, and B. Brun, J.Electroanal.Chem., submitted.
16. N. Lakshiminarayanaiah, in: "Transport Phenomena in Membranes," Academic Press, New York - London, p.130 (1969).

PREPARATION OF MICROPOROUS MEMBRANES BY PHASE INVERSION PROCESSES

H. Strathmann

Fraunhofer-Gesellschaft IGB
Nobelstr. 12
7000 Stuttgart 80, FRG

INTRODUCTION

The majority of todays membranes used in microfiltration, ultra-filtration or reverse osmosis consists of a symmetric or asymmetric microporous structure, which may in case of the asymmetric structure carry a more or less dense skin on the surface. The membranes may differ considerably in their structure, their function and in the way they are produced[1]. Preparation procedures for the different membrane types are described in patents and publications in detailed recipes, which are deeply rooted in empiricism. Superficially the preparation technique of a microporous polyethylene tube made by extrusion seems to have very little in common with the preparation of an asymmetric skin-type reverse osmosis membrane made by the precipitation technique described by Loeb and Sourirajan[2]. A more detailed analysis, however, reveals that the formation of both membrane types is determined by the same basic mechanism and governed by a process referred to by Kesting[3] as phase inversion. In fact, basically all polymeric ultrafiltration and reverse osmosis membranes and the majority of the microfiltration membranes, no matter how different their structures and their mass transport properties may be, are made by the phase inversion process. This process involves the conversion of lipid homogeneous polymer solutions of two or more components into a two-phase system with a solid, polymer-rich phase forming the rigid membrane structure and a liquid, polymer-poor phase forming the membrane pores. Phase separation can be achieved with any polymer mixtures, which forms, under certain conditions of temperature and composition, a homogeneous solution and separates at a different temperature or composition into two phases. To obtain a microporous medium such as a membrane requires both phases to be coherent. If only the solid phase is coherent and the liquid phase incoherent, a closed-cell foam structure will be obtained. If the solid phase is incoherent, a polymer powder will be obtained instead of a rigid structure.

The phase separation technique, however, is not only applicable to polymer solutions, it may also be utilized to prepare microporous membranes from glass mixtures and metal alloys as demonstrated in Figure 1, which shows the structure of a) a microporous asymmetrically structured glass membrane, b) a symmetrically structured microfiltration membrane prepared from regenerated cellulose esters and c) an asymmetrically structured reverse osmosis membrane made from polyamid. Although the structures, the

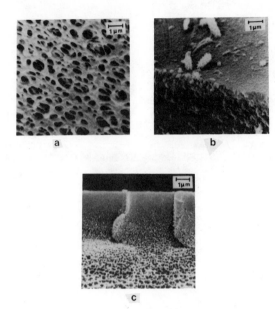

Fig. 1. Scanning electron micrographs of phase inversion membranes:
a) asymmetric microporous glass membrane, b) symmetric microfiltration membrane, c) asymmetric reverse osmosis membrane.

properties, and the preparation and procedures seem superficially to have
nothing in common, they are governed by the phase separation process and
are determined by thermodynamic and kinetic parameters, such as the
chemical potentials and diffusivities of the individual components and the
Gibb's free energy of mixing of the entire system. Identification and
description of the phase separation process is the key to understanding
the membrane formation mechanism, a necessity for optimizing membrane
properties and structures.

PHENOMENOLOGICAL DESCRIPTION OF MEMBRANE PREPARATION BY PHASE SEPARATION

Three different techniques are used for the preparation of state of
the art synthetic polymeric membranes by phase inversion: 1) thermogelation
of a two or more component mixture, 2) evaporation of a volatile solvent
from a two or more component mixture and 3) addition of a nonsolvent to a
homogeneous polymer solution. All three procedures may result in symmetric
microporous structures or in asymmetric structures with a more or less
dense skin at one or both surfaces suitable for reverse osmosis, ultrafiltration or microfiltration. The only thermodynamic presumption for all
three preparation procedures is that the free energy of mixing of the
polymer system under certain conditions of temperature and composition is
negative; that is, the system must have a miscibility gap over a defined
concentration and temperature range[4].

1. Thermogelation of a Two-Component Mixture

A microporous system may be obtained by thermogelation of a two
component mixture. The principle of the process is illustrated schematically in Figure 2, which shows a phase diagram of a two component mixture as
a function of temperature. At a specific temperature, the mixture forms a
homogeneous solution for all compositions but at a lower temperature shows
a miscibility gap over a wide range of compositions. The points P and S of
the diagram represent the pure components P and S respectively, and points

on the line P-S describe mixtures of these two components. If a homogeneous mixture of the composition X_P at a temperature T_1, as indicated by the point A in Figure 2, is cooled to the temperature T_2, as indicated by point B, it will separate into two different phases, the compositions of which are indicated by the points B' and B". The point B' represents the phase rich in component P and the point B" the phase rich in component S. The lines B'-B and B"-B represent the ratio of the amounts of the two phases in the mixture. If one phase is either still liquid while the other phase has already been solidified as this is the case in some polymer-solvent mixtures[5], or if one component can be dissolved as this is the case in some glass or metal alloys[6], a microporous system is obtained, where the liquid or dissolved phase forms the membrane pores and the solid phase the rigid membrane structure. Thermogelation is readily applied for the preparation of glass and metalmembranes and the extrusion of microporous polyethylene or polypropylene tubes now are widely used in cross-flow microfiltration.

2. Evaporation of a Volatile Solvent From a Three-Component Polymer Solution

This method was one of the earliest used in making microporous membranes[7] and has subsequently been applied by Kesting for making reverse osmosis membranes in his dry or complete evaporation process[8]. The principle of the process can be illustrated with the aid of a three-compound mixture consisting of a polymer, a volatile solvent and a third component, which by itself is a nonsolvent for the polymer. This three-component mixture is completely miscible over a certain composition range and contains a miscibility gap over another composition range as indicated in Figure 3, which presents an isothermic phase diagram of the three-component mixture. The corners of the triangle represent the pure components. Boundary lines between any two corners represent mixtures of two components, and any point within the triangle represents a mixture of all three components. Within a certain compositionally defined range of thermodynamic states, all three components are completely miscible, whereas in a different range – the miscibility gap – the system decomposes into two distinct phases. If the volatile solvent is completely evaporated from a homogeneous mixture of 10% polymer, 60% solvent, and 30% nonsolvent, as indicated by point A in Figure 3, the composition of the mixture will change from that represented by point A to that represented by point B. At point B the system consists of only two components: polymer and nonsolvent. Since this point is situated within the miscibility gap, the system is separated into two phases: a polymer-rich phase indicated by point B'

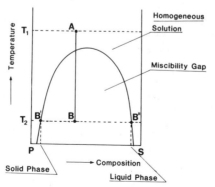

Fig. 2. Schematic diagram showing the formation of a microporous membrane by thermal gelation of a polymer solution exhibiting a miscibility gap at certain conditions of temperature and composition.

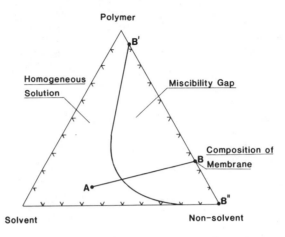

Fig. 3. Schematic diagram showing the formation of a microporous membrane
by evaporation of a solvent from a three-component polymer mixture
exhibiting a miscibility gap at certain conditions of temperature
and composition.

forming the rigid structure, and the phase B" forming the liquid filled
pores of the membrane.

3. Addition of a Nonsolvent to a Homogeneous Polymer Solution

This technique is widely used today for the preparation of symmetric
microfiltration membranes as well as for manufacturing asymmetric "skin-
type" ultrafiltration or reverse osmosis membranes[9]. The preparation
procedure can again be rationalized with the aid of a three-component
isothermic phase diagram shown schematically in Figure 4.

The phase diagram of the three-component mixture shows a miscibility
gap over a large range of compositions. If a nonsolvent is added to a
homogeneous solution consisting of polymer and solvent, as indicated by the
point A on the solvent-polymer line, and if the solvent is removed from the
mixture at about the same rate as the nonsolvent enters, the composition of
the mixture will change following the line A - B. At point C, the composition of the system will reach the miscibility gap and two separate phases
will begin to form a polymer-rich phase represented by the upper boundary
of the miscibility gap and a polymer-poor phase represented by the lower
boundary of the miscibility gap. At a certain composition of the three-
component mixtures, the polymer concentration in the polymer-rich phase
will be high enough to be considered as solid. This composition is represented by point D in Figure 4. At this point the membrane structure is
more or less determined. Further exchange of solvent and nonsolvent will
lead to the final formation of the membrane, the porosity of which is
determined by point B, which represents the mixture of the solid polymer-
rich phase and the liquid phase which is virtually free of polymer and
solvent as represented by points B' and B" respectively. While thermogelation can also be used for the preparation of metal and glass alloy
membranes, are the solvent evaporation and nonsolvent precipitation procedures limited to the preparation of polymer membranes.

The thermodynamic description of the formation of microporous systems
by means of the phase diagrams, as illustrated in Figure 2 to 4, is based
on the assumption of thermodynamic equilibrium. It predicts under what
conditions of temperature and composition a system will separate into two
phases and the ratio of the two phases in the heterogeneous mixture.

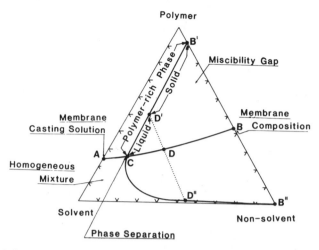

Fig. 4.   Schematic diagram showing the formation of a microporous membrane
          by addition of a non-solvent to a homogeneous polymer solution.

Related to the membrane formation procedure, the thermodynamic description
predicts the overall porosity that will be obtained at specified states.
However, no information is provided about the pore sizes, which are determined by the spatial distribution of the two phases. Equilibrium thermodynamics is not able to offer any explanation about structural variation
within the membrane cross-section; that is, whether the membrane has a
symmetric or asymmetric structure or a dense skin at the surface. These
parameters are determined by kinetic effects, which depend on system
properties such as the diffusivities of the various components in the
mixture, the viscosity of the solution and the chemical potential gradients
which act as driving forces for diffusion of the various components in the
mixture. Because these parameters change continuously during the phase
separation, which constitutes the actual membrane formation process, no
transient states of equilibrium will be achieved. Especially in polymer
systems, frozen states will often be obtained that are far from equilibrium
and that can be stable for long time periods, on the order of years. The
chemical potential and diffusivities of the various components in the
system, and their dependencies on composition, temperature, viscosity, etc,
are difficult to determine by independent experiments and therefore are not
readily available. This makes a quantitative description of the membrane
formation mechanism nearly impossible. A qualitative description, however,
which allows rationalization of the membrane formation and correlation of
the various preparation parameters with membrane structures and properties,
is possible.

EXPERIMENTAL PROCEDURES

     The experimental procedure described in this paper are concentrated on
the preparation of membranes from various polymer-solvent systems by
precipitation in a nonsolvent only. The membranes are then characterized
in terms of their transport properties and structures. Furthermore, the
three-component phase diagrams are determined for various polymer-solvent-
precipitant systems.

1.   Preparation of Phase Inversion Membranes by Non-Solvent Precipitation

     The preparation of phase inversion membranes is described in detail
recipes in the literature [9-11] and various patents.

a) Description of Relevant Preparation Parameters. In the preparation procedure of phase inversion membranes several significant parameters determining the structure and properties of the membrane can be identified:

1. The polymer and its concentration in the casting solution
2. The solvent or solvent system
3. The precipitant or precipitant system
4. The precipitation temperature.

In addition to these parameters several other procedures may affect the properties of the membrane, for example, an annealing step by which a membrane may be shrunk due to heat treatment or an evaporation step during which part of the solvent or solvents are evaporated from the surface of the film prior to precipitation.

b) Polymers, Solvents, Precipitants and Procedure Used in the Membrane Preparation Process. The majority of the membrane discussed in this paper are prepared from three different polymers; that is, cellulose acetate E-393-3 manufactured by Eastman Chemical Corporation, polysulfone PS 1700 made by Union Carbide and an aromatic polyamide made by DuPont under the trade name Nomex. Dimethylsulfoxide (DMSO), dimethylacetamide (DMAc), dimethylformamide (DMF), n-methylpyrrolidone (NMP) and acetone served as solvents for the preparation of the casting solutions. Several additives such as benzene, tetrahydrofurane (THF), formamide and lithium chloride were used in the casting solution[12]. Water, methanol, glycerole, formic acid and mixtures of these components were used as precipitants. The polymer concentration in the casting solution was varied between 5 and 25 wt % polymer.

2. Characterization of the Membranes in Terms of Structure and Transport Properties

The membranes are generally characterized in terms of their mass transport properties, that is their transmembrane fluxes and molecular weight cut-off, by filtration tests using hydrostatic pressure differences between 5 and 100 bar. The membrane structures are studied by scanning electron micrographs.

3. Rate of Polymer Precipitation During Membrane Formation

The rate of precipitation of the polymer during the membrane formation process was determined by precipitating films of different thickness and by an optical microscope using a technique described in the literature[13].

EXPERIMENTAL RESULTS AND DISCUSSION

1. General Observations

In this part of the paper several general observations concerning membrane structures, filtration properties and preparation procedures are described.

a) The Membrane Structure. Using scanning electron microscope techniques, four different rather typical structures shown in Figures 5a to 5d can be observed with phase inversion membranes. Photograph a) shows a cross-section of a microfiltration membrane prepared from a cellulose nitrate solution by precipitation with water vapor in a humidity controlled environment. The membrane shows a "sponge"-like structure with no skin on the bottom or top surface. Photograph b) shows a cross-section of a

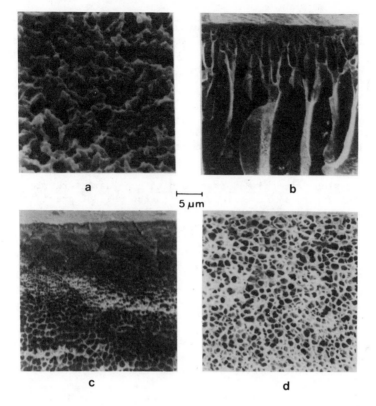

Fig. 5. Scanning electron micrographs of membrane cross-sections with
typical structures: a) symmetric microporous membrane without a
skin; b) asymmetric membrane with a "finger"-type structure and a
dense skin at the surface; c) asymmetric membrane with "sponge"-
type structure, a dense skin at the surface and a porous substructure with increasing pore diameters from the top to the
bottom side of the membrane; d) asymmetric membrane with a
"sponge"-type structure, a dense skin at the surface and a porous
structure with uniform pore size distribution over the entire
cross-section.

typical ultrafiltration membrane prepared from a polyamide solution and
precipitated in water by immersing the polymer solution into a water bath.
The membrane shows a "finger"-type structure with a dense skin at the
surface and large pores penetrating the entire membrane cross-section.
The pores increase in diameter from the top to the bottom side. Photograph
c) shows the cross-section of a typical reverse osmosis membrane prepared
from a polyamide solution and precipitated in water by immersing the
polymer solution into a water bath. The membrane shows a sponge-type
structure with a dense skin at the surface and a porous structure underneath with increasing pore diameter from the top to the bottom side of the
membrane. The only differences in the preparation procedures of the membranes shown in photograph b) and c) are the polymer concentration and the
precipitation temperature. Photograph d) shows the cross-section of a
reverse osmosis membrane prepared from a polyamide solution and precipitated in a water-solvent mixture. The membrane shows a sponge-type
structure with a dense skin at the surface and a porous structure underneath with a relative uniform pore size distribution over the entire
cross-section. Membranes with this type of structure can usually be dried
without changing their mass-transport properties.

b) <u>The Polymer, Polymer Concentration and the Membrane Structure</u>. The scanning electron micrographs of Figures 6a to c show the cross-section of three membranes with nearly identical structures and ultrafiltration properties listed in Table 1 prepared from three different polymers; that is cellulose acetate, polyamide and polysulfone, by precipitation in a water bath. The scanning electron micrographs of Figures 7a to 7c show the cross-sections of membranes made from one polymer-solvent system; that is polyamide in DMAC, with different polymer concentrations. These membranes show completely different structures and filtration properties which are listed in Table 2. The results of Figures 6 and 7 indicate that the same type of membrane can be prepared from various polymers and that from one polymer various types of membranes can be made.

c) <u>Precipitation Rate and Membrane Structure</u>. Characteristic membrane structures can generally be correlated with the rate of precipitation of the polymer solution. Figure 8 shows two series of photographs taken during the precipitation process under an optical microscope. The magnification and the time interval at which the photographs were taken is the same in both cases. The rate of precipitation slows down as the precipitation front moves further into the casting solution. The photographs also show that the finger-type structure precipitates much faster than the sponge-type structure.

There are various simple means by which the rate of precipitation can be altered. An increase in the polymer concentration of the casting solution generally results in an decrease in the precipitation rate and a transition from a finger-type structure to a sponge-type structure. Figure 9 shows the structure of membranes prepared from a polyamide-NMP casting

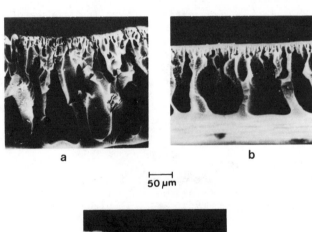

a   b

50 μm

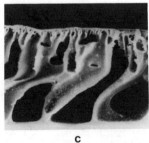

c

Fig. 6. Scanning electron micrographs of membrane cross-sections prepared from three different polymer-solvent systems by precipitation in water: a) 12% cellulose acetate in DMAc; b) 12% polyamide in DMSO; c) 12% polysulfone in DMF.

Table 1. Filtration Properties and Porosities of Membranes Prepared from Different Polymer-solvent Systems by Precipitation in Water

Polymer	Filtration rate* (cm/s)	Rejection** (%)		Membrane porosity (%)
		γ-globulin	bov.albumin	
Cellulose acetate in DMAc	3.5×10^{-3}	99	98	80
Polyamide in DMSO	2.1×10^{-3}	97	72	82
Polysulfone in DMF	1.9×10^{-3}	96	80	83

* The filtration rates were determined with DI-water at 2 bar and room temperature.
**The rejections were determined at 2 bar and room temperature with solution of 1% γ-globulin and bov. albumin.

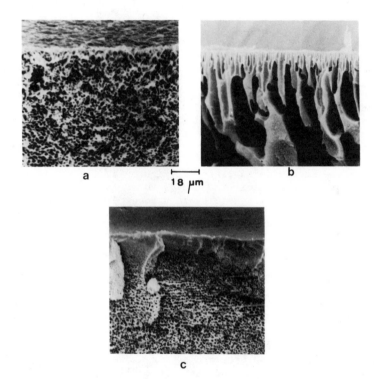

18 μm

a   b

c

Fig. 7. Scanning electron micrographs of membrane cross-sections prepared from polyamide solutions of different concentration and by different precipitation procedures: a) 10% polyamide in DMAc precipitated by water vapor; b) 10% polyamide in DMAc precipitated in a water bath; c) 22% polyamide in DMAc precipitated in a water bath.

solution of various polymer concentrations by precipitation in water at room temperature. In Table 3 the precipitation rates and the filtration properties of these membranes are summarized. The same effect can be achieved using different fluids as precipitants as indicated in Figure 10 and Table 4 where the structures and precipitation rates of different

Table 2. Filtration Properties and Porosities of Membranes Prepared from a
Polyamide-DMAc-water System

Polymer-concentration	Precipitant	Porosity (%)	Filtration rate* (cm/s)	Rejection**	Structure
10% polyamide in DMAc	water vapor	89	2×10^{-1} at 2 bar	0% for Dextran 100	symmetric "sponge"-type
10% polyamide in DMAc	water	79	1.8×10^{-3} at 2 bar	88% for bov.albumin	asymmetric "finger"-type
22% polyamide in DMAc	water	71	8×10^{-5} at 100 bar	99% for $MgSO_4$	asymmetric "sponge"-type

* The filtration properties were determined with deionized water and
  solutions of 1% solids at room temperature.
**The molecular weight of Dextran 100 was approximately 2 million Dalton.

membranes prepared from a solution of 15% polyamide in DMAc by precipitation in different fluids are shown. Going from formic acid to glycerin in
the precipitation bath, the rate of precipitation decreases drastically and
the structure changes from a finger- to sponge-type structure. Various
solvents or solvent mixtures may lead to the same results[12]. One
possible means of changing the rate of precipitation rather drastically is
achieved by addition of the nonsolvent from the vapor phase. This leads to
membranes with a more or less symmetric sponge-type structure with rather
large pore diameters as demonstrated in Figure 5a. In all cases studied
during this work fast precipitation leads to a finger-type structure and
slow precipitation generally results in a sponge-type structure.

2. Thermodynamic and Kinetic Fundamentals of Phase Separation Processes

The three-component phase diagram shown schematically in Figure 4 is
the description of an equilibrium state. It reflects the conditions under
which a multicomponent mixture is either stable as a homogeneous phase or
decays into two separate phases. In macromolecular systems, however,
equilibrium is frequently never reached and the phase separation is largely
governed by kinetic parameters, which are rather difficult to determine for
systems with more than two components. For simplicity, the basic thermodynamic and kinetic relations of the phase separation process are discussed
for a binary system. These basic relations, however, are also valid for
multicompound systems and can therefore be applied to rationalize the
membrane formation mechanism.

a) Thermodynamic Description of a Binary System With Limited Miscibility.
The thermodynamic state of a system with limited miscibility can be described in terms of the free energy of mixing. At constant pressure and
temperature, three different states can be distinguished:

1. A stable state with a homogeneous solution in a single phase, which is
   thermodynamically determined by [4,14]:

$$\Delta G > 0 \text{ and } \left(\frac{\delta \mu_i}{\delta X_i} \right)_{P,T} > 0. \tag{1}$$

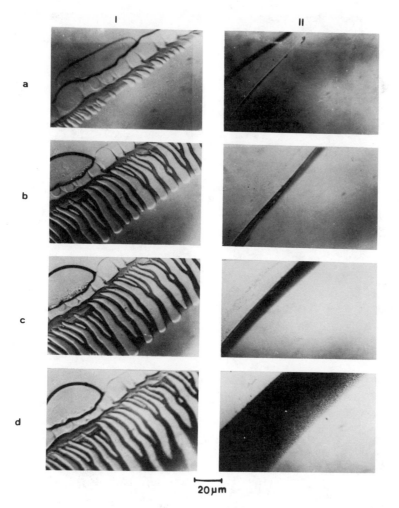

I   II

a

b

c

d

|————| 20μm

Fig. 8. Photo-micrographs of the casting solution/precipitant interphase
at (a) the beginning of the precipitation and after (b) 12 sec.,
(c) 24 sec. and (d) 5 min. Series I giving a "finger"-type
structure and series II giving a "sponge"-type structure.

2. An unstable state where the homogeneous solution separates spontaneously into two phases in equilibrium. This state, which is always
located within the miscibility gap, is thermodynamically determined
by:

$$\Delta G < 0 \text{ and } \left(\frac{\delta \mu_i}{\delta X_i} \right)_{P,T} < 0. \tag{2}$$

3. An equilibrium state given by the phase-boundary composition and
thermodynamically determined by:

$$\Delta G = 0 \text{ and } \left(\frac{\delta \mu_i}{\delta X_i} \right)_{P,T} = 0. \tag{3}$$

Here, ΔG is the free enthalpy of mixing, μ_i is the chemical potential
of component i, and X_i is its mole fraction.

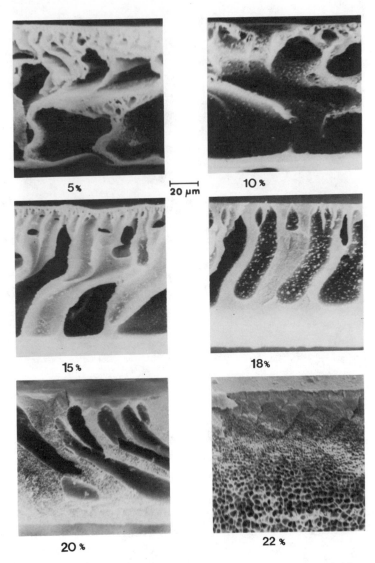

Fig. 9. Scanning electron micrographs of membrane cross-sections prepared from various polyamide concentrations in NMP by precipitation in water at room temperature.

b) Kinetic Description of a Binary System With Limited Miscibility. The kinetic interpretation of a system with limited miscibility can be expressed in terms of the diffusion coefficient D of the system. At constant pressure and temperature, again three different states can be distinguished[4]:

(1) A stable state with the homogeneous solution in one phase which is kinetically determined by:

$$D > 0. \tag{4}$$

(2) An unstable state, where the homogeneous solution decays spontaneously into two different phases. This state is determined by:

$$D < 0. \tag{5}$$

(3) An equilibrium state given by the phase-boundary composition. Here the diffusion coefficient becomes zero.

$$D = 0. \tag{6}$$

The physical significance of these three states becomes clear when one realizes that the diffusion coefficient is defined by Fick's law in terms of a concentration gradient as the driving force. However, the actual driving force for the flux of a component is not the gradient in its concentration but the gradient in its chemical potential. It is possible to produce situations where the gradients in concentration and chemical potential are of different sign. In this case the diffusion coefficient defined by Fick's law will be negative.

The diffusion coefficient can be related to the chemical potential driving force by[14]:

$$D_i = B_i \left(\frac{\delta \mu_i}{\delta X_i} \right)_{P,T} = 0. \tag{7}$$

Here D_i is the diffusion coefficient of component i, μ_i is its chemical potential, and X_i its mole fraction. B_i is a mobility term, which is always positive. Introducing the chemical potential of a nonideal solution, which is given by:

$$\mu_i = \mu_i^o + RT \ln X_i + RT \ln f_i^s . \tag{8}$$

into Equation (7) leads to:

$$D_i = \frac{B_i RT}{X_i} \left(1 + \frac{\delta \ln f_i^s}{\delta \ln X_i} \right). \tag{9}$$

Here μ_i^o is a standard potential, f_i is an activity coefficient referring to the pure phase, R is the gas constant, and T is the absolute temperature. The last term in Equation (9) determines whether the diffusion coefficient is positive, negative or zero according to these possibilities:

$$\frac{\delta \ln f_i^s}{\delta \ln X_i} > -1, < -1, \text{ or } = 1 . \tag{10}$$

Equations (8) to (10) indicate that, if for any reason the activity coefficient of a component i in a binary solution is raised to the extent that the product $f_i X_i$ becomes larger than unity, phase separation will occur and the component i will flow from an area of lower concentration into an area of higher concentration. Thus the flux of the component i is against the concentration gradient but i follows the chemical potential gradient.

In summary:

(1) The driving force for any mass flux is not the concentration gradient but the gradient in the chemical potential.

(2) In phase separation processes, the components are transported against their concentration gradient because the activity coefficient is increased to such an extent that the product of $f_i X_i$ is larger than unity as a consequence.

127

Table 3. Rates of Precipitation and Filtration Properties of Membranes
Prepared from Polyamide–NMP Solution of Various Polyamide Concentrations by Precipitation in Water at Room Temperature

Polymer-concentration (%)	MgSO₄	Rejection* Cytochrome C	bov. albumin	Filtration rate (cm/s × 10⁴)	Porosity (vol. %)	Precipitation rate (s)
5**	0	0	10	56	91	32
10**	0	43	84	32	85	40
15**	8	92	100	9	81	52
18***	75	100	100	18	79	83
20***	90	100	100	4	77	142
22***	98	100	100	1.6	76	212

*   The rejection was determined with solutions of 1% solids
** applied pressure 5 bar
***applied pressure 100 bar.

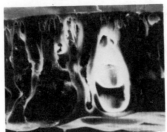
Formic acid

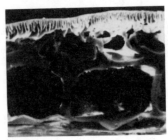
Methanol

50 μm

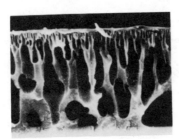
Water

Acetone

Acetic anhydride

Glycerol

Fig. 10. Scanning electron micrographs of membrane cross-sections prepared
from a casting solution of 15% polyamide in DMAc precipitated at
room temperature in various precipitation media.

Table 4. Rates of Precipitation and Structures of Membranes Prepared from a Casting Solution of 15% Polyamide in DMAc by Precipitation in Different Media

Precipitant	Rate of precipitation (sec)	Membrane structure
Formic acid	32	"finger"-type structure
Methanol	39	"finger"-type structure
Water	50	"finger"-type structure
Acetone	110	"sponge"-type structure
Acetic anhydride	234	"sponge"-type structure
Glycerol	667	"sponge"-type structure

The activity coefficient of a component can be changed for example by changing the temperature or the composition of the mixture. A typical example is the so-called "salting -out" effect, where the activity of the salt in an aqueous solution is raised by adding an organic solvent to such an extent that the salt precipitates from the solution.

3. Discussion of Membrane Structures and Properties Based on Thermodynamic Relations of Phase Separation.

Most of the structural differences of phase inversion membranes can be rationalized by the basic thermodynamic and kinetic relations of phase separation.

a) The Precipitation Process. Before discussing the details of the formation of symmetric or asymmetric finger- or sponge-type membrane structures it is useful to describe the concentration profiles of the casting solution components through the precipitating membrane. Figure 11 shows the concentration profiles of polymer, solvent and precipitant at some intermediate time during the precipitation of a polymer film cast on a glass plate. This figure shows the composition of the casting solution at the point of precipitation (C), the point of solidification (D) and the final membrane composition (B) taken from the precipitation pathway in Figure 4.

During precipitation the casting solution can be divided into three layers. Travelling from the glass plate towards the precipitation bath these layers are:

(1) The casting solution layer: This is the layer closest to the glass plate, and has a composition similar to the original casting solution. Little solvent has diffused out, and little precipitant has diffused into this layer.

(2) The fluid polymer layer: This layer lies between the point of precipitation on one side and the point of solidification on the other side. In this layer the casting solution divides into a polymer-rich and a polymer-poor phase. The composition of the components in both phases are shown in Figure 11. Assuming the phases are in equilibrium, their compositions can be related by the tie lines in the phase diagram, Figure 4. At the point of precipitation, the precipitated polymer contains still a high solvent concentration and a low precipitant concentration; it is therefore quite fluid. The polymer nearest the precipitation bath has been precipitated longer, and has lost solvent and gained more precipitant; its viscosity is therefore higher. Thus, the viscosity of the precipitated polymer climbs from

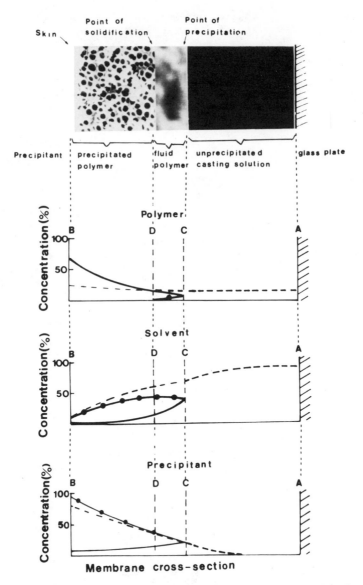

Fig. 11. Schematic diagram of the concentration profiles of polymer, solvent and precipitant through a precipitating membrane. --- total concentration; —— concentration in the polymer-rich phase; -•- concentration in the polymer-poor phase.

the point of precipitation C until it becomes almost a solid at the point of gelation, D. During this time bulk movement of the precipitated polymer takes place to form the matrix of the final membrane.

(3) The solid polymer layer: In this layer, the solid, polymer-rich phase undergoes continuous desolvation, and the system composition changes from D to B. It is the shrinkage or syneresis of the solid polymer accompanying this composition change that produces stresses in the polymer. Because the polymer is solid, these stresses cannot be as easily relieved by bulk movement of polymer as in the fluid polymer layer. Instead, the polymer structure either slowly undergoes creep to relieve the stress, or, if the stress builds up too rapidly to be dissipated by creep, the polymer matrix breaks in weak spots.

Two different techniques have been employed for the precipitation of membranes from a polymer casting solution. In the first method, the precipitant is introduced from the vapor phase. In this case the precipitation is slow, and a more or less homogeneous structure is obtained without a dense skin on the top or bottom side of the polymer film. This structure can be understood when the concentration profiles of the polymer, the precipitant and the solvent during the precipitation process are considered. The significant feature in the vapor-phase precipitation process is the fact that the rate-limiting step for precipitant transport into the cast polymer solution is the slow diffusion in the vapor phase adjacent to the film surface. This leads to uniform and flat concentration profiles in the film. The concentration profiles of the precipitant at various times in the polymer film are shown schematically in Figure 12.

Because of the flat concentration profiles, the solution at virtually the same time over the entire film cross section, and no macroscopic gradients of activity or concentration of the polymer are obtained over the film cross section. On a microscopic scale, however, because of thermal molecular motions, there are areas of higher and lower polymer concentration, which act as nucleation centers for polymer precipitation. These microscopic areas of higher polymer concentration are randomly distributed

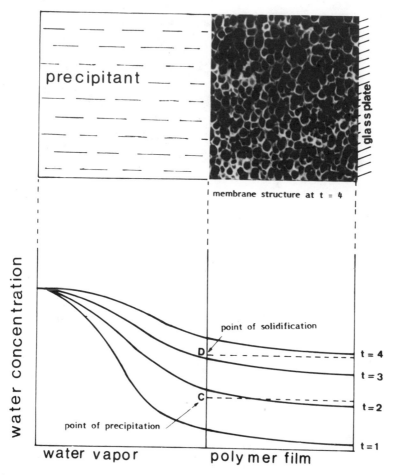

Fig. 12. Concentration profiles of the precipitant in the casting solution at various times during the formation of a symmetric structured membrane.

throughout the cast polymer film. Therefore, a randomly distributed polymer structure is obtained during precipitation. This structure is also shown in Figure 13 in the form of a scanning electron microscope picture of the cross section of a symmetric membrane obtained with a vapor phase precipitant.

In the second membrane preparation procedure the precipitant is added to the casting solution by immersing the cast polymer film in a bath of the precipitation fluid. In this case the precipitation is rapid, and a skinned membrane structure is obtained. This structure, and especially the skin formation, can again be understood by considering the concentration profiles of the polymer, the solvent and the precipitant during the precipitation process. These profiles over the cross section of the cast polymer film are shown schematically in Figure 13. The most important feature in immersion precipitant are the steep concentration and activity gradients of all components obtained at the polymer solution – precipitation medium interface. Because of the activity gradients, the transport of the polymer during phase separation at the interface is no longer random, but directed into the casting solution.

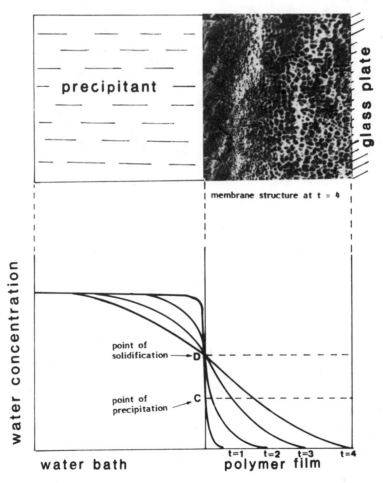

Fig. 13.   Concentration profiles of the precipitant in the casting solution at various times during the formation of an asymmetric skin type membrane.

When the cast polymer film is immersed into the precipitation bath, solvent leaves and precipitant enters the film. At the film surface the concentration of the precipitant soon reaches a value resulting in phase separation. In the interior, however, the polymer concentration is still far below the limiting concentration for phase separation. Phase separation therefore occurs initially at the surface of the film, where due to the very steep gradient of the polymer chemical potential on a macroscopic scale, there is a net movement of the polymer perpendicular to the surface. This leads to the formation of the membrane skin. This skin then hinders further precipitant flux into and solvent out of the casting solution and thus becomes the rate-limiting barrier for precipitant transport into the casting solution, and the concentration profiles in the casting solution interior becomes less steep. Thus, once the precipitated skin is formed, the same situation in the sublayer is obtained as in a membrane precipitated from the vapor phase, and a structure with randomly distributed pores is formed.

b) Skin Type Membranes With "Sponge"- and "Finger"-like Structures. In skin-type membranes two characteristic structures shown in Figure 5 are obtained. One is a sponge-like structure and the other is a finger-like substructure underneath the skin.

The formation of the sponge-structured membranes can be easily rationalized utilizing the description of the precipitation process given above. With finger-structured membranes the formation process is more complex and cannot entirely be described by the thermodynamic and kinetic arguments of phase separation processes. Other phenomena such as syneresis, shrinkage and stress relaxation in the precipitated polymer also play an important role. The formation of finger-structured membranes is conveniently divided into two steps: the initiation, and the propagation of fingers. The formation of the skin is identical with that of the sponge-structured membranes. However, as the result of syneresis, shrinkage stress in the solid polymer skin cannot be relieved by creep relaxation of the polymer and the homogeneous layer ruptures. The points at which the skin has been fractured form the initiation points for the growth of the fingers. Once a finger has been initiated, shrinkage of the polymer causes it to propagate by draining the freshly precipitated polymer at the bottom of the finger to the side of the finger. This is schematically shown in Figure 14 which indicates the growth of a "finger" at various times. Within a finger the exchange of solvent and precipitant is much faster than through the unfractured skin, and the precipitation front moves much faster within a finger than in the casting solution bypassed between fingers. This solution is protected from immediate exposure to the precipitant by a layer of precipitated polymer. Precipitation therefore occurs much slower and a sponge-like structure is formed between the fingers. Typical finger-like structure membranes almost always have a bottom skin. The bottom skin may be caused by the adhesion of the casting solution to the glass plate, preventing the last polymer fluid at the bottom of the finger from moving to the sides of the finger. This fluid polymer thus solidifies in place sealing off the finger.

4. Different Parameters and Their Effect on Membrane Structure and Performance

The mechanism for the formation of symmetric or asymmetric microporous membranes outlined above allows many of the variables of the membrane preparation procedures to be rationalized.

a) The Selection of the Polymer-Solvent-Precipitant Systems. The precipitant and the solvent used in membrane preparation determine both the activity coefficient of the polymer in the solvent-precipitant mixture and

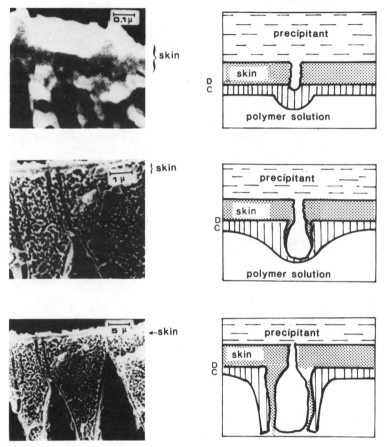

Fig. 14.   Schematic diagram of finger formation at various times during
precipitation.

the concentration of polymer at the point of precipitation and solidification.   The activity of the polymer in the mixture with the solvent and
precipitant, however, determines the rate of precipitation and thus the
structure and properties of the final membrane.

The effect of additives to the casting solution or precipitant on the
membrane structure can also be explained by changes of the activity coefficients of the polymer, the solvent or the precipitant.

b) The Effect of the Polymer Concentration in the Casting Solution on the
Membrane Structure.   A low polymer concentration in the casting solution
tends to precipitate in a finger structure, while high polymer concentrations tend to form sponge-structured membranes.   The effect of polymer
concentration on membrane structures can be explained by the initiation
and propagation of fingers.   Higher polymer concentration in the casting
solution produces a higher polymer concentration at the point of precipitation, which will thus tend to increase the strength of the surface layer
of polymer first precipitated, and tend to prevent initiation fingers.   The
increasing viscosity of the casting solution has the same effect.

Although most aspects of the formation of asymmetric skin type membranes can satisfactorily be rationalized by applying the basic thermodynamic and kinetic laws of phase separation processes, there are other
parameters, such as surface tension[16], polymer relaxations[17], solvent

loss by evaporation[18] etc., which are not directly related to the phase separation process, but nevertheless will have a strong effect on the membrane structure and properties.

REFERENCES

1. H. K. Lonsdale, J.Membr.Sci., 10:81 (1982).
2. R. E. Kesting, "Synthetic Polymeric Membranes," McGraw-Hill, New York (1971).
3. S. Loeb and S. Sourirajan, Adv.Chem.Serv., 38:117 (1962).
4. R. Haase, "Thermodynamik der Mischphasen," Springer Verlag, Berlin (1956).
5. R. Schnabel, Proceed. 5th Intern. Symp. Fresh Water From the Sea, 4:409 (1976).
6. A. J. Castro, US Patent, 4 247 498 (1981).
7. R. Zsigmondy and C. Carius, Chem.Ber., 60B:1047 (1927).
8. R. E. Kesting, J.Appl.Polym.Sci., 17:1771 (1973).
9. H. Strathmann, "Trennung von Molekularen Mischungen mit Hilfe Synthetischer Membranen," Steinkopff, Darmstadt (1979).
10. S. Loeb, in: "Desalination by Reverse Osmosis," U. Merten, ed., M.I.T. Press, Cambridge, Mass. (1966).
11. S. Manjikian, Ind.Eng.Chem.Prod.Res.Develop, 6:23 (1967).
12. H. Strathmann, Habilitationsschrift, Universität Tübingen (1981).
13. M. A. Frommer, R. Matz, U. Rosenthal, Ind.Eng.Chem.Prod.Res.Develop., 10:193 (1971).
14. J. Stauff, "Kolloidchemie," Springer Verlag, Berlin (1960).
15. R. Haase, "Thermodynamik der irreversiblen Prozesse," Steinkopff Verlag, Darmstadt (1963).
16. G. B. Tanny, J.Appl.Polym.Sci., 18:2149 (1974).
17. D. M. Koenhen, M. H. V. Mulder, C. A. Smolders, J.Appl.Polym.Sci., 21:199 (1977).
18. J. E. Anderson and R. Ullman, J.Appl.Phys., 44:4303 (1973).

CHITOSAN HOLLOW FIBERS: PREPARATION AND PROPERTIES

F. Pittalis and F. Bartoli

Assoreni
Monterotondo
Italy

SUMMARY

The preparation and properties of dialysis and/or haemodialysis chitosan hollow fibers are described. In particular spinning conditions are described in which hollow fibers showing good dialytic and ultra-filtrative properties are obtained starting from chitosan solutions in acetic acid. The transport properties of these fibers have been studied making use of model substances such as urea, uric acid, creatinine, inulin and vitamin B_{12}. The obtained results are discussed and compared with those obtained starting from commercial products such as cuprophan hollow fibers.

INTRODUCTION

Chitosan is a natural polymer prepared by complete or partial deacetylation of chitin which is a main component of the shell of crustaceans such as crab, lobster and the like[1].

The presence of amino groups in the base repetitive unit of chitosan (Figure 1) determines the reactivity of this polymer with organic and/or inorganic acids with formation of water soluble salts. The so obtained solution can be used for the preparation of films and membranes by evaporation or alkali coagulation procedures[2]. The excellent dialytic properties of chitosan membranes have stimulated a wide interest for a possible use of this material in dialytic or ultrafiltrative processes. In particular a considerable attention has been devoted to the development of methods by which hollow fibers could be conveniently obtained.

In this communication we describe the preparation and properties of chitosan hollow fibers obtained by a wet spinning procedure in which an alkali coagulating bath is used as the outside quench medium and a gas mixture containing ammonia and air or nitrogen is fed to the inner bore of the fiber. This mixture provides both the central quench medium and the necessary support for the nascent fiber during the subsequent steps of the preparation.

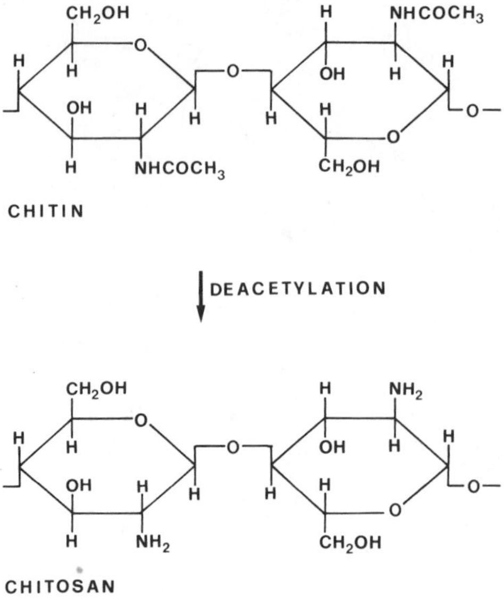

Fig. 1. Deacetylation scheme of chitin to chitosan

MATERIALS AND METHODS

A commercial chitosan (Kurifix, Kurita, Japan) and an EniChimica
(Milano, Italy), product have been used for the preparation of the fibers.
Both samples had a molecular weight of about 600,000–800,000, a deacetylation degree of 67–70% and an ash content of 2.5%. Chitosan was dissolved
in 0.3 M acetic acid at a concentration of 3% w/w. This solution, showing
a viscosity of about 1200 cP, was extruded through a tube-in-orifice
spinneret and a gas mixture containing ammonia and air 60/40 v/v was fed to
the central bore of the spinneret. The external quench medium was NaOH 1M.
After the coagulating bath the fiber was driven by two pairs of godets
where it was flushed with water and dried at 60–70°C. The dried fiber was
then collected in form of skeins.

The so obtained fibers have been tested for their transport properties and compared with commercial products such as cuprophan hollow fibers.
To this purpose small reactors have been prepared, each containing 50
fibers whose length was 30 cm. These reactors have been used for diffusion
and solvent permeability tests. Burst pressure tests have also been
performed.

Diffusion Measurements

Diffusion measurements have been performed at 37°C making use of model
substances such as urea, sodium chloride, uric acid, creatinine, vitamin
B_{12} and inulin. Solutions of these substances were recycled inside the
fibers and the amount of the diffused molecules was measured in the external recycling solution. The amount of solution of each tested substance
was 100 ml and that of the external solution was 300 ml; both solutions
were recycled at a rate of 800 ml/h. Diffusion apparatus is shown in
Figure 2. Diffused substances were measured in the external solution by
conductometric means in the case of sodium chloride and by an enzymatic or
colorimetric method in the case of urea, uric acid, creatinine, vitamin B_{12}
and inulin[3].

Solvent Permeability Measurements

Measurements of hydraulic permeability[4] have been made at 37°C and at pressure differences of 100, 200 and 300 mmHg making use of the apparatus shown in Figure 3. During these tests a solution of blue dextran 2000 was circulated inside the fibers so that a possible leak could be detected.

Burst Pressure Tests

Burst pressure tests have been performed pressurizing the fibers with nitrogen. A pressure decrease due to the burst of same fibers was indicated by a pressure gauge connected to the fibers.

RESULTS

Some physical parameters of the tested chitosan fibers, prepared starting both from the commercial and EniChimica chitosan, are shown in Table 1 together with the corresponding parameters of commercial cuprophan

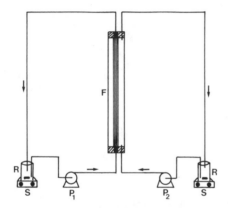

Fig. 2.   Diffusion apparatus. P_1, P_2 pumps; F fibers; R recycle reservoir; S magnetic stirrer.

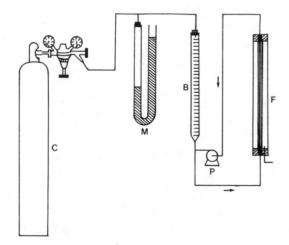

Fig. 3.   Hydraulic permeability apparatus.   C air cylinder; M pressure gauge; B burette; P pump; F fibers.

139

fibers. It can be seen that gelled chitosan fibers are characterized by a very high water content which undergoes an irreversible decrease upon drying. However dried chitosan fibers show a significant water uptake which is about four times higher than that of cuprophan fibers. This effect accounts for the highly hydrophilic nature of the chitosan polymer and should eventually lead to a high solvent permeability for these fibers.

Figures 4 and 5 show some scanning electron microscope pictures of chitosan fibers prepared starting from the EniChimica polymer. An homogeneous sponge-like structure is shown, as is normally observed in dialysis and haemodialysis membranes. The same structure has been observed in chitosan fibers obtained from the Kurifix polymer.

Diffusion Measurements

Diffusion measurements have been performed at different concentrations of the model substances in the feed solutions (from 50 mg/l to 1 g/l).

Table 1. Characteristics of Tested Fibers

Fiber	Φext μm	Wall thickness μm	Gelled fiber H_2O content g H_2O/g pol.	Dried fiber residual H_2O mg H_2O/g pol.	Dried fiber H_2O uptake g H_2O/g pol.	Count denier g/9 Km
EniChimica	450	15	22	130	2.5	144
Kurifix	450	15	23	130	2.4	153
Cuprophan	250	18	–	120	0.6	126

Fig. 4. Chitosan hollow fiber. Structure.

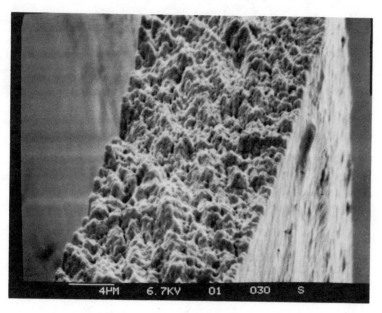

Fig. 5. Chitosan hollow fiber. Structure.

These measurements allowed us to verify the first Fick's law in the case of the tested fibers[5]:

$$J = - D \frac{\delta C}{\delta X} \qquad (1)$$

where J is the flux of the permeant molecule expressed in $mg/\acute{c}m^2.sec$, D is the diffusion coefficient expressed in cm^2/sec and $\delta C/\delta X$ is the derivative of the concentration of the diffusing substance with respect to the X axis perpendicular to the membrane plane. In the case of very thin membranes formula (1) can be approximated as:

$$J = - \frac{D}{d} \Delta C \qquad (2)$$

where a gradient has replaced a concentration differential and d is the membrane thickness. Formula (2) has been used to calculate the diffusion coefficients of the model substances which are shown in Table 2.

The reported data demonstrate a good permeability of chitosan fibers, in particular in the case of uric acid, vitamin B_{12} and inulin. In addition no significant difference can be observed between chitosan fibers prepared from Kurifix chitosan and EniChimica chitosan.

Solvent Permeability Measurements

Hydraulic permeability measurements performed at pressure differences of 100, 200, and 300 mmHg are reported in Table 3. The obtained results show that chitosan fibers are 3-4 times more permeable to water than cuprophan fibers. This result agrees with the very high water uptake that we observed for the dried chitosan fibers.

Burst Pressure Tests

Chitosan fibers prepared both from Kurifix and from EniChimica chitosan showed to withstand pressure differences up to 600 mmHg.

Table 2. Diffusion Coefficients

Fiber	NaCl cm^2/sec	Urea cm^2/sec	Uric acid cm^2/sec	Creatinine cm^2/sec	Vitamin B_{12} cm^2/sec	Inulin cm^2/sec
EniChimica	$2.2 \cdot 10^{-6}$	$2.1 \cdot 10^{-6}$	$7.0 \cdot 10^{-7}$	$3.1 \cdot 10^{-6}$	$2.8 \cdot 10^{-7}$	$1.7 \cdot 10^{-7}$
Kurifix	$2.5 \cdot 10^{-6}$	$2.4 \cdot 10^{-6}$	$6.5 \cdot 10^{-7}$	$3.2 \cdot 10^{-6}$	$2.9 \cdot 10^{-7}$	$1.7 \cdot 10^{-7}$
Cuprophan	$2.7 \cdot 10^{-6}$	$2.7 \cdot 10^{-6}$	$5.9 \cdot 10^{-7}$	$3.1 \cdot 10^{-6}$	$1.8 \cdot 10^{-7}$	$1.7 \cdot 10^{-7}$

Table 3. Hydraulic Permeability

Fiber	$\Delta P=100$ mmHg $1/h \cdot m^2$	$\Delta P=200$ mmHg $1/h \cdot m^2$	$\Delta P=300$ mmHg $1/h \cdot m^2$
EniChimica	2.0	4.2	6.2
Kurifix	2.0	4.1	6.0
Cuprophan	0.5	1.1	2.2

CONCLUSIONS

The tests we performed indicate that chitosan can be used for the preparation of hollow fibers. The good results obtained for diffusivity and water permeability suggest that these fibers could be used in dialysis and/or haemodialysis appliances.

In particular the spinning procedure we described seems to be interesting. A significant advantage of this procedure is the possibility to feed to the central bore of the fiber a gaseous phase which acts as a quenching medium and as a support for the structure of the fiber. In addition it is possible to avoid the use of oils such as isopropyl myristate or 2-ethyl hexanol normally used as filling liquids for the preparation of hollow fibers.

Acknowledgements

The authors are indebted to Dr. R. Camanzi for his scanning electron microscope pictures of the fibers and to Mr F. Brunacci for his skillful assistance in the preparation and testing of the fibers.

REFERENCES

1. R. A. Muzzarelli, "Chitin," Pergamon Press, Oxford (1977).
2. B. C. Averbach, R. B. Clark, Processing of Chitosan Membranes, MIT Sea Grant Program 78-14, Cambridge, Mass. (1978).
3. R. J. Henry, D. C. Canon, J. W. Winkelman, "Clinical Chemistry: Principles and Technics," Harper & Row, New York (1974).
4. E. Klein, F. Holland, A. Lebeuf, A. Donnaud, J. K. Smith, Transport and mechanical properties of hemodialysis hollow fibers, J.Membr. Sci., 1:371 (1976).
5. J. Crank and G. S. Park, "Diffusion in Polymers," Academic Press, London (1968).

RESEARCH ON THE PREPARATION OF A NEW TYPE OF

POLYARYLSULFONE MEMBRANE FOR ULTRAFILTRATION

S.-J. Han, K.-F. Wu, G.-X. Wu and Y.-B. Wang

Institute of Environmental Chemistry
Academia Sinica Beijing
China

SUMMARY

This paper is written in relation to the introduction of a new type of
polyarylsulfone membranes for ultrafiltration (UF) purposes. This new
membrane material is a condensation-polymerized product of phenolphthalein
and dichlorodiphenylsulfone (PDC).

For preparation of the membrane it is required to achieve a sufficiently high degree of polymerization. The character of the PDC polymer
material is quite good, including thermostability at higher temperature,
better chemical resistance and an unique hydrophilic property. Good
permeablility performance has also been shown for the UF membranes made of
PDC polymer.

In this paper details of the preparation conditions and performances
of the membrane itself have been discussed.

INTRODUCTION

In recent years membrane ultrafiltration has gained increasing attention as a simple and convenient technique for concentrating, purifying
and fractionating solutions of low-to-high molecular-weight solutes. Both
laboratory and industrial applications of UF are becoming a reality.

The cellulosic polymers, aromatic polyamides, and polysulfones are by
far the most important membrane materials established in the RO/UF technology of today. The polysulfone materials have long been used for making
porous supports for composite RO membranes. Now they are increasingly
being used for making membranes for direct use in RO/UF application. In
view of their outstanding chemical and hydrolytic stability and higher
mechanical strength, aromatic polysulfones have tremendous potentiality as
RO/UF membrane material. Of the aromatic polysulfone family of polymers,
three kinds of materials are commercially available under trade names Udel,
Radel and Victrex. There are still other kinds of polysulfones which were
prepared in laboratories.

This paper is written in order to introduce a new type of polyarylsulfone membranes for UF. This new material is a condensation-polymerized

143

Table 1. Some Properties of PDC Polymer*

Some properties at high temperature:	
glass temperature (Tg), °C	260
decomposition temperature, °C	465
weight losses, 400 °C, 9 hrs	3%
200 °C, 210 days	no change
Hydrolytic decomposition properties at high temperature:	
200°C, 300 days, Tg	no change
150°C, 200 hrs, strength	no change
Reduced viscosity $\eta_{sp/c}$	0.45 - 0.60

*These data were given by Dr. Liu Ke-Jing, Zhang Hai-Chun (Institute of Applied Chemistry, Academia Sinica, Changchun, China).

product of phenolphthalein and dichlorodiphenylsulfone (PDC), the molecular structure of which is as follows:

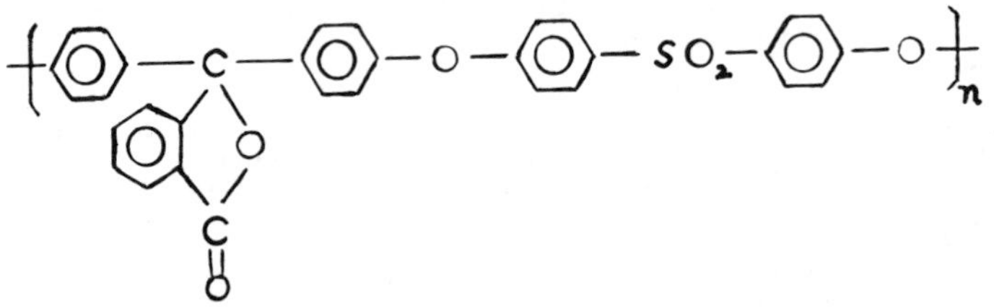

Compared with other kinds of polyarylsulfones, PDC polymer has much better thermal and hydrolytic stability and hydrophilic property.

In this paper details of the preparation conditions, membrane performances and morphological structures of the UF membranes made of PDC polymer have been discussed. Good membrane performances have been shown by the experimental results.

Experimental Methods and Reagents

The characteristics of PDC polymer are shown in Table 1. Pure commercial chemical reagents were used in the experiment. Membranes were prepared as flat sheets by Loeb-Sourirajan technique. The solution for membrane casting consisted of PDC polymer, additive and solvent. DMF was used as solvent and water as a gelation medium. Permeability and retention of UF membranes were measured on a flat sheet UF cell at operating pressure of 3 kg/cm^2 and temperature of 25°C. Flux of pure water (F) was obtained from Equation (1)

$$F = \frac{V}{S \, t} \text{ ml cm}^{-2} \text{ hr}^{-1} \tag{1}$$

V: volume of pure water permeating through the membrane

t: time of ultrafiltration

S: effective area of membrane.

Retention (R) was measured using bovine serum albumin solution and R was obtained from Equation (2)

$$R(\%) = (1 - \frac{c}{c_o}) \ 100\%$$
(2)

c_o : concentration of the solution before ultrafiltration

c : concentration of the permeate solution.

The concentrations of protein were determined by UV spectrophotometer at 280 nm (wave length). Morphological structure of membranes was observed by using optical microscope and electronic microscope techniques.

RESULTS AND DISCUSSIONS

Effect of Different Kinds of Additive Reagents

Table 2 shows performances of UF membranes prepared with range of additive reagents. Using the additive reagents presented in Table 2, flux of membranes are much larger, retentions all above 90%. These indicated that these additive reagents are appropriate for preparing PDC UF membranes with higher porosity and more uniform pore size.

Effect of the Contents of Additive Reagents

The effects of the contents of PEG-1540 and PVP in the casting solution on performances of membranes are shown in Figure 1 and Figure 2. It is apparent from Figure 1, that the flux increased and the retention decreased with the increasing content of PEG-1540. In the case of PVP it is quite different from that of PEG-1540. As shown in Figure 2, the flux decreased and retention slightly increased with increasing content of PVP.

Effect of Polymer Concentration

The effects of PDC polymer concentration in the casting solution are shown in Figure 3. In all the cases, the flux decreased and the retention increased with increasing content of PDC polymer. This result is similar to those obtained by Masato Nishimura et al.[1] and Han Shi-Jing, Liu Ting-Hui et al.[2] with PS(Udel). This shows that the surface pore size of membranes decreased with increasing content of PDC polymer in the casting solution.

Table 2. Effect of Additive Reagents

Additive reagents	$F(ml/cm^2,hr.)$	$R(\%)$
Butanol (BtoH)	145	95.6
Ethylene glycol monomethyl ether (EGME)	152	96.1
Polyethylene glycol-400 (PEG-400)	175	94.2
Polyethylene glycol-1540 (PEG-1540)	181	90.2
Polyvinyl pyrolidone (PVP)	176	91.6
Diethylene glycol (DEG)	152	94.1
Dioxane	158	96.2
None	92	95.1

PDC: 13% (wt.)   additive: 8% (wt.)

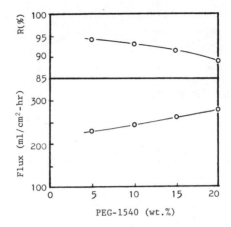

Fig. 1.　Effect of PEG-1540 contents

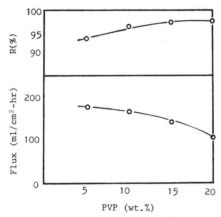

Fig. 2.　Effect of PVP contents

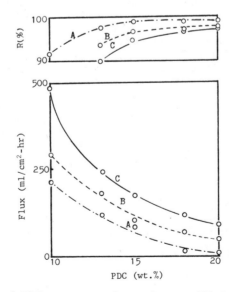

Fig. 3.　Effect of PDC concentration.　A. no additive; B. PVP as an additive; C. PEG-1540 as an additive.

Effect of Evaporation Time

Evaporation time is an important parameter which can influence the performances of UF membranes. The flux slightly decreased and the retention slightly increased with prolonged evaporation time (see Figure 4). This can be explained by the fact that the amount of evaporated solvent increased with prolonged evaporation time and hence more dense surface layer of membrane is formed.

Effect of Gelation Temperature

The flux increases and retention decreases with increasing of gelation temperature (see Figure 5). This phenomenon may be due to the fact that the pore size on the membrane surface increases with the increase of gelation temperature. Therefore, membranes with different pore sizes can be prepared by changing gelation temperature.

Effect of Relative Humidity

Membranes were prepared under different relative humidities. It was found that the difference of humidities does not cause significant variation in flux and retention of membranes (see Table 3). This differs from the case as for polysulfone Udel[2].

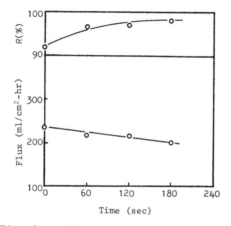

Fig. 4.   Effect of evaporation time.

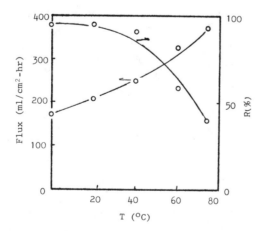

Fig. 5.   Effect of gelation temperature.

Optical micrographs of PDC UF membranes are displayed in Figure 6 and Figure 7. From Figure 6 and Figure 7, it is shown that PDC UF membranes have a dense surface layer and finger-like structure of support layer with more uniform pore size and higher porosity.

CONCLUSION

From the experimental results, the following conclusions are made:

- PDC polymer is a better material for preparation of UF membranes. The PDC UF membranes have higher flux.
- PVP, PEG, BtoH, EGME, DEG and dioxane are all suitable additive reagents for making PDC UF membranes.
- The average pore size of PDC membrane can be changed by varying PDC concentration in casting solutions or by varying gelation temperatures.
- At room temperature, evaporation time has only a little influence on flux and retention, and relative humidity does not have any significant influence on the performance of PDC membranes.
- From the optical microscope observation, it has been shown that PDC UF membranes mentioned in this paper have typical finger-like structure with more uniform pore size and higher porosity.

Table 3. Effect of Relative Humidity

Relative humidity %	F ml/cm^2,hr.	R%
30	213	98.2
45	245	96.1
60	228	96.9
75	249	96.9
90	236	94.6

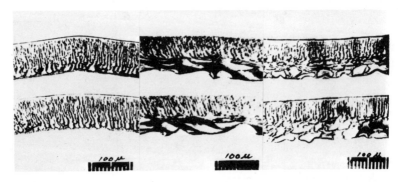

Fig. 6. Optical micrographs of cross-section structure of PDC UF membranes (a. no additive; b. PVP as an additive; c. PEG-1540 as an additive).

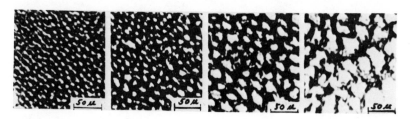

Fig. 7. Optical micrographs of pore distribution at different level of support layer of PDC UF membrane (PVP as an additive).

Acknowledgement

The authors are grateful to Dr. Liu Ke-Jing and Zhang Hai-Chun, Institute of Apply Chemistry, Academia, Sinica, Changchun, China, for supplying the PDC polymer.

REFERENCES

1. M. Nishimura, T. Muro, and Y. Tsujisaka, Preparation and performance of polysulfone membrane for ultrafiltration, Kobunshi Ronbunshu, 34:713 (1977).
2. Han Shi-Jing, Liu Ting-Hui, Wu Kai-Fen, Xu Yong-Sheng, Wang Guo-Yi, and Li Shu-Shen, Research on polysulfone membrane for ultrafiltration, J.of Environ.Scie., 1, No.6, 14 (1980).

OUTER SKINNED HOLLOW-FIBERS-SPINNING AND PROPERTIES

J. M. Espenan and P. Aptel

Laboratoire de Chimie-Physique et Electrochimie
(CNRS, LA 192), Université Paul Sabatier
31062 Toulouse, France

SUMMARY

Asymmetric hollow fibers with the skin on the outside of the fiber
were prepared. A wet-dry spinning technique was employed, using a tube-
in-orifice spinneret. Highly permeable hollow fibers were obtained by
addition of a pore-forming agent in the spinning dope, associated with
the use of a solvent of the polymer as core liquid. Using a dope mixture
composed of polysulfone/Triton X100/dimethylformamide, the principal
variables of the spinning process were studied.

INTRODUCTION

In a series of previous papers[1-3], our attention was focused on
the spinning of symmetric polysulfone hollow fibers with the skin on the
inside, using a wet-dry spinning technique. We came to the conclusion
that rheological phenomena play a prominent part in fixing the final
properties of the fiber[3]. The spinning number, an empirical dimensionless coefficient was introduced to estimate the variation of the
hydraulic permeability coefficient as a function of the spinning process
variables.

Our purpose here is to present a method for the preparation of outer
skinned ultrafiltration hollow fibers which have the following properties:

- Outer diameter less than 1000 µm

- Hydraulic permeability higher than 10^{-11}m/s.Pa

- Resistance to collapse higher than 5 MPa

For a given system: spinneret/spinning dope/core liquid/coagulant bath
we will also give some correlations between the properties of the fiber
(outer and inner diameters, hydraulic permeability) and the spinning
variables (spinning dope and core fluid velocities, air-gap distance).

PRINCIPLES OF THE PREPARATION OF HIGHLY PERMEABLE OUTER SKINNED HOLLOW FIBERS

That the formation of asymmetric membranes can be explained in terms of phase separation phenomena in polymer solutions is now widely accepted[4-5]. This analysis shows that for a given polymer-solvent system, a top layer will be formed by immersion of the collodion in a strong coagulant and that the permeability of the membrane will be increased by decreasing the polymer concentration in the solution (formation a less dense top layer and of a more porous substructure). Another well known method for increasing the permeability of an asymmetric membrane is to add a pore forming agent[1-2-6].

The preparation method for asymmetric hollow fibers is closely related to that used for flat-sheet membranes. However one of the major differences is that for hollow fibers we have the possibility of changing the properties of the membranes by varying the nature of the core fluid. This means for instance that if a strong coagulant is used a dense inner skin will be obtained.

Because our purpose was to obtain highly permeable outer skinned hollow fibers, it was decided to combine the effects of the addition of a pore forming agent to polymer solution of relatively low concentration and the use of a solvent of the polymer as core fluid. The solvent present in the inside channel of the nascent fiber has in fact three roles to play:

- it keeps the inner channel open

- it avoids the formation of an inside skin

- it diffuses into and through the dope solution lowering its polymer content and reducing the rate of precipitation, these two effects being both favorable to the formation of a more porous structure.

EXPERIMENTAL

Materials

The following polymer materials were used: polysulfone Udel 3500 from Union Carbide; polyvinylidene fluoride Foraflon 1000 LD from Péchiney Ugine-Kuhlman and polyvinylpyrrolidone K25 from Fluka - Triton X100 and analytical grade solvents were also purchased from Fluka.

Spinning Line

The spinning line is shown schematically in Figure 1. The spinning solution consisting of a viscous, degassed and filtered polymer solution is forced into a coaxial tube spinneret while a core liquid is delivered through the inner tube.

The spinneret is maintained at a distance h from the coagulation bath. While passing through this air gap the extrudate is elongated. A variable speed take-up roll collects the fiber.

The spinneret has an external radius of 0.75 mm, the outer and inner radii of the tube were 0.3 and 0.2 mm respectively.

The dope solution was forced to the spinning jet under nitrogen pressure. The flow rate Q_s, was controlled gravimetrically with an accuracy of 5%. The core liquid is delivered by means of a linear pump.

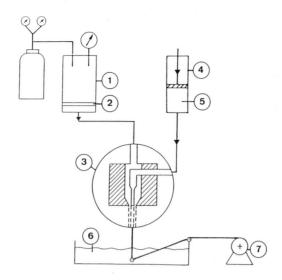

Fig. 1. Fiber spinning system: (1) spinning solution reservoir;
(2) filter; (3) spinneret; (4) linear pump; (5) core liquid;
(6) coagulation bath; (7) variable speed take-up roll.

All the experiments were performed at room temperature, and the coagulation
bath was maintained at 23°C.

Fiber Characterization

Inner and outer diameters of the hollow fibers were estimated by means
of both an optical microscope and a scanning electron microscope SEM.
Cross-sections of hollow fibers were prepared for the SEM by breaking the
fibers at the temperature of liquid nitrogen. The specimen was then coated
with silver and scanned on the JEOL JSM U3, at 9–12 kV accelerating voltage. Outer diameters were also measured by means of a micrometer.

The dimensions of the fibers are not strictly constant with length and
are known with an error of 12%.

The hydraulic permeability coefficient (Lp) of the fibers was measured
on small modules consisting of 10–50 fibers encased in a polymethacrylate
tube 20 cm long, with 1.1 cm inside diameter. The module was sealed at
each end with an epoxy resin, one end was cut to open the hollow core of
the fibers so as to allow the filtrate to be collected. Two ports were
opened at the both ends of the fiber housing in such a way as to allow the
circulation of the feed stream. Lp was calculated from measurements at
25°C of the mass flow rate obtained at various average transmembrane
pressure differences. The precision of Lp (15%) is strongly dependent on
the accuracy of the value of the membrane area. This area was taken as
equal to the external active surface of the fibers.

RESULTS AND DISCUSSION

Influence of Additives and Nature of the Core Fluid

To illustrate the cogency of the principles stated above, we have
reported some typical data in Table 1 showing the influence of pore-forming
additive and the nature of the core fluid on the hydraulic permeability of
the fibers. Comparison of fibers Nos. 03 and 9 shows that the addition of

Table 1. Effect of Dope and Core Fluid Compositions on the Hydraulic
Permeability of Hollow Fibers. PSf: Polysulfone; PVP: Polyvinylpyrrolidone; PVDF: Polyvinylidene fluoride; Triton X100;
DMF: N,N-Dimethylformamide; DMA: N,N-Dimethylacetamide. Water
was used as coagulating fluid.

Fiber number	Dope composition (wt.fraction)			Core fluid	L_p $(10^{-10}$ m/s.Pa)
	Polymer	Additive	Solvent		
03	PSf (0.29)		DMF (0.71)	DMF	< 0.01
9	PSF (0.29)	Triton (0.22)	DMF (0.49)	DMF	3.4
BK3	PSf (0.29)	Triton (0.22)	DMF (0.49)	DMF (0.9) Water(0.1)	0.46
BK4	PSf (0.29)	Triton (0.22)	DMF (0.49)	DMF (0.7) Water(0.3)	< 0.01
P5	PSf (0.24)	PVP (0.24)	DMA (0.52)	DMA	0.8
C1	PVDF (0.28)	Triton (0.18)	DMF (0.54)	DMA	2.6

a surfactant in the spinning dope has a strong influence on the permeability. An increase in water content in the core fluid results in a
sharp decrease in the permeability (Fibers BK 3-4). Fibers P5 and C1 also
exhibit a high permeability. These results mean that the principles we
suggested to get highly permeable hollow fibers could be applied to various
polymer/additive systems.

A cross-section of fiber No.9 is shown in Figure 2a. Under the outer
skin, is a sponge pore structure and then a finger-like pore structure. It
should be noted that the macrovoids are cylindrical over a large part of
their length and are open on the inner surface of the fiber. The holes
which are formed on the inner surface are regularly dispersed and their
radii are identical (roughly 10 μm in diameter). The photomicrograph shown
in Figure 2b shows a resemblance to the cross-section of the ionotropic gel
membranes developed by Thiele.

Influence of the Spinning Variables

Table 2 shows the range of variation of the parameters studied and the
corresponding range of variation of the properties of the hollow fibers.
For all these experiments, the spinning dope was composed of PSf (weight
fraction 0.29). Triton X100 (weight fraction 0.22) and DMF (weight
fraction of 0.49); water was used as the coagulation fluid and DMF as the
core fluid.

Effect of Q_c. A first series of data concerns the effect of the flow
rate of DMF which is delivered through the needle. As more and more DMF is
forced into the core of the fiber, the internal diameter increases while

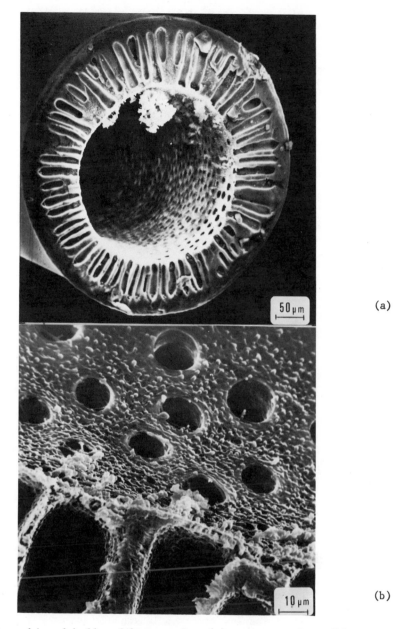

(a)

(b)

Fig. 2. Outer skinned hollow fiber no. 9. (a) cross-section; (b) enlarge-
ment of the inside surface showing pores of nearly uniform
diameter.

the outer diameter does not change significantly (Figure 3a). At the same
time, Lp, the hydraulic permeability coefficient rises (Figure 3b), starting from very low values for Q_c lower than 10^{-2} cm^3/s. Cross -sections of
fibers are shown in Figure 3c. The SEM photographs clearly indicate the
increase in the inner diameter which results from an increase in Q_c (fiber
6 to fiber 9). The outer parts of the cross-sections are very similar and
the effect of increasing the flow rate of DMF seems to be to dissolve
progressively the internal part of the fiber.

Table 2. Range of Variation of the Spinning Variables and of the
Properties of the Hollow Fibers

Parameter

Spinning solution : flow rate, Q_s (cm^3/sec)

$$0.04 < Q_s < 0.28$$

Core fluid : flow rate, Q_c (cm^3/sec $\times$ 10^2)

$$0.5 < Q_c < 5.5$$

Air gap : h (cm)

$$0.6 < h < 4$$

Take-up velocity, v_t (cm/sec)

$$83 < v_t < 300$$

Properties of the hollow fibers :

Hydraulic permeability coefficient, Lp (m/sec.Pa $\times$ 10^{10})

$$0.02 < Lp < 3.4$$

Outer diameter, OD (μm)

$$300 < OD < 900$$

Inner diameter, ID

$$1.5 < OD/ID < 7.2$$

Effect of Q_s The results concerning the influence of the flow rate of
the spinning solution on the dimensions and the permeability of the fibers
are shown on Figures 4a-4c. The outer diameter increases, the permeability
decreases, while the inner diameter does not change when the extrusion flow
rate increases. The microphotographs are very similar to the previous
ones. Note that the magnification is decreasing from fiber No.11 to fiber
No.13, so the decrease in the inner diameter is only apparent.

Effect of h   By increasing the air gap, the emerging filament is
more and more strongly drawn out and as a result, both diameters decrease
(Figure 5a) when h reaches 4 cm, the inner channel of the fiber closes
up (Fiber 5). Figure 5b indicates that the hydraulic permeability coefficient stays at a high value. The finger-like porous structure of the
fibers changes significantly with the length of the air gap (Figure 5c).
Clearly, the number of macrovoids decreases when h increases (Fiber 2 to
Fiber 5), but the size of the holes in the inner surface is constant
(15 μm).

Discussion

It appears difficult to present a general correlation relating Lp to
the three spinning variables we have studied. However if we plot Lp versus
the thickness, e, of the wall of the hollow fibers, data points fall
roughly on a straight line (Figure 6). Then a general trend is a decrease
in Lp when e increases. Simple calculations have shown that this variation
of Lp is not due to a significant pressure drop in the inner channel or in
the porous part of the wall containing the macrovoids. So, it appears that
the resistance to the mass transfer is in fact limited to the outer skin

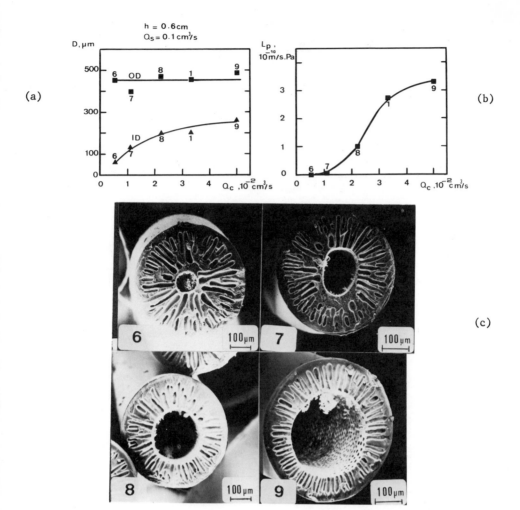

Fig. 3. Effect of the flow rate, Q_C, of the core fluid for constant values
of Q_S (0.1 cm³/s) and h (0.6 cm). (a) Outer diameter (OD) and
inner diameter (ID) vs. Q_C. (b) hydraulic permeability coefficient (Lp) vs. Q_C. (c) SEMs of cross-sectional views of hollow
fibers no. 6-9. Each fiber is quoted by its number on curves and
photographs.

and to the sponge porous structure: the correlation relating Lp to e, must
not be interpreted as an effect of a variation of the resistance of the
overall thickness.

Considering now the residence time of the filament in the air gap, and
assuming that an increase in the resistance of the outer skin could result
from an increase of the evaporation time or from a decrease of the water
vapor inflow, we have plotted the variation of Lp versus the residence
time. Any correlation cannot be deduced from this analysis. We must point
out that the very short residence time (5-10 ms) could explain that no
clear effect have been observed.

In our opinion, the correct explanation for the variation of Lp vs. e
would take into account the diffusion of the core fluid into and through

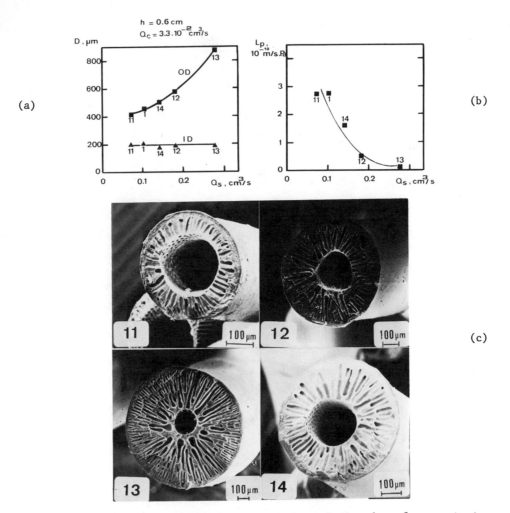

(a) (b) (c)

Fig. 4. Effect of the flow rate, Q_s, of the spinning dope for constant
values of Q_c (0.033 cm³/s) and h (0.6 cm). (a) Outer diameter
(OD) and inner diameter (ID) vs. Q_s. (b) hydraulic permeability
coefficient (Lp) vs. Q_s. (c) SEMs of cross-sectional views of
hollow fibers no. 11-14. Each fiber is quoted by its number on
curves and photographs.

the polymer dope before coagulation occurs. The diffusion phenomena create
a concentration profile lowering the polymer content up to the outer
surface of the fiber. The polymer content decreases all the more rapidly
that the filament is thinner.

Lastly, a correlation relating the outer diameter to the drawing ratio
is shown in Figure 7. For a given value of v_t/v_s, the outer diameter can
be estimated by the empirical equation (corr. coef. = 0.946):

$$OD(\mu m) = 1439 \left(\frac{v_t}{v_s} \right)^{-0.427}$$

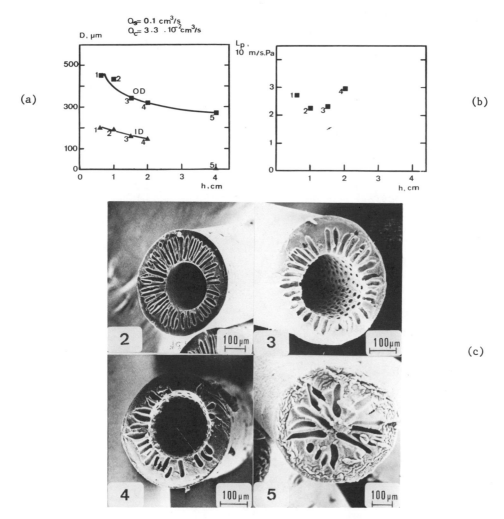

Fig. 5. Effect of the length of the air gap, h, for constant values of Q_S (0.1 cm³/s) and Q_C (0.033 cm³/s). (a) Outer diameter (OD) and inner diameter (ID) vs. h. (b) hydraulic permeability coefficient (Lp) vs. Q_S. (c) SEMs of cross-sectional views of hollow fibers no. 2-5. Each fiber is quoted by its number on curves and photographs.

CONCLUSION

This work has shown that highly permeable outer skinned hollow fibers can be spun using a wet-dry technique. The key features of the process are:

1. Addition of a pore forming agent in the spinning dope, associated with the use of a solvent of the polymer as core liquid.
2. A correct balance between the flow rate of the core liquid and the drawing ratio.

Acknowledgements

The authors are grateful to Institut Nationale de Recherche Chimique Appliquée and Association Nationale de la Recherche Technique for financial support through contract No. 01381.

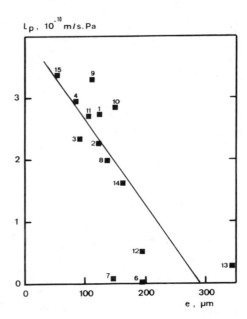

Fig. 6. Hydraulic permeability coefficient vs. fiber-wall thickness, e. Each fiber is quoted by its number. It must be pointed out that all the fibers have a rejection coefficient nearly 1 for poly-ethyleneglycol 20.000.

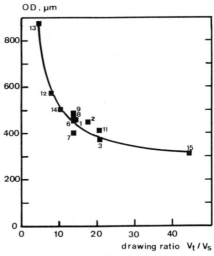

Fig. 7. Outer diameter vs. drawing ratio. Each fiber is quoted by its no.

REFERENCES

1. P. Aptel, F. Ivaldi et J. P. Lafaille, Fibres creuses en polysulfone. Préparation et propriétés de transfert en ultrafiltration, C.R.Acad.Sci., Ser. II, 293:681 (1981).
2. P. Aptel, F. Ivaldi and J. P. Lafaille, Development of asymmetric hollow fibers. Spinning process and transport properties in ultrafiltration, Proceedings of the 2nd World Congress of Chemical Engineering, Montreal, 4:191 (4-9 Oct. 1981).

3. P. Aptel, N. Abidine, F. Ivaldi, and J. P. Lafaille, Polysulfone
    hollow fibers: Effect of spinning conditions on ultrafiltration
    properties, J.Membrane Sci. (to be published).
4. H. Strathmann and K. Koch, The formation mechanism of phase-inversion
    membranes, Desalination, 21:241 (1977).
5. J. G. Wijmans, J. P. B. Baaij, and C. A. Smolders, The mechanism of
    formation of microporous or skinned membranes produced by immersion
    precipitation, J.Membrane Sci., 14:263 (1983).
6. I. Cabasso, Hollow-fiber membranes, in: "Kirk-Othmer, Encyclopedia of
    Chemical Technology," Vol.12, Wiley, New York, pp.492-517, third
    edition (1980).

FACTORS AFFECTING THE STRUCTURE AND PROPERTIES

OF ASYMMETRIC POLYMERIC MEMBRANES

A. Bottino, G. Capannelli, and S. Munari

Institute of Industrial Chemistry
University of Genoa, Corso Europa 30
16132 Genoa, Italy

SUMMARY

Ultrafiltration polyvinylidene fluoride based membranes were prepared by the phase-inversion process. The morphology, the performance and mechanical properties of the membranes were investigated in relation to some preparation variables such as casting solution composition (solvent and additives in the solution), exposure or evaporation conditions (time and exposure temperature), coagulation conditions (additives in the coagulation bath and temperature of the bath). Morphology of the membranes was examined with a scanning electron microscope. Ultrafiltration performance was evaluated with dextran solutions. Burst pressure resistance was chosen as criterion for mechanical properties.

INTRODUCTION

Many polymeric membranes used today in separation processes are composed of a rather sophisticated asymmetric structure, consisting of a very thin selective layer on a highly porous thick sublayer. This type of membranes (asymmetric membranes) is classically produced by the so-called phase-inversion process, which is a multistep process involving: solution preparation, film casting, partial solvent evaporation, coagulation and sometimes heat post-treatment. Morphology, performance (flux, rejection) and mechanical properties of the membranes are strictly related to the above preparation step. Basic research on the relationship between conditions of preparation and ultimate properties of membranes become of fundamental importance for their optimization.

In this work the morphology, the performance and mechanical properties of several polyvinylidene fluoride based membranes are reported in relation to some preparation variables in the phase-inversion process such as casting solution composition (solvent and additives in the solution), exposure or evaporation conditions (time and exposure temperature), and coagulation conditions (additives in the coagulation bath and temperature of the bath).

EXPERIMENTAL

Materials

Two polymers, polyvinylidene fluoride (PVDF) and sulfonated poly-vinylidene fluoride (PVDFS) were considered for membrane preparation. The former was a commercial product (Foraflon 1000-HD Ugine Kuhlmann), the latter was obtained by sulfonation of the former[1]. Reagent grade, N,N-dimethylformamide (DMF), N,N-dimethylacetamide (DMA), N-methyl-2-pyr-rolidone (NMP) and dimethylsulfoxide (DMSO) were used as solvents in the casting solution. Polyoxyethylenglycols (PEG) with average molecular weight ($\bar{M}w$) of 600, 3000, 10000 and 35000 respectively were used as additives in the casting solutions. Inorganic salts ($ZnSO_4 \cdot 7H_2O$, $NaNO_3$) were used as additives in the coagulation bath. Dextran with $\bar{M}w$ of 40000 was used as solute of the feed solution in ultrafiltration tests.

Membrane Preparation

Membranes were prepared by the phase-inversion process. Polymer solutions were cast at room temperature (20°C) on a glass plate with a knife. After exposure in air at temperature ranging from 20 to 140°C for a period varying between 30 and 420 seconds the plate was immersed in the coagulation bath composed of pure water or salt solutions. The temperature of coagulation bath was varied between 3 and 70°C. The membrane were removed from the bath and leached overnight under running water before testing.

Membrane Morphology Studies

The membranes were dried by using a Balzers Union critical point dryer, frozen in liquid nitrogen and fractured to obtain cross sections. The specimens were coated with gold by using a Balzers Union sputtering device and examined in a Cambridge 250 K scanning electron microscope.

Ultrafiltration Performance

The membranes were tested in a laboratory-scale pilot plant with 1000 ppm dextran feed solution. Operating conditions were: temperature 40°C, average pressure 200 KPa, recirculation rate 5 m/s. Details on the plant and membrane module are reported in another paper[2]. Dextran in permeate and concentrate was determined colorimetrically[3].

Burst Pressure Evaluation

Burst pressure resistance of membranes was chosen as criterion for mechanical properties. Burst pressure was evaluated submitting a membrane specimen to increasing water pressure and measuring the pressure to crack the membrane.

RESULTS AND DISCUSSION

Effect of Solvent

The morphology of the membrane is deeply affected by the type of the solvent of the dope. This fact can be clearly seen in Figure 1 which reproduces scanning electron micrographs of cross sections of membranes obtained from PVDF solutions in different solvent. All the membranes exhibit structures with more or less finger-like shaped cavities. Because of the different interactions between polymer, solvent and non solvent the size and the shape of cavities and the consequent overall thickness of the membranes are different by changing the solvent.

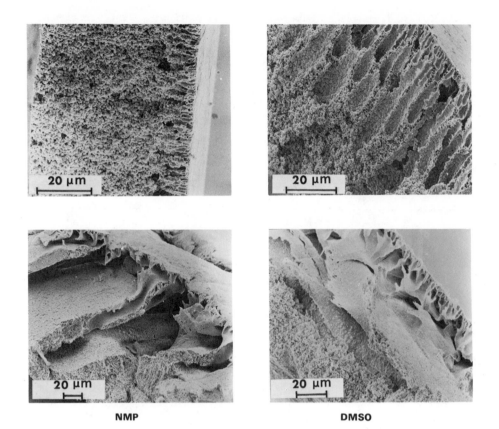

NMP   **DMSO**

Fig. 1.   Scanning electron micrographs of cross sections of membranes cast
           from 15 wt % PVDF solutions in different solvents.   Exposure con-
           ditions: 30 sec at 20°C.   Coagulation bath: water at 3°C.

        Among the solvents used DMF produces membranes with lower thickness
and smaller cavities.   The cavities are close to the dense layer and extend
into the globular inner sublayer for different depths depending on the
solvent quality.   The membrane performances reported in Table 1, show that
the rejections remain practically constant, but there is a wide change in
fluxes.   The more compact membranes exhibit lower fluxes and there results
suggest that at least in some experimental conditions (high precipitation
rate) the rejection depends mainly from the skin layer that is instantaneously formed after the immersion of the dope in the coagulation bath.   In
these cases the flux depends mainly on the hydraulic resistance of the
membranes cross section and the larger are the cavities, the higher are the
fluxes.

Effect of Additives in the Casting Solution

        Figure 2 reproduces scanning electron micrographs of cross sections
of membranes obtained from PVDF solutions containing different amounts of
PEG 600.   All the membranes exhibit finger-like structures.   Increasing
the amount of PEG 600 the finger-like cavities become progressively bigger
and extend for about 50% of the thickness of the membrane when 10 g of
additive are added to 100 g of casting solution.

Table 1. Performance of Membranes Cast from 15 wt % PVDF Solutions in Different Solvents. Exposure Conditions: 30 sec at 20°C. Coagulation Bath: water at 3°C.

Solvent	Flux $1/m^2$ d	Rejection %
DMF	150	75
DMA	210	70
NMP	2400	78
DMSO	5700	81

The effect of the $\overline{M}w$ PEG on the overall morphology of the membrane is shown in Figure 3. An increase of $\overline{M}w$ of the additive causes a further increase of the size of the cavities which penetrate the entire cross section when 10 g of PEG 35000 are added to 100 g of casting solution. An infrared analysis of the casted membrane shows that a certain amount of the PEG is entrapped in the polymeric matrix. The amount of entrapped PEG increases as the $\overline{M}w$ increases. Prolonged rinsing in hot water does not change significantly the amount of the PEG entrapped. DSC analysis shows that the PEG is mainly dispersed in the PVDF network as the melting peak of the former is not present.

The effect of additional amounts of PEG 600 and PEG 3000 in the casting solution on the performance of the membranes is shown in Figure 4. As it can be seen an increase of PEG concentration causes an increase of the flux while the rejection remains practically constant. On the same figure is also reported the performance of the reference membranes obtained by adding to the casting solution an additional amount of DMF instead of PEG. The flux of the membrane is also strongly affected by the $\overline{M}w$ of PEG. However an increase in the $\overline{M}w$ causes a detrimental effect on rejection as can be seen in Figure 5. An addition of PEG 600 or PEG 3000 in the casting solution leads to a decrease in the burst pressure of the membranes as can be seen in Figure 6. When up to 5 g of PEG 600 or PEG 3000 are added to 100 g of casting solution, the burst pressure of the resulting membranes is about 60% lower than the burst pressure of the corresponding reference membranes.

From the above figure no relevant effect of the $\overline{M}w$ of PEG on the burst pressure can be foreseen. This fact was further confirmed with membranes prepared by casting solutions containing PEG of higher $\overline{M}w$ (10000 and 35000 respectively). The effect of an additional amount of PEG and PVDFS casting solution on the morphology, ultrafiltration performance and burst pressure of the membranes is shown in Figures 7-9. The membranes prepared from binary mixtures of PVDFS and DMF (i.e. without additives) exhibit a structure with cavities considerably bigger than the membranes obtained from binary mixtures of the unmodified polymer. At the same time PVDFS yields membranes with higher flux, lower rejection and burst pressure. By adding PEG 3000 in PVDFS casting solution a further increase in the size of cavities (Figure 7), and flux (Figure 8) is obtained while the rejection and burst pressure (Figure 9) are lowered.

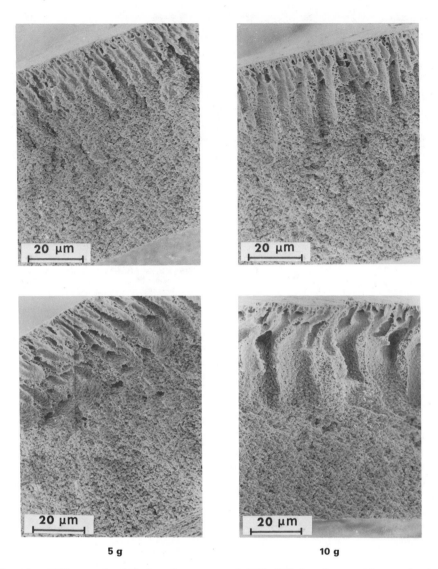

5 g   10 g

Fig. 2.   Effect of additional amount of PEG 600 in the casting solution
          (g PEG/100 g casting solution) on the morphology of membranes.
          Casting solution: 15 wt % PVDF in DMF. Exposure conditions:
          30 sec at 20°C. Coagulation bath: water at 3°C.

The obtained results can be explained as follows. As it is well known
the formation of macrovoids and finger-like cavities occurs at high ratio
between the flow of nonsolvent (JNS) inwards the cast film and the flow of
solvent (JS) outwards. Because of its extreme affinity to water the presence of PEG in the cast dope increases the ratio JNS/JS. This increase
depends on the amount of PEG in the dope and, under the same amount, is
higher, as lower is the diffusion coefficient of PEG, i.e. as higher is
its $\overline{M}w$.

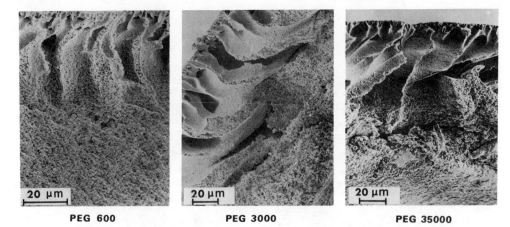

PEG 600   PEG 3000   PEG 35000

Fig. 3.   Effect of $\bar{M}w$ of PEG added to the casting solution (10 g PEG/100 g casting solution) on the morphology of membranes. Casting solution: 15 wt % PVDF in DMF. Exposure conditions: 30 sec at 20°C. Coagulation bath: water at 3°C.

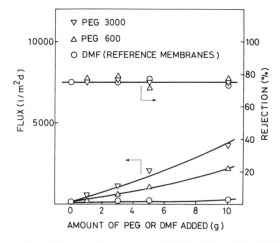

Fig. 4.   Effect of additional amount of PEG 600, PEG 3000 and DMF in the casting solution (g PEG or DMF/100 g casting solution) on the performance of membranes. Casting solution: 15 wt % PVDF in DMF. Exposure conditions: 30 sec at 20°C. Coagulation bath: water at 3°C.

It has been shown in Figure 5 that the increase of the PEG $\bar{M}w$ and consequently the amount of entrapped polymer causes a rejection reduction. It could be easy to argue that membranes with largest cavities present more defects in the thin layer and therefore reduction in mechanical properties and rejection, but as the PEG acts also as a plasticizer on the PVDF a different effect could be present also. The right answer is given only by direct evaluation of the density and distribution of the pores and experiments on this subject are in progress in our lab. The addition of PEG to the sulphonated PVDF (PVDFS) membranes produces a less relevant effect as the sulphonated polymers by themselves increase the ratio JNS/JS and therefore change the membranes morphology.

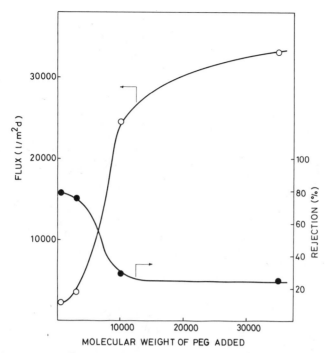

Fig. 5. Effect of $\overline{M}w$ of PEG added to the casting solution (10 g PEG/100 g casting solution) on the performance of membranes. Casting solution: 15 wt % PVDF in DMF. Exposure conditions: 30 sec at 20°C. Coagulation bath: water at 3°C.

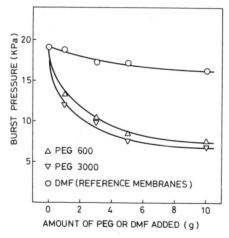

Fig. 6. Effect of additional amount of PEG 600, PEG 3000 and DMF in the casting solution (g PEG or DMF/100 g casting solution) on the burst pressure of membranes. Casting solution: 15 wt % PVDF in DMF. Exposure conditions: 30 sec at 20°C. Coagulation bath: water at 3°C.

Effect of Exposure Conditions of the Casting Solution

The effect of time and exposure temperature of the PVDFS casting solution on Flux and Rejection of the resulting membranes is shown in Figures 10 and 11. Because of the insignificant evaporation due to the

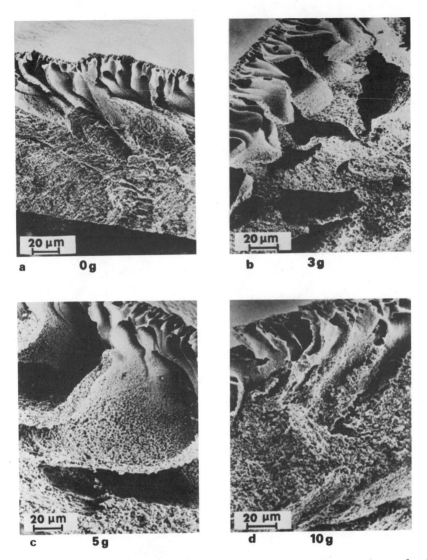

Fig. 7. Effect of additional amount of PEG 3000 in the casting solution
(g PEG/100 g casting solution) on the morphology of membranes.
Casting solution: 20 wt % PVDFS in DMF. Exposure conditions:
30 sec at 20°C. Coagulation bath: water at 3°C.

high boiling point of the solvent, no apparent effect of exposure time at
20°C on the performance of the membranes was found. Increasing the exposure temperature evaporation of the solvent takes place and exposure
time can play an important role: the flux decreases (Figure 10) while the
corresponding rejection increases (Figure 11). At 140°C, after 420 seconds
exposure time a compact transparent homogeneous film is obtained. Obviously increasing time and exposure temperature the burst pressure of the
membrane also increases as can be seen in Figure 12.

Evaporation conditions of the casting solution can affect the morphology of the membranes as can be observed as an example from scanning
electron micrographs reported in Figure 13. The structure of the membrane

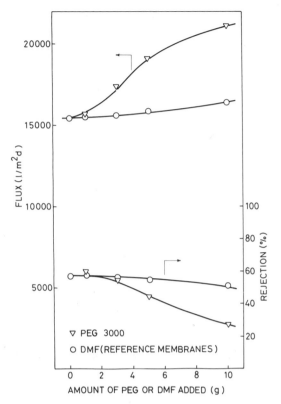

Fig. 8. Effect of additional amount of PEG 3000 and DMF in the casting
solution (g PEG or DMF/100 g casting solution) on the performance
of membranes. Casting solution: 20% PVDFS in DMF. Exposure
conditions: 30 sec at 20°C. Coagulation bath: water at 3°C.

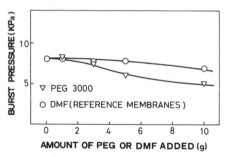

Fig. 9. Effect of additional amount of PEG 3000 and DMF in the casting
solution (g PEG or DMF/100 g casting solution) on the burst
pressure of membranes. Casting solution: 20% PVDF in DMF.
Exposure conditions: 30 sec at 20°C. Coagulation bath: water
at 3°C.

reproduced in Figure 13a is quite similar to that shown in Figure 7a. This
confirms the fact that exposure time at 20°C has no influence on the
characteristics of membranes.

This effect is simply explained in terms of increase of polymer concentration in the top layer of the cast film, and as a consequence a more
dense and thicker skin in the resulting membrane is obtained.

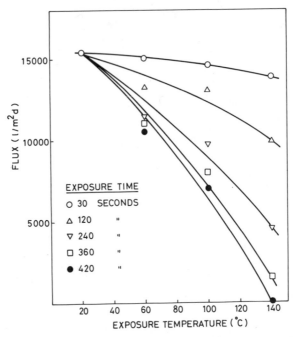

Fig. 10. Effect of exposure condition on the flux of the membranes. Casting solution: 20 wt % PVDFS in DMF. Coagulation bath: water at 3°C.

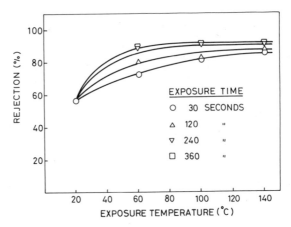

Fig. 11. Effect of exposure condition on the rejection of the membranes. Casting solution: 20 wt % PVDFS in DMF. Coagulation bath: water at 3°C.

Effects of Additives in the Coagulation Bath

Figure 14 reproduces scanning electron micrographs of membranes prepared by immersion of PVDFS casting solution in salt solutions containing 25 wt % $NaNO_3$ (Figure 14a) or $ZnSO_4 \cdot 7H_2O$ (Figure 14b). Comparison of Figure 7a with Figure 14 shows that by adding inorganic salts to a water bath membranes of different morphological structure can be obtained. The membrane obtained by immersion in $NaNO_3$ solution exhibit an entirely globular structure while the membrane obtained by immersion in $ZnSO_4 \cdot 7H_2O$ solution presents an intermediate structure between the extremes shown in Figure 7a and Figure 14a. The different structures shown in Figure 14

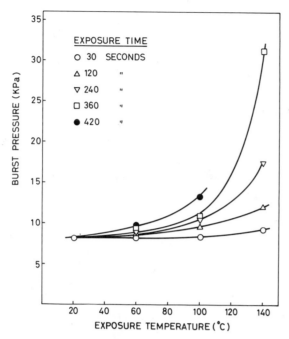

Fig. 12. Effect of exposure condition on the burst pressure of the
membranes. Casting solution: 20 wt % PVDFS in DMF. Coagulation
bath: water at 3°C.

could be mainly related to the different molar concentration of salt in the
coagulation bath. The effect of additional amounts of NaNO$_3$ in the water
bath on the performance of the membranes is shown in Figure 15. An increase in salt concentration causes a strong decrease of the flux and a
substantial improvement of the corresponding rejection.

The obtained results depend on the influence of the salts dissolved in
the coagulation medium on the ratio JNS/JS. The most important phenomena
involved are:

a) higher precipitation rate of the membrane skin and subsequent reduction of the possibility for the nonsolvent penetration inside the
   membrane;
b) driving force reduction due to the change of the chemical potential.

As from the higher resolution SEM observation is not evident. An
increase on the skin thickness and the overall membrane thickness decreases; the decrease of the JNS/JS ratio due to the change of the chemical
potential looks to be the most important effect.

Effect of Temperature of Coagulation Medium

In a earlier paper[4] we studied the effect of temperature of coagulation bath on morphology of PVDFS membranes. This study has been extended
to the unmodified polymer and the results of microscopic investigations and
burst pressure evaluations are reported in Figure 16. The figure shows
that increasing the temperature of water bath the formation of an alveolar
structure, formed by closed cells separated from polymeric walls, takes
place. This change in morphology is coupled with a strong increase of
burst pressure. Also this phenomenon can be explained in terms of kinetic
effects as the values of JNS and JS are greatly enhanced by temperature.

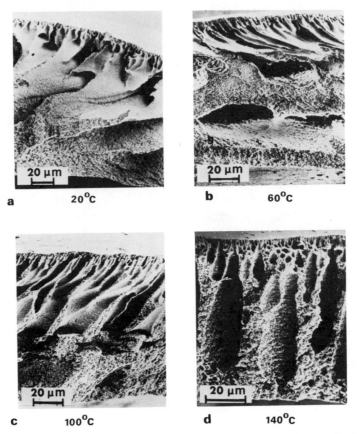

Fig. 13. Effect of exposure temperature on the morphology of the
membranes. Casting solution: 20 wt % PVDFS in DMF. Exposure
time: 360 sec. Coagulation bath: water at 3°C.

CONCLUSIONS

In conclusion, the experimental data strongly support the idea that
the ratio JNS/JS is largely controlling the membranes morphology and performances. The phase inversion process is mainly controlled by kinetic
phenomena and the additives both in the dope and in the coagulation medium
change the driving force in the boundary line.

Transmembranes fluxes are highly dependent on the overall membrane
resistance, the rejection is less influenced by the change of the preparation parameters, owing to the fact that the skin formation takes place
instantaneously before that the exchange phenomena become important.

The investigation of the skin properties requires the knowledge of
pore density and distribution and in this direction efforts must be made in
order to clarify the mechanism of membranes formation.

Further explanation of the same results obtained in this work can be
found in several papers of membrane literature[5-32].

| a | NaNO₃ | b | ZnSO₄·7H₂O |

Fig. 14. Effect of inorganic salts in the coagulation bath on the morphology of membranes. Casting solution: 20 wt % PVDFS in DMF. Exposure conditions: 30 sec at 20°C. Salt concentration in the bath: 25 wt %. Temperature of the bath: 3°C.

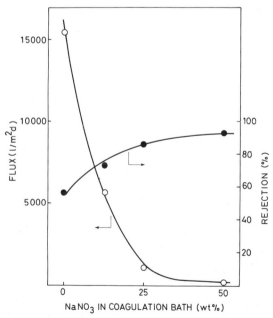

Fig. 15. Effect of NaNO₃ concentration in the coagulation bath on the performance of the membranes. Casting solution: 20 wt % PVDFS in DMF. Exposure conditions: 30 sec at 20°C. Temperature of the bath: 3°C.

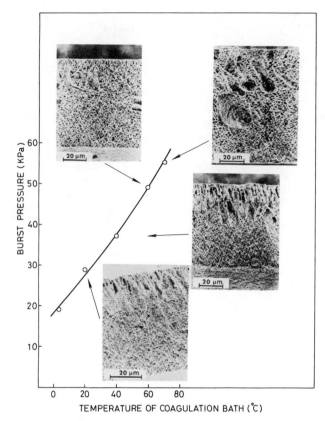

Fig. 16. Effect of temperature of coagulation bath on the burst pressure and morphology of the membranes. Casting solution: 15 wt % PVDF in DMF. Exposure conditions: 30 sec at 20°C. Coagulation bath: water.

Acknowledgement

This work was financially supported by the Consiglio Nazionale delle Ricerche (CNR) (Programmi finalizzati Chimica Fine e Secondaria - Sottoprogetto Membrane), Rome (Italy).

REFERENCES

1. F. Vigo, G. Capannelli, C. Uliana, and S. Munari, Modified polyvinylidene fluoride (PVDF) membranes suitable for ultrafiltration purpose application, La Chimica e l'Industria (Italy), 64:74 (1982).
2. A. Bottino, G. Capannelli, A. Imperato, and S. Munari, Ultrafiltration of hydrosoluble polymers. Effect of operating condition on the performance of the membranes, J.Membrane Sci., in press.
3. M. Dubois, K. A. Gilles, J. K. Hamilton, P. A. Rebers, and F. Smith, Calorimetric method for determination of sugars and related substances, Anal.Chem., 28:350 (1956).
4. A. Bottino, G. Capannelli, and S. Munari, Effect of coagulation medium on properties of sulfonated polyvinylidene fluoride membranes, J.Appl.Polym.Sci., submitted for publication.
5. R. E. Kesting, Synthetic polymeric membranes, McGraw Hill, New York (1971).

6. M. A. Frommer, R. Matz, and U. Rosenthal, Mechanism of formation of reverse osmosis membranes, Ind.Eng.Chem.Prod.Res.Develop, 10:193 (1971).
7. H. Strathmann, P. Scheible, and R. W. Baker, A rationale for the preparation of Loeb-Sourirajan-type cellulose acetate membranes, J.Appl.Polym.Sci., 15:811 (1971).
8. M. A. Frommer and D. Lancet, The mechanism of membrane formation: membrane structures and their relation to preparation conditions, in: "Reverse Osmosis Membrane Research," H. K. Lonsdale and H. E. Podall, eds., Plenum Press, New York (1972).
9. M. A. Frommer and R. M. Messalen, Mechanism of membrane formation. VI. Convective flows and large voids formation during membrane preparation, Ind.Eng.Chem.Prod.Res.Develop., 12:328 (1973).
10. M. T. So, F. R. Eirich, H. Strathmann, and R. W. Baker, Preparation of asymmetric Loeb-Sourirajan membranes, Polym.Lett., 11:201 (1973).
11. G. B. Tanny, The surface tension of polymer and asymmetric membrane formation, J.Appl.Polym.Sci., 18:2149 (1974).
12. O. Kutowy and S. Sourirajan, Cellulose acetate ultrafiltration membranes, J.Appl.Polym.Sci., 19:1449 (1975).
13. H. Strathmann, K. Kock, P. Amar, and R. W. Baker, The formation mechanism of asymmetric membranes, Desalination, 16:179 (1975).
14. L. Broens, D. M. Koenhen, and C. A. Smolders, On the mechanism of formation of asymmetric ultra- and hyperfiltration membranes, Desalination, 22:205 (1977).
15. M. Guillotin, C. Lemoyne, C. Noel, and L. Monnerie, Physicochemical processes occurring during the formation of cellulose diacetate membranes. Research of criteria for optimizing membrane performance. IV. Cellulose diacetate-acetone - organic additive casting solutions, Desalination, 21:165 (1977).
16. D. M. Koenhen, M. H. V. Mulder, and C. A. Smolders, Phase separation phenomena during the formation of asymmetric membranes, J.Appl. Polym.Sci., 21:199 (1977).
17. H. Strathmann and K. Kock, The formation mechanism of phase inversion membranes, Desalination, 21:241 (1977).
18. D. Tirrell and O. Vogl, Regular copolyamides. IV. Characterization of membrane morphology, J.Polym.Sci.Polym.Chem.Ed., 15:1889 (1977).
19. R. D. Sanderson and H. S. Pienaar, A morphological approach to desalination in hyperfiltration, Desalination, 25:281 (1978).
20. M. Sugimara, M. Fujimoto, and T. Uragami, Effect of casting solvent on permeation characteristics of polyvinylidene fluoride membranes, Polym.Prepr.Am.Chem.Soc.,Div.Polym.Chem., 20:999 (1979).
21. I. Cabasso, Practice aspects in the development of a polymer matrix for ultrafiltration, in: "Ultrafiltration Membranes and Applications," A. R. Cooper, ed., Plenum Press, New York (1980).
22. C. Lemoyne, C. Friederich, J. L. Halary, C. Noel, and L. Monnerie, Physicochemical processes occurring during the formation of cellulose diacetate membranes. Research of criteria for optimizing membrane performance. V. Cellulose diacetate-acetone-water-inorganic salt casting solutions, J.Appl.Polym.Sci., 25:1883 (1980).
23. C. A. Smolders, Morphology of skinned membranes: a rationale from phase separation phenomena, in: "Ultrafiltration Membranes and Applications," A. R. Cooper, ed., Plenum Press, New York (1980).
24. T. Uragami, M. Fujimoto, and M. Sugihara, Studies on synthesis and permeabilities of special polymer membranes. 24. Permeation characteristics of poly(vinylidene fluoride) membranes, Polymer, 21:1047 (1980).
25. H. Bokhorst, F. W. Altena, and C. A. Smolders, Formation of asymmetric cellulose acetate membranes, Desalination, 38:349 (1981).
26. T. Uragami, Y. Naito, and M. Suginara, Studies on syntheses and permeability of special polymer membranes. 39. Permeation character-

istics and structure of polymer blend membranes from poly(vinyl-idene fluoride) and poly(ethylene glycol), Polym.Bull., 4:617 (1981).

27. T. Uragami, M. Fujimoto, and N. Suginara, Studies on syntheses and permeability of special polymer membranes. 27. Concentration of poly(styrene sulphonic acid) in various aqueous solutions using poly(vinylidene fluoride) membranes, Polymer, 22:240 (1981).

28. T. Uragami, Y. Ohsumi, and M. Sguginara, Studies on syntheses and permeability of special polymer membranes. 40. Formation conditions of finder-like cavities on cellulose nitrate membranes, Desalination, 37:293 (1981).

29. W. Pusch and A. Walch, Membrane structure and its correlation with membrane permeability, J.Membrane Sci., 10:325 (1982).

30. T. C. Shen and I. Cabasso, Ethyl cellulose anisotropic membrane, in: "Macromolecular Solutions," R. B. Seymour and G. A. Stahl, eds., Pergamon Press, New York (1982).

31. J. G. Wismans, J. P. B. Baais, and C. A. Smolders, The mechanism of formation of microporous or skinned membranes produced by immersion precipitation, J.Membrane Sci., 14:263 (1983).

32. Xu Xie-Quing, Additives for the preparation of polyvinylidene-based ultrafiltration membranes, especially for hemodialysis and hemo-osmometer, Desalination, 48:79 (1983).

SEPARATION CHARACTERISTICS OF ULTRAFILTRATION MEMBRANES

G. Jonsson and P. M. Christensen

Instituttet for Kemiindustri
Technical University of Denmark
DK-2800 Lyngby, Denmark

The selectivity of ultrafiltration membranes is determined primarily by the ratio between the hydrodynamic diameter of the solute and the apparent pore diameter. The retention characteristics of a given UF membrane are usually presented as the retention versus the molecular weight of different macromolecules, and the molecular weight cut-off is said to be the M_w-value which is almost totally rejected.

Most ultrafiltration data are so highly influenced by concentration polarization that a determination of the membrane transport properties is very uncertain. Often, it is found that the flux is independent of pressure and that the retention decreases with pressure[1], which is inconsistent with all transport theories that consider only the membrane.

In an earlier paper[2] different methods for determining the selectivity of reverse osmosis membranes were investigated. A combined viscous flow and frictional model was found to give the best correlation with the experimental data. This so-called finely-porous model has been presented in detail[3]. It was shown to give a good description of retention data for rather tight UF membranes having M_w cut-off values in the range 1000-6000 daltons.

EXPERIMENTAL

The ultrafiltration experiments were performed in the reverse osmosis loop with test cell I without sectioning as described elsewhere[4]. Commercial cellulose acetate membranes, DDS 600, were used. The membranes were pressurized at 10 atm maximal pressure for 16 hours before use, to ensure stability. Two sets of measurements were done:

1) At constant pressure level, the circulation velocity was varied in 7 steps from 156 to 348 cm/s.
2) At constant circulation velocity (250 cm/s), the pressure was varied in 10 steps from 10 to 1 atm.

In all experiments the bulk solution consisted of 0.1% of the macromolecule in distilled water. Solute concentrations of the feed and permeate were determined refractometrically with a differential refractometer, Waters model R-403.

179

Measurements were carried out with the following macromolecules:

1) Fractionated and well characterized dextrans from Pharmacia, Uppsala, Sweden:

   a) Dextran FDR   (M_w = 5,200, M_n = 3,500)
   b) Dextran T 10 (M_w = 9,300, M_n = 5,700)
   c) Dextran T 20 (M_w = 22,300, M_n = 15,000)

2) Polyethylenglycols from Merck, Schuchardt, Germany:

   a) PEG 1,000 (M_w =   950 - 1,050)
   b) PEG 2,000 (M_w = 1,900 - 2,200)
   c) PEG 4,000 (M_w = 3,500 - 4,500)
   d) PEG 6,000 (M_w = 5,000 - 7,000)
   e) PEG 10,000 (M_w = 8,500 - 11,500)
   f) PEG 20,000 (M_w $\cong$ 20,000)

3) Polyvinylpyrrolidon from Fluka AG, Switzerland:

   a) PVP K15 (M_w $\cong$ 10,000)

For each macromolecule some of the permeate samples were concentrated to between 0.5 - 1.0% using a reverse osmosis batch cell equipped with a HR 98 membrane, which totally rejects all fractions of the macromolecules. These concentrated permeate samples were analyzed by HPLC using a Spectra Physics model SP 8000 equipped with a Chrompack TSK G 3000 SW column (30 cm x 7.5 mm). A solution of 0.1% Na_2SO_4 was used as eluent at a flow rate of 0.5 ml/min at 30°C. Sample size was 100 μl.

RESULTS AND DISCUSSION

The mass transfer coefficient k_i, for this test cell has earlier been determined for low molecular weight solutes[4]. However, due to the very high polarization for ultrafiltration data, K_i is determined by the same method for some of the macromolecules, too.

From the film-theory model the following relation can be derived[5]:

$$\ln\left(\frac{1-S}{S}\right) = \ln\left(\frac{1-R}{R}\right) + \frac{J_v}{k_i} \tag{1}$$

In turbulent flow, K_i is normally found to be proportional to the flow velocity, u, raised to the power 0.8 according to the relation:

$$\frac{k_i d_h}{D_i} = A \cdot Re^{0.8} \cdot Sc^{1/3} \tag{2}$$

Consequently, a plot of $\ln(1-S)/S$ versus $J_v/u^{0.8}$ should give a straight line intersecting at $\ln(1-R)/R$ and with a slope equal to $(AD_i^{2/3}/d_h^{0.2} \nu^{0.467})^{-1}$. This is shown in Figure 1 for PEG 4000 at 4 different pressure levels. With increasing pressure the true retention, R, increases, but the slope remains constant. Knowing the true retention, the mass transfer coefficient can be calculated from Equation (1). This was done for three of the macromolecules and the K_i-values corresponding to u = 250 cm/s is given in Table 1. For the other solutes k_i was calculated from Equation (2) knowing the constant A from the slope in Figure 1. The diffusion coefficients were calculated from the Stokes-Einstein relation:

$$D_i = kT/(6 \pi \eta r_i) \tag{3}$$

where radius of the different macromolecules has been estimated from Figure 2, which is based on literature values[6,7,8].

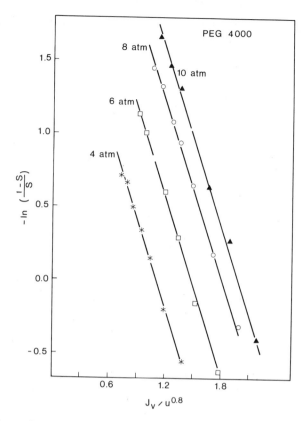

Fig. 1. Data for PEG 4,000 at varying circulation velocity for estimation of the mass transfer coefficient.

Figures 3 and 4 show the observed retention versus the permeate flux for three PEG's and dextrans, respectively. Due to polarization the curves go through a maximum at which point the increase in true retention is balanced by the increase in concentration polarization. The flux value where the maximum is situated depends mainly on the retention level and the mass transfer coefficient. This special phenomenon may result in such a situation that the observed retention of a bigger molecule will be lower than for a smaller molecule. This is seen in Figure 3, where the curves for PEG 20,000 and PEG 6,000 cross each other at high flux values. In Figures 3 and 4 the true retentions calculated from Equation (1) using the k_i-values from Table 1 are further shown as dotted lines. Now a steady increase in retention is seen approaching the maximal retention, R_{max}, at infinite flux.

According to the final-porous model the true retention is given by the relation:

$$\frac{C_m}{C_b} = \frac{1}{1-R} = \frac{b}{K} + \left(1 - \frac{b}{K}\right) \exp\left(-\frac{t\lambda}{\varepsilon}\frac{J_v}{D_i}\right) \tag{4}$$

Due to the high correlation between the two parameters b/k and $t\lambda/\varepsilon$ in Equation (4) only few of the experimental data-sets give reasonable values of the estimated parameters from a non-linear parameter estimation[9].

Table 1. Diffusion Coefficients Estimated from Literature Data and
Experimental Determined Mass Transfer Coefficient for the
Different Macromolecules

Solute	D_i $(10^{-6}$ cm/s$)$	k_i^* $(10^{-3}$ cm/s$)$
PEG 1,000	2,65	(4,27)
PEG 2,000	2,03	(3,20)
PEG 4,000	1,46	2,23
PEG 6,000	1,15	(2,09)
PEG 10,000	0,852	1,67
PEG 20,000	0,566	(1,23)
Dextran FDR	1,44	(3,29)
Dextran T 10	1,13	2,74
Dextran T 20	0,762	(2,04)
PVP K 15	0,946	(2,40)

* values in parenthesis are calculated from eq. (2).

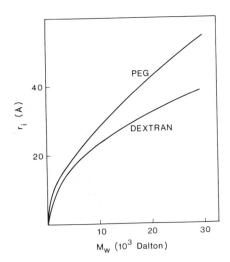

Fig. 2. Solute radius versus molecular weight estimated from literature
values.

However, assuming that $t\lambda/\varepsilon$ is a membrane constant independent of the
solutes, all the data-sets can be analyzed together, with b/k being constant for each solute. In this way $t\lambda/\varepsilon$ was estimated to 10^{-4} cm and the
solute parameters, b/k, and the maximal retention, $R_{max} = 1 - k/b$, is given
in Table 2.

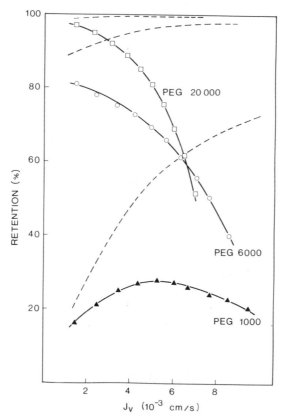

Fig. 3. Observed retention versus permeate flux for three different PEGs. The dotted curves are the true membrane retention calculated from Equation (1).

In Figure 5 the experimental determined R-values for PEG 6000 are shown as a function of the permeate flux. Further the theoretical curve calculated from Equation (4) using the parameters from Table 2 is shown. This gives a quite good agreement, although there is a tendency that the experimental data increase more steeply than the theoretical. This difference, which is a common trend for all the solutes might be due to a variation in the pore size of the membrane and therefore in the b/k-values. This change in shape is also responsible for the problems with the parameter estimation for the individual data-sets, giving very small $t\lambda/\varepsilon$-values and correspondingly high b/k-values.

In Figure 6 two other methods for estimating R_{max} are shown. According to Pusch[10] a linear relation exists between $1/R$ and $1/J_v$, from which R_{max} can be determined by extrapolation to zero. For all the solutes this gives, however, R_{max}-values above 100%, as demonstrated in Figure 6. In another method, which is not theoretical justified, R_{max} is determined by extrapolation of R versus $1/J_v$ to zero. As the linearity is not found at low pressures, the values of R_{max} shown in Table 2 were estimated by linear regression of the experimental data in the pressure region 6-10 atm. For some of the solutes this gives quite a good agreement with the R_{max}-values determined from Equation (4); however, some of the values are unrealistically high. In Figure 6 the theoretical curves calculated from Equation (4) are further shown as dotted lines, from which it can be seen that both curves level out when $1/J_v \to o$.

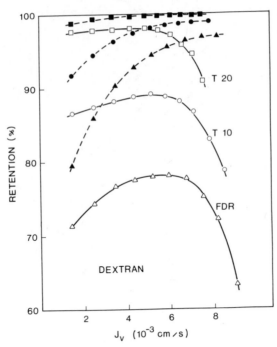

Fig. 4. Observed retention versus permeate flux for three different dextrans. The dotted curves are the true membrane retention calculated from Equation (1).

Table 2. Estimated Values of b/K and R_{max} from Equation (4) Using a Nonlinear Parameter Estimation

Solute	b/K ($t\lambda/\varepsilon = 10^{-4}$cm)	$R_{max} = 1 - {}^K/b$ (%)	R_{max}^* (%)
PEG 1,000	7,99	87,48	87,13
PEG 2,000	20,07	95,02	95,58
PEG 4,000	42,7	97,66	>100
PEG 6,000	69,5	98,56	99,56
PEG 10,000	141,8	99,29	99,12
PEG 20,000	419,0	99,76	99,89
Dextran FDR	46,3	97,84	>100
Dextran T 10	103,7	99,04	99,91
Dextran T 20	603,8	99,83	99,95
PVP K 15	35,02	97,14	>100

* Estimated by linear regression of R vs. $1/J_v$ in the pressure range 6-10 atm.

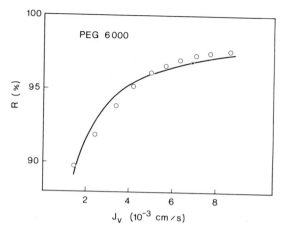

Fig. 5.   Comparison of true retention data. The curve drawn in is
calculated from Equation (4) using the data from Table 2.

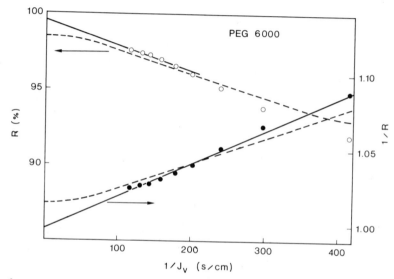

Fig. 6.   Comparison of true retention data. The lines drawn in are the
calculated regression lines, whereas the dotted curves are
calculated from Equation (4).

Figure 7 shows the elution curves for a mixture of PEG standards and a
dextran T 10, respectively. In addition, the permeate from a dextran T 10
solution is shown as the dotted curve. Comparison with the elution curve
for the bulk solution, a clear shift to a lower molecular weight can bee
seen. Figure 8 shows two similar comparisons of the bulk and permeate
solution for PEG 20,000 and PEG 4,000, respectively. For the high molecular weight solute a shift is again seen, whereas the low molecular
solute is unchanged. The elution curve for PEG 20,000 shows further that
there are some impurity of PEG 10,000 in this substance. Comparing the top
heights of the concentrated permeate and the bulk solution it can be seen
that PEG 10,000 is decreased in concentration relative to PEG 20,000. This
is in agreement with the situation shown in Figure 3, where the retention
of PEG 20,000 falls below that of PEG 6,000 at high pressure. In Figure 9
the calibration curves for PEG and dextran are given as $\log(M_w)$ versus
elution volume. From these curves it is found that M_w is reduced from

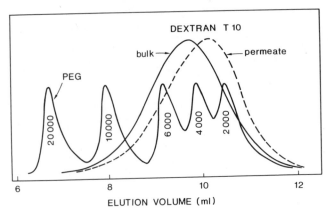

Fig. 7. Elution curves for a mixture of PEG standards. Further a comparison of the bulk and permeate for dextran T 10.

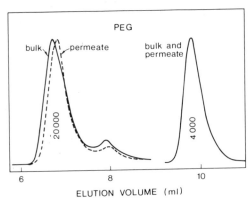

Fig. 8. Comparison of the bulk and permeate for PEG 20,000 and PEG 4,000.

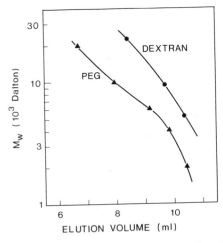

Fig. 9. Calibration curves for PEG and dextran.

20,000 to 18,500 for PEG 20,000 and from 9,300 to 6,300 for dextran T 10 by the permeation through the membrane.

Figures 10 and 11 show cut-off curves at three different pressure levels for PEG and dextran, respectively. At low pressure the retention level for the two different macromolecules is quite similar, but with increasing pressure the retention for PEG decreases much, whereas dextran shows a maximum at 6 atm and then decreases a little. This different behavior seems mainly to be caused by a difference in the solution properties of the two macromolecules. As seen from Figure 2, PEG has a higher hydrodynamic volume than dextran of the same molecular weight and so also a lower diffusivity. Therefore concentration polarization is more severe for PEG than dextran so that the retention decreases much faster. This phenomenon is verified in Figure 12, where the cut-off curve is given as the intrinsic value, R_{max}, versus M_w. Now PEG, dextran and PVP fall on almost the same curve, showing that the separation characteristics of the membrane itself are primarily determined by the molecular weight of the solute.

The retention data have further been correlated with Equations for steric exclusion and frictional interaction with the pore walls given by Ferry[11] and Haberman and Sayre[12], respectively:

$$K = 2(1-\alpha)^2 - (1-\alpha)^4 \tag{5}$$

and

$$b = \frac{1-2.105 \ \alpha + 2.0865 \ \alpha^3 - 1.7068 \ \alpha^5 + 0.72603 \ \alpha^6}{1-0.75857 \ \alpha^5} \tag{6}$$

where $\alpha = r_i/r_p$.

In Figure 12, $R_{max} = 1-k/b$ has been calculated for PEG for three different pore radii and shown as a function of M_w determined from Figure 2. At low M_w the experimental data are close to the curve for $r_p = 20$ Å, but above $M_w = 2,000$ they increase much slower than expected from Equations (5) and (6). At $M_w = 5,000$ they cross the curve for $r_p = 30$ Å, approaching the curve for $r_p = 60$ Å at $M_w = 20,000$, where R_{max} is close to unity. This may indicate that the membrane has an average pore radius around 30 Å, but a pore size distribution from 20 Å to 60 Å with quite few pores in the upper range.

In Figures 10 and 11 similar curves have been calculated for the observed retention in the following way: From Equations (5) and (6) the parameter b/K is calculated for different r_i and r_p values. Then the true retention, R, is calculated from Equation (4) at different pressure levels (corresponding to given J_v-values) assuming $t\lambda/\varepsilon = 10^{-4}$ cm and D_i determined from Equation (3). Finally, the observed retention is calculated from Equation (1) knowing the mass transfer coefficient from Equation (2) at u = 250 cm/s, and M_w from Figure 2. Again, for $M_w < 2,000$ the experimental data are quite close to the calculated curve for $r_p = 20$ Å. At $M_w = 5,000$ the experimental data at 2 and 10 atm cross the calculated curves for $r_p = 30$ Å and they seem to approach the 60 Å curve for M_w-values above 20,000. Thus the similarity in the position of the calculated curves relative to the experimental data in Figures 10 and 12 seems to confirm that the membrane has a pore size distribution between 20 Å and 60 Å for r_p.

CONCLUSION

Taking careful consideration of the concentration polarization the true membrane retention was found in reasonable agreement with the

187

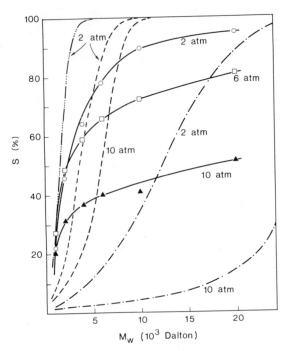

Fig. 10. Observed retention versus molecular weight for PEG at three
different pressure levels. The dotted curves are calculated
from Equations (1)-(6) for three different pore radii: r_p = 20 Å
$(-\cdots-)$, r_p = 30 Å $(---)$ and r_p = 60 Å $(-\cdot-)$.

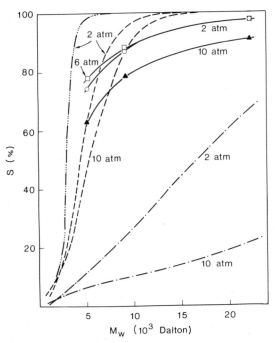

Fig. 11. Observed retention versus molecular weight for dextran at three
different pressure levels. The dotted curves are calculated
from Equations (1)-(6) for three different pore radii: r_p = 20 Å
$(-\cdots-)$, r_p = 30 Å $(---)$ and r_p = 60 Å $(-\cdot-)$.

188

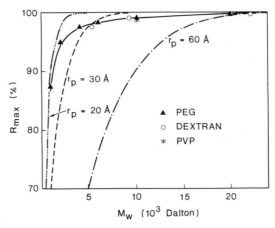

Fig. 12. The maximal retention determined from Equation (4) versus
molecular weight for all the macromolecules investigated.
The dotted curves are calculated from Equations (5) and (6)
for three different pore radii.

finely-porous membrane model. The estimated R_{max}-values show an unambig-
uous dependence of the molecular weight for the three kinds of macromolecules investigated. However, the experimental cut-off curves depend so much
on the operating condition (pressure, circulation velocity) and the solution properties of the macromolecules that a distinct M_w cut-off value for
UF membranes is very uncertain. Using Equations relating the ratio r_i/r_p
to the distribution coefficient and the friction factor, it is concluded
that the membrane has an average pore radius of about 30 Å, but a pore size
distribution from 20 Å to 60 Å with quite few pores in the upper range.

NOMENCLATURE

b	– friction parameter	R	– true retention
C_b	– bulk concentration	R_{max}	– maximal retention
C_m^b	– concentration at membrane	Re	– Reynold number
	surface	S	– observed retention
D_i	– diffusion coefficient	Sc	– Schmidt number
d_h^i	– hydraulic diameter of	t	– tortuosity factor
	membrane channel	T	– temperature
J_v	– volume flux	u	– circulation velocity
K	– distribution coefficient	ε	– fractional pore area
k_i	– mass transfer coefficient	η	– viscosity
k^i	– Boltzmann constant	λ	– skin layer thickness
r_i	– solute radius	ν	– kinematic viscosity
r_p^i	– pore radius		

REFERENCES

1. W. F. Blatt, A. Dravid, A. S. Mickaels, and L. Nelsen, in: "Membrane
   Science and Technology," J. E. Flinn, ed., Plenum Press, New York
   (1970).
2. G. Jonsson, Desalination, 24:19 (1978).
3. G. Jonsson and C. E. Boesen, Desalination, 17:145 (1975).
4. G. Jonsson and C. E. Boesen, Desalination, 21:1 (1977).
5. G. Jonsson and C. E. Boesen, in: "Synthetic Membrane Processes,"
   G. Belfort, ed., Academic Press, Orlando (1984).

6. K. A. Granath, J.Colloid.Sci., 13:308 (1958).
7. K. A. Granath and B. E. Kvist, J.Chromatog., 28:69 (1967).
8. H. A. Ende, in: "Polymer Handbook," J. Brandrup and E. H. Immergut, eds., Interscience Publishers, New York (1967).
9. A. J. Barr, J. H. Goodnight, J. P. Sall, and J. T. Helwig, "A User's Guide to SAS 79," SAS Institute Inc., Raleigh, North Carolina (1979).
10. W. Pusch, Ber.Bunsenges.Physik.Chem., 81:269 (1977).
11. J. D. Ferry, J.Gen.Physiol., 20:95 (1936).
12. W. L. Haberman and R. M. Sayre, "Motion of Rigid and Fluid Spheres in Stationary and Moving Liquids Inside Cylindrical Tubes," David Taylor Model Basin Report No.1143, Department of the Navy (1958).

SEPARATION OF AMINOACIDS BY CHARGED

ULTRAFILTRATION MEMBRANES

Shoji Kimura and Akiyoshi Tamano

Institute of Industrial Science
University of Tokyo, 7-22-1 Roppongi
Minatoku, Tokyo 106, Japan

INTRODUCTION

An ultrafiltration membrane, that has a fixed charge, is interesting
for practical applications, because it has two variables, namely a pore
size and a charge density. So the membrane can be used as an ordinary
ultrafiltration membranes, and it can at the same time reject inorganic
salts by the effect of the electric charge. From this nature this membrane
is considered to be applicable for the purification of secondary sewage
effluent, which contains both organic solutes and dilute inorganic salts.

It may also be possible to separate solutes having a charge and those
without charge. From this nature this membrane is considered applicable
for the separation of various solute, that have different iso-electronic
points.

Membranes made of sulfonated polysulfone have been developed and used
in our previous works, in which manufacturing procedure and their ultrafiltration nature were presented[1]. Rejections of inorganic solutes were
measured and their transport parameters were explained theoretically by
using the relation of Donnan equilibrium and the appropriate form of
activity coefficient for the counter ion in the membrane[2,3].

In this investigation results of separation of different aminoacids by
this membrane is reported.

EXPERIMENTALS

Membranes

Sulfonated polysulfone(SPS) was obtained by sulfonation of polysulfone
(P 1700, Union Carbide), the charge density of which is 1.6 meg/g at
maximum. Tubular membranes were cast from solutions of SPS/N-methyl-
2pyrrolidone/LiNO$_3$. Details of casting procedure were written in the
literature[1].

Characterization Of Membranes

Ultrafiltration experiments using various molecular weight poly-ethylene glycol(PEG) were performed and it was found that a membrane designated as M_1 had a fractional molecular weight cut-off value of about 10,000. This membrane was used throughout this work. It's pore radius was also estimated as 1.87 nm.

Rejections of various inorganic salts were also measured. From these data reflection coefficients and solute permeabilities were obtained. Based on the theoretical considerations developed before[2] osmotic co-efficients of the membrane charge were estimated from reflection coef-ficients. Finally solute permeabilities were calculated and were found to be well coincident with experimental data.

Rejection Of Aminoacids

Among three types of aminoacid following acid were used in this experiment.

As basic acid:

L-ornithine hydrochloride (molecular weight 167.6, iso-electric point 9.70), L-lysine hydrochloride (m.w. 182.7, i.e.p. 9.59), L-arginine hydrochloride (m.w. 210.7, i.e.p. 11.15).

As acidic acids:

sodium L-glutamate (m.w. 187.1, i.e.p. 3.22), sodium L-aspartate (m.w. 173.1, i.e.p. 2.77).

As neutral acids:

glycine (m.w. 75.1, i.e.p. 5.97), L-isoleucine (m.w. 131.2 i.e.p. 5.94).

First rejections of single acid were measured and transport parameters were determined. Also osmotic coefficients were estimated as before. Next the effect of pH on the rejection was measured by using typical three types of acids, and the relation of dissociation and rejection was presented. Lastly mixtures of two types of acids were separated by the membrane. Experimental conditions were as follows for all experiments: flow rate 4 1/min, pressure 0.3-0.9 mpa, temperature 25°C.

RESULTS

Rejection of Single Aminoacid

Rejections of three types of aminoacids are shown in Figure 1, 2 and 6, where it is seen that an acidic aminoacid is rejected very well and a neutral acid is not rejected much, while rejections of a basic acid are about 50% without pH adjustment. A basic aminoacid seemed to be absorbed in the membrane and flux decreased to some extent.

In Table 1, experimentally determined values of reflection coef-ficients (σ) and permeabilities (P) were listed. In Table 2 osmotic coefficients calculated from reflection coefficients (ϕ) are listed and permeabilities calculated using these osmotic coefficients (Pcal) are compared with experimental values P. It is seen that agreement is reasonable.

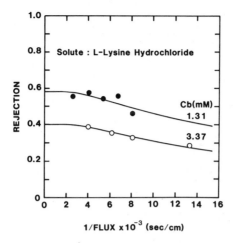

Fig. 1. R vs. $1/J_v$ plot of L-lysine hydrochloride.

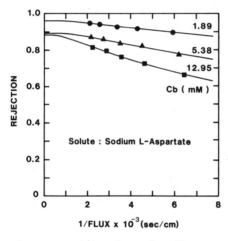

Fig. 2. R vs. $1/J_v$ plot of sodium L-aspartate.

Table 1.   Reflection Coefficients and Permeabilities

Solutes	C_b(mM)	σ(-)	$P \times 10^4$(cm/s)
L-ornithine hydrochloride	1.68	0.67	0.2
	5.73	0.7	0.95
L-lysine hydrochloride	1.31	0.59	0.55
	3.37	0.4	0.55
L-arginine hydrochloride	1.19	0.6	0.58
	3.53	0.23	0.68
sodium L-glutamate	1.30	0.95	0.14
	4.10	0.87	0.23
	13.20	0.86	0.44
sodium L-aspartate	1.89	0.96	0.16
	5.38	0.89	0.32
	12.95	0.88	0.6

Table 2. Osmotic Coefficients and Permeabilities

Solutes	C_b(mM)	σ(-)	ϕ(-)	$P_{cal} \times 10^4$	$P \times 10^4$(cm/s)
L-ornithine hydrochloride	1.68	0.67	0.007	0.4	0.2
	5.73	0.7	0.03	0.96	0.95
L-lysine hydrochloride	1.31	0.59	0.005	0.65	0.55
	3.37	0.4	0.009		0.55
L-arginine hydrochloride	1.19	0.6	0.0046	0.57	0.58
	3.53	0.23	0.0073	0.4	0.68
sodium L-glutamate	1.30	0.95	0.016	0.22	0.14
	4.10	0.87	0.016	0.57	0.23
	13.20	0.86	0.055	0.69	0.44
sodium L-aspartate	1.89	0.96	0.025	0.20	0.16
	5.38	0.89	0.025	0.56	0.32
	12.95	0.88	0.06	0.64	0.6

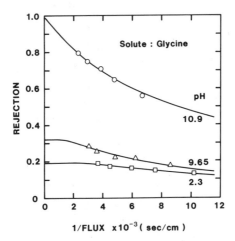

Fig. 3. Effect of pH on R of glycine.

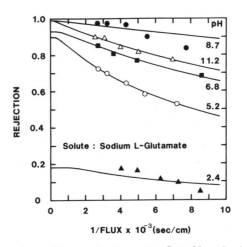

Fig. 4. Effect of pH on R of sodium L-glutamate.

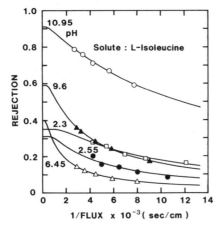

Fig. 5.   Effect of pH on R of L-isoleucine.

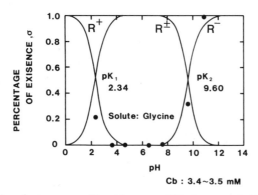

Fig. 6.   Relation between reflection coefficient and dissociation of glycine.

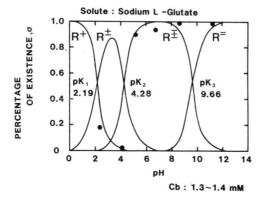

Fig. 7.   Relation between reflection coefficient and dissociation of sodium L-glutamate.

Effect of PH on Rejection

Effect of pH on rejection of glycine, sodium L-glutamate and L-isoleucine are shown in Figure 3, 4 and 5, respectively, while their correspondence to the form of acids are shown in Figures 6, 7 and 8 respectively. It is clearly seen that the rejection of neutral form is low, and rejections increase at both sides of isoelectric point.

Rejection Of Mixed Acids

Rejections of a mixed solution of sodium L-aspartate and L-isoleucine is shown in Figures 9 and 10. In Figure 9 the concentration of L-isoleucine is kept constant, while the concentration of sodium L-aspartate is changed. In Figure 10 concentrations are changed vice versa. Rejections of a mixed solution of L-isoleucine and L-ornithine hydrochloride is shown in Figure 11.

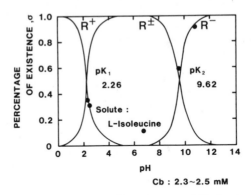

Fig. 8. Relation between reflection coefficient and dissociation of L-isoleucine.

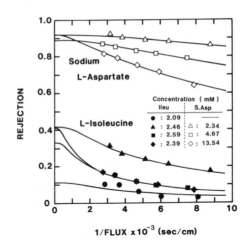

Fig. 9. Rejection from a mixed solution of sodium L-aspartate and L-isoleucine; concentration of L-isoleucine constant.

From these Figures it is seen that there are some effect of mixing on rejections, but it seems possible to separate amino-acids having different iso-electric points by a charged ultrafiltration membrane.

CONCLUSION

SPS membrane, having a charge density of 1.6 meq/g and a fractional molecular weight cut-off of 10,000, can separate low molecular weight solutes by its charge. In this work aminoacids, whose molecular weight from 75 to 200, were used as solutes and it is shown that their rejection can be changed by pH adjustment and their separation can be done by the difference of iso-electric points.

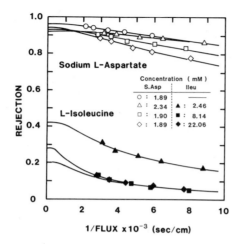

Fig. 10. Rejections from a mixed solution of sodium L-aspartate and
L-isoleucine; concentration of sodium L-aspartate constant.

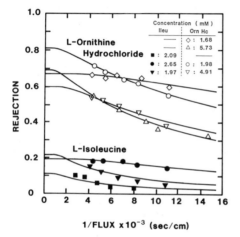

Fig. 11. Rejections from a mixed solution of L-ornithine hydrochloride and
L-isoleucine.

REFERENCES

1. I. Jitsuhara and S. Kimura, J.Chem.Eng.of Japan, 16:389 (1983).
2. I. Jitsuhara and S. Kimura, J.Chem.Eng.of Japan, 16:394 (1983).
3. S. Kimura and I. Jitsuhara, Desalination, 46:407 (1983).

PREDICTING FLUX–TIME BEHAVIOR OF FLOCCULATING

COLLOIDS IN UNSTIRRED ULTRAFILTRATION

R. M. McDonogh, C. J. D. Fell and A. G. Fane

School of Chemical Engineering and Industrial Chemistry
University of New South Wales
Kensington 2033, Australia

ABSTRACT

A model for the unstirred ultrafiltration of flocculating colloids is presented. The model allows for the growth of colloidal flocs above the membrane and the tendency to flocculate is related back to the zeta potential of individual particles. By calculating the deposited floc size from flocculation kinetics and particle proximity from double layer theory, it is possible to use the Carman-Kozeny equation to determine the specific resistance of successively deposited layers. This in turn allows the progress of the ultrafiltration to be predicted. Calculated flux declines and final fluxes from the model agree well with those measured experimentally for spherical silica colloid, r = 12.5 mm, in the range $0 \leq$ zeta potential ≤ 35 mV.

INTRODUCTION

Investigation of the dominant mechanisms prevailing in the ultra-filtration of colloids is a necessary precursor to the adoption of ultra-filtration as a pretreatment method for colloid contaminated waters destined for reverse osmosis.

Previous studies on the ultrafiltration of colloids have been carried out in a cross flow mode. Blatt et al.[1] and Porter[2] have both commented from experimental observation that conventional ultrafiltration theory based on the convective back transport of particles may not strictly apply to colloidal systems. Similarly, Henry[3] has found that the dependence of flux on shear rate is greater in colloidal cross flow filtration than is predicted by the thin film back transport model.

The general conclusion from all of this work, as well as that of Kraus[4], Madsen[5] and Trettin and Doshi[6] is that the ultrafiltration of a colloid in a cross flow mode is controlled by two separate mechanisms, one due to shear initiated convective back transport of colloid particles (this may be assisted by the 'tubular pinch' effect) and one due to the colloid chemistry of the particles being ultrafiltered.

By working in an unstirred batch cell, the effects of convective back transport are removed and the influence of colloid chemistry can be investigated directly. In an earlier study[7] we have reported on the unstirred ultrafiltration of non-flocculating colloids and have shown that the retentate permeability can be predicted on a theoretical basis.

In the present paper we develop a model for the unstirred ultra-filtration of colloids that can flocculate. This model allows the prediction of flux-time curves which can be compared with those obtained experimentally. The model can also be used to predict the effect of colloid particle zeta potential on the retentate cake permeability.

THEORY

Colloids dispersed in solution are not thermodynamically stable, and will coalesce (flocculate) if collision takes place as a result of Brownian motion such that the repulsive energy due to particle charge is exceeded.

For the case where there is no repulsion (so called 'fast' coagulation conditions) every collision leads to coagulation. Von Smoluchowski[8] has shown that the rate of disappearance of primary particles (J) is then given by:

$$J = - \frac{dv}{dt} = - 8\pi Dv^2 R_o \qquad (1)$$

Here v, D and R_o are the particle concentration, diffusivity and critical interparticle spacing respectively.

For more slowly coagulating systems where there is appreciable inter-particle repulsion, Fuchs[9] has modified Equation (1) by introducing a correction factor W which accounts for particle potential energy (V). Equation (1) now becomes:

$$J = 8\pi Dv^2 R_o /W \qquad (2)$$

and W is given by:

$$W = R_o \int_{R_o}^{\infty} \exp \left(\frac{V}{kT} \right) \frac{dR}{R^2} \qquad (3)$$

with R_o being $2r$, with r the radius of the particle.

The total number of particles at any time (v_t) can be found by integration of Equation (2):

$$v_t = \frac{v_o}{1 + 8\pi Dv_o R_o t/W} \qquad (4)$$

If the particles are assumed spherical, particle diffusivity can be expressed as

$$D = \frac{kT}{6\pi \eta R_o} \qquad (5)$$

so that:

$$v_t = \frac{v_o}{1 + \dfrac{4kTv_o t}{3\eta w}} \qquad (6)$$

Equation (6) has been shown[10] to apply reasonably over the whole of the flocculation process. It thus becomes possible to use Equations (3) and (6) to obtain a population-time history of a flocculation and to calculate the volume average floc size as a function of time according to Equation (7)

$$R_{floc,t} = R_o \left(\frac{v_o}{v_t}\right)^{\frac{1}{3}}$$

(7)

Calculation of Permeability of Retentate Layer

In an earlier publication[7] we have shown the permeability of the retentate layer in unstirred colloid ultrafiltration can be calculated from a knowledge of the particle size and zeta potential. The interaction of double layers on adjacent particles leads to inter-particle pressure $(P_{d\ell})$:

$$P_{d\ell} = 2 \eta^\circ kT(\cosh(Z_i e\psi/kT) - 1)$$

(8)

which is counterbalanced by the applied hydraulic pressure. By evaluating ψ from particle charge and attractive potential, it is possible to calculate particle spacing and then voidage (ε) and to hence calculate retentate permeability (in terms of specific resistance (α)) from the Carman-Kozeny equation

$$\alpha = \frac{180}{4r^2} \cdot \frac{(1-\varepsilon)}{\rho\varepsilon^3}$$

(9)

Computer Model

The computer model uses Equations (6) - (9) to estimate the changing specific resistance of the retained layers during unstirred ultrafiltration.

The program (flowsheet given in Figure 1) examines events occurring from t=o, when the solution is first prepared, and considers initially the flocculation in the bulk solution before the ultrafiltration run is commenced at $t=t_i$. Depending on the proclivity of the colloid to coalesce, consideration of events during this time can be important.

After the commencement of ultrafiltration (i.e. beyond $t=t_i$), the program steps forward in time increments of Δt, calculating the flux according to Equation (10):

$$J = \frac{\Delta P}{\eta(R_{mem} + R_{cake})}$$

(10)

Where R_{mem} is the membrane resistance (determined from solvent flux) and R_{cake} is the total retentate resistance calculated for the previous time increment.

The program examines coagulation occurring in the bulk solution during Δt, as well as considering (for completeness) the extent of coagulation taking place in the high concentration polarised layer immediately above the retentate cake. The concentration profile in this layer is calculated (as recommended by Trettin and Doshi[6]) using the instantaneous flux and Equation (5). In practice it is found that the short residence time in this polarised layer is insufficient to contribute greatly to coalescence in the system.

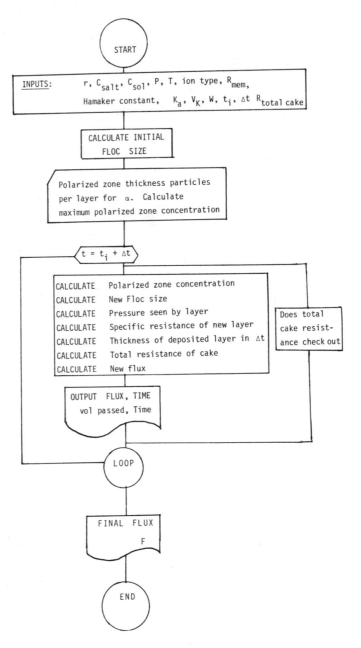

Fig. 1. Flowsheet of computer model.

From the flux and a knowledge of the prevailing particle size (Equation 7), it is possible to calculate the thickness of the retentate layer deposited during Δt by:

$$W_t = \Delta t \cdot \frac{Jc_o}{\rho(1-\varepsilon)} \tag{11}$$

and the specific resistance of this layer by application of Equation (9).

202

It is then possible to sum the resistances of the various retentate layers already laid down and to arrive at an overall retentate resistance (R_{cake}) to be used for the next time step.

Assumptions made in the above scheme are, (i) that flocs once deposited as retentate undergo no further coagulation, and (ii) that the bulk colloid concentration remains approximately constant (a reasonable assumption in view of the relatively small quantity of permeate that is removed).

Values of the parameters used in the model were:

$$\Delta t = 0.36 \text{ s}$$
$$\rho = 2400 \text{ kg m}^{-3}$$
$$\Delta P = 100 \text{ kPa}$$
$$r = 12.5 \text{ nm}$$

η and ε are calculated according to Reference 11. The Hamaker constant (used in Equation (8), was taken as 0.4.

EXPERIMENTAL

The experimental protocol adopted has been previously described[7].

The colloid used in each experiment was Syton CX 30 ($r = 12.5$ nm) supplied by Monsanto Ltd. It was filtered before use to remove any aggregates.

To prepare a colloid solution having particles with a specific zeta potential, concentrated, filtered colloid solution was diluted to 0.3 wt.% colloid by the addition of a chosen salt solution (see Table 1). These salt solutions were themselves prepared from Analar grade reagents and were filtered through both 0.2 μm Millipore and Amicon PM30 membranes to remove any possible contaminants. The time at which the salt solution was added to the concentrated colloid was noted for use in subsequent flocculation calculations.

Zeta potentials of the colloid particles were measured using a Micromeritics Mass Transport Zetameter, such that specific measurements differed by less than 5%.

Amicon PM 30 membranes (impermeable to colloid) were used in all ultrafiltration studies. A new membrane was used for each run, and was prepared according to the manufacturer's specification.

Table 1. Zeta Potential and Calculated Final Floc Sizes

Salt Type	Concentration $\times 10^{+2}$M	Zeta potential mV mV	Calculated final floc size nm
$CaCl_2$	10.02	0.5	299
$MgCl_2$	0.98732	2	63.8
$CaCl_2$	1.3604	4	43.2
NaCl	3.4223	8	26.1
$CaCl_2$	0.3401	10	18.3
KCl	2.6824	14	17.3
NaCl	1.7112	23	13.1
NaCl	0.8556	31	12.5
Water	–	32	12.5

The experimental equipment consisted of an unstirred cell which was fed from a feed reservoir through a Millipore filter. Transmembrane pressure was applied by a nitrogen cylinder. The cell was initially charged with the colloid solution and flux was then monitored after pressure had been applied by determining the rate of permeate collection with a Shimadzu electronic balance. The accuracy of such flux measurements was ±2%.

RESULTS

Flux-Time Curves

Figures 2 and 3 show typical experimental flux-time curves and also flux-time curves predicted using the model.

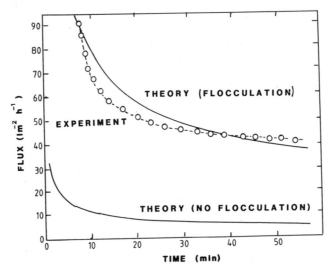

Fig. 2.   Flux-time plot for zeta potential of 0.5 mV. Continuous lines show predicted fluxes for model with and without flocculation considered.

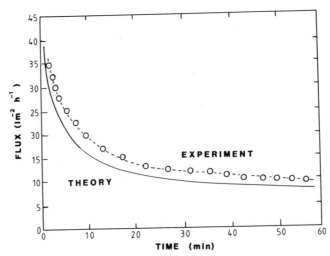

Fig. 3.   Flux-time plot for zeta potential of 31 mV. Continuous line represents model output. For this case no significant flocculation occurs.

At low zeta potential (Figure 2), flux drops away quickly (as expected) with increasing permeate passage. Predictions from the model with flocculation accounted for (as indicated) are in reasonable agreement with experiment, although the model predicts that flux initially falls less quickly than is experimentally observed. When no allowance is made for flocculation (lower curve), predicted flux values are much lower than those experimentally observed. This is because the model then predicts that deposited unflocculated particles, having a low zeta potential, do not repel each other strongly and are hence more closely spaced. The result is a less permeable deposited cake and a predicted lower flux.

At high zeta potential (Figure 3), predictions from the model are again in reasonable agreement with experiment, but are generally a little lower. The model predicts that flux initially drops away more quickly with time than is experimentally observed. For the case of high zeta potentials, predictions from the model are unchanged if allowance is made for flocculation as significant flocculation is unlikely within the time frame of the experiment.

Depending on the zeta potential of particles in the colloid solution, flux-time curves for all experiments performed showed similar behavior to those in Figure 2 and 3, with the model incorporating flocculation providing reasonable estimates of experimental fluxes.

T/V Versus V Curves

Filtration Theory suggests that the progress of an unstirred experiment follows Equation (12):

$$\frac{t}{V} = \frac{R_{mem} \, \eta}{\Delta P . A_{mem}} + (\frac{\alpha \, c \, \eta}{2 \, A_{mem}^2 \Delta P}) \, V \tag{12}$$

for constant specific resistance (α), corresponding to an incompressible cake, this equation predicts a linear relationship between T/V and V.

At high zeta potentials, the T/V versus V plots were straight lines, as previously reported[7], with a correlation coefficient of 0.990 or better.

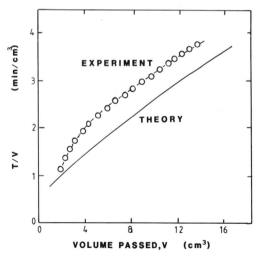

Fig. 4. T/V versus V plot for zeta potential of 8 mV. Continuous line is model prediction.

At low zeta potentials, experimental plots were strongly curved at low values of V as shown in Figure 4. Although the model incorporating flocculation does predict some curvature in the T/V versus V plot, it clearly does not predict the rapidly changing specific resistance in the early section of a run. However, the slopes of the experimental and predicted curves are similar later in the run, indicating that the model is then predicting correct specific resistances.

Final Flux

Figure 5 shows experimental final flux values for all runs compared with those computed by the model. Predicted values are in reasonable agreement, but generally lie slightly lower than those observed experimentally. Notably, predicted fluxes are high for low zeta potentials where flocculation can, and does, occur.

Specific Resistance

Figure 6 shows experimental specific resistance values (calculated from T/V versus V plots) compared with those predicted by the model. Having regard to the attendant error in measuring zeta potentials, agreement is encouraging. Most particularly, the model predicts the low specific resistances experimentally observed at low zeta potentials.

Predicted values of specific resistance at high zeta potentials are in close agreement with those earlier calculated[7] by a procedure which treated the deposited cake as having a uniform specific resistance. In the present model the contributions to total cake resistance from each layer within the cake are summed.

DISCUSSION

As originally conceived, the purpose of the present study was to determine whether an earlier model for predicting the specific resistance of the cake deposited in unstirred colloid ultrafiltration could be extended to encompass the possibility of colloid flocculation. It was considered that flocculation could account for the relatively high fluxes obtained when colloids having a low zeta potential were ultrafiltered.

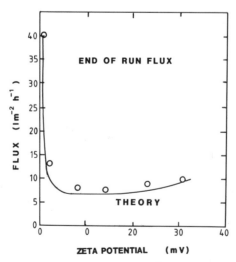

Fig. 5. Experimental end of run fluxes (circles) compared with predicted flux for various zeta potentials.

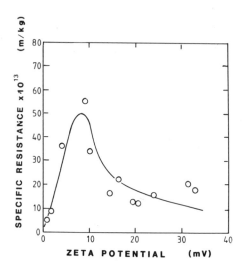

Fig. 6. Experimental specific resistance (circles) compared with model
predicted specific resistances for various zeta potentials.

In developing the model, close attention was paid to ensuring that the
likelihood of coagulation in the concentration boundary layer immediately
above the deposited cake was accounted for. It was reasoned that coagulation would be markedly enhanced in this high concentration zone. In the
event, it was found that most coagulation took place in the bulk solution.
For example, Table 1 gives an indication of the extent of coagulation
occurring over the experimental run at various zeta potentials. When
account is taken of this coagulation, the experimentally observed changes
in cake permeability with zeta potential are explained.

The strength of the present model lies in the fact that it does not
use adjustable parameters. The predicted fluxes or specific resistances
are based on measurable properties of the colloidal system. The calculated
specific resistance is sensitive to the value of zeta potential used and is
to a lesser extent dependent on other factors such as Hamaker constant (mid
range value used), temperature and transmembrane pressure. The fact that
it is possible to arrive at reasonable estimates of final flux using this
approach is encouraging and suggests that it is mechanistically sound.

It remains to be resolved why the T/V versus V curves show strong
curvature early in the experimental run. It is likely that this has to do
with the structure of the initial layers of colloid deposited, and this
will be the subject of future investigation as will the extension of the
study to other colloids.

From the practical viewpoint the results suggest that in considering
the use of ultrafiltration as a pretreatment process for waters containing
colloidal contaminants, it may be possible to optimise the addition of
flocculant so as to make the most advantageous change to particle zeta
potential and hence deposited cake resistance.

CONCLUSION

The present model for the unstirred ultrafiltration of colloids gives
predictions for final flux and deposited cake specific resistance which are
in encouraging agreement with experiment.

Most importantly, by incorporating the possibility of colloid floc-culation during an experimental run, the model can explain the low specific resistances obtained at low zeta potentials and can predict the effect of zeta potential on the unstirred ultrafiltration of colloids.

Acknowledgement

The authors wish to acknowledge support from the Water Research Foundation of Australia and from MEMTEC Ltd. R. McDonogh is a Commonwealth of Australia postgraduate scholar.

NOMENCLATURE

A_{mem}	– membrane area (m^2)	t	– time (s)
c	– colloid concentration ($kg\ m^{-3}$)	T	– temperature (K)
D	– colloid diffusivity ($m^2 s^{-1}$)	v	– concentration of particles per unit volume (m^{-3})
e	– electronic charge (C)		
J	– ultrafiltration flux ($\ell m^{-2} h^{-1}$)	V_t	– concentration of particles at time t
k	– Boltzman's constant (JK^{-1})		
ΔP	– transmembrane pressure (Nm^{-2})	Z_i	– valence
$P_{d\ell}$	– osmotic pressure due to double layer (Nm^{-2})	α^1	– permeability ($m\ K_s^{-1}$)
		ϵ	– voidage
r	– radius of particle (m)	η	– viscosity (Nsm^{-2})
R_o	– critical interparticle spacing (m)	K	– reciprocal Debye length (m^{-1})
$R_{floc,t}$	– interparticle spacing (floc diameter at t (m)	ρ	– density ($kg\ m^{-3}$)
		χ	– double layer potential

REFERENCES

1. W. F. Blatt, A. Dravid, A. S. Michaels, and L. Nelson, Solute polar-ization and cake formation in membrane ultrafiltration: Causes, consequences and control techniques in "Membrane Science and Technology," J. E. Flinn, ed., Plenum Press, New York, p.47 (1970).
2. M. C. Porter, Concentration polarization with membrane ultra-filtration, Ind.Eng.Chem.Prod.Res.Dev., 11:234 (1972).
3. J. D. Henry Jr., Cross-flow filtration, in: "Recent Development in Separation Science," Vol.2, CRC Press, Cleveland, p.205 (1972).
4. K. A. Kraus, Cross-flow filtration and axial filtration, paper presented at the 29th Annual Purdue Industrial Waste Conference, LaFayette, IN., May (1974).
5. R. E. Madsen, "Hyperfiltration and Ultrafiltration in Plate-and-frame Systems," Elsevier, New York (1977).
6. D. R. Trettin and M. R. Doshi, Ultrafiltration in an unstirred batch cell, Ind.Eng.Chem.Fundam., 19:189-194 (1980).
7. R. M. McDonogh, C. J. D. Fell, and A. G. Fane, Surface charge and permeability in the ultrafiltration of non-flocculating colloids, J.Membrane Sci., (accepted for publication).
8. M. von Smoluchowski, in: "Colloid Science," H. R. Kruyt, ed., Elsevier, Amsterdam, 1:278-279 (1952).
9. N. Fuchs, in: "Colloid Science," H. R. Kruyt, ed., Elsevier, Amsterdam, 1:283-285 (1952).
10. H. R. Kruyt and A. E. Arkel, in: "Colloid Science," H. R. Kruyt, ed., Elsevier, Amsterdam, p.295 (1952).
11. "Chemical Rubber Handbook," 60th Edn., CRC Press, p.F-51 (1980).

SEPARATION CHARACTERIZATION OF ULTRAFILTRATION MEMBRANES

Gun Trägårdh and Karin Ölund

University of Lund, Division of Food Engineering
P.O. Box 50, S-230 53
Alnarp, Sweden

At the division of Food Engineering at the University of Lund work is going on to find a standard method for characterizing UF-membranes. The aim is to find a method which will make it possible to compare membranes from different manufacturers. Today, this is very difficult since information about the membranes often is based on experiments using different test molecules at different operating conditions.

The work presented here is a comparison of separation qualities for two different types of membranes. They are both manufactured by the DDS company in Denmark and the reported cut off is 20,000 daltons for both membranes. One of them is a polysulfone (PS) membrane type GR 61 PP. The other one is a membrane made of cellulose acetate (CA) type 600.

The method used is based on the analysis of retentate and permeate from membrane filtration experiments. The analysis used are gel permeation chromatography (GPC) and laser light scattering measurements. The distribution of mole weights in permeate and retentate obtained by GPC-analysis, enables the calculation of the retention of different mole weights. Laser light scattering measurements complete the mole weight distribution results by giving the size of the particles calculated from diffusion coefficients by the Stoke-Einstein equation. In the experiments presented here a test solution of 0.5% Dextran (Pharmacia Fine Chemicals) with an average mole weight of 10,000 daltons has been used.

THE UF-EQUIPMENT

The test loop used for the characterization experiments contains a magnetic drive sealless gear pump, a thermocouple for temperature registration, a membrane module, a pressure gauge, a flow meter and a heat exchanger. The membrane module is shown in detail in Figure 1.

The inlet and the outlet of the module have been chamferred in order to avoid sharp edges which could disturb the flow profile at the membrane surface. For the same reason the geometry of the inlet is identical to that of the membrane flow channel for a distance of at least forty hydraulic diameters. The membrane area is 20×100 mm^2. Five different channel heights can be chosen between 1 and 5 mm by inserting PTFE-rods of very exact heights. The PTFE-rods go all the way through the inlet channel and the module.

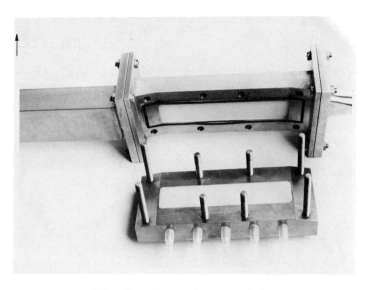

Fig. 1.   The membrane module.

THE EXPERIMENTAL METHOD OF THE MEMBRANE TESTS

Before starting a UF-experiment, distilled and membrane filtration water is circulated in the test loop until steady state is reached. This is done at the same operating conditions as will be used with the test solution afterwards. The circulation velocity is regulated by the pump and nitrogen gas is used to adjust the pressure independently of circulation velocities. The permeate which is lost from the test loop during an experiment is immediately replaced by permeate from a pressure vessel in connection to the circuit. When steady state is reached a test solution of Dextran is ultrafiltrated at five circulation velocities, from 5 m/s to 1 m/s. This corresponds to Reynolds numbers of 20,000 – 4,000 at a channel height of 2 mm. Permeate is collected at each circulation velocity at constant permeate fluxes.

ANALYSIS OF SAMPLES FROM THE ULTRAFILTRATION EXPERIMENTS

Gel Permeation Chromatography

The samples collected during ultrafiltration are analyzed by gel permeation chromatography, where particles are separated according to size and detected by a differential refractometer. In this way the mole weight distribution can be found and the retentions of different mole weights can be estimated.

Light Scattering Measurements

To find the correlation between mole weight and particle size for Dextran, five calibration fractions of known average mole weights supplied by Pharmacia Fine Chemicals, were measured by light scattering. In this work the light source was a Spectra Physics 164-06, 2 W argon ion laser. The spectrometer and the compact autocorrelator/microcomputer unit used belong to the Malvern system 4600, including soft ware for the system.

THEORIES

The observed retention $R_{obs} = \dfrac{C_b - C_p}{C_b}$, frequently used when describing membrane characteristics, does not correspond to the real conditions, since the concentration at the membrane surface is higher than that in the bulk due to concentration polarization. Another approach is to use the true retention[1].

$$R = \frac{C_m - C_p}{C_m} \tag{1}$$

The film model equation

$$\frac{C_m - C_p}{C_b - C_p} = \exp(J_v/k) \tag{2}$$

and the relation

$$k \propto U^a \tag{3}$$

give together

$$\ln\left(\frac{1 - R_{obs}}{R_{obs}}\right) = \ln\left(\frac{1-R}{R}\right) + const. \frac{J_v}{U^a} \tag{4}$$

From this relation the true retention for different mole weights can be determined indirectly by extrapolation to infinite circulation velocity.

To find the true retention versus circulation velocity without taking the mole weight distribution into account, the film model Equation (2) and the Deissler Equation (5) are used.

$$Sh = \frac{k\,D_h}{D} = A\,Re^a\,Sc^b \tag{5}$$

From the Deissler equation the mass transfer coefficient k is calculated for different circulation velocities. The concentration c_m at the membrane is then calculated from the film model equation and the true retention is found from Equation (1).

RESULTS AND COMMENTS

It was found that the PS-membrane and the CA-membrane had different separation qualities although the cut off reported by the manufacturer is 20,000 daltons for both membranes.

In Figure 2 chromatograms show the mole weight distribution for the test solution and for permeates from both membranes, collected at two different circulation velocities. At a velocity of 5 m/s, the permeate from a CA-membrane has a more narrow mole weight distribution than the permeate from a PS-membrane. This effect is lost at lower circulation velocities as shown for 1 m/s. It should also be noticed that the mole weight distributions in permeates are more affected by circulation velocities for CA-membranes than for PS-membranes.

The difference between the two membranes in observed retention as a function of mole weight and hydrodynamic radius is shown in Figure 3 and Figure 4. The average observed retention values for three experiments on each membrane are shown in Table 1. The standard deviations for these experiments are about 5%. Another point is that there seemed to be more variations between different pieces of PS-membranes, than for CA-membranes.

It would be interesting to try a correction for concentration polarization where the true retention R_{true} is found. Therefore, we assume that the film model equation and the Deissler equation are valid in our concentration ranges. The constants used in the Deissler Equation (5) are a = 0.875 and b = 0.25. The true retention as a function of circulation velocity is presented in Figure 5 for the PS-membrane and in Figure 6 for the CA-membrane.

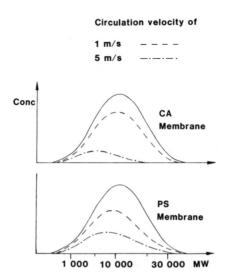

Fig. 2. Mole weight distribution of test solution and permeates at circulation velocities of 1 m/s and 5 m/s.

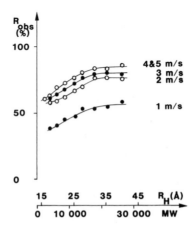

Fig. 3. Observed retention versus mole weight and hydrodynamic radius for a polysulfone membrane.

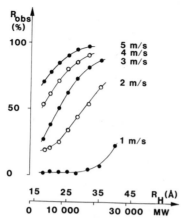

Fig. 4. Observed retention versus mole weight and hydrodynamic radius for a cellulose acetate membrane.

Table 1. R_{obs} at Different Circulation Velocities and Mole Weights

MW	PS membrane R_{obs} at different circulation velocities			CA membrane R_{obs} at different circulation velocities		
	1 m/s	3 m/s	5 m/s	1 m/s	3 m/s	5 m/s
5 000	39	58	63		36	72
10 000	44	70	73		59	88
15 000	49	76	78		75	94
20 000	51	82	83		86	96

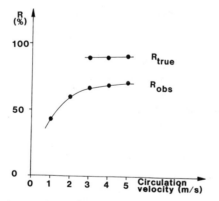

Fig. 5. True retention and observed retention versus circulation velocity for a polysulfone membrane.

It may be discussed to what extent the film model Equation (2) and the Deissler Equation (5) are valid for this type of calculation. The physical properties of the test solution are varying with the solute concentration

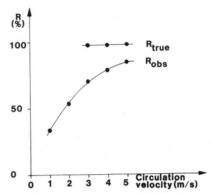

Fig. 6. True retention and observed retention versus circulation velocity
for a cellulose acetate membrane.

in the polarized layer. This fact affects especially the viscosity and the
diffusivity and is not considered in the equations. However, working with
a low concentration of the solute the error is diminishing. It has to be
checked especially that the calculated concentration at the membrane
surface (C_m) is below the gel concentration if the type of solute used has
gelling characteristics.

NOMENCLATURE

A, a, b – constants

C_b – solute concentration in bulk (kmol/m³)

C_m – solute concentration at membrane surface (kmol/m³)

C_p – solute concentration in permeate (kmol/m³)

D – diffusion coefficient (m²/s)

D_h – hydraulic diameter (m)

J_v – volume flux (m³/m², s)

k – mass transfer coefficient (m/s)

R – true retention $1 - \dfrac{C_p}{C_m}$

R_H – hydrodynamic radius (m)

R_{obs} – observed retention $1 - \dfrac{C_p}{C_b}$

Re – Reynolds number

Sc – Schmidt number

Sh – Sherwood number

U – circulation velocity (m/s)

REFERENCE

1. S. Nakao and S. Kimura, Analysis of solutes rejection in ultrafiltration, J.Chem.Eng.Japan, 14:32 (1981).

INFLUENCE OF SURFACTANTS ON THE TRANSPORT BEHAVIOR

OF ELECTROLYTES THROUGH SYNTHETIC MEMBRANES

D. Laslop and E. Staude

Institut für Technische Chemie
Universität – Gesamthochschule Essen
Postfach 103 764, D-4300 Essen 1

INTRODUCTION

Soon after the hyperfiltration of saline feed became a feasible process for the production of fresh water, in addition to the investigation of the permeation of salts through membranes also the behavior of surfactants in this separation process likewise claimed basic interest. The goal of investigation of these substances can be split into two parts. On one hand the degree of rejection of the surfactants using various types of membranes is the object of the experiments. On the other hand the mutual influence between surfactants and other feed ingredients on the permeation through the membrane was placed in the forefront.

Using cellulose acetate (CA) membranes the behavior of nonionic and anionic surface active agents was examined[1, 2]. In both cases the rejection of the surfactants was very high, and the volume flow decreased with increasing concentration. As with the solutions of polyoxyethylated nonionic detergents the presence of sodium chloride in the feed did not alter remarkably the transport behavior[1]. More details can be derived from the work dealing with the inherent characteristics of the surface active substances. Thus, the critical micelle concentration (CMC) of the surfactant in the feed plays a significant role in as much as a liquid membrane above the CMC is formed at the feed solution/membrane interface. In this case the transport resistance is increased[3]. For these experiments modified Loeb-type CA membranes were used, the surfactant solutions were prepared from polyoxyethylene-nonylphenoles and poly(vinylalkyl) ethers. Increasing hydrophilicity of the various surfactants increases the permselectivity of the membrane system. Modified Loeb-type membranes annealed at different temperatures were utilized for the investigation of transport phenomena of anionic, cationic, and nonionic surfactants. The influence of the CMC on the behavior of anionic and cationic sufactants was negligible, whereas for the nonionic surfactants a considerable increase in the rejection above the CMC was observed[4]. The concentration of the surfactants was in the range of 10^{-5} to 10^{-3} mol/l.

The influence of surfactant/electrolyte system on the transport behavior of CA membranes was studied using plating rinse water containing chromium salts and tetraethylammonium perfluoroctanolsulfonate (FT 248), especially developed for chromic plating bath[5]. With increasing surfactant concentration and also rising electrolyte concentration a weighty drop of both volume flow and electrolyte rejection was observed. This drop indicated the concentration where the CMC was reached.

From these results it can be derived that both structure and charge of surface active agents are responsible for rejection and water permeability. In those cases where in the concentration polarization layer the CMC is reached an additional effect is caused, which depends on the nature of the surfactant. This layer or even "liquid" membrane adjacent to the membrane determines the volume flow by changing the water binding capacity[3]. Naturally this must have a feedback on salt retention, which is enhanced by increasing water binding. In addition, there is a concentration dependent interaction between surfactants and electrolytes which influences the volume flow. But this was measured using one special type of surfactant only.

The following results are presented which were obtained using solutions of anionic, cationic, and nonionic sufactants and chlorides of mono- and divalent electrolytes.

EXPERIMENTALS

The experiments were carried out in a closed loop hyperfiltration apparatus which was described previously in detail[6]. The working pressure was 40.7 ± 0.4 bars and the temperature 293K ± 0.1. Two types of CA membranes (trade name Nadir), manufactured and kindly supplied by Kalle, Niederlassung der Hoechst AG, Wiesbaden, were used. Their characteristics are (measured by using 0.1M NaCl at 40 bars): CA-10 R = 29%, J = 2100 $1/(m^2 d)$ and CA-93 R = 82%, J = 920 $1/(m^2 d)$.

For preparing the solution the following detergents were used: sodium dodecylbenzenesulfonate, SDS (trade name Texapon L 100, Henkel GmbH, Dusseldorf); sodium 1-octanesulfonate, OSS, and benzalkonium chloride, BAC (purchased from Fluka GmbH, Neu Ulm); tetraethylammonium perfluoroctanolsulfonate, FT 248) (purchased from M & T GmbH, Stuttgart); and polyoxyethylene-polyoxypropylene block copolymers, PL 64 (trade name Pluronic PL 64, distributed by Serva GmbH, Heidelberg). As electrolytes for these solutions $NiCl_2$ and NaCl were used in two different concentrations.

The volume flow was measured gravimetrically, and the concentrations of the surfactants were determined by colorimetry using methylene blue and methyl red as indicators. The nonionic surfactant concentration was measured by the speed of phase separation of a surfactant containing water/ethyl methyl ketone solution using eosin as a contrast medium. The concentration of chloride was measured by potentiometry.

EXPERIMENTAL RESULTS

Anionic Surfactants

Using SDS it can be derived clearly from Figure 1 that the volume flow is strongly influenced by the surfactant and the electrolyte concentration, respectively. The valency of the ions is dominant. In this Figure the relative volume flow J_i/J_o (J_i = solution volume flow, J_o = pure water volume flow) using the CA-93 membranes are drawn. The same is valid for the CA-10 membranes. In both cases the volume flow drops sharply at a distinct surfactant concentration. But in the latter one it can be noted that in consequence of the high permeability through the membrane the surfactant gel layer is built up at lower substrate concentration, and it

is more intense in its efficiency as a secondary membrane. This is deriv-
able from the J_i/J_o values, which are smaller than those for the CA-93
membrane. The same result was found in the absence of any electrolyte.

This strongly marked decrease of the volume flow coincides with a
shift of the chloride rejection (Figure 2). For high nickel chloride
concentrations using the better desalting membranes the chloride rejection
decreases, whereas it increases with those membranes exhibiting a low
rejection. For low nickel chloride concentrations one could assume the
same behavior at higher surfactant concentration. But this could not be
realized experimentally because of the fouling layer on the membrane built
up by Ni/surfactant complexes. In the former case this layer more drastically reduces the water transport than the electrolyte transport, consequently, the rejection of chloride becomes smaller. In the case of the
low rejection membrane the fouling layer resistance rises. This layer acts
as the real membrane and is more dense than the CA membrane.

The rejection of the surfactant as a function of its concentration is
shown in Figure 3. As may be seen from the Figure the rejection is always

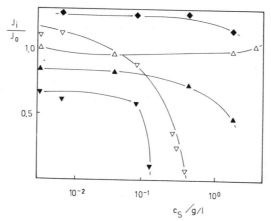

Fig. 1.   Relative volume flow (J_i/J_o) as function of SDS concentration c_S.
Membrane CA-93, Feed: $NiCl_2$ ▽, $NaCl$ △, open symbols c(electrolyte)
0.005M, filled symbols c(electrolyte) 0.1M, without electrolyte ◆.

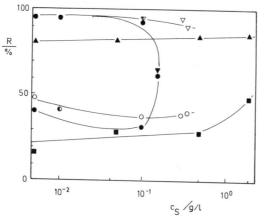

Fig. 2.   Chloride rejection R as function of SDS concentration c_S. Open
symbols c(electrolyte) 0.005M, filled symbols c(electrolyte) 0.1M;
$NiCl_2$ ○ CA-10, ▽ CA-93; $NaCl$ □ CA-10, △ CA-93.

less when using solutions containing electrolytes compared with those free from any electrolyte. It increases with increasing solute concentration and reaches its highest value at micelle formation which occurs at a lower surfactant concentration with nickel chloride in the feed than with sodium chloride. Opposite to the sodium chloride/SDS system in the nickel chloride/SDS system Ni/SDS salt precipitates in the concentration polarization layer. There is a transformation of the dynamic gel layer into a partial crystalline inhomogeneous scaling layer, which allows permeation by crystal defects. This can be seen from the chloride rejection and the SDS rejection. According to Figure 3 there is a negligible rejection of SDS at low concentration when the sodium chloride content is high, whereas at low sodium concentration the rejection is high and increases with increasing SDS concentration. The reason is the increasing solubility of the surfactant at a high electrolyte/surfactant ratio, this effect does not occur at low ratios.

In contrast to SDS the fluorosurfactant FT 248 is considerably more hydrophobic, thus the absorption onto the membrane polymer is much more pronounced. Also the influence of the sodium chloride on the volume flow

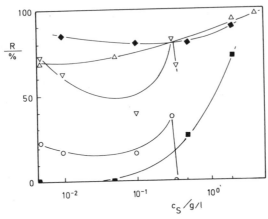

Fig. 3. SDS rejection R as function of the concentration c_S $NiCl_2$: ○ CA-10, ▽ CA-93, c: 0.005M; NaCl: ■ CA-10, c: 0.1M, △ CA-93, c: 0.005M, without electrolyte ◆ CA-93.

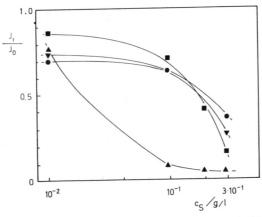

Fig. 4. Relative volume flow J_i/J_0 as function of FT 248 concentration. Electrolyte concentration 0.1M; $NiCl_2$ ● CA-10, ▼ CA-93, NaCl ■ CA-10, ▲ CA-93.

Table 1. Comparison of Anionic Surfactant Solutions Using CA-10 Membranes

Concentration of surfactant	SDS 0.1M NiCl$_2$			FT 248 0.1M NaCl		
	$\frac{J_i}{J_o}$	R_{Cl^-} %	$J_i c_{perm}$ M/(m^{2}d)	$\frac{J_i}{J_o}$	R_{Cl^-} %	$J_i c_{perm}$ M/(m^{2}d)
0.005	0.649	41	226	-	-	-
0.01	0.586	41	245	0.860	20	195
0.1	0.506	32	233	0.708	11	168
0.15	0.114	62	26	-	-	-
0.2	-	-	-	0.415	13	108
0.3	-	-	-	0.126	38	31

at the high concentration is changed (Figure 4). In contrast to the SDS containing solution the solubility of the two electrolytes is reversed. At high sodium chloride concentration the fluorosurfactant precipitates before the CMC is reached. In this system the sodium cation plays the same role as the nickel cation in the SDS containing solution. This can be seen from Table 1. This Table also contains data for the chloride flow through the membrane, calculated by $J_i c_{permeate}$ (M/(m^{2}d)). The values of the chloride rejection for both the highly and the poorly concentrated sodium chloride solutions tend to converge, but this effect is not as strongly pronounced as with the SDS system. The rejection of the surfactant shows the normal behavior, it rises with increasing concentration.

Cationic Surfactant

In this cationic surfactant the hydrophilic nitrogen atom is tetrahedrally surrounded by the hydrophobic parts of the molecule. The volume flow is moderately reduced at higher surfactant concentration. Analogous to this the chloride rejection increases with increasing surfactant concentrations. The salts in the solution influence the transport of the surfactant through the membrane, sodium chloride is more effective than nickel chloride. Using the low desalting membrane a negative rejection of BAC is observed indicating that there seems to be an interaction of substrate and membrane polymer. This is perhaps comparable with the dissolving of distinct organic molecules in the membrane phase[7].

Nonionic Surfactant

The effect of the pluronic PL 64 on the permeation of water or chloride through the membrane is extremely small. The gel layer on the membrane seems actually to act as liquid membrane, its high hydrophilicity causes no pronounced resistance for hydrophilic substances.

DISCUSSION

A comprehensive representation of the various types of behavior of the electrolyte/surfactant systems can be received by comparing the corresponding values as a function of the membrane characteristics.

Figure 5 shows the experimental values of the surfactant rejection using sodium chloride as electrolyte for the low desalting membranes. The

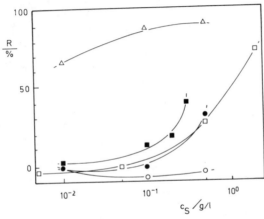

Fig. 5. Rejection of surfactants as function of surfactant concentration in the presence of 0.1M NaCl. Membrane CA-10; SDS □, OSS o, FT 248 ■, BAC ●, PS 64 △.

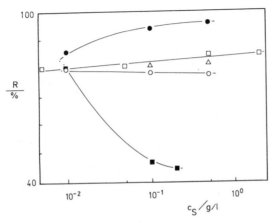

Fig. 6. Rejection of chloride as function of surfactant concentration in the presence of 0.1M NaCl. Membrane CA-93; SDS □, OSS o, FT 248 ■, BAC ●, PL 64 △.

results are shown for 0,1 M NaCl solution. At low surfactant concentrations the rejection is poor, and OSS as a substance with very small surface active qualities permeates very well the membrane. Rising concentration increases the rejection except in the case of nonionic surfactant. The same was observed using the membrane annealed at a higher temperature, but at the same sodium chloride concentration as before. As to the chloride rejection the summary is given in Figure 6. Without remarkable influence on the permeability are nonionic and those anionic surfactants which build gel layer-like micelles. As pointed out, the strong hydrophobic/hydrophilic FT 248 does not belong to these substances. On the other hand the gel layer of the cationic surfactant intensifies the chloride rejection on account of its fixed charge.

When solutions containing electrolytes and surface active substances are hyperfiltrated three different effects have to be considered which occur in the concentration polarization layer. They correspond to the various regions in the phase diagram of these surfactants. In the first case there are separate surfactant molecules which facilitate the permeation of both the water and the surfactant. The behavior is changed when electrolytes are added to the surfactant solution. As may be derived from the results the volume flow decreases particularly when the open membrane type is used. The normal influence of the surfactant on the volume flow which is the consequence of the surface tension effect at the membrane/ liquid boundary layer is now possibly covered over by the change of the water structure caused by the electrolytes. A competitive transport within the concentration polarization layer occurs, which can result in a facilitated transport of the smaller surfactants through the membrane at low concentration. This surfactant effect becomes smaller with increasing surfactant concentration. Using higher rejection membranes the surfactant as well as the electrolyte concentrations in the concentration polarization layer increases and also the back diffusion rises.

The second effect arises when the CMC is reached which is noticeable by increasing surfactant rejection; contrary to this the volume flow as well as the chloride rejection are affected only slightly. The transition to the third situation is continuous, the transport resistance is now exerted by the completely developed gel layer (liquid membrane) or by the cristalline fouling layer adjacent to the membrane. The volume flow is greatly hampered, additionally, the values for the rejection of surfactants as well as for the electrolytes increase because of decreasing volume flow. This can be seen from the columns in Table 1, indicated as $J_i c_{perm}$.

Individual effects of the structure of the surfactants may influence the above mentioned effects. Thus BAC may interact with the membrane polymer because of its positively fixed charge. But, the steric hindrance plays a role, the charge is shielded and its rejection is comparable low as it was found for the CA-10 membrane. According to this, the rejection of the chloride ion is high, and the volume flow is low.

For technical application the results indicate that the rejection of surfactants can be increased in the presence of electrolytes when the solutions concerned are hyperfiltrated, for the electrolytes cause the wellknown decrease of the CMC.

REFERENCES

1. S. Sourirajan and A. F. Sirianni, Ind.Eng.Chem.Prod.Res.Develop, 5:30 (1966).
2. H. Hindin, P. J. Bennett, and S. S. Narayanan, Water Sewage Works, 116:466 (1969).
3. R. E. Kesting, W. J. Subcasky, and J. D. Paton, J.Colloid Interface Sci., 28:156 (1968).
4. C. Kamizawa and S. Ishizaka, Bull.Chem.Soc.Jpn., 45:2967 (1972).
5. E. Staude, W. Assenmacher, and H. Hoffmann, Desalination, in press.
6. E. Staude and W. Assenmacher, Desalination, 49:215 (1984).
7. W. Pusch, H. G. Burghoff, and E. Staude, Proc.Int.Symp.Fresh Water Sea 5th, 4:143 (1976).

EFFECTS OF ANTIFOAMS ON CROSS-FLOW

FILTRATION OF MICROBIAL SUSPENSIONS

K. H. Kroner, W. Hummel, J. Völkel and M. -R. Kula

Gesellschaft für Biotechnologische Forschung mbH
Mascheroder Weg 1
D-3300 Braunschweig, F.R.G.

INTRODUCTION

Membrane processes such as ultrafiltration and cross-flow filtration will be of increasing importance in downstream processing of biologically active proteins from microbial sources[1], especially under conditions of containment in manufacturing, which is expected from extensive use of recombinant-DNA technology. Recently we began investigations into cross-flow filtration for the recovery of enzymes[2].

At present the major problem in applying this technique for harvesting and separation of microbial cells from spent media is the rapid fouling of membranes during operation. Fouling is due to membrane polarization and formation of sub-layers of particles and other compounds present in fermentation broth. One of the identified agents contributing to the phenomenon is the presence of antifoams, which are used at variable levels in fermentation, to reduce foam problems. Antifoams are generally hydrophobic materials and tend to concentrate on interfaces. They adsorb or coat on the membrane surface, thereby reducing the flux rates and often also the retention of macromolecules, e.g. proteins will be increased. Some basic studies of the effects of antifoams during cross-flow filtration of some influencing parameters will be presented in this paper.

ANTIFOAM CHARACTERIZATION AND SELECTION

Antifoams are chemicals which reduce the propensity of a solution to foam. For aqueous based fermentation fluids, the antifoam is generally a slightly water soluble or water insoluble surface active agent. It functions by reducing electrical double layer repulsion or by eliminating the surface elasticity of the foam.. The main effect is that the antifoams go to the air-water interface and reduce the stability of any foam which forms[3,4]. In order to be used in a fermentation process, the antifoam must not inhibit organism growth or product formation, and should not interfere with measurement techniques, e.g. pH or dissolved oxygen probes. Another point which is quite important but sometimes ignored is that the antifoam should not interfere with subsequent purification procedures.

In modern bacterial fermentations the most commonly used antifoams are polypropylene glycols of molecular weight 2000, copolymers of polypropylene

223

and polyethylene glycols with different composition and silicon oil emulsions, based on methyl siloxanes. From studies we have carried out in our laboratory during cultivation of Brevibacterium spec. in a corn steep medium, we selected three antifoams which do not interfere with bacterial growth and product formation and which were used for the following studies:

1. PPG 2000 (polypropylene glycol, CWH Hüls)
2. Ucolub N115 (polypropylene – polyethylene glycol, Brenntag)
3. Silicon oil emulsion (Tegosipon T52, Goldschmidt)

EFFECTS OBSERVED ON CROSS-FLOW FILTRATION OF MICROBIAL SUSPENSIONS

The main effect of antifoams is the reduction of the transmembrane flux during cross-flow filtration of fermentation broths due to membrane fouling, both with microporous as well as ultrafiltration membranes. Figure 1 shows for example, the influence of about 0.1% of the Ucolub N115 antifoam on performance during filtration of Escherichia coli fermentation broth. The filtration was performed in a tube type cross-flow module with a 0.3 μm microporous polypropylene membrane, from Membrana. The flux rates are plotted against time during concentration of the broth. One can see that the flux rate with antifoam is reduced to nearly 60% of the value of the control experiment. Also, the initial flux decrease was even more drastic and took place within a few minutes. This example is not singular but reflects our general experience with the antifoam effect, whereby flux decrease varies from case to case. The degree of fouling is a function of the type and amount of antifoam, the type of membrane, the organism and other components present in the broth. Generally flux reduction up to 50% are observed in the presence of moderate antifoam concentrations. This can be seen from Table 1, where a Brevibacterium spec. culture broth with corn steep was filtered in a stirred cell, using various membranes and the antifoam mentioned. The results show a great variety for all membranes and antifoams used. The polyglycols show the greatest effect, whereas the silicone oil emulsion was not so active as a fouling agent, which can be attributed to the fact, that most of the silicone oil will be attached to the cells.

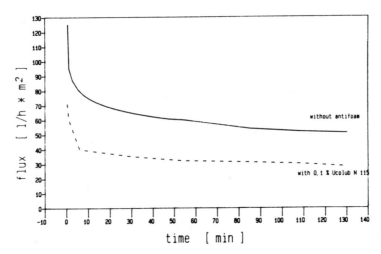

Fig. 1. Influence of antifoam on cross-flow filtration performance. Escherichia coli fermentation broth; tube type membrane: PP 0.3 μm (Membrana), V = 2 m/s, Δρ = 1 bar, temperature: 22°C.

Table 1. Influence of Antifoams on Cross-flow Performance Filtering <u>Brevibacterium Spec.</u> Broth - Different Types of Membranes in a Stirred Cell*[1]

membrane	initial flux 1/h m^2	remaining flux *[2]					
		PPG 2000		Ucolub N 115		Silicon	
		0.01 %	0.1 %.	0.01 %	0.1 %	0.01 %	0.1 %
HVLP 0.45 µm (Millipore)	72.4	60	53	56	50	93	89
PP 0.2 µm (Membrana)	51.9	83	73	92	84	94	92
Membrafil 0.22 µm (Nucleopore)	55.2	85	73	90	84	78	75
PA 0.45 µm (Sartorius)	41.9	82	67	82	72	90	88
PC 0.4 µm (Nucleopore)	23.3	-	63	-	59	-	85
PSVP (10^{6}d) (Millipore)	17.9	90	75	-	-	-	84

*[2] % of initial rates (30 min values)
*[1] $\Delta \rho$ = 0.5 bar, stirrer tip speed 2 m/s, 45 cm^2
medium: corn steep; phenylalanine, phosphate

The second interfering effect of antifoams is to increase the retention values of proteins. It is known, that proteins can be retained by microporous membranes due to fouling phenomena[2], and antifoams seem to aggravate this problem. Figure 2 shows four examples of enzymes, three extracellular and one intracellular (solubilized by disruption of the cells), which were markedly changed in their retention by the addition of polypropylene glycol during diafiltration in a tube type polypropylene module with 0.3 µm cut-off, filtering real fermentation broths, with recycling of the filtrate stream. To avoid time effects each point was determined after 30 minutes of diafiltration. The three extracellular enzymes, a protease, an amylase and pullulanase, generally appear to be less affected compared to the intracellular α-glucosidase, but in the case of the latter, the initial retention was even higher because of the presence of disrupted cells. Apart from the possibility of more specific interaction of single enzymes with the antifoam[5], there seems to be a increasing retention with increasing molecular weight of the proteins (protease 30 000 dalton - pullulanase 150 000 dalton). This indicates that the effective pore size comes down to that of ultrafiltration membranes with cut-offs between 50 - 150000 dalton ($\sim$10 nm). Even small amounts of about 0.1% of PPG are enough to change the retention values in the range of approximately 20-30%.

SOME BASIC STUDIES OF INFLUENCE PARAMETERS

Hydrodynamic Conditions

To understand further, the interaction of antifoams on membranes, we carried out experiments with an antifoam alone to avoid the complex situation in real broths. The antifoam chosen was PPG 2000, which is a viscous liquid, forming droplet suspensions or emulsions at low concentrations.

Figure 3 shows the influence of the liquid velocity on the flux rate as well as the retention of PPG. This experiment was carried out in a

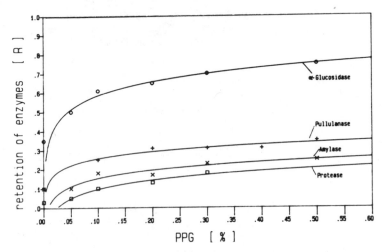

Fig. 2. Influence of antifoam on the retention of enzymes. Fermentation
broths: Bacillus licheniformis (protease), Bacillus subtilis
(amylase), Klebsiella pneumoniae (pullulanase); disrupted
Saccharomyces cerevisiae α-glucosidase tube type membrane:
PP 0.3 µm (Membrana). V = 2 m/s, Δρ = 0.9 bar, T = 22°C.

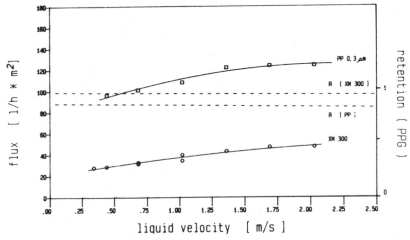

Fig. 3. Influence of liquid velocity of flux and antifoam retention.
Spiral flow cell (SFC); membranes: XM-300 (Amicon), PP 0.2-0.3 µm
(Membrana); Δρ = 1 bar, T = 25 °C, 1% PPG 2000 (Concentrations of
antifoam are measured by turbimetry).

spiral flow cell (SFC) with two different types of membranes. The first
was a microporous polypropylene membrane with an average pore size of about
0.2 - 0.3 µm, and the second was an ultrafiltration membrane XM-300, from
Amicon, with about 300 000 dalton cut-off. One can observe a slight
increase of the flux rates of both membranes with increasing velocity,
whereas there is only minor change in retention in both cases. The slopes
of a standard log-log plot for both membranes gave similar values of 0.3
and 0.35, respectively, which reflect mass transfer under laminar flow
conditions, when considering the gel polarization model, although the
nominal Reynold's numbers are between 2600 and 12000 (Calculated for
water). The action of shear on the polarization layer seems to be of the
same order for both membranes, but the absolute values of the flux rates

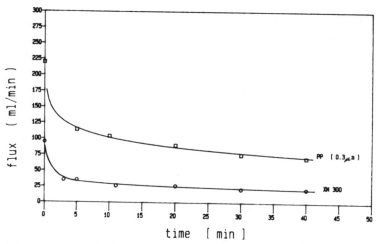

Fig. 4. Time effects of flux decay with antifoam. SFC; membranes: XM-300
(Amicon), PP 0.2-0.3 μm (Membrana); V = 2 m/s, Δρ = 1 bar,
T = 25°C, 0.1% PPG 2000.

are quite different. For the more open membrane one can assume that there
is a superimposed penetration of the pores by the antifoam with some
breakthrough of antifoam, dependant on the differential pressure and the
shear velocity. The more dense UF-membrane should act as a total barrier
for the dispersal of antifoam and therefore polarization should take place
immediately. If this assumption holds there should be a difference in the
time behavior of the membranes. This was illustrated generally in Figure
4, where the actual flux rate is plotted against time, during a diafiltration procedure with 0.1% PPG in the spiral flow cell. The UF-membrane
decreases in flux within a few minutes and then remains nearly constant,
whereas the flux of the polypropylene membrane drops over a longer time
period. This variation in time behavior, which could also be further
changed under different experimental conditions, is one of the major
drawbacks when studying membrane antifoam interactions. In general, we
have used only 30 minutes values for comparison during these experiments.

The influence of differential pressure on flux and antifoam retention
can be seen from Figure 5. An open microporous membrane, type HVLP, with
average pore size of 0.45 μm (Millipore) and the XM-300 membrane were
compared, for diafiltration in the SFC in the presence of 1% PPG at a
liquid velocity of about 2 m/s. For the XM-300 membrane, a curve found
typically with gel-polarization is observed, with a slight increase in
retention. For the more open membrane, a quite drastic change in flux
rates with differential pressure can be seen, with some incosistencies in
the course of the curve between 0.2 - 0.5 bars which may be due to
penetration of the antifoam coat into the pores. The difference between
the asymptotic flux rate values of the membranes is very large and besides
the degree of coating, the parameter to explain these differences seems to
be the mean average pore size. This assumption is confirmed by our
experiments with different membranes, as can be seen from Figure 6. Eleven
different membranes in two different experimental set-ups were examined.
The flux decay is clearly a function of the average pore size. With
increasing pore size the retention of PPG decreases with a reduction of the
concomitant loss in flux. From the variation of flux for membranes of
equal pore size it appears that membrane material is also of importance. A
critical pore size around 10-30 nm exists below which total retention is
observed.

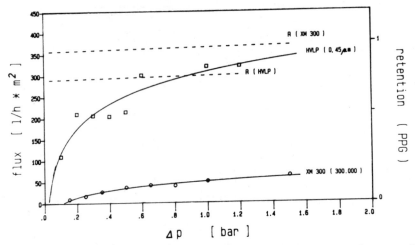

Fig. 5. Flux as function of differential pressure. SFC; membranes: XM-300
(Amicon), HVLP 0.45 µm (Millipore); V = 2 m/s, T = 25°C, 1% PPG
2000.

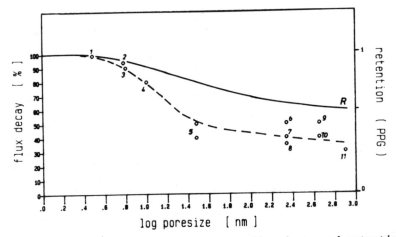

Fig. 6. Influence of average pore size on flux decay and retention of PPG.
SFC and stirred cell; V = 1 m/s, T = 25°C, Δρ = 1 bar, 0.1% PPG;
membranes: XM-50 (Amicon) (1), XM-100 (Amicon) (2), PTHK (Milli-
pore) (3), XM-300 (Amicon) (4), PSVP (Millipore) (5), PP 0.2 µm
(Membrana) (6), Membrafil 0.22 µm (Nucleopore) (7), PC 0.2 µm
(Nucleopore) (8), HVLP 0.45 µm (Millipore) (9), PA 0.4 µm
(Sartorius) (10), CN 0.8 µm (Sartorius) (11). (Pore sizes of
the UF-membranes were estimated according to the method of
Sarbolouki[6]).

Physico-chemical Parameters

Besides the hydrodynamic conditions, environmental parameters such as
antifoam concentrations, pH, ionic strength and temperature determine
further the rate of antifoam-caused fouling. The predominant effect is
clearly due to the amount of antifoam used.

Figure 7 shows the flux decay and retention of PPG as a function of
the concentration of PPG for a membrane of type PSVP, with 1,000,000 dalton
cut-off, from Millipore. This example reflects the generally observed

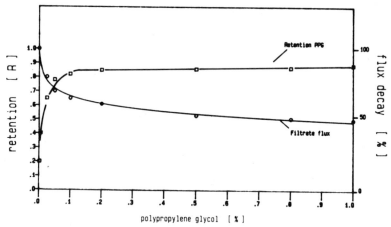

Fig. 7. Influence of antifoam concentration on flux decay and antifoam
retention. SFC; membrane: PSVP 10^6 Dalton (Millipore); V = 1 m/s,
$\Delta\rho$ = 1 bar, T = 26°C.

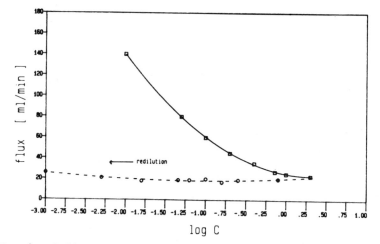

Fig. 8. Influence of antifoam concentration on flux - redilution
experiment. SFC; membrane: PP 0.2-0.3 μm (Membrana);
V = 2 m/s, $\Delta\rho$ = 1 bar, T = 25°C (PPG 2000 as antifoam).

concentration effect. With increasing concentration the retention of PPG
also increases and the flux comes down asymptotically to about 50% of
initial value at 1% of PPG. Most of the change in membrane performance is
observed at low concentrations of around 0.1% PPG and any further changes
are only slight (time effects). Relevant concentrations for PPG in fermentation broths of complex media are in the range of up to 1%, therefore
strong effects on membrane performance are to be expected.

Figure 8 shows the results which were obtained during a redilution
experiment, which demonstrates that antifoam caused fouling of a microporous membrane, here polypropylene with 0.2 - 0.3 μm, is concentration
dependant in forward direction with accumulation of PPG, but not reversible
in principle after redilution. This experiment was carried out in a
SFC-module with approximately 2 m/s liquid velocity at 1 bar differential
pressure. An increase in flux was observed only, when the critical solubility concentration of the PPG was reached (approximately 0.005% at 25°C),
due to a washout effect.

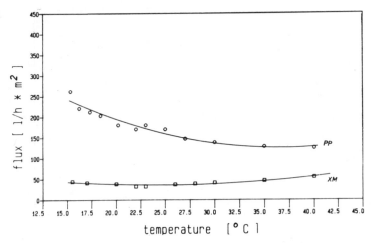

Fig. 9.   Influence of temperature on flux. SFC; membranes: XM-300
          (Amicon), PP 0.2-0.3 μm (Membrana); V = 1.5 m/s, Δρ = 1 bar;
          1% PPG 2000.

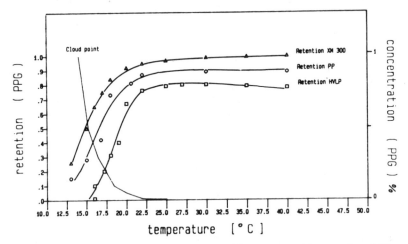

Fig. 10. Influence of temperature on antifoam retention and antifoam
          solubility (indicated by the "cloud point"). SFC; membranes:
          XM-300 (Amicon), PP 0.2-0.3 μm (Membrana), HVLP 0.45 μm
          (Millipore); V = 1 m/s, Δρ = 1 bar; 1% PPG 2000.

        Another important factor dealing with the performance of cross-flow
filtration is the temperature during filtration, due to changes in the
critical solubility of the antifoam in water. Figure 9 shows the flux
rates for a XM-300 membrane and a microporous polypropylene membrane as a
function of the temperature. For the UF-membrane it could be expected that
the flux rate goes through a minimum, due to solubility effects in the cold
(<20°C) and higher diffusion rates with increasing temperature. But for
the microporous membrane this behavior is not observed. The filtration
rate decreases with increasing temperature. This was the striking effect,
compared to normal filtration behavior. It seems to be that the temperature effect is counteracted by the superimposed time effect. It may be
that at higher temperatures the penetration of antifoam into the pore
structure becomes easier, because of decreasing viscosity, for example, and
therefore clogging will be accelerated. This striking temperature effect
can be observed also with fermentation broths. For example, we have seen

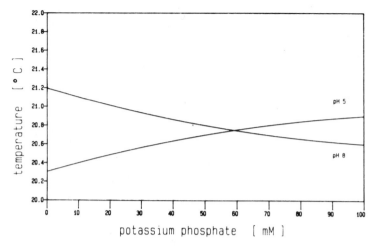

Fig. 11.

that the filtration rate of a broth of a Bacillus licheniformis cultivation
will be improved by about a factor of 2 by lowering the temperature down to
5°C[5].

Figure 10 shows the relation between retention and the critical
solubility of the PPG in function of the temperature for three different
membranes the data obtained from the experiments in the SFC. There is a
good correlation between retention values and the critical solubility at a
given temperature, which indicates that mainly the disperse form of the
antifoam will be retained. From this experiment one can also derive some
assumptions about the mechanism of antifoam fouling: The finely dispers
antifoam droplets will be transported by convective flux to the surface of
the membrane, where they coalesce and spread over the membrane, with
similar mechanisms acting at liquid-gas interfaces. Adsorption then takes
place with time by removing water, but this should depend on membrane
material, the type of antifoam and other factors. The thickness of the
layer will then be controlled by the hydrodynamic conditions. In our
opinion this outline is more realistic for dense membranes with smooth
surfaces (UF-membranes); for more open membranes (microporous types), one
should consider, that spreading could be hindered by the roughness and
structure of the surface. On the other hand, coalescence takes place,
locally or inside the pores and penetration of the pore structures should
be expected to depend on the hydrodynamic conditions, which, in our
opinion, explains the time effects observed and the lower retention compared to UF-membranes.

The influence of other environmental parameters such as pH or ionic
strength, may have further effects on antifoam fouling, due to changes in
antifoam solubility, as well as on adsorption and interaction kinetics
between membranes, antifoams and other media components, such as proteins.

The critical solubility of PPG, for example, is a function of pH and
ionic strength, with increasing ionic strength at pH 8 the solubility
decreases, as indicated by the decrease of the "cloud point". For pH 5 a
reverse effect was observed. Due to changes in retention of antifoam this
will lead to changes in the flux rates, but will not be discussed here in
more detail. For the experimental trials which were performed here we used
a moderate ionic strength of 10 mM phosphate buffer with a pH of about 6.5.

CONCLUDING REMARKS

The presence of antifoam in fermentation broths interferes with membrane processes. These studies give a rough indication of the interactions within membrane processes in the downstream processing of enzymes. The situation is often more complex in real fermentation broths, since the spent fermentation media contain other hydrophobic or surface active substances e.g. lipids from the feedstocks or the organism themselves, and high membrane fouling rates will result from a "cummulative effect". More basic research is needed, for example into the characteristics of broths with respect to their fouling rates on membranes.

Not withstanding the disadvantage of fouling during operation, membrane processes offer the chance of complete removal of such substances from the filtrates prior to subsequent purification steps, due to the high retention of "lipid" substances.

Some careful thought should be given to the experimental procedures to study the antifoam effects, since the experimental set-up can influence the resultant data, sometimes drastically. For example, the ratio of installed membrane area to total broth volume, the material properties of machine parts and total wetted area as well as the form of addition of the anti-foams have an important effect on the reproducibility of results, especially when comparing laboratory experiments with pilot scale or process scale data. A membrane area of total volume ratio of 1-2 m^2/100 l generally seems to be well suited in our opinion. Cleaning procedures as well as improvements to diminish the antifoam effects, such as the use of small amounts of detergents or prior removal of lipids by adsorption on activated carbon or silica gel, for example, show some advantages too, but care should be given with respect to the biological active substances, which are processed, as well as secondary effects on subsequent operations.

Acknowledgements

We should like to thank Membrana Corp. (Enka), Wuppertal, for their gifts of some membrane materials.

REFERENCES

1. A. S. Michaels, Membranes in biotechnology - current status and future prospects, in: "4th Symp. on Synthetic Membranes in Sci. and Technology," Tübingen (1983).
2. K. H. Kroner, H. Schütte, H. Hustedt, and M. -R. Kula, Cross-flow filtration in the downstream processing of enzymes, Process Biochem., 19:67 (1984).
3. J. Bryant, Anti-foam agents, in: "Methods in Microbiology," J. R. Norris and D. W. Ribbons, eds., Academic Press, London (1970).
4. M. J. Hall, S. D. Dickinson, R. Pritchard, and J. I. Evans, Foams and foam control in fermentation processes, in: "Progress in Industrial Microbiology," Vol.12, O. J. D. Hochenhull, ed., Churchill Livingstone, London (1973).
5. K. H. Kroner, unpublished results.
6. M. N. Sarbolouki, A general diagram for estimating pore size of ultrafiltration and reverse osmosis membranes, Sep.Sci.Technol., 17:381 (1982).

YEAST CELL ENTRAPMENT IN POLYMERIC HYDROGELS:

A KINETIC STUDY IN MEMBRANE REACTORS

M. Cantarella, V. Scardi, A. Gallifuoco*, M. G. Tafuri*
and F. Alfani*

Istituto di Fisiologia Generale, University of Naples
Via Mezzocannone 8, 80134 Naples, Italy
*Istituto di Principi di Ingegneria Chimica, University of
Naples, P.le Tecchio, 80125 Naples, Italy

ABSTRACT

The preparation of hydrogel films with immobilized yeast cells and
their use as membrane in a reactor is discussed. The permeate flow rate,
which is initially high, decreases with operational time and reaches an
asymptotic value which linearly depends on the applied pressure in the
investigated range (0.1-4 atm). The rate of flux decay was shown to be a
linear inverse function of time and of a decay constant K.

The yeast invertase activity present in the membrane was monitored and
the experimental values of the Michaelis constant, k_m = 60 mM, and the
activation energy, E_a = 7367 cal/mol, show that enzyme kinetics is close to
that observed with free cells.

The operational thermostability was measured in the membrane reactor
at 45°C and the invertase activity is significantly high for a prolonged
time.

INTRODUCTION

This work deals with the immobilization of yeast cells in a polymeric
carrier and their characterization in a membrane reactor. Several reasons
suggest the use of this type of reactor in bioconversion studies. In fact,
it was shown that on laboratory scale they represent a very useful tool for
performing kinetic tests under well-defined operational conditions, such as
temperature, pH and ionic strength, in a similar way to other reactor
configurations, but in addition they allow to monitor for long period
reaction rate at constant substrate and biocatalyst concentrations. The
ultrafiltration membrane reactor is the only continuous reactor in which
this conditions can be satisfied, when working with a soluble catalyst.
Therefore, rates of thermal inactivation can be easily evaluated[1] and,
whenever it occurs, inhibitory effects can be detected and the inhibition
mechanism can be identified[2].

On industrial scale the use of membrane reactors was suggested mainly
because soluble catalyst can be reused. Moreover, biocatalyst immobilization in capillary ultrafiltration systems offers the possibility to get

high productivity since a high active surface-to-volume ratio can be easily achieved.

Former studies on enzymatic membrane reactors have been extended in order to investigate on the kinetic behavior of immobilized cells. Experiments were performed with Saccharomyces Cerevisiae, a yeast rich in endo- and exo-enzymes, and employed in industrial processes. The invertase activity was tested mainly because of its significant content in the yeast cell and of the difficulty to immobilize successfully the purified enzyme.

In the recent years attention has been devoted to catalyze bioconversion with cells instead of purified enzymes since it allows reduction in the process costs. In fact, enzyme purification is no more necessary; multienzymatic reactions are made possible owing to the presence of a variety of enzymes in the cell; the conformation of cell-bound enzymes is generally stable owing to the favorable microenvironment.

However, for industrial purposes it is more convenient to immobilize the cells in order to achieve a high cell density and a reduced cell growth which ensure high reaction yield per unit reactor volume. Moreover, the process layout is simplified since viscosity is reduced and product separation is made easier.

Among the possible different immobilization techniques the entrapment in hydrogels of polyhydroxyethylmethacrylate was selected. This polymeric carrier presents favorable conditions for the biocatalyst since it retains water up to 70% by weight and its biocompatibility is sufficiently good. The polymerization can be carried out even at refrigerator temperature and does not cause pH shift. Reaction time, depending on the water content and the expected degree of polymerization, ranges between few minutes to 2-3 hours; all other conditions, such as monomer and initiator concentrations, are not so drastic to induce a marked biocatalyst denaturation[3]. Finally, hydrogels have good mechanical properties which suggest their use in different reactors. The present paper particularly describes the preparation of hydrogel enzyme-films and their use as membrane in a bioreactor.

EXPERIMENTAL

Materials and Methods

The experiments were carried out with lyophilized Saccharomyces cerevisiae prepared from a single commercial lot of yeast. The monomer, 2-hydroxyethylmethacrylate (HEMA), and the crosslinking agent, ethylendimethacrylate (EDMA) were both from Rohm Gmbh (Germany) and BDH (England) respectively and were purified before use. The initiators of the polymerization reaction, $(NH_4)_2S_2O_8$ and $Na_2S_2O_5$, were from Bio Rad (USA) and I.C.N. (USA) respectively. All other chemicals were pure grade reagents from usual sources.

Product concentration was determined on 1 ml aliquot of the reaction assayed for reducing sugars by the Nelson colorimetric method (4), using D-glucose as a standard.

The following polymers have been used as molecular weight standards for rejection measurements: polyethyleneglycol (M.W. 4,000), PEG, from Merck (F.R.G.), polyvinylpyrrolidone (M.W. 15,000), PVP/K15, from EGA-Chemie (F.R.G.), polyvinylpyrrolidone (M.W. 44,000), PVP/K44, from BDH (England).

Preparation of the Membranes

All the membranes tested in this study were prepared according to the following standard procedure. A fixed amount (4.5 mg dry wt.) of _Saccharomyces cerevisiae_ were suspended in water (15 ml), and 6g of HEMA were mixed with 0.5% w/v of EDMA; both solutions were maintained in N_2 atmosphere to remove oxygen traces, then they were mixed under stirring and the polymerization was started with the addition of 0.3% w/v of ammonium persulphate and 0.3% w/v of sodium metabisulphite based on organic phase. Then, the homogeneous solution was introduced in a mould prepared with two perfectly flat glass plates sealed with rubber and silicone. Polymerization was carried out at 25°C for 1.5 - 3 hours, in the absence of air. The resulting membrane (1 mm thick) was washed several times at refrigerator temperature to remove the cells not entrapped. The recovery of initial cell concentration in the different preparations tested during this study was above 90%.

Reactor System

The ultrafiltration reactor consists of an unstirred, flat membrane cell (4.5 cm i.d., 8 ml overall internal volume). The reactor is fed by a nitrogen pressurized reservoir. System temperature is controlled within ±0.5°C by total immersion in a thermostatic water bath. Samples are collected by means of an automatic fraction collector which permits measurements of permeate flow rate. A more detailed description of the experimental layout is reported in [5].

RESULTS AND DISCUSSION

Membrane Performances

Firstly the permeate flow rate was measured as function of the applied pressure and time. Experiments were performed with distilled water and at constant temperature, 45°C. Only minor variations of the flux were observed at lower temperature up to 25°C. In Figure 1 a plot of permeate flow rate Q versus process time is reported. An initial very rapid decay was observed, which lasted roughly 10 hours, it was followed by a second phase during which the flux slowly approaches to an asymptotic value. The data of Figure 1 refer to an intramembrane applied pressure equal to 0.3 atm, but similar behavior was also observed at lower and higher pressure fields. However, depending on the value of ΔP the length and the shape of the transient period tend to vary as discussed later on.

In the course of the run the pressure was released twice for 10-15 minutes each time, then applied again. During these short periods a partial recovery of the initial flux is achieved and after a time-lag the fluodynamic behavior of the membrane tends to that observed prior of the disturbance. Both magnitude and length of the phenomenon depend on the instantaneous physical properties of the membrane which in turn are function of the time elapsed since the beginning of the run.

Several attempts were made to identify the rate equation of flux decay. Time exponential and hyperbolic functions do not adequately fit the experimental data. The best correlation was found with the following rate equation.

$$Q - Q_{AS} = \frac{Q_o - Q_{AS}}{1 + KT_p}$$

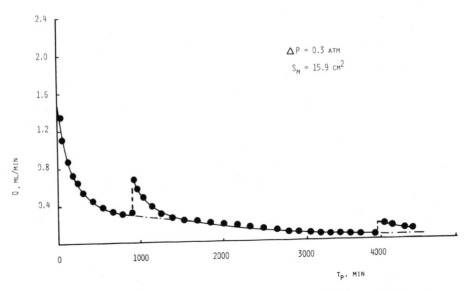

Fig. 1. Permeate flux vs. process time. The dotted lines indicate pressure release.

where Q_o and Q_{AS} are the initial and asymptotic values of the flux, and k is a flux decay constant. This can be evaluated by plotting $1/(Q - Q_{AS})$ versus T_p (see Figure 2). The intercept on the ordinate allows accurate determination of Q_o whereas $k = - 1/T*_p$, $T*_p$ being the intercept on the abscissa. The fitness of the experimental data was satisfactory in the whole range of explored applied pressure (0.1 - 4 atm) and the results confirm that k depends on both pressure and membrane conditions, i.e. previous pressure cycles, relaxation time. A detailed investigation is now in progress.

Moreover, the asymptotic values of permeate flow rate were plotted (Figure 3) versus applied pressure; a linear behavior was observed up to the maximum explored value, 4 atm. Membrane permeability measured by the slope of the straight line was $2.02 \cdot 10^{-3}$ ml/min atm cm^2.

In a different set of experiments the ultrafiltration cell was fed with solutions of commercial polymers at different molecular weights and the membrane cut-off was identified.

The experimental apparatus consisted of a pressurized reservoir, filled with distilled water, and a multiport valve, which enables the injection of 10 ml of polymer solution (10^{-5}M for PVP/k15 and PVP/k44 and 10^{-2}M for PEG 4,000) without perturbing the pressure distribution and the flow regime within the cell. Once the permeate flux was stabilized, the polymer was injected, and its concentration in the effluent samples was measured spectrophotometrically at 220 nm for PVP and at 260 nm for PEG.

PVP/k15 and PEG 4,000 freely flow throughout the membrane whereas PVP/k44 is totally rejected.

Kinetic Characterization

The invertase activity of the yeast cells entrapped in the membrane was assayed at different substrate concentrations and temperatures in order to determine the enzyme activity recovery and the kinetic parameters of the reaction rate equation.

236

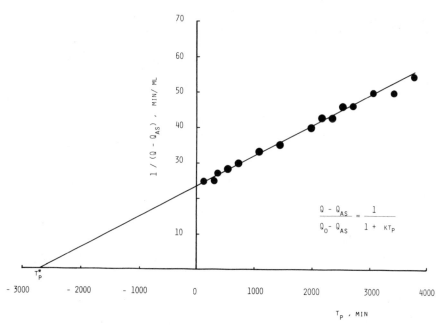

$$\frac{Q - Q_{AS}}{Q_0 - Q_{AS}} = \frac{1}{1 + KT_P}$$

Fig. 2.   Linear inverse plot of permeate flux for the determination of the flux decay constant $K = 1/T_P^*$.

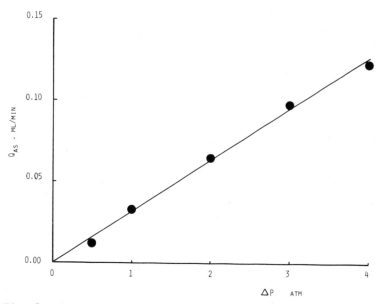

Fig. 3.   Asymptotic permeate flux as function of applied pressure.

In a first set of experiments the reactor was kept at 45°C and was fed with sucrose solutions in 50 mM Na-acetate buffer, pH=5, at different concentrations from 40 to 300 mM. The product concentration and flux were monitored for at least 50 hours and the reaction rate was measured at steady-state conditions according to the following mass-balance equation:

$$R = \frac{Q_{AS} \, C_p}{N_c}$$

237

Experiments have been performed using different membranes prepared under the same experimental conditions previously detailed.

By inspection of Figure 4 some conclusions can be reached. The dotted line represents the kinetic behavior of free cells in a batch reactor. The immobilized cell kinetic obeys a Michaelis-Menten rate equation, and k_m is slightly larger than that for soluble cells. This increase, which might indicate a reduced affinity of the enzyme for the substrate, may also be determined by the presence of limited mass-transfer resistance. In fact, in this case substrate concentration in the bulk and in the vicinity of the active site would not be the same, the second being the smaller and dependent on the diffusion coefficient in the polymeric membrane. Therefore, the values for substrate concentration, as reported in Figure 4, would be apparent.

Secondly, the enzyme activity recovery varies in the different preparations even at a standard procedure. It was noted, under the polymerization condition adopted, that the observed loss of activity mainly depends on the time of polymer formation and this tends to change in the limits previously reported without reasonable justification.

However, the enzymatic activity of the cells in the hydrogels is satisfactory, and a recovery between 37.5% and 71% was observed in this study.

The pressure of mass-transfer resistance in the membrane is confirmed by the values of the activation energy, E_a, for immobilized and free cells. Figure 5 shows a typical Arrhenius plot from which, an activation energy of 7367 cal/mol for the membrane entrapped cells and of 8990 cal/mol for the free cells have been evaluated via a least square method.

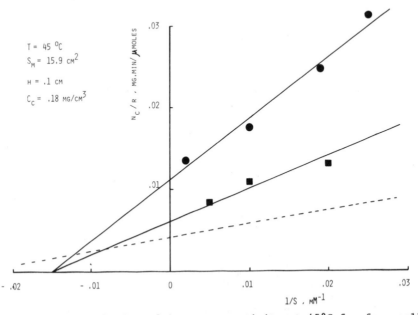

Fig. 4. Lineweaver-Burk plot of invertase activity at 45°C for free cell (dotted line) and immobilized S. cerevisiae in the membrane: (●) polymerization time 3 h at 25°C, (■) polymerization time 1.5 h at 25°C.

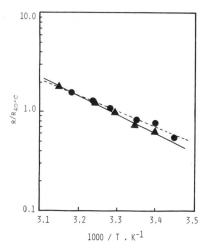

Fig. 5. Arrhenius plot of dimensionless invertase activity for free and
immobilized yeast cells: (▲) free cells, $R_{40°C}$=.775 μmol/mg min
(●) immobilized cells, polymerization time 3 h at 25°C, $R_{40°C}$=
.278 μmol/mg min.

Particles were also taken off from the membrane film, reduced to a
constant size (20-30 mesh), according to the procedure described elsewhere[3], and assayed in a batch reactor for determining the effect of pH
on reaction rate. A broader range of maximum reaction rates on pH variation was determined in comparison with the free cells without observing
shift in the optimum pH value[3].

The invertase thermostability of the yeast cells entrapped in the
membrane was measured at 45°C, working at about 1% substrate conversion
and feeding into the reactor a constant concentration of sucrose (50 mM in
50 mM Na-acetate buffer, pH=5). Under these conditions the reaction rate
is directly proportional to the instantaneous concentration of active
enzyme. The rate of sucrose hydrolysis slowly decreases following an
exponential decay with process time. During an initial time-interval,
which lasts roughly 80 h, the flux across the membrane decreases, and
therefore under these unsteady-state conditions, the determination of
reaction rate is masked by the dynamic behavior of the reactor. However,
invertase activity was monitored for prolonged time, 900 h, and from the
slope of log R versus time an half life of 1108 h was calculated, almost
identical to that observed with free cells.

CONCLUSIONS

Encouraging results have been obtained during this exploratory study
dealing with the immobilization of cells in hydrogels. The material in the
form of film is suitable for application in a membrane reactor. Its
characterization indicates good mechanical resistance and interesting
properties of flow even at reduced pressure.

On the other hand the results of the investigation on the kinetic
properties of immobilized cells indicate that membranes with a satisfactory
recovery of the initial activity can be prepared and that, according to the
data on invertase activity, only minor modifications in the rate equation
are observed.

In fact, the affinity of the enzyme towards the substrate and the activation energy remain almost unchanged thus indicating that during the immobilization procedure enzyme configuration is not altered, and mass-transfer limitations to reactant and product transport in the hydrogel film are not relevant.

The behavior of the membranes is not yet totally defined, and experiments are in progress to measure accurately the rejection towards species at different molecular weights and the dependence of permeate flow rate and kinetic properties on the concentration of crosslinking agent, on the nature of chemical species employed as redox agents and on the other polymerization conditions.

Acknowledgements

The work was supported by the Italian Department of Public Education (MPI), Tematica "Reattoristica Biochimica".

NOMENCLATURE

C_c — yeast cell concentration in the membrane, mg/cm^3

C_p — product concentration, $\mu mol/ml$

E_a^p — invertase activation energy, cal/mol

H — membrane thickness, cm

K — flux decay constant, min^{-1}

K_m — Michaelis constant, mM

N_c^m — yeast cell amount, mg

ΔP — applied pressure, atm

Q — permeate flux, ml/min

Q_o — initial flux, ml/min

Q_{AS} — asymptotic flux, ml/min

R — sucrose conversion rate, $\mu mol/min$

S — sucrose concentration, mM

S_M — membrane surface, cm^2

T — temperature, °C

T_p — process time, min or h

T_p^* — inverse of flux decay constant, min

REFERENCES

1. F. Alfani, M. Cantarella, and V. Scardi, Use of a membrane reactor for studying enzymatic hydrolysis of cellulose, J.Membr.Sci., 16:407 (1983).
2. M. Cantarella, A. Gallifuoco, V. Scardi, and F. Alfani, Enzyme stability and glucose inhibition in cellulose saccharification, in Ann. N.Y. Acad. Sci., J. Hitchcock, ed., in press (1984).
3. M. Cantarella, C. Migliaresi, M. G. Tafuri, and F. Alfani, Immobilization of yeast cells in hydroxyethylmethacrylate gels, Applied Microb.and Biotechnol., in press (1984).
4. N. Nelson, A photometric adoption of the Samogyi method for the determination of glucose, J.Biol.Chem., 153:375 (1944).
5. V. Scardi, M. Cantarella, L. Gianfreda, R. Palescandolo, F. Alfani, and G. Greco, Enzyme immobilization by means of ultrafiltration techniques, Biochimie, 62:635 (1980).

PERFORMANCE OF WHOLE CELLS POSSESSING CELLOBIASE ACTIVITY

IMMOBILIZED INTO HOLLOW FIBER MEMBRANE REACTORS

A. Adami, C. Fabiani,* M. Leonardi,* M. Pizzichini*

Cattedra di Microbiologia Industriale
Univ. degli Studi di Milano, Italia
*Enea, Divisione Chimica, Lab. Chimica Applicata
Cre Casaccia, Roma, Italia

Cells of Hansenula henriici possessing specific activity (in the 0.0017-0.065 U/mg/min. range) towards cellobiose, have been immobilized into Amicon hollow-fiber modules, used as ultrafiltration devices. The performance of the resulting membrane enzyme reactor (REM) was tested with cellobiose solutions (0.25-1% in buffer at pH=4,4) as substrate. More than 98% of the cells resulted immobilized in the module and a good retention of hydrolytic activity was observed.

During long time experiments (260 hours of continuous operation) a small reduction of cellobiase activity was recorded. The reactor performance greatly depends both on the number of total enzyme units immobilized into the cartridge (different amount of total cells) and on the specific activity of the used cells.

Batch experiments, simulating some of the working conditions of the membrane reactor, have been performed, and the results compared with those obtained with the membrane reactor. A simplified mathematical model was used to obtain the kinetic parameters of the membrane reactor.

INTRODUCTION

Immobilization of whole cells in continuous reactors has some advantages if compared with the immobilization of purified enzymes. Savings in enzyme separation and purification, the increase in enzyme half life and stability, the presence of possible cofactors[1] (the enzyme remains in its native state) are some of these reasons.

Immobilization into hollow fiber membrane bioreactors offers a great number of engineering potentialities[2,3]. This technique allows for a quick and effectual entrapment of the cells into the membrane support with packing densities higher than in usual cell suspension[4].

In this work yeast cells of Hansenula henriici having β-glucosidase activity[5,9] were immobilized into the shell side of Amicon hollow fiber membranes and the continuous hydrolysis of cellubiose has been studied. The hollow fiber module was used in the ultrafiltration mode with the substrate solution flown from the shell side into the fiber lumen. This kind of immobilization is one of the different loading techniques[10,11,12] offered by an hollow fiber reactor. The kinetic behavior of batch and hollow fiber reactor operated in similar conditions were compared.

MATERIALS AND METHODS

A yeast strain of the genus Hansenula selected and identified as Hansenula henriici has been used in all experiments. The main characteristics of this yeast are: whitish colonies, global cells, lateral budding; only pseudohyphae can be formed; sexual reproduction present with typical shaped ascospores; urea hydrolysis negative. Fermentation absent. Positive utilization of cellobiose, D-ribose, D-xylose, L-rhamnose, trehalose, glycerol, erythritol, ribitol, xylitol, D-mannitol, citrate, salicin, arbutin, methanol, ethanol, nitrate, ethylamine. Negative utilization of: D-galactose, L-sorbose, L-arabinose, D-arabinose, sucrose, maltose, melibiose, lactose, raffinose, melezitose, inulin, starch, galactitol, inositol, D-gluconate, DL-lactate, succinate. Growth at 42°C positive.

The cells were prepared either as frozen samples in buffer solution (Na citrate-citric acid, 0.025 M pH 4,4) or as freezed dried samples. Specific activity on cellobiose resulted to be in the range 0.017 U/mg/min and 0.065 U/mg/min in previously optimized conditions of reaction, following the method of Shewale and Sadana[13]. Before use a certain amount of cells was defrosted and resuspended in the same buffer solution in order to check the activity values.

Cellular concentration was expressed as yeast cell dry weight/volume (mg/ml).

β-glucosidase activity was obtained by measuring the μ moles of glucose produced by a certain amount of cells (mg) in 10 minutes at 50°C. The activity was expressed as μ moles of glucose produced per mg of cells per minute: U/mg/min.

The cellobiose (from Sigma) solutions had a sugar concentration in the range of 0.1-4% (w/v) in buffer citrate solution (0.025 M, pH=4.4).

Glucose concentration was determined by means of the Beckman Glucose Analyzer 2 according to the enzymatic oxidation

$$\beta\text{-D-glucose} + O_2 \xrightarrow[H_2O]{\text{glucose oxidase}} \text{gluconic acid} + H_2O_2$$

Amicon Hollow Fiber Modules

Two different modules H1P100-43 and H1P30-43 were used with a cut-off of 100,000 and 30,000 respectively. The polysulphone hollow fibers characteristics are: 20.3 cm length, 2 mm outer and 1.1 mm inner diameters. The volume of the sponge region is 26 cm^3.

RESULTS AND DISCUSSION

Batch Hydrolysis

Cellobiose batch hydrolysis experiments were performed to evaluate preliminarly the km and V_{max} kinetic parameters of the yeast β-glucosidase. Two sets of experiments were made at 50°C in the standard buffer solution: 1) static batch experiments with variable cell/substrate concentration ratios and reaction times; 2) shaken batch experiments.

Typical Michaelis–Mentes plots obtained with 6 different cellobiose concentrations in static (a) and dynamic (b) batch hydrolysis conditions are shown in Figure 1. From these data the usual Lineweaver-Burk plots allows for the evaluation of the km and V_{max} kinetic parameter (Figure 2). In the static conditions cellobiose solutions with concentrations in the 0.003 M (roughly 0.1% w/v) 0.06 M (roughly 2% w/v) range react with 60 mg of cells (dry weight) per cm^3 of solution at 50°C and pH=4.4 for ten minutes. The values of km and V_{max} obtained were of 0.0263 M and 0.0444 U/mg/min respectively. In the shaken batch the concentration of cellobiose varied in a wider range from 0.006 M (roughly 0.2% w/v) to 0.117 M (roughly 4% w/v). The cell concentration was 10 mg (dry weight) per cm^3 of reaction solution. Temperature, pH and reaction time were the same as above. Values of 0.0556 M for km and 0.2222 U/mg/min for V_{max} were obtained.

The comparison of the two V_{max} values (five fold higher in shaken batch) suggest that dynamic conditions as those present in a membrane continuous reactor should favor cell performances. Moreover in an hollow fiber reactor operated in ultrafiltration mode the continuous product recovery from the reaction volume (the fiber wall) should avoid product inhibition.

This negative effect is reflected in the data of Figure 3 where the glucose produced in shaken batch conditions (mg/ml) as a function of time at different cell/substrate ratios R and at two different initial cellobiose concentrations is reported.

R (=mg of cells/mg of substrate) varies from 0.086 to 4.3 for a cellobiose concentration of 1% (w/v) in the reaction mixture. For a cellobiose concentration of 0.5% (w/v) the range of cells/substrate ratios we studied was of 0.86 to 8.6. A plateau has been reached (after about 3 hours) for almost all experimental conditions which means that the hydrolysis rate decrease with time as glucose is produced. This negative effect should not be present in the hollow fiber factor.

Hollow Fiber Membrane Reactor

The scheme of the hollow fiber bioreactor is shown in Figure 4. Hydraulic characterization of the cartridges was carried out by measuring the flow rate as a function of the applied pressure difference ΔP in three

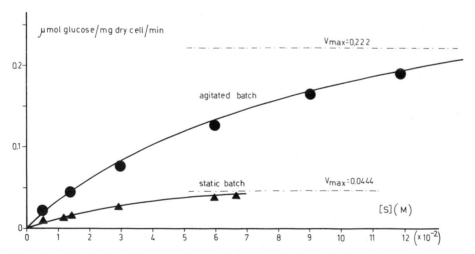

Fig. 1. Michaelis–Menten plots of <u>Hansenula henriici</u> β–glucosidase in agitated batch and static batch conditions.

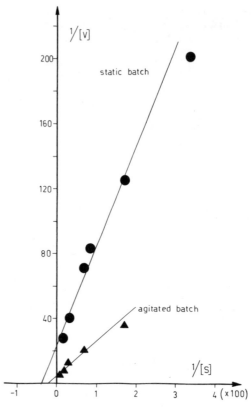

Fig. 2. Lineweaver-Burk plots of <u>Hansenula henriici</u> β-glucosidase in
agitated batch and static batch conditions.

different conditions. In the first test distilled water was flown through
the unloaded membrane and the obtained values of flow rates and corresponding ΔP were then compared with those obtained with cellobiose solution
(1%w/v) as test liquid. The test was repeated using the same cartridge
loaded with cells (1.5g of dried cells) and the above cellobiose solutions.
Loading conditions were the same that we used in the continuous hydrolysis
experiments. It has been experimentally tested that more than 98% of the
cells is actually immobilized by a single operation of ultrafiltration.

     As expected, the volume flow through the hollow fiber (shell side →
lumen side) gradually increases in both cases with the different applied
liquid pressure. The pressure increases with the liquid viscosity (curves
a and b, Figure 5) and with a given solution is greater when the cartridge
was loaded with cells (curves b and c). As shown in Figure 5, a ΔP=1.5 atm
with 1% cellobiose results in a permeate flow of 0.77 cm^3 sec^{-1}, in the
unloaded membrane and 0.2 cm^3 sec^{-1} in the loaded cartridge. The reduced
volume permeability observed with loaded fibers reflects both on the
operating condition (a higher pressure is needed to maintain constant
fluxes) and in the reactor performance (the contact time between cells and
substrate is increased).

     In a representative continuous hydrolytic test (Figure 6) in the
bioreactor (70 hours, cellobiose 1%) the pressure increases in order to
maintain a constant permeation flow is shown. This effect could limit the
life time of the hollow fiber reactor. To balance this pressure increase
it is possible to gradually increase the feed pump flow rate (curve b).

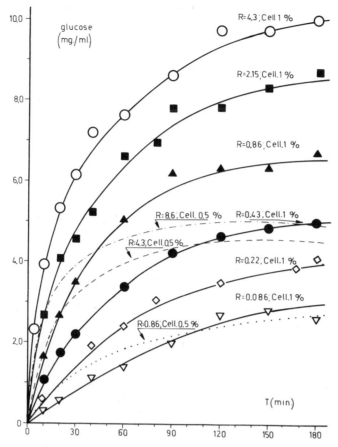

Fig. 3. Glucose (mg/ml) produced as a function of hydrolysis time, at different R (cell/substrate ratio) values and at two different initial cellobiose concentrations (0,5 and 1% w/v).

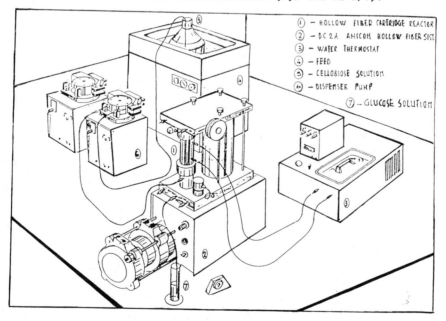

Fig. 4. Experimental apparatus.

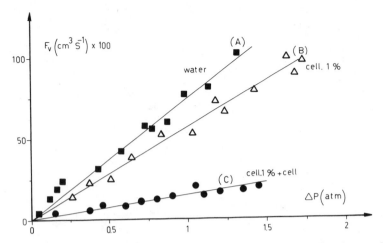

Fig. 5. Hydraulic permeability of Amicon hollow-fiber membrane H1P30-43 (F_v) as a function of the applied ΔP in three different working conditions: (A) distilled water through unloaded membrane, (B) cellobiose solution (1% w/v) through unloaded membrane, (C) cellobiose solution (1% w/v) through cell loaded membrane.

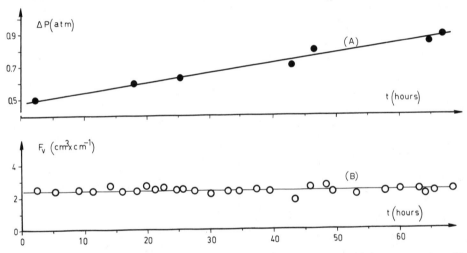

Fig. 6. Time dependence of applied ΔP pressure (A) and (B) permeation flow as a function of time.

However in some continuous hydrolysis test thermophilic mould growth was observed which contributes to the pressure increase. Nevertheless when this did not occur it was possible to reach a maximum of 236 hours of continuous hydrolysis.

Continuous hydrolysis experiments were performed on the loaded cartridge by flowing at a constant rate (ΔP control) the substrate solution (heated at constant temperature of 50°C) from the shell side to the lumen side i.e. in U.F. mode. The module was heated at the same temperature, using a water jacket. The flow rate of the peristaltic pump feeding the cellobiose solution varied into a range of 2-3 ml/min in order to have a high residence time into the reactor and to maximize the hydrolytic yield. The permeate was collected and its volume was measured at fixed time

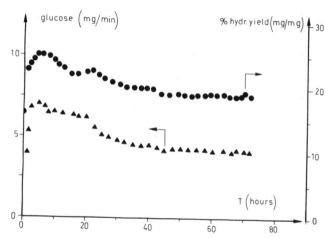

Fig. 7. Hydrolysis trend in continuous reactor; velocity of glucose production (mg/min) and hydrolytic yield (mg of glucose/mg of initial cellobiose) as a function of time.

intervals in order to calculate the flow rate. On each collected fraction the glucose produced was measured. To get a picture of the decrease of β-glucosidase activity and its stability at long times the reactor was run to a maximum of 236 hours. A typical curve for the hydrolysis of cellobiose to glucose in the continuous reactor is shown in Figure 7. Both glucose production (mg/min) and conversion (mg of produced glucose/mg of initial cellobiose) rates as a function of time are reported. A maximum value of glucose production is reached in about 5 hours followed by a stabilization and then a decreasing trend, almost constant glucose production rate. This behavior was general for all the hydrolysis tests performed whose data are reported in Table 1.

The maximum hydrolysis yield was roughly 25% and decreased to about 20% during the stabilization period, after 50-70 hours. During this period the glucose production rate amounts to 4.5 mg/min. In Table 1 our continuous hydrolysis tests in H.F. membrane reactor are compared in relation to the medium hydrolysis yield and to the activity value. In particular this comparison can be done for experiments 1, 2, 4 at the same specific activity (0.017 U/mg/min) and for No.3 and 5 at a higher specific activity (0.030 U/mg/min).

In Figure 8 we report a medium hydrolysis yield as a function of total cell units immobilized into the cartridge. The yield is higher with lower substrate concentrations. On the other hand, the maximum rate in the production of glucose is reached only with higher cellobiose concentrations. These results allow to chose the most suitable substrate concentration in order to have better hydrolytic yield or higher glucose production rates. The amount of the cell loaded in the membrane reactor can cause an increase in the flow pressure through the fibers especially in long lasting experiments. From this point of view it is better to increase the cellular specific activity rather than the cell quantity immobilized in the H.F. membrane reactor.

A Simplified Model

The reported experimental hydrolysis data have been used for evaluating the kinetic k_m and V_{max} parameters in the reactor to compare them with those obtained in batch conditions.

Table 1. Comparison of Experimental Parameters in Amicon H1P100-43 Hollow Fiber Membrane Reactor

Test Number	1	2	3	4	5
Dry cells (mg) of Hansenula henriici	2100	936	860	912	1720
Cell activity at the production (U/mg/min)	0.017	0.017	0.030	0.017	0.030
Total units in cartridge	35.7	15.9	25.8	15.5	51.6
Cellobiose solution % (w/v)	0.25	0.40	0.50	1.00	1.00
Cellobiose solution (M)	0.007	0.012	0.015	0.029	0.029
Continuous hydrolysis period (hours)	16.5	133.5	71.5	53.5	71.5
Feed medium flow rate (ml/min)	1.8	3.0	2.4	2.7	2.2
Total volume of permeate (1)	1.8	25.0	10.0	8.7	9.6
Residence time in reactor	10.0	6.0	7.5	6.7	8.2
Medium value of produced glucose (mg/ml)	0.31	0.25	0.94	0.57	2.19
Maximum value of produced glucose (mg/ml)	0.38	0.34	1.12	0.93	2.72
Activity medium value (U/mg/min)	0.00015	0.0045	0.0146	0.0093	0.0155
Activity maximum value (U/mg/min)	0.0018	0.0061	0.0174	0.0152	0.0193
Medium hydrolysis yield (M glucose/M cellobiose)	0.25	0.12	0.35	0.11	0.02

Working conditions: temperature 50°C, pH 4.4 in Na citrate-citric acid buffer solution 0.025 M.

For this purpose many models at different levels of complexity are available for immobilized enzyme reactors (Rony, Waterland, etc). In these models different operating conditions are assumed which reflect onto the obtained mathematical solutions. Models based on diffusion processes have complicated solutions. But it has been demonstrated[14] that even in these cases a kinetic control predominates when the square of Thiele module is less than 0.1. Therefore, for the sake of simplicity, we have assumed a plug flow model sketched in Figure 9, in which the membrane containing the immobilized cells act as a barrier for the substrate solution volume flow. This flow is produced by an applied pressure difference (ΔP) between the external (shell-side) hollow fiber surface (at the high pressure volume) and the internal (lumen side) surface (at low pressure).

As small ΔP values have been used, the observed compaction of the gelled dynamically formed enzyme membrane[15] is disregarded and the whole sponge region of the fiber wall (26.0 cm^3) is assumed as reaction space.

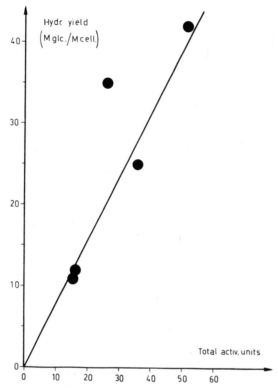

Fig. 8. Hydrolytic yield (M glucose/M Feed cellobiose) as a function of total enzymic units immobilized in the reactor.

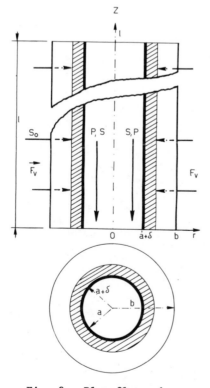

Fig. 9. Plug-flow scheme.

Other assumption are:

- each fiber has the same behavior
- constant radial volume flow following Hagen-Poiseuille Equation[16]
- flat concentration profiles in any layer of the system.

These hypotheses, even if strict, lead to a simple equation still containing all the useful operating parameters which may effect the reactor yield.

The mass balance at the lumen surface ($\Sigma^i = 2 \pi a.1$ being a the lumen, and 1 the fiber length) is

$$F_v \frac{dS}{dr} = - R\Sigma^i \tag{1}$$

with F_v volume flow rate ($cm^3 s^{-1}$), S=substrate concentration, R=the Michaelis Menten reaction rate in the hollow fiber reactor.

$$V = \frac{V^*_{max} \cdot S}{km^* + S} \tag{2}$$

The condition $\Sigma^i \gtrless \Sigma^e$ (Σ^e external fiber surface) is assumed for small wall thickness ($\delta \ll a$). Integration of Equation 1 over the reaction volume gives with the help of (2) and the boundary conditions

$S=S_o$ for $r=a+\delta$

$S=S$ for $r=a$

$$-km^* \ln(1-X) + S_o X = \frac{V_{max} \cdot \Sigma \cdot \delta}{F_v} \tag{3}$$

with $X = \frac{S_o - S}{S_o}$

The $(\Sigma\delta)/F_v$ ratio correspond to a space time ν and Equation (3) can be written

$$-km^* \ln(1-X) + S_o X = \nu V^*_{max} \tag{4}$$

From Equation 3 (or 4) the km* and V_{max} kinetic parameters (for a given cell concentration $V^*_{max} = [E].K$) can be obtained by a plotting XS_o vs.$\ln(1-X)$. This has been done in Figure 10 with the experimental X values obtained from conversion plots (see Figure 7). For four substrate concentrations (0.25%, 0.4%, 0.5%, 1.0%) the X data were calculated in the nearly constant part of the plot. From the linear plot of Figure 10 km* = 0.024M and V_{max} = 0.030 U/mg/min. These values are very similar to those calculated from batch experiments (Table 1). These results seem to indicate that in the assumed experimental conditions the process is kinetically controlled. However it must be mentioned that these values may depend on the applied ΔP. In fact for higher ΔP the compaction of the gelled membrane can be expected followed by a reduction in the reaction volume. In hyperfiltration experiments gelled dynamically formed membranes are compacted to a thickness even of 100 Å at high pressures. The observation of this effect on gelled cell membrane could be a way to evaluate the effective membrane thickness.

The presence of a gelled membrane is essentially shown by the increased resistance to the volume flow (decreased volume permeability) already discussed. In terms of Hagen-Poiseuille law

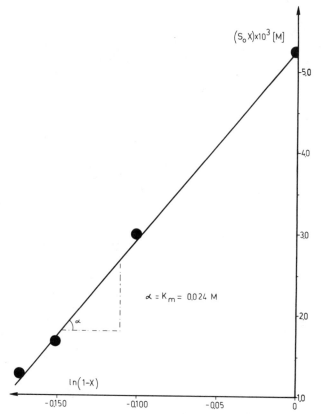

Fig. 10. Evaluation of kinetic hollow fiber parameter according to
equation 3 (or 4).

$$F_v = - Q \cdot \frac{\Delta P \cdot \Sigma}{\delta} \qquad (5)$$

with $\dfrac{Q}{\delta} = \dfrac{G \varepsilon^r p^2}{8 \mu \delta}$ = permeability and $\varepsilon = n \pi r_p^2$ = porosity

(r_p = mean pore diameter, μ = solution viscosity, G = geometric factor,
n = number of fibers). A change of volume permeability for a given
solution in the presence (Q_1) or absence (Q_2) of the gelled membrane can
be analyzed by the ratio

$$\frac{Q_1}{Q_2} = \frac{\varepsilon_1}{\varepsilon_2}$$

i.e. the gelled membrane can reduce the porosity of the fiber as a consequence of an obstructive effect. In the used Amicon cartridge the porosity
of loaded fiber is reduced by 70%:

$$\frac{Q_1}{Q_2} = \frac{0.15}{0.57} = 0.26$$

CONCLUSION

A comparison between the batch hydrolysis conditions and those obtained in the continuous reactor is not possible. However we can assume

that agitated batch better simulate the hydrodynamic conditions of the continuous reactor than the static system. We then compare the two different hydrolysis rates in two similar tests. We choose therefore a cellobiose concentration of 0.5% (w/v), a value of the ratio weight of dry cells/weight of substrate of 8.6 in batch conditions and of 9.5 in hollow fiber reactor and a cellular specific activity of 0.003 U/mg/min for both of them. The medium value of glucose production rate in the continuous reactor (2.2 mg of glucose/min) indicates an activity retention of 37% if compared with the hydrolysis rate in agitated batch conditions (Figure 3).

All performed tests whose results are reported in Table 1 follow an hydrolysis curve as those reported in Figure 7 (glucose production rate or conversion rate). The maximum hydrolytic yield expressed as mg ratio achieved in the discussed tests was 25% decreasing at 20% in a 50-70 hours time period (Table 1). In this time interval the glucose production amounts to 4.5 mg/min. An analysis of the reported data (Table 1) shows that comparing tests 1, 2, 4 (same cell specific activity 0.017 U/mg/min) and 3.5 (higher specific activity 0.030 U/mg/min), the hydrolysis yield rises increasing the total cell units in the cartridge and/or decreasing the substrate concentration. These results allow for the choice of suitable substrate concentrations in order to obtain higher hydrolytic yield or alternatively higher glucose production rates. An increase in the amount of cells used to load the continuous reactor can produce a reduction in flow conditions which may be negative in long time experiments. Therefore a higher specific activity could be more useful than a higher reactor cell content. The simplified plug-flow model used allows for a simple calculation of km* and V*$_{max}$ in the bioreactor and may be used as a test for effective porosity in loaded hollow fibers and for the compaction of the gelled membrane dynamically formed. However more experimental results are needed to prove its effective usefulness for a hollow fiber cell reactor used in ultrafiltration conditions.

REFERENCES

1. F. B. Kolot, New trends in yeast technology-immobilized cells, Process Biochem., 15(7):2 (1980).
2. L. R. Waterland et al., Enzymatic catalysis using asymmetric hollow fiber membranes, Chem.Eng.Commun., 2:37 (1975).
3. In Ho Chim and Ho Nam Chang, Variable-volume hollow-fiber enzyme reactor with pulsatile flow, AIChE J., 29(6):910 (1983).
4. D. S. Inloes et al., Ethanol production by Saccharomyces cerevisiae immobilized in hollow-fiber membrane bioreactors, Appl.Environ. Microbiol., 46(1): 264 (1983).
5. N. Moldoveanu and D. Kluepfel, Comparison of β-glucosidase activities in different Streptomyces strains, Appl.Environ.Microbiol., 46(1):17 (1983).
6. H. Yoshioka and S. Hayashida, Relationship between carbohydrate moiety and thermostability of β-glucosidase from Mucor miehei YH-10, Agric.Biol.Chem., 45(3):571 (1981).
7. S. R. Parr, Some kinetic properties of the β-glucosidase (cellobiase) in a commercial cellulase product from Penicillium funiculosum and its relevance in the hydrolysis of cellulose, Enzyme Microb. Technol., 5:457 (1983).
8. K. Ohmine et al., Kinetic study on enzymatic hydrolysis of cellulose by cellulase from Trichoderma viride, Biotechnol.Bioeng., 25:2041 (1983).
9. J. Woodward and A. Wiseman, Fungal and other β-D-glucosidases-their properties and applications, Enzyme Microb.Technol., 4:73 (1982).
10. P. R. Rony, Multiphase catalysis II. Hollow fiber catalysis, Biotechnol.Bioeng., 13:431 (1971).

11. P. R. Rony, Hollow fiber enzyme reactors, J.Amer.Chem.Soc., 94:23 (1972).
12. L. R. Waterland et al., A theoretical model for enzymatic catalysis using asymmetric hollow fiber membranes, AIChE J., 20:50 (1974).
13. J. G. Shewale and J. C. Sadana, Cellulose and β-glucosidase production by a basidiomycete species, Can.J.Microbiol., 24:1204 (1978).
14. D. B. Van Dongen, D. O. Cooney, Hydrolysis of Salicin by β-glucosidases in a hollow fiber membrane reactor, Biotechnol.Bioeng., 19:1253 (1977).
15. E. Drioli, "Recent Development in Sep. Science," Vol.III part B, p.343, CRC, Press Inc., Cleveland, Ohio (1977).
16. Sun-Tak Hwang, K. Kammermeyer, "Membranes in Separation," A. Weissberger, ed., Chap.3, John Wiley and Sons Inc., New York (1975).

RECENT DEVELOPMENTS OF ULTRAFILTRATION IN DAIRY INDUSTRIES

J. L. Maubois

Laboratoire de Recherches de Technologie Laitière I.N.R.A.
65 rue de Saint-Brieuc
35042 Rennes Cedex, France

The growing use of membrane ultrafiltration technique in the world dairy industry, during the last decade, can be illustrated by evolution of installed membrane areas: 300 m² in 1971 with most of the first industrial equipment being built in New Zealand for treating casein whey, 80,000 m² in 1982 (Figure 1) and probably, more than 100,000 m² nowadays to our knowledge. All milk producing countries now, have industrial UF plants either for treating whey, or for concentrating milk or both. Reasons for such a growth, which can be simultaneously qualified as slow, if it is taken into account the technico-economical advantages brought about by this separation technique, but a fast growth in the particularly traditional character of the dairy industry, must be considered:

- biochemical characteristics of milk and its by-products (mostly whey). Indeed, molecular sieving of these liquids realized by UF does not require, in first approach, use of UF membranes with a sharp cut-off because of the existing discontinuity in molecular sizes of milk components. There is a gap between the molecular size of lactose (0.2 nm) and the one of the smallest whey protein (2.5 x 4.0 nm for the α-Lactalbumin monomer, Swaisgood, 1982). The deep knowledge of properties and biochemical characteristics of milk components was also certainly one of the main factors of this sharp growing use of UF. It was relatively easy for biochemists, microbiologists and technologists to define optima parameters of carrying out this technique.
- Successive progresses realized in matters of UF membranes and equipments. Three membrane generations has been developed during the last ten years: the first one made mostly from cellulose acetate is now completely abandoned because of its pH (3 to 8) and temperature (50°C maximum) limitations. The second one made from sulphonic polymers represents the main part of the world total membrane area. The third one, made from inorganic materials, essentially zirconium oxyde and amorphous carbon, but some pilot scale ceramic equipments are now running, was mostly developed and marketed in France, where it represents more than 20% of the total national surface.

The four configurations of UF membrane equipments (plate and frame-tubular, hollow-fiber, spiral-wound) are present in the dairy industry with their advantages and their disadvantages regarding performances (expressed in liters of treated product or collected permeate per hour and per m² of membrane), investment cost (1500 to 3500 US $/m² of surrounded membrane),

255

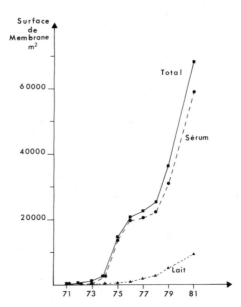

Fig. 1. Evolution of membrane areas in the world dairy industry according to data of De Boer and Hiddink (1980) and new data collected from Abcor, Alfa Laval, Dorr Oliver, Ladish Triclover, Pasilac, Rhône-Poulenc and S.F.E.C. (Maubois and Brulé, 1982).

energy consumption (0.2 KWH/m² for hollow fiber conception; 0.5 KWH/m² for plate and frame, spiral wound equipments; 0.7 to 1.0 KWH/m² for tubular conception), membrane life (manufacturers warranties vary from 1 to 3 years), floor-space and easiness and cost of cleaning. Recent UF dairy installations take in account the pseudo plastic rheological behavior of milk and whey UF retentates at high protein contents (Goudédranche et al., 1980 - Figure 2). The proposed solutions, besides substitution of volumetric pumps to centrifugal pumps, are the shortening of membranes length as marketed by Romicon or a totally new design of membrane support plate (37 module of Pasilac).

Recent Development of UF Application in Dairy Industry

For illustrating the last developments of use of UF in the dairy industry, we have chosen three recent applications which have been recently scaling up.

The first one is related to cheesemaking. Progresses in matters of UF equipment (mineral membranes and new design of finishing units) have led to two new application fields for so called M.M.V. processes (Maubois et al., 1969) (Figure 3): the making of fresh acid cheeses and of semi hard cheeses. Indeed, thanks to the very strong mechanical resistance characteristics of mineral membranes and/or to the mastering of tangential flow speed of UF retentate along the membrane, we were able (Mahaut et al., 1982) to restudy ultrafiltration of coagulated milk as proposed by Stenne (1973) in replacement of curd centrifugation for making fresh cheeses as Quarg. The obtained results concerning UF fluxes (Figure 4) were very surprising. Initial fluxes when milk curd (renneted milk brought to pH 4.5 through lactic acidification) is ultrafiltrated were at least 60% higher than those observed when normal fluid milk (pH 6.7) is treated with the same equipment. As shown by curves in Figure 4, the flux decrease is stopped around concentration 2:1 and even a slight flux increase related to increase of drop pressure is noticeable above concentration 3:1. Such a

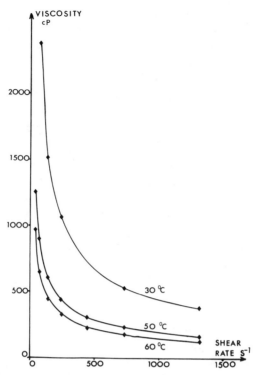

Fig. 2. Variation of viscosity of a highly concentrated liquid precheese
(T.S.: 44.4%; Nx6,38 = 20.4%) versus shear rate (Goudédranche
et al., 1982).

behavior in permeation flux curve must be related to a highly porous
texture of polarization layer. In opposition to what happens during
ultrafiltration of fluid milk where polarization layer is progressively
built by a coalescence of dense globular proteins, this polarization layer
during ultrafiltration of milk curd must be constituted by a weak network
of paracasein chains which, moreover, have no electric charges. High
porosity of this network is confirmed not only by the high permeation flux
but also by the lower retention of small whey proteins, lower retention
which has led the membrane manufacturer to develop a tightened membrane for
this application. In our opinion, ultrafiltration of milk curd is very
promising for many cheese varieties. It can be said that five industrial
units are already running in Germany for making Quarg.

We were also able to reproduce with excellent industrial feasibility
conditions, liquid precheeses having a protein content higher than 21g
p. 100 (Goudédranche et al., 1980). Such precheeses constitute a cheese
base from which an extraordinary variety of cheeses can be made either
similar to traditional cheeses or entirely new according to the nature and
the amount of additives (lactic starters - ripening enzymes - spices,
herbs; or by mixing one precoagulated precheese with another one just
renneted and differently inoculated, etc.). In this field, only applications to unripened cheeses such as Mozarella have been quickly developed.
Industrial scale up of ripened cheeses such as Gouda or St Paulin varieties
is slower because of the tradition of cheesemakers in spite of tremendous
technico-economical advantages: a 19% increase of cheese yielding capacity
from milk, total suppression of curd work in vats, and also the need to
develop and to build a new dose-molding equipment.

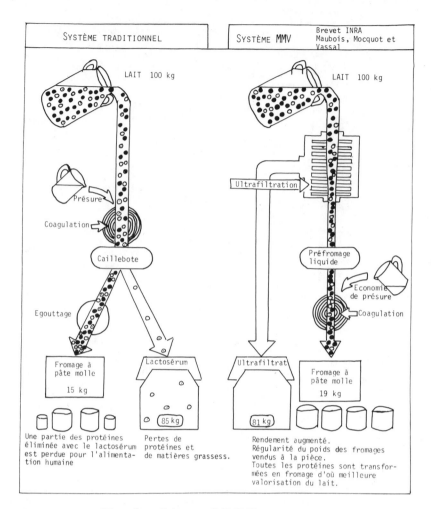

Fig. 3.   Schema of M.M.V. process.

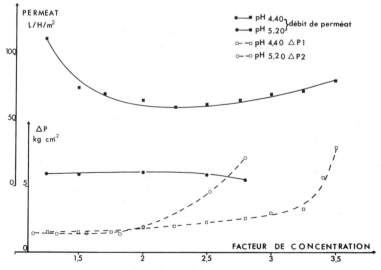

Fig. 4.   Permeate flux observed during ultrafiltration of curdled milk at
          pH 4.6 (Mahaut et al., 1982).

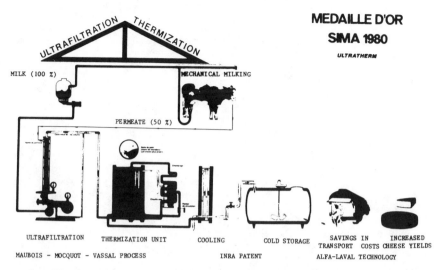

MEDAILLE D'OR
SIMA 1980
ULTRATHERM

MILK (100 %)   MECHANICAL MILKING

PERMEATE (50 %)

ULTRAFILTRATION   THERMIZATION UNIT   COOLING   COLD STORAGE   SAVINGS IN   INCREASED
                                                                  TRANSPORT COSTS CHEESE YIELDS

MAUBOIS - MOCQUOT - VASSAL PROCESS   INRA PATENT   ALFA-LAVAL TECHNOLOGY

Fig. 5.   Schema of ultrafiltration-thermization at farm level.

The second development that we wish to present during this lecture is
the implementation of membrane ultrafiltration treatment combined with a
thermization of UF retentate in milk producing farms. This program which
is at the end of the R and D step was originated by the following observations:

a) Because of generalization of milk refrigeration in bulk tanks and
therefore the less frequent collections that this derives, the nature of
contaminant flora of milk was strongly modified during the last fifteen
years. Instead of having mesophilic acidifying germs, the dominant flora
is psychrotrophic for most of the actual milk collected all over the world.
But, for this type of microorganisms, if the cells are easily destroyed by
the heat treatments classically used in the dairy industry, the same cells
have the inopportune property to produce very heat stable enzymes (pro-
teases and lipases), so heat stable that even an UHT treatment does not
destroy their activity (Alichandis and Andrews 1977). Activity of these
enzymes leads to organoleptic defects (off flavors - rancidity - gelation)
in diazy products (consumption milk - cheese - butter) when total count of
psychrotrophic germs has reached the level of one million cells per ml.
Such a level is easily obtained when raw milk is stored two or three days
at the farm and waits one or two more days at the plant before receiving
its first heat treatment. As simultaneously proposed by Zall (1980) it
seemed to us that a possible solution to this strong degradation of milk
bacteriological quality was to realize heat treatment as near to the
milking machine as possible, known as the main contamination part of the
farm equipment.

b) Except for consumption milk, human beings are not very well using
the main milk components water and lactose since they are doing husbandry.
For example, in France, 70% of lactose produced by dairy cows is not
directed for human consumption (Fauconneau, 1980). So, the following
question arises: why the need to cool, to store, to collect and to treat
these components with more and more energy consuming techniques, when a
feasible technique exists in ultrafiltration, for separating them and when
the best user of these same components is the dairy cow, as well demonstrated by Schingoethe (1976) and Thivend (1977).

259

Fig. 6. New prototype of farm ultrafiltration-thermization equipment for treating continuously milk produced by 35 to 60 cows.

In a collaboration program with the firms Alfa-Laval and Electricité de France, funded by E.E.C. and Brittany public authorities, three generations of equipments, progressively automatized, carrying out the process line described in Figure 5, were designed, built and tested. The last one (Figure 6) was multiplied to five units and installed in five milk producers having a dairy herd between 35 to 100 dairy cows. Milk going out of a milking machine is continuously ultrafiltrated until the volume reduction (around 2) required for obtaining the desired protein content in retentate. This protein content is defined by a refractometer inserted in the UF recirculation loop. The permeate is immediately returned as drink to the milked cows. The retentate is continuously thermized (65°C to 72°C – 15s), cooled and directed to the bulk tank. Bacteriological results obtained with this equipment are confirmed by those described by Bénard et al. (1981): average total mesophilic count after eight months of running twice a day, seven days a week around 15,000 CFU/ml in eight cumulated milkings. Such results authorize a collection twice a week. The Biochemical quality of the milk collected is totally restored: no lipolysis, and no proteolysis can be detected. From the first economical studies which have been conducted on this subject, economical studies based on an investment cost of 10,000 US dollars for the equipment, it can be said that the net profit of use of the proposed treatment when Emmental cheese is made will be around 0.6 US cent/liter of normal milk (Floriot et al, 1984).

If this application is spread out, the whole dairy economy will be modified. Indeed, beyond general improvement of quality of milk collection and resulting products, all capacities for collecting, transporting, storaging and transformation at dairy plants (pumps, plate heat-exchangers, cheese vats, evaporators, spray driers...) will be doubled. Composition of milk and whey powders will be modified (their nutritional value will be increased) and the financial support of the dairy market must be revised.

The last application of membrane ultrafiltration that we want to evocate is related to membrane biotechnology. The enzymatic membrane

reactor technology is evidently the most flexible and the most convenient
fragmentation technique for biological fluids. Indeed, enzyme confinement
in free state in a small volume, as the one represented by the retentate
loop and membrane porosity, offers the possibility to simulate in vitro
physiological conditions as those existing, for example, in intestinal
tractus. Reactions can be continuously or sequentially realized with a
very high enzyme substrate ratio in non-prohibitive conditions, the enzyme
being recovered. Multienzymes reactors can also be envisaged provided that
physicochemical parameters of optimum enzymatic activities were consistent.
Finally, beyond continuous removal of hydrolyzed products as soon as they
are split from the substrate, the membrane can act by its own porosity
characteristics on the size of hydrolyzed fragments.

In collaboration with medical researchers, we have applied this
membrane biotechnology to the fragmentation of milk proteins (Figure 7).
The purpose of this research was to simulate in the reactor the different
steps of intra intestinal hydrolysis in order to obtain the particular
peptidic sequences of amino-acids which are induced from intestinal
receptors hormonal and enzymatic secretions. Such sequences are required
for the nutrition of people who suffer from carcinogenic diseases, entero-
colite illnesses or congenital malformations. Two processes have been
developed, carrying out hydrolysis of milk proteins formerly purified with
pancreatic enzymes: One from whey proteins (Maubois et al., 1979) another
from casein (Brulé et al., 1980). From these products, has been prepared
complete liquid food with lipides and glucides for clinical tests in
reanimation rooms. These trials were so successful that an industrial
production of several ten tons has been started. Indeed, it appears that
peptidic sequences obtained from milk proteins are promoting either a
partial regeneration of missing intestinal segments or a compensatory
hypertrophy of remaining intestinal segments (Mendy, 1984).

But, whatever was the interest of these peptidic hydrolysates for
reanimation of so badly affected people, it is very little in comparison
to the potential which they are opening. Indeed, some of the peptides contained in the mixture coming out from the enzymatic membrane reactor have

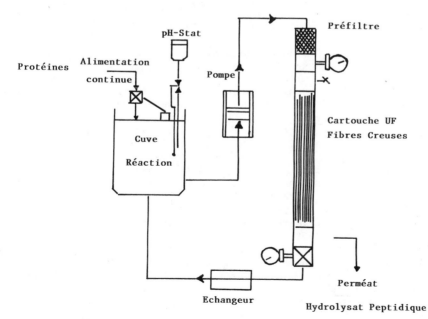

Fig. 7. Schema of an enzymatic membrane reactor.

now been identified as similar to opiates and/or acting as mediators of these exorphines. So, isolation and purification of milk exorphines is envisageable for preparing natural drugs acting on sleep regulation, appetite, growing and even insulin secretion. Other peptides could have other immuno or hormo stimulating action on vital organs (brain, heart etc.) or even on the humans psychism (Guillemin, 1984). It is, maybe, all that the drug industry of the third millenary can derive from membrane biotechnology.

REFERENCES

Alichandis, E., and A. T. Andrews, 1977, Some properties of the extra-cellular protease produced by the psychrotrophic bacterium Pseudomonas fluorescens strains, Biochimica et Biophysica Acta, 485:424-433.

Benard, S., Maubois, J. -L., and A. Tareck, 1981, Ultrafiltration-thermisation du lait à la production: aspects bactériologiques, Lait, 61:435-457.

Brule, G., Roger, L., Fauquant, J., and Piot, M., 1980, Procédé de trai-tement d'une matière première à base de caséine, contenant des phosphocaséinates de cations bivalents. Produits obtenus et applications, Brevet francais, No.80 02 280.

Brule, G., Roger, L., Fauquant, J., and Piot, M., 1980, Procédé de trai-tement d'une matière à base de caséine, contenant des phospho-caséinates de cations monovalents ou leurs dérivés. Produits obtenus et applications, Brevet francais, No.80 01 281.

Fauconneau, G., 1976, Symposium IBM "Soyons prêts pour les années 1980," Hamburg, RFA.

Floriot, J. -L., 1984, Evaluation économique prévisionnelle de l'impact de l'insertion de l'ultrafiltration-thermisation à la ferme sur la filière Emmental, Etude D.G.R.S.T., No.81 C 1375.

Goudedranche, H., Maubois, J. -L., Ducruet, P., and Mahaut, M., 1980, Utilization of the new mineral UF membranes for making semi-hard cheeses, Desalination, 35:243-258.

Guillemin, R., 1984, Les Neuropeptides, Biofutur, (24), p.7.

Mahaut, M., Maubois, J.-L., Zink, A., Pannetier, R., and Veyre, R., 1982, Eléments de fabrication de fromages frais par ultrafiltration sur membrane de coagulum de lait, Technique Laitière, 961:9-13.

Maubois, J.-L., Mocquot, G., and Vassal, L., 1969, Procédé de traitement du lait et des sous-produits laitiers, Brevet francais, No.2 052 121.

Maubois, J.-L., Roger, L., Brule, G., and Piot, M., 1979, Hydrolysat enzymatique total des protéines de lactosérum. Obtention et applications, Brevet francais, No.79 16 483.

Mendy, R., 1984, (cité par Rajnchapel-Messai J.), Fragmentation des protéines laitières, Biofutur, 24:60-61.

Schingoethe, D. J., 1976, Whey utilization in animal feeding: a summary and evaluation, J.Dairy Science, 59(3):556-570.

Stenne, P., 1973, Procédé de fabrication d'aliments protéiques, notamment de fromages, Brevet francais, No.2 232 999.

Swaisgood, H. E., 1982, Chemistry of milk protein, in: "Developments in Dairy Chemistry - I. Proteins," P. F. Fox, ed., Applied Science Publishers Ltd., 1-59.

Thivend, P., 1977, Utilisation de l'ultrafiltration de lactosérum dans l'alimentation du ruminant, Economie Agricole, (2):25-27.

Zall, R. R., 1980, Can cheese making be improved by treating milk on the farm? Dairy Industries Intern., 45(2):28-29, 31-48.

ELECTRODIALYSIS OF DILUTE STRONTIUM CATIONS

IN SODIUM NITRATE CONCENTRATED SOLUTIONS

Claudio Fabiani and Massimo De Francesco

Dip. TIB, Divisione Chimica, Lab. Chimica Applicata
Cre - Casaccia
Roma, Italia

Membrane processes may be advantageous in the treatment of low activity nuclear wastes, but often their usefulness is reduced because of the presence in the waste solution of inactive salts at high concentrations. Electrodialysis of 0.5 M $NaNO_3$ solution containing Sr^{2+} at very low concentration (ppm) is discussed as a decontamination method for Sr^{90}. Electrodialysis was performed with commercially available membranes (Nafion 125 cationic from DuPont and RAI R-5035 anionic from Raipore Co.) at different current densities in the 8-40 mA/cm^2 range. The dependence of sodium and strontium ions transport numbers and separation factors on operating conditions is considered. The effect on separation efficiency of the presence of EDTA as a Sr complexing agent is discussed.

INTRODUCTION

Electrodialysis is a potentially useful method for treating nuclear waste solution but no extended applications of this method are found in nuclear technology[1]. The narrow range of useful salinities, the sensitivity of membrane to fouling and operational problems associated with mounting and dismantling the cell stack are the main drawbacks of this separation technique. However the continuous character of the process, the higher life-time of electrodialysis membranes as compared with those used in reverse osmosis and the availability of electrical power in nuclear plants are all factors which deserve much attention.

The selective separation of radionuclides at trace levels from inactive solution components at medium or high concentrations is often an unresolved problem of nuclear waste chemical technology. In these cases many separation methods which have proved to be useful for dilute solutions can not be used any more. For instance colloidal TiO_2 is a good and inexpensive adsorbent for Sr^{90} but the presence of high concentrations of Na^+ strongly reduces the decontamination effect[2].

In the present work the electrodialysis is examined as a method for treating concentrated low activity nuclear waste solutions. A typical solution simulating a real nuclear waste will be considered (Table 1).

EXPERIMENTAL

Materials

All chemicals used to prepare standard solutions (NaNO$_3$ 0.5M, Sr
5 ppm as Sr(NO$_3$)$_2$ pH=10 adjusted with NaOH) were as received purity grade
reagents from Carlo Erba (Italy). The cation selective membrane Nafion 125
from DuPont and the anion selective R-5035 membrane from RAI corporations
were used. The RAI membrane is a typical membrane for electrodialysis.
The Nafion membrane has been extensively used as separator in electrolysis
cells, batteries and fuel cells. This membrane, not specifically tailored
for electrodialysis, was chosen because its chemical, transport and structural properties are known. The main features of the membrane couple are
collected in Table 2. Before use membranes were both equilibrated in the
standard solution for at least 24 hours.

Table 1. Simulating and Real Low Activity
Nuclear Waste Solutions (tail-end
solution S11 of the Cs$^{134+137}$, Sr^{90}
decontamination process of a Magnox
type storage pond[2])

	Tail-end S11 solution	Simulating solution
Temperature (°C)	29 - 30	30
pH	8 - 9	10
Na NO$_3$ %	3 - 4	4 (0.5 M)
Chemical composition	(µCi/l)	(ppm)
Sr^{90}	3 . 10^{-1}	5
Cs$^{134+137}$	5 . 10^{-2}	
Co^{144}	5 . 10^{-3}	
Cr^{51}	3 . 10^{-3}	
Ru^{106}	3 . 10^{-3}	
Nb^{95} + Zr^{95}	2 . 10^{-3}	
Co^{60}	3 . 10^{-4}	

Table 2. Membranes Characteristics.

	NAFION 125		RAI R-5035	
Membrane	Perfluorinate Sulfonic Acid Cationic exchange		Quaternized Vinylbenzylamine grafted strong anionic exchange	
Tickness (cm)	0.025		0.020	
Resistance (ohm cm^2)	(0.5 NaCl)	13-14	(0.6 KCl)	13-16
Water content (% Dry basis)		15-20		23
Permselectivity	KCl 1N/3N	70 mV	KCl 0.5N/1.0 N	82
Transport number			t$_-$	0.85
	t+ (NaCl 0.4M, 25°C) 0.94 (5)		t$_-$ (NiCl$_2$ 0.42M, 25°C) 0.70 (5)	
K$_H^M$ ion exchange constant (3)	K$_H^{Na}$ = 1.22; K$_H^{Sr}$ = 4.24			
Diffusion Coefficient (4)	D$_{Na}$ = 9.44 10^{-7} cm^2 S^{-1}			
Diffusion Coefficient water (4)	D$_w$ = 2.65 10^{-6} cm^2 S^{-1}			
Electoosmotic Coefficient	β (NaCl 0.4 M) = 130 cm^3 F^{-1} (5)		β (NiCl$_2$ 0.4M) = 118 cm^3 F^{-1} (5)	

Electrodialysis Unit and Experimental procedure. The electrodialysis
laboratory apparatus (Figure 1) is a modified Bell II unit supplied from
Berghof (Tubingen, West Germany). The cell stack is formed by six membrane
couples of 37 cm² surface area for each membrane and with channels of 2 mm
between each couple. Most of the experiments were performed by flowing an
equal volume of standard solution (2000 cm³) in both the concentrated and
diluted compartments at a rate of 1400 cm³/min.

Both solutions were thermostated at 30 ± 0.1°C and their pH and
conductivity values were continuously measured (in two thermostated
measuring cells) and recorded. The applied stack voltage was also continuously measured. The changes of concentration of Na⁺ in both solution
were obtained from conductivity data by means of the calibration curve of
Figure 2. These values correlate very well with those measured by flame
photometry at four check points in the $NaNO_3$ concentration range used to
build the conductivity plot. The Sr^{2+} concentration was measured on small
aliquot (less than 1 cm³) of both solutions by using a Varian atomic
adsorption spectrophotometer model 975.

Electrodialysis was performed in the 0-40 mA/cm² current density range
during 5-6 hours. Two sets of experiments at 0.5 and 1.0 amperes were done
to test the effect of EDTA as Sr^{2+} complexing agent. EDTA was added to the
standard solution up to a molar ratio EDTA/Sr of two.

RESULTS AND DISCUSSION

Flow rates as high as 3 cm s⁻¹ minimize concentration polarization
effects[6] and current densities of 30-40 mA/cm² could be used. However
the practical absence of polarization was checked by recording the pH of

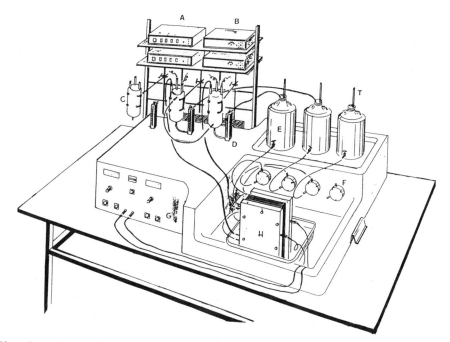

Fig. 1.   Electrodialysis Laboratory Test Unit: (A) pH-meters, (B)
          conductivity meters, (C) measuring cells, (D) flowmeters, (E)
          electrode rinse, diluate and concentrated solutions, (H) cell
          stack, (G) power supply.

both diluate and concentrate solutions. As shown in Figure 3 concentrate
solutions maintain their initial value around pH-10 while the pH of the
diluate solutions decreases slowly at high current densities.

Therefore the potential applied to the cell stack increases in order
to maintain constant the imposed current density (Figure 4). Also the
voltage increase is a consequence of the increase in membrane resistance
due to the decrease in electrolyte concentration in the dilute solution.
In fact Nafion membranes in 0.01 - 1.0M alkali hydroxide solutions have an
electrical resistance lower than that of corresponding aqueous solutions.
However this resistance increases of two orders of magnitude (1-100 Ohm cm)
as the electrolyte concentration is reduced from 1 to 0.01 M [7]. On the
contrary the membrane selectivity practically remains ideal since the
counterion (Na$^+$) transport number keeps above 0.95 [7]. The small polarization effect observed through the pH of the diluate and the applied
voltage is therefore a consequence of the desalting process of the standard

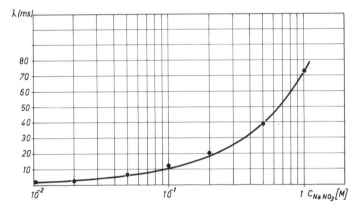

Fig. 2. Conductivity-concentration calibration curve for the standard type
solutions (ph = 10,NaNO$_3$ varying concentrations). T = 30°C.

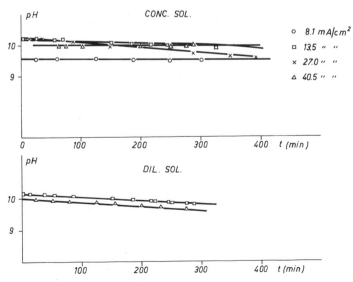

Fig. 3. pH variation in concentrated and diluted solutions during
electrodialysis experiments.

266

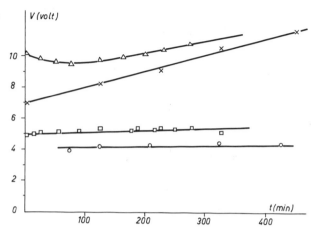

Fig. 4. Applied cell voltage during electrodialysis: 8.1(o), 13.5(□), 27.0(x), 40.5(△) mA/cm².

feed solution. In fact the Na/Sr molar ratio in the feed solution is initially $8 \cdot 10^3$ and electrodialysis produces a simple concentration on NaNO$_3$. The flux of the salt as referred to the unit area of the stack unit cell can be described by Equation (6).

$$J_s = (\tau_c - \tau_a) \, i/F - | \, (D_c/d_c) + (D_a/d_a) \, | \, \Delta C \tag{1}$$

being τ_c, τ_a the Hittorf's transport numbers of the cation through the cationic and anionic membranes respectively; Δ C concentration difference between concentrated and dilute solutions and i (mA cm^{-2}) the current density. The $(\tau_c - \tau_a)$ difference represents the net transport number for the cation and is related to the coulomb efficiency η_c

$$\eta_c = (\tau_c - \tau_a) \, 100F/i \tag{2}$$

in the absence of back diffusion effects. These negative effects are taken into account through the last terms equation 1. D_c, D_a are the salt diffusion coefficients in the two membranes of thickness d_c and d_a. However it can be shown that back diffusion is actually negligible at least for small time intervals when ΔC is small. By using the data reported in Table 2 the value of the back-diffusion terms result to be of the order of 10^{-8} moles cm^{2}s^{-1} for Na at i=27 mA cm^{-2} and t=400 min. This evaluation has been done by assuming for Na diffusion through the anionic and cationic membranes a D_a/D_c ratio proportional to the membrane water content. The back diffusion contribution was therefore disregarded.

The net transport number for cations can be obtained from the plots of Figures 5 and 6 where electrolyte concentration changes in the diluate and concentrated solutions are reported as a function of time t. The electro-osmotic effect should also be considered. The volume flow in an electro-membrane process is due both to the drag effect on the water molecules and to the transport of hydration water of ionic species[8,9]. Volume flow J_v can be accounted for by means of the Equation (6).

$$J_v = (\beta_c + \beta_a) \, i/F \tag{3}$$

being β_c, β_a (cm^{3}F^{-1}) the electroosmotic coefficients of cationic and anionic membranes respectively. In Equation (3) the water diffusion contribution has been disregarded. From Equations (2) and (3), the concentration of the cation in the concentrated solution at time t is given by

267

$$C(t) = \frac{m_o + n\ (\tau_c - \tau_a)\ (I/F)\ t}{V_o + n\ (\beta_c + \beta_a)(I/F)\ t} \qquad (4)$$

being m_o the initial moles in the solution of volume V_o (2000 cm^3) ($C_o = m_o/V_o$ initial molar concentration); n the number of cell units (6 in the present case) and I the electrical current in amperes (i=I/A being A the membrane surface). From Equation (4).

$$\Delta C(t) = C(t) - C_o = \frac{(n/V_o)\ [(\tau_c - \tau_a) - C_o (\beta_c + \beta_a)]\ (I/F)\ t}{1 + (n/V_o)(\beta_c + \beta_a)\ (I/F)\ t} \qquad (5)$$

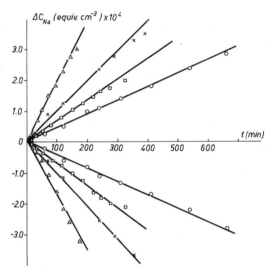

Fig. 5. Na$^+$ concentration variations in feed and concentrated solutions at different current densities: 8.1(o), 13.5($\square$), 27.0(x), 40.5($\triangle$) mA/cm^2.

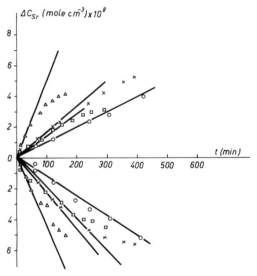

Fig. 6. Sr^{+2} concentration variation in feed and concentrated solutions at different current densities: 8.1(o), 13.5($\square$), 27.0(x), 40.5($\triangle$) mA/cm^2.

268

Table 3. Net Transport Numbers of Na^+ and Sr^{2+} and Na/Sr Selectivity.

I (A)	i (mA cm^{-2})	C_{Na} x 10^3 (mole cm^{-3})	α_{Na} x 10^8 (mole cm^{-3}S^{-1})	$(\hat{\tau}_c - \hat{\tau}_a)$	C_{Sr} x 10^6 (mole cm^{-3})	α_{Sr} x 10^{10} (mole cm^{-3}S^{-1})	$(\hat{\tau}_c - \hat{\tau}_a)_{Sr}$	P_{Sr}^{Na}
0.3	8.1	0.48	0.75	0.92	0.053	0.018	0.021 10^{-2}	0.48
0.5	13.5	0.52	1.17	0.88	0.064	0.040	0.027 10^{-2}	0.40
1.0	27.0	0.60	1.67	0.69	0.062	0.045	0.016 10^{-2}	0.45
1.5	40.5	0.47	2.75	0.74	0.058	0.083	0.019 10$^-$	0.48

At small t values it is $(n/V_o)(\beta_c + \beta_a)(I/F)\ t \ll 1$ for high current densities and therefore Equation (5) reduces to

$$\Delta C(t)=(n/V_o)\ [(\tau_c+\tau_a)-C_o(\beta_c+\beta_a)]\ (I/F)\ t \qquad (6)$$

which allows for the evaluation of the net transport number of the cations $(\tau_c - \tau_a)$ from the initial slope of the ΔC vs. t curves at constant I. The electroosmotic contribution can be computed from the data of Table 2. The results of these calculations on the experimental data reported in Figures 5-6 for the concentrated solutions are collected in Table 3.

The Na effective transport number at small current densities agrees quite well with the literature values reported in Table 2. By increasing the current density the transport number decreases (Figure 7). The electroosmotic correction is 10-15% of the net transport number. On the contrary due to the very low Sr concentrations the net transport number for this cation is very small and can be taken as a constant in the used

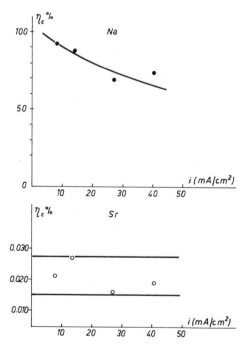

Fig. 7. Net cation transport numbers reported as coulomb efficiencies for Na^+ and Sr^{+2}.

current density range. In Table 3 the selectivity of the separation
process defined as

$$P^{Na}_{Sr} = \frac{(\tau_c - \tau_a)_{Na} \; C_{oSr}}{(\tau_c - \tau_a)_{Sr} \; C_{oNa}} \qquad (7)$$

is also reported. The P values are practically constants in the used
current density range. These low values reflect the higher ion-exchange
equilibrium constant for divalent alkaline earth cations with respect to
monovalent cations in Nafion type membranes (Table 2). In fact Nafion is
not an univalent selective membrane as those currently used in desalination
processes. These kind of membranes have $P^{Na}_{Divalent}$ selectivities in the
range 3–5[6] and they should produce a high separation between Na and Sr
during electrodialysis.

The initial Na/Sr molar ratio of $\sim 8 \cdot 10^3$ after electrodialysis (t=2
hours) changes to $\sim 7 \cdot 10^3$ (Figures 5, 6). However an effective desalting
effect is obtained as a consequence of the higher coulomb efficiency of the
Na transport.

The initial 0.5M concentration in $NaNO_3$ is reduced to 10^{-2}M after
about 6 hours at 27.0 mA/cm^2 current density. In the same time interval
the Sr concentration in the diluate decreases to $2 \cdot 10^{-6}$M (0.15 ppm) which
corresponds to a concentration reduction factor (decontamination factor of
active solutions) of 97%. This is an interesting result for decontamination processes of low activity wastes.

Effect of Sr Complexing Agents

In Figures 8–9 the results obtained by adding EDTA to the standard
$NaNO_3$ 0.5M, Sr 5 ppm, pH=10 solution are presented. These experiments were
made to prove the effectiveness of a strong complexing agent to confine the
simulating radionuclide element (Sr) in the feed (diluate) solution during
the electodialysis process. Experiments were performed at two current
densities 13.5 and 27.0 mA cm^{-2} and at a constant EDTA/Sr=2 molar ratio.
The standard solution plus EDTA was circulated in the dilute compartment of
the stack while in the concentrated compartment a $NaNO_3$ 0.5M pH=10 solution
was used. The Sr concentration C(ppm) vs. t plot at 13.5 mA/cm^2 shows no
particular effects in the feed (diluate) compartment. The Sr concentration
increases slowly as a consequence of the electroosmotic water transport.

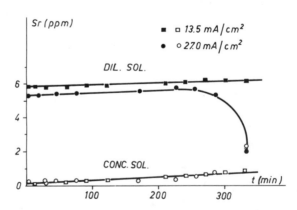

Fig. 8. Variation of Sr concentration in diluate and concentrated
solutions from a feed solution containing EDTA. EDTA/Sr=2 molar
ratio at 13.5($\square$,$\blacksquare$) and 27.0($\circ$,$\bullet$) mA/cm^2.

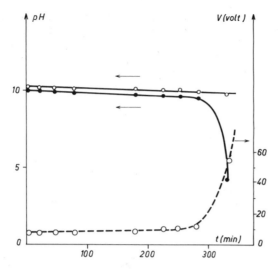

Fig. 9. pH and applied cell voltage during electrodialysis of the
solutions of Figure 8.

After 6 hours the Na concentration rises to 0.7M (curve in Figure 5) while
the Sr concentration remains below 1 ppm (Figure 8). These results show
that EDTA is a very good complexing agent to retain all the Sr in the feed
(waste) solution during the desalting process. The NaNO$_3$ concentration is
reduced from 0.5M to 0.25M (50%) after five hours while Sr is practically
confined in the feed solution. At higher current density (27.0 mA/cm^2) the
same results are obtained during the first five hours of electrodialysis.
For longer times a strong Na$^+$ concentration polarization occurs because the
sodium concentration decreases in the diluate at a very small level and
practically the whole NaNO$_3$ has been transferred into the concentrated
solution (Figure 5). In these conditions OH$^-$ ions become competing transported species as referred to NO$^-_3$ and the pH of the feed (diluate)
solution falls to acidic values (pH=4.13, Figure 9). At these pH values
the EDTA-Sr complex is not formed and Sr^{+2} participate to the transport
process as it is shown by the sudden decrease of Sr concentration in the
diluate solution. The relevant desalting effect in the feed (standard)
solution determines an increase in the solution and membrane resistances
and the applied voltage increases above 50 Volts (Figure 9). These effects
limit the useful current density or, conversely, the electrodialysis time.

CONCLUSION

This study is a preliminary analysis of electrodialysis as a desalting
process for low activity nuclear wastes (Table 1) when these wastes are so
concentrated that usual separation techniques like ion-exchange and adsorption, can not be used for decontamination.

Electrodialysis with non univalent-cation selective membranes, as
Nafion 125, produces an effective reduction in electrolyte concentration
(NaNO$_3$ concentration decreases from 0.5 to 0.01 M). A 97% decrease in the
concentration of the species simulating the radionuclide (Sr for Sr^{90}) is
also observed. This effect is due to the very low P^{Na}_{Sr} selectivity (0.40-
0.48). The use of specific univalent cation selective membranes which show
a sodium/divalent selectivity in the range 2.5 - 5 should greatly improve
these results.

271

The presence of complexing agents as EDTA determines a quite complete retention of the Sr^{+2} species during the desalting process. In this case polarization effects due to the strong dilution of the initial solution must be controlled by using small current densities or small electro-dialysis times. The use of EDTA makes electrodialysis an interesting method as a pre-treatment of low active wastes even with non specific univalent cation selective membranes like Nafion.

REFERENCES

1. R. G. Gutman, Membrane Processes, in: "Radioactive Waste: Advanced Management Methods for Medium Active Liquid Waste," Horwood Academic Pub., Switzerland, Chap.4 (1981).
2. C. Fabiani, M. DeFrancesco, Strontium separation with UF membranes from dilute aqueous solutions, J.Memb.Science, submitted for publication.
3. L. Bimbi, M. DeFrancesco, and C. Fabiani, Elettrodialisi. Caratterizzazione di una coppia di membrane per la separazione di ioni metallici mediante elettrodialisi.Sistema Ni^{+2}/Na^+., ENEA technical report RT/CHI(83), 2 Roma, Italy (1983).
4. H. L. Yeager and B. Kipling, Ionic diffusion and ion clustering in a perfluorosulfonate ion-exchange membrane, J.Phys.Chem., 83:1836 (1979).
5. A. Steck and H. L. Yeager, Water sorption and cation exchange selectivity of a perfluorinated ion exchange polymer, Anal.Chem., 52:1215 (1980).
6. T. Nishiwaki, in: "Industrial Processing with Membranes," R. E. Lacey and S. Loeb, eds., John Wiley Sons Inc., New York, Chap.6 (1972).
7. G. Scibona, C. Fabiani and B. Scuppa, Electro chemical behaviour of Nafion type membranes, J.Memb.Science, 16:37-50 (1983).
8. B. R. Bresslau and I. F. Miller, A hydrodynamic model for electro-osmosis, Ind.Eng.Chem.Fund., 10:554 (1971).
9. C. Fabiani, G. Scibona, and B. Scuppa, Correlation between electo-osmotic coefficient and hydraulic permeability in Nafion membranes, J.Memb.Science, 16:51-61 (1983).

ION-SELECTIVE MEMBRANES FOR REDOX-FLOW BATTERY

H. Ohya, K. Emori, T. Ohto and Y. Negishi

Department of Chemical Engineering
Yokohama National University
Hodogayaku Yokohama 240, Japan

SUMMARY

For the purpose to measure voltage drop across the separating membranes of redox-flow battery, cells were made of transparent poly-vinyl chloride. Their dimensions are 35 mm in inner diameter and 50 mm in length. To obtain voltage on the surface of each side of membrane, the measured voltages in the cell were extrapolated to the surface.

Area resistivities of membrane during charge and discharge, R_c, R_d are calculated by the following equations, using membrane voltage difference E_{mc} and E_{md}, and membrane potential E_m measured at the condition of open circuit, $R_c = (E_{mc}-E_m)/i_c$, $R_d = (E_m-E_{md})/i_d$ where i_c and i_d are current densities during charge and discharge. Values of R_c and R_d are obtained with several commercially available ion-exchange membranes.

Mobilities of hydrogen ion and chloride ion in the membrane are calculated from the data of resistivity of membrane, fixed charge in the membrane, and concentration of redox ions in the solutions, assuming Donnan equilibrium.

INTRODUCTION

Secondary battery which uses aqueous redox system has been known since end of last century. But little attention has been paid of it, since its electro motive force is low and its weight is heavy because of aqueous solution. But it has become important to be used as a power storage of electricity system. Redox-flow secondary battery (in short as "battery") started to get revaluation again[1,2,3 etc]. Particularly NASA has carried out a lot of research announcement[4-11]. In Japan, it has picked up to be one of the theme of "Moon-Light Project" of Ministry of International Trade and Industry, and at Electrotechnical Laboratory, vigorous research is in progressing.

The battery is constructed from redox-couple, membrane as separator, electrodes etc. As for the membrane, it's characteristics may differ corresponding with redox-couple and many kind of membranes were recommended and investigated. Battel Research Institute used H^+ ion permeable membrane but NASA used anion exchange membranes. Electrotechnical Laboratory uses cation exchange membranes which are all different types.

Table 1. Basic Properties and Area Resistivity during Charge and Discharge
for Several Ion Exchange Membranes

Ion Exchange Membranes	Transfer-ence Number* [-]	IEC** [meq/g-wet membrane]	Membrane Thickness [mm]	Area Resistivity $[\Omega.cm^2]$ (1)	(2) Charge	Discharge
Cation Exchange membranes						
CMVR	.985	1.94	.15	6.69	1.6	1.6
C-2T	.982	1.63	.15	6.00	5.5	1.1
C-4T	.981	1.86	.15	5.66	3.7	0.5
C-7T	.977	1.86	.15	6.17	2.5	3.2
C-58T	.971	1.79	.15	5.97	4.1	2.2
CL-25T	.988	1.63	.15	8.52	4.2	2.2
Naffion-125	-	-	-	-	$2.0^{(3)}$	$2.0^{(3)}$
Anion Exchange						
ACS	.994	2.14	.15	8.12	3.7	3.8
AVS-4T	.991	2.06	.15	7.78	3.1	2.5
ACH-45T	-	-	-	-	3.2	2.2

* Based on membrane potential defference between aqueous solutions of potassium chloride of which concentrations are 0.1 and 0.01 N
** Equilibrated with aqueous solution of 0.01 N NaOH or HCl.
(1) In the aqueous solution of 35 410 mq/l KCl, at 28°C.
(2) In the aqueous solution of 1N HCl and 1M $FeCl_2/CrCl_3$.
(3) In the aqueous solution of 2N HCl and 1M $FeCl_2/CrCl_3$.

In this presentation, it was adopted by the same redox-couple as NASA
and Electrotechnical Laboratory are using, aqueous hydrochloride solution
of chloride of iron and chromium. Various kind of ion-exchange membrane
were used to investigate their electrical behavior.

Measurement of Electrical Resistance of Ion Exchange Membrane

Ion exchange membrane. General properties of ion exchange membranes
used are shown in Table 1. Transport numbers in membrane were determined
from membrane potential generated between aqueous solutions of KCl 0.1N and
0.01N, and their values are between 0.971 to 0.994 which means the used
membranes have good quality in permselectivity. The ion exchange capacity
measured by ordinary method is approximately 2 meg/mg wet membrane. Area
resistivity for anion membrane is approximately 8 $\Omega \cdot cm^2$ and 6 $\Omega \cdot cm^2$ for
cation. As far as resistivity concerns, cation membrane seems suitable for
this purpose.

Measuring cell. A set of measuring half cells were made of transparent PVC pipe to measure electrical resistance of ion exchange membrane
which is placed between redox solutions as shown in Figure 1. The concentration of the solution in each half cell was kept constant by a magnetic stirrer at 166 r.p.m. Carbon plate, 4 mm in thickness was used as
an working electrode. The current density measured on the surface of the
electrode was considerably small compared with reported data. Therefore
polyvinyl plates with 4 mm thickness were inserted in the slant line part
as shown in Figure 1, and effective membrane was reduced to 34°, 10^{ϕ}, 7.9^{ϕ}
and 5% so as to increase current density on the surface of membrane.

Experimental Apparatus. As it is shown in Figure 2, the experimental
apparatus was composed of measuring half cells, mini cell which was
developed at Electro Technical Laboratory (effective membrane are of 18 mm
x 100 mm), two 200 ml Erlenmeyer flasks: one which was used as a reservoir

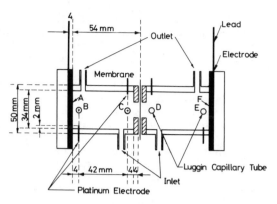

Fig. 1.   Voltage measuring cells with membrane.   Each cell has two Luggin
capillary tubes and two platinum electrodes.

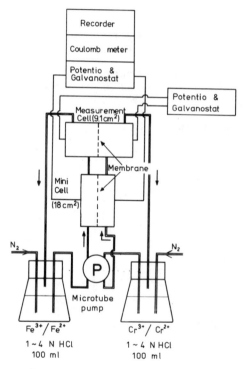

Fig. 2.   Experimental set-up with micro cell and measuring cell.

for $FeCl_2/FeCl_3$ in aqueous HCl solution and other for $CrCl_2$ $CrCl_3$ and two
microtube pump (Eyela Mp 3 supplied from Tokyo Rikagaku.   Tygon tube was
used to connect components.   Potentiostat/Galvanostat HA301 and HA211
(Hokuto Denko) were used to charge and discharge to the mini cell and
measuring two half cells.   Almost all charge and discharge occurred on mini
cell, so Coulomb meter HF201 (Hokuto Denko) was attached to measure the
depth of charge here.

Potential distribution measurement.   (1) Measuring point: Platinum
rods and Rugin's capillaries were inserted in point B,C,D,E as shown in
Figure 1.   Working electrode on anode side was at A and cathod side was

at F. On the Rugin's capillary half cell side, HCl solution with the
concentration equivalent to concentration of the redox solution was filled
in and connected to Ag/AgCl electrode half cell by saturated KCl salt
bridge. HCl solution was flowed in to purge redox species, when the color
in the capillary become conspicuous. (2) Open circuit potential distribution : In Figure 3 are shown two potential distributions referring to
point A which were measured using Ag/AgCl electrode and platinum electrodes
at open circuit. When the circuit is open, the measured potential at
points B and C, and points D and E should be equal. Difference between the
two pair should be equal to membrane potential. Substantial difference
between Ag/AgCl electrodes and platinum was not observed. So, the measurements were carried out mainly using platinum electrodes and Ag/AgCl electrode is inserted at only point c to measure single electrode potential.
(3) potential distributions during charge and discharge: As an example,
measured potential distribution during charge and discharge are shown in
Figure 4. In each half cell the potential gradient is observed. Potential
at each membrane surface is obtained by extrapolating the curve to the
surface, assuming linear gradient. Potential difference between the both
side of membrane during charge and discharge, Emc and Emd can be calculated.

Experimental Procedure. Ion exchange membrane was installed between
measuring half cells and soaked in aqueous HCl solution (ordinary 1N) for
about 3 hours. Then, the cells were washed with pure water for 5 minutes
and introduced at each side, aqueous $CrCl_3$ or $FeCl_2$ solution (1.0∿1.25 M)
of particular HCl concentration. Then measurement of potential distribution was carried out at the conditions of the constant electric current
charge or constant resistance discharge. Also periodically the circuit was
open to measure membrane potential.

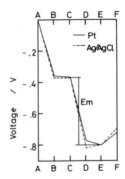

Fig. 3. Voltage distribution in the cell at open circuit with Ag/AgCl
electrodes. Refer point name to Figure 1.

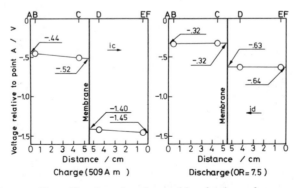

Fig. 4. Voltage distribution in the cells during charge and discharge.

Experimental Results

Data processing method. Using anion exchange membrane ACS, an example of experimental result during charge and discharge at the condition of depth of charge close to zero is shown in Figure 5. After start of charge, membrane potential E_m and membrane potential difference E_{mc} showed increasing tendency but after about one hour it became constant. But value of difference $(E_{mc} - E_m)$ remained constant about 0.05. Area resistivity during charge R_c, defined as in the following equation, is about $4\,\Omega \cdot cm^2$.

$$R_c = \frac{E_{mc} - E_m}{i_c} \qquad (1)$$

When switching from charge to discharge, the continuity of membrane potential is still existing. But potential difference between membrane E_{md} becomes smaller than E_m. The value of $(E_m - E_{md})$, as shown in Figure 5, decreases with time. But current density i_d is also decreasing and area resistivity of membrane R_d, defined as in the following equation, is shown almost constant at about $3.5\ \Omega \cdot cm^2$.

$$R_d = \frac{E_m - E_{md}}{i_d} \qquad (2)$$

R_c and R_d of membranes. When concentration of redox species is 1.25M, R_c at current density $127A/m^2$ and R_d over constant resistance 7.5Ω for various membranes are shown on Table 1. Resistance of charging is usually larger, and R_c for C-47 is about seven times larger than R_d. But there are such CMVR, Nafion 125 and ACS which show no difference.

Effect of concentration levels of HCl. In order to lessen the resistance in solution and membrane, concentration of HCl in aqueous solution is increased. Figure 6 shows the change in resistance of membrane when the concentration of HCl is increased from 1N to 4N. R_c of CMVR membrane in 1N is $1.6\Omega \cdot cm^2$ but decreased to $0.5\Omega \cdot cm^2$ in 4N. R_c of ACS membrane also decreases from 3.7 to $2.0\Omega \cdot cm^2$.

Effect of current density. When current density is increased from 127 to 204 and $509\ A/m^2$ with concentration of 1M of redox species and 1N of HCl solution, R_c keeps almost constant. But when the redox concentration is increased to 1.25M, R_c at lower current density increases slightly. Actually it might be said that current density has no effect on R_c and R_d.

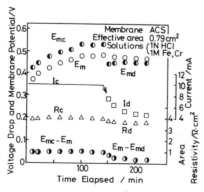

Fig. 5. Charge of voltage drop across the membrane E_{mc} and E_{md}, and membrane potential E_m with time.

DISCUSSION

Ionic concentration in membrane and electric conductivity of membrane. If it is possible to assume that Donnan's equilibrium of ions between ion exchange membrane and solution prevails, the ionic concentrations in cation exchange membrane for H^+ ion and Cl^- ion may be expressed as follows[14]:

$$\bar{C}_{H^+} = \frac{1}{2} \left(\sqrt{\bar{C}_X^2 + 4\, C_H\, C_{Cl^-}} + \bar{C}_X \right)$$

$$(3)$$

$$\bar{C}_{Cl^-} = \frac{1}{2} \left(\sqrt{\bar{C}_X^2 + 4C_H C_{Cl^-}} - \bar{C}_X \right)$$

Concentration of chloride ion in the solution C_{Cl^-} is the sum of that derived from HCl and mean of that from $FeCl_2$ and $CrCl_3$ assuming completely dissociated. Concentration of fixed dissociable group in the membrane $\bar{C}_X$ is assumed equal to ion exchange capacity IEC. On Table 2 are shown the measured values of membrane area resistivity R which are shown in Figure 6, membrane electric conductivities κ which are obtained using

Table 2. Electric Conductivity of Membranes and Concentration of Hydrogen Ion and Chloride Ion in Membranes, Assuming that Concentration of Fixed Charge Equals to Ion Exchange Capacities Shown on Table 1

		Membrane: C M V R				A C S			
Normality	[N]	1	2	3	4	1	2	3	4
R	[$\Omega \cdot cm^2$]	1.50	.85	.55	.50	3.75	2.50	1.96	1.50
κ	[mS/m]	1.00	1.76	2.73	3.00	.40	.60	.76	1.00
C_{H^+}	[$kmol/m^3$]	1	2	3	4	1	2	3	4
C_{Cl^-}	[$kmol/m^3$]*	3.5	4.5	5.5	6.5	3.5	4.5	5.5	6.5
$\bar{C}_{H^+}$	[$kmol/m^3$]**	3.40	4.41	5.42	6.42	1.08	2.11	3.13	4.14
$\bar{C}_{Cl^-}$	[$kmol/m^3$]**	1.03	2.04	3.05	4.05	3.22	4.25	5.27	6.28

* Concentration of chloride in the solution is summation of that due from hydrochloride acid and means of that from ferrous chloride and chromium chloride.

** With equation (4)

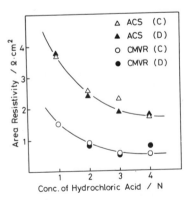

Fig. 6. Decrease of electric resistivity with increase of concentration of hydrochloric acid.

Equation (4) and ionic concentration in the membrane calculated from Equation (3)

$$\frac{1}{\kappa} = \frac{R}{d} \tag{4}$$

where d is thickness of membrane.

Mobilities of hydrogen ion and chloride ion in membrane. The electric conductivity is generally given as in the following equation

$$\kappa = \sum_i \kappa_i = F\sum_i z_i u_i \bar{c}_i$$

where κ_i, z_i, u_i, $\bar{c}_i$ are electric conductivity, valence number including sign, mobility and concentration in membrane of i th component respectively.

Assuming mobile ions in the membrane are only hydrogen ion and chloride ion, κ may be expressed as follows

$$\kappa = F(u_H^+ \bar{c}_H^+ - u_{Cl}^- \bar{c}_{Cl}^-) \tag{5}$$

The mean values of u_H^+ and u_{Cl}^- can be obtained from κ, $\bar{c}_H^+$ and $\bar{c}_{Cl}^-$ in Table 2. The values of u_H^+ and u_{Cl}^- for CMV and ACS are calculated by the least square method as follows.

For CMV $u_H^+ = 2.69 \times 10^{-5}$, $u_{Cl}^- = -3.66 \times 10^{-5}$

For ACS $u_H^+ = 1.04 \times 10^{-4}$, $u_{Cl}^- = -0.91 \times 10^{-4}$ [cm^{2}v^{-1}s^{-1}].

Acknowledgement

This research was supported by a grant-in-Aid of Scientific Research (No.56045052,57045040) of The Ministry of Education, Science and Culture Japan. The mini cell was supplied from Electrotechnical Laboratory of MITI.

NOMENCLATURE

C : concentration [kmol. or keq./m^3]
d : thickness of membrane [mm]
E_m : membrane potential difference [V]
E_{mi} : voltage drop across membrane [V]
i : current density [A/m^2]
R : electric resistivity [Ω.cm^2]
κ : electric conductivity [mS/m]

Superscript
− : inside membrane
Subscript
c : during charge cycle
Cl^- : chloride ion
d : during discharge cycle
H^+ : hydrogen ion
x : fixed charge density

REFERENCES

1. W. Vielstich, "Fuel Cells," Wiley, p.400 (1970).
2. K. D. Beccu, Zur Deckung des zukünftigen Spitzenbedarfs elektrischer Energie mittels elektrochemischer Energiespeicher, Chemie.Ing. Techn., 46:95 (1974).
3. L. H. Thaller, "Redox Flow Batteries," Proc. Symp. on Load Levelling, Atlanta, p.353 (Oct.1977).
4. R. E. Lacey and D. R. Cowson, "Development of Anion-selective Membranes," NASA CR-134932 (Oct.1975).
5. P. R. Prokopins, "Model for Calculating Electrolytic Shunt Path Losses

in Large Electrochemical Energy Conversion Systems," NASA TMX-3359 (April 1976).

6. S. S. Alexander and R. B. Hodgdon, "Anion Permselective Membranes," NASA CR-135316 (1977).

7. S. S. Alexander, R. B. Hodgdon, and W. A. Waite, "Anion Permselective Membranes," NASA CR-159599 (Mar.1979).

8. L. H. Thaller, "Redox Flow Cell Energy Storage Systems," NASA TM-79143 (1979).

9. M. A. Reid, R. F. Gahn, J. S. Ling, and J. Charleston, "Preparation and Characterization of Electrodes for the NASA Redox Storage Systems," NASA TM-82702 (Sept.1980).

10. N. H. Hagedorr and L. H. Thaller, "Redox Storage Systems for Solar Applications," NASA TM-81464 (Sept.1980).

11. J. Giner and K. Gahill, "Advanced Screening of Electide Couples," NASA CR-159738 (Feb.1980).

12. K. Nozaki, H. Kaneko, A. Negishi, and T. Ozawa, Research and development of redox flow cells in electrotechnical laboratory, Denki Kagaku, 51:189 (1983).

13. T. Berzins, "Electrochemical Characterization of Nafion Membranes in Chlor-Alkali Cells," Electrochemical Society Fall Meeting, Atlanta, U.S.A. (Oct.1977).

14. T. Hanai, "Membrane and Ion," Kagakudogin, Kyoto.

ZERO GAP MEMBRANE CELL AND SPE CELL TECHNOLOGIES

VS. CURRENT DENSITY SCALE UP

A. Nidola

Oronzio De Nora
Impianti Elettrochimici S.p.A.
Milan, Italy

ABSTRACT

Both zero gap and SPE* cells technologies, in the field of chlor-alkali process, have been studied and compared vs. current density scale up.

The zero gap cell technology includes a permselective bilayer membrane with metallic electrodes, activated with suitable electrocatalysts for the wanted reactions, contacting the membrane itself.

The SPE cell technology includes a permselective bilayer membrane contacting the metallic anode, $DSA^{(R)}$, activated with suitable electro-catalysts for chlorine evolution, whereas the cathode electrode is made directly on the cathode side of the bilayer membrane in order to form a membrane bonded cathode system.

The SPE cell configuration compared with a zero gap cell achieves:

- better current density distribution

- lower or nil cathode bubble effect and lower polarization of caustic conc. even at high current density and high caustic strength

- longer membrane life and

- lower cell voltage with more or less the same caustic faraday efficiency

Examples of V/A characteristics for both technologies are given, detailed and commented on in the paper.

---

*The term 'SPE' (solid polymer electrolyte), which has been used worldwide since 1969 to indicate an electrolytic cell equipped with an 'activated' permselective membrane, should be replaced by the new term 'M & E' (membrane bonded electrode) technically more appropriated and pertinent for the matter. Therefore in the text - in order to avoid misunderstandings the 'old term SPE' shall be always written together with the 'new term M & E' according to this form: SPE (M & E).

Up-to-date, in the field of the chlor-alkali process made with membrane cells, three technologies can be considered:

1) narrow gap technology
2) zero gap technology and
3) membrane bonded electrode technology

The narrow gap shows great limitations mainly regarding the mechanical stability of the membrane. In other words this means that inert and sophisticated spacers must be inserted between the membrane and the electrodes in order to prevent or to keep under control membrane deformations which are undesired from the point of view of the current density distribution (cell performance). On the other hand the presence of spacers reduces the effective active area of the membrane with the consequence increase of the current density on the working surface of it.

Another possibility is to have an asymmetric narrow gap i.e. located only from one side for example at the cathodic one by keeping the membrane directly in contact with the anode through the application of a certain hydrostatic pressure.

However, even in this case, the pressure gap must be limited otherwise a penalty in the caustic faraday efficiency can take place due to the increase of the caustic back diffusion. Through these simple considerations the narrow gap cell is a technology based on a lot of delicate compromises. As consequence, at the present time, it seems to be less attractive compared to both the zero gap and the SPE (M & E) systems, the two technologies which are the object of this presentation[1].

The presentation shall follow these lines:

a) Brief description of the two technologies: zero gap and SPE (M & E).

b) Technical characteristics including:

1. Electric field
2. Polarization of concentration
3. Bubble effect
4. Catalytic activity of the electrodes

c) Electrochemical performances vs. current density and caustic strength scale ups.

d) Break down voltage.

BRIEF DESCRIPTION OF THE TECHNOLOGIES

Brief definitions and descriptions of the two technologies:

i) zero gap and
ii) SPE (M & E)

are given, before entering in detail about their characteristics and performance, in order to give a better understanding of this comparison.

Zero Gap Systems

A zero gap system consists of an electrolytic cell equipped with a permselective bilayer membrane directly contacting the two electrodes; particularly the anodic membrane layer contacts the anode (DSA$^{(R)}$) whereas the cathodic membrane layer contacts the cathode (DSC$^{(R)}$). The two electrodes - one of the peculiarities of the ODN cell design - are not fixed but removable; in other words they are located between the permselective membrane and the current leads in a proper way in order to allow satisfactory electric contacts during cell operation and an easy mechanical removal during cell disassembly for inspection and recoating if necessary. Through this description the components of a zero gap cell include:

- permselective bilayer membrane

- removable electrodes

- fixed current leads

according to the model shown below in Figure 1.

SPE (M & E)

An SPE (M & E) system consists of a bilayer permselective membrane having the electrocatalysts, anodic and/or cathodic, directly applied on its surface. From this point of view the SPE (M & E) represents a very compact electric unit. Through this definition the components of an SPE (M & E) system include:

A permselective bilayer membrane having a) on the anodic surface an electrocatalyst suitable for promoting the anodic reaction of the chlorine evolution and to prevent the parasitic reaction of the oxygen evolution and, b) on the cathodic surface an electrocatalyst able to favor hydrogen evolution.

A nickel cathodic current lead on the cathodic side consisting of two parts, the former contacting the activated membrane highly flexible and having a large density of electric contacting points whereas the latter, in the form of a rigid structure, far from the membrane.

A titanium anodic current lead on the anodic side consisting as for the previous case, of two parts, one directly contacting the activated membrane elastic enough to assure a satisfactory electric contact distribution whereas the second one far from the membrane consisting of a rigid structure.

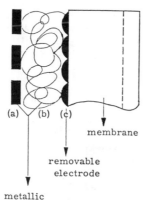

(a)   (b)   (c)

→ membrane

→ removable
electrode

→ metallic
current
lead system   (a) rigid massive structure

(b) flexible thin structure

Fig. 1. Zero gap cell: schematic view

283

A schematic view of an SPE (M & E) system is given in Figure 2. An alternative to the SPE (M & E) system described above includes an hybrid system wherein the membrane bonded electrode concerns the cathode side or y whereas the anode one consists of a metallic anode, having a high active surface contacting the membrane in a proper way.

TECHNICAL CHARACTERISTICS

Descriptions and Comparisons

The technical characteristics involved for the comparison concern four aspects which are fundamental in the economy and the classification of the two proposed technologies. They include:

1. Electric field
2. Polarization of concentration phenomena
3. Bubble effect and
4. Catalytic activity of the electrodes

Electric field[1]. Graphic examples given in Figure 3 are evidence that the distribution of the electric field varies through the cell cross section - mainly across the membrane - vs. the cell design or the electrode configuration. Through this simple scheme the role played by both the electrode geometries and the membrane-electrode gap can be seen.

α) schematic cross section view

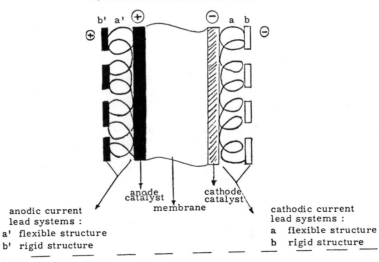

anodic current
lead systems :
a' flexible structure
b' rigid structure

cathodic current
lead systems :
a flexible structure
b rigid structure

β) schematic mass balance picture

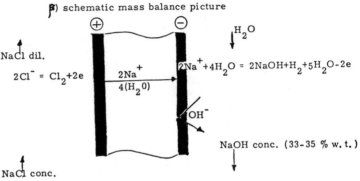

Fig. 2. SPE (M & E) system.

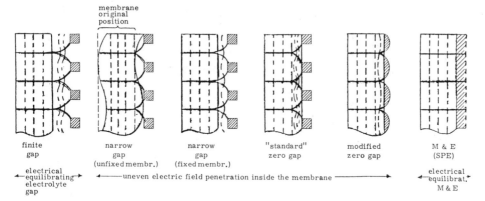

membrane
original
position

finite	narrow	narrow	"standard"	modified	M & E
gap	gap	gap	zero gap	zero gap	(SPE)
	(unfixed membr.)	(fixed membr.)			

electrical
←equilibrating→
electrolyte
gap

←————uneven electric field penetration inside the membrane ————→

electrical
←equilibrat.→
M & E

Fig. 3. Electric field profiles vs. membrane cell configurations.
□ membrane; ▨ electrode (cathode); --- equipotential line;
—— electric line.

For example, across the membrane, the electric field has the tendency
to become only in two opposite cases, i.e.:

- high gap cell and
- zero gap cell provided that the electrodes are formed or applied on
  the membrane surfaces in order to have an SPE (M & E) system.

In all the other combinations, including the zero gap cell, both
traditional and modified, an uneven current density distribution takes
place. Regarding the electric field profile around the electrodes the best
situation is observed again with the M & E system whereas finite gap becomes critical as for electrodes having a finite surface are used. Through
the graphic analyses made and commented on above a classification regarding
the electric field profile vs. cell design written in the decreasing order
of its uniformity degree is tentatively the following:

SPE (M & E) > modified zero gap $\geq$ finite gap > zero gap >

> narrow gap > narrow gap
(fixed membrane) (unfixed membrane)

Polarization of concentration[1]. The polarization of concentration –
which is an electrochemical phenomenon dependent on current density –
discussed in this section considers only the cathode side which is more
critical than the anodic side. However all the concepts and the models
taken into consideration for the cathode side can be extrapolated to the
anode one. In Figure 4 are drawn the profiles concerning:

- electric transfer number of [OH^-] and the
- concentration (activity) of caustic soda

vs. five cell geometries, i.e. finite gap, narrow gap (fixed membrane),
zero gap (both traditional and modified) and SPE (M & E).

According the the models drawn above, the electric number equations in
the three cell sections for 1 [F] passing are [2] [3];

anolyte	$t\ Na^+ + t\ Cl^- = 1;\ t\ Na^+ \cong t\ Cl^-$
membrane	$\bar{t}\ Na^+ \cong 1;\ t\ OH^- \cong 0$
catholyte	$t\ Na^+ + t\ OH^- = 1;\ t\ OH^- \cong 4\ t\ Na^+$
and the cathode reaction	$H_2O = \tfrac{1}{2}H_2 + OH^- - 1\ F$

285

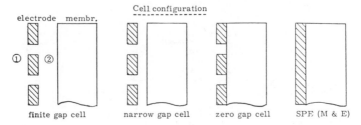

Fig. 4. Caustic soda polarization of concentration – Polarization profiles distribution.

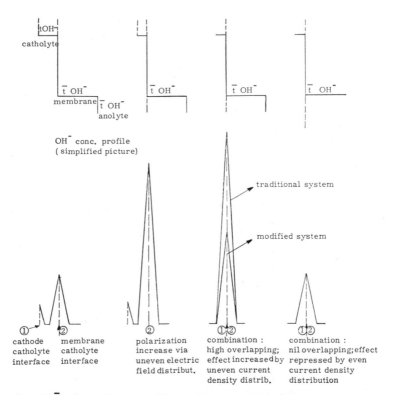

Fig. 5. OH⁻ electric transfer number profile (simplif. picture).

seems evident that the best situation, wherein the caustic polarization is kept as low as possible, is obtained by using an SPE (M & E) system. In the case of zero gap technology a higher polarization, even for the modified version, arises owing the uneven current density distribution both through the membrane and on the cathode.

Bubble effect. The bubble effect is a phenomenon depending on many variables such as:

- membrane type (equivalent weight and structure)
- cell design and electrode configuration
- caustic strength
- current density
- operation time (aging phenomena) etc.

By using treated membranes i.e. membrane having the surface contacting the electrolytes coated with hydrophilic compounds, as for example TiO_2, SiO_2, ZrO_2, Nb_2O_5 etc., the bubble effect or [4] in other words, the hydrogen gas sticking, is cancelled or drastically reduced.

Particularly, as an example, if the cathodic side is considered, oxide treated membranes, in practice, do not show any bubble effect phenomenon even if the cathode is contacting the membrane itself provided that the electrolysis is carried at typical conditions used today (NaOH - 32-35%, temperature 90°C, current density 2-3 kA/m^2).

However, the bubble effect, which disappeared under the above conditions, emerges again when operating under more severe conditions for example by increasing current density (>5 kA/m^2) and/or caustic strength (>40%).

This tendency is enhanced by working time indicating that, in some way, phenomena related to the membrane may cause variation in the surface tension which is the parameter ruling the sticking of the bubble gases to the organic polymer.

electrode membrane

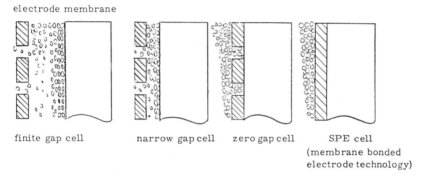

finite gap cell   narrow gap cell   zero gap cell   SPE cell
                                                               (membrane bonded
                                                               electrode technology)

Fig. 6.   Bubble gas (bubble effect) distribution.

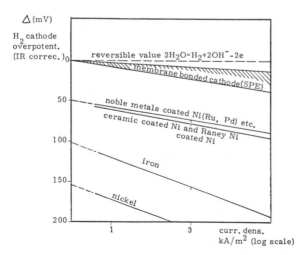

Fig. 7.   Catalytic activity for hydrogen evolution reaction vs. cathode type and geometry (cathodic polarization curves)[6]. Working conditions: NaOH 32%; Temp. 90°C.

Through this information, which shall be detailed later on, graphic models on the bubble effects vs. cell technologies can be tentatively given in Figure 6 indicating that, again the SPE (M & E) system - due to its structure - shows, in practice, a nil ohmic drop penalty.

Catalytic activity of the electrodes. Also in this case, the last of this chapter, the cathode side only - in order to simplify the presentation - is put under focus. However all the concepts and the experiences taken into consideration for the cathode side can be, in some cases, extrapolated to the anode one.

In Figure 7 are plotted the catalytic activities, concerning the reaction

$$e + H_2O = \tfrac{1}{2}H_2 + OH^-$$

expressed in terms of hydrogen overpotential (IR corrected) vs. different cathodes (types and geometries).

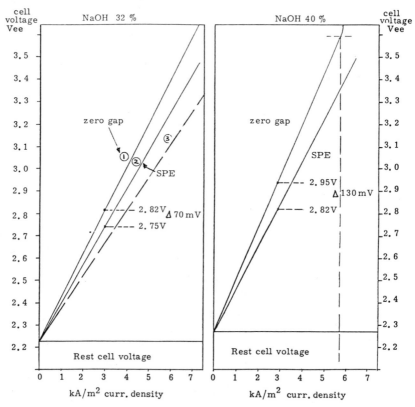

Fig. 8.   SPE (M & E) cell technology vs. zero gap cell technology V/A characteristics vs. caustic strength. Line 1 - zero gap (modified system); line 2 - SPE (M & E); line 3 - practical limit. Working conditions:

NaOH 32%		NaOH 40%
conc. anolyte	190 g/l	190 g/l
pH anolyte	≈ 3.5	3.5 – 4.0
temperature	90°C	90°C
operation time	250 hrs	250 hrs

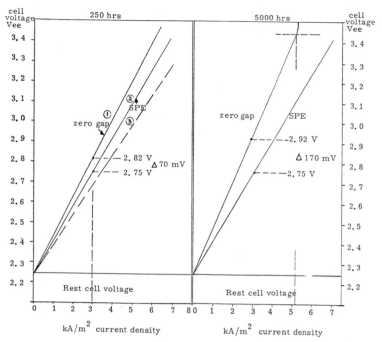

Fig. 9. SPE (M & E) cell technology vs. zero gap cell technology –
V/A characteristics vs. operation time. Line 1 – zero gap
(modified system); line 2 – SPE (M & E); line 3 – practical limit.
Working conditions:

	250 hrs	5000 hrs
conc. anolyte	190 g/l	190 g/l
pH anolyte	≅ 3.5	3.5 –
caustic strength	32%	32%
temperature	90°C	90°C

Table 1. Electrochemical Terms Involved in the Cell Break Down Voltage

Electrochemical terms	zero gap (modified system)	SPE (M & E)
Anode side	Anolyte -gas and liquid (ohmic drop)	the same
	Anode -coating (Tafel and polariz. of conc.)	the same
	-structure (ohmic drop)	the same
Cathode side	Catholyte -gas and liquid (ohmic drop)	absent
	-coating (Tafel and polariz. of conc.)	the same
	-structure (ohmic drop)	the same
Membrane	-Donnan potential	the same
	-inner resistivity	the same
	-polar. of conc.	the same
	-bubble effect	absent

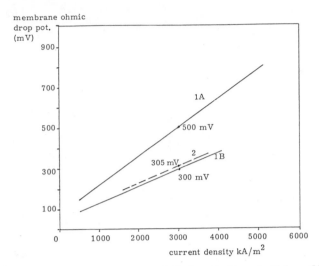

Fig. 10. Membrane ohmic drop potential (excluding bubble effect contributions (gas sticking)) vs. current density and membrane type – (inner resistivity + Donnan potential). 1A – untreated old generation; 1B – untreated new generation; 2 – treated new generation. Working conditions: Anolyte – conc. 175 g/l – pH 3.0; catholyte – NaOH 32%; temperature 90°C.

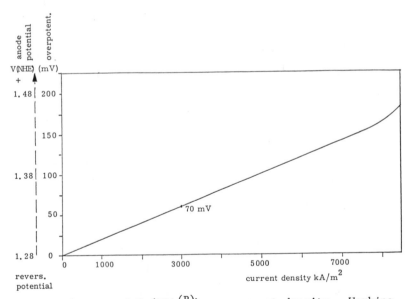

Fig. 11. Anode potential (DSA $^{(R)}$) vs. current density. Working conditions: Anolyte – conc. 175 g/l – pH 3.0; temperature 90°C.

The analyses of the cathode polarization curves (Tafel's lines) made in NaOH 32%, at 90°C, provide evidence of the superiority of the membrane bonded cathode in respect to all the others. The main explanation for the large gap between the M & E (cathode) and all the others investigated LHOC (low hydrogen overpotential cathodes) is related to the surface area. In practice in the case of the M & E the cathode surface is so large, dis-

persed and uniform, that the determining step of the hydrogen evolution
overall reaction is given by recombination or dimerization[5]

$$H_{AD} + H_{AD} = H_2$$

which is, in practice, current density independent. On the contrary for
the other cases, wherein a finite surface is used, the determining step of
the hydrogen evolution is given by

$$H + H^+ + e = H_2$$
or
$$2H^+ + 2e = H_2$$

which both are current density dependent[5].

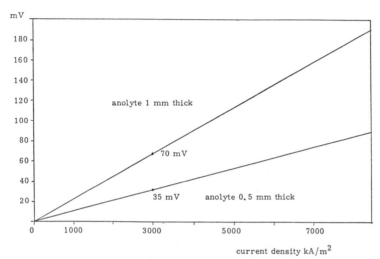

Fig. 12.   Anolyte ohmic drop vs. current density and catholyte thickness –
           (gas free). Anolyte – conc. 175 g/l – pH 3.0; temperature
           90°C; anolyte thickness – 0.5 mm, 1.0 mm.

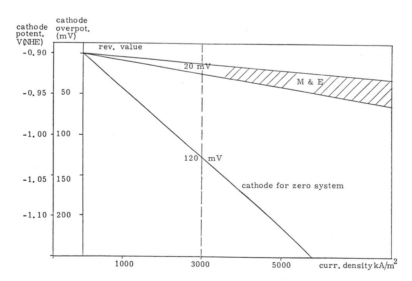

Fig. 13.   Cathode potential vs. current density. Caustic strength – 32%;
           temperature   90°C

ELECTROCHEMICAL PERFORMANCES VS. CURRENT DENSITY AND CAUSTIC SODA STRENGTH: SCALE-UPS FOR DIFFERENT OPERATION TIMES

Examples of V/A characteristics vs:

- current density scale-up
- caustic strength increase and
- operation time

for zero both gap (modified system) and SPE (M & E) cell technologies are given in this chapter.

Through the analyses of the experimental data summarized in Figures 8 and 9 the following remarks can be tentatively made:

i) SPE (M & E) cell technology shows V/A characteristics having a slope less than the one observed on zero gap (modified system) cell technology over all the explored range of current density (0-7 kA/m^2).

ii) The voltage gap between the two compared technologies increases more and more vs. current density scale-up, caustic soda strength increase and operation time indicating that SPE (M & E) technology becomes more attractive in the range of high current density and high caustic soda strength.

iii) The V/A characteristics related to SPE (M & E) remain straight lines over all the range of experimental current densities, caustic soda strength and operation time.

iv) The V/A characteristics related to zero gap (modified system) have the tendency to increase the slope vs. current density scale-up, caustic soda strength and operation time. The change in the slope, for the case of the zero gap (modified system) cell technology might be interpreted in terms of additional ohmic drop penalties arising on the membrane under certain working conditions and probably ascribed to a complex combination of bubble effect (hydrogen gas sticking) and caustic polarization of concentration phenomena.

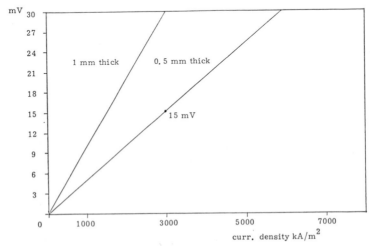

Fig. 14.   Catholyte ohmic drop vs. current density and catholyte thickness - (gas free). Working conditions: Catholyte strength - 32%; temperature - 90°C.

CELL BREAK DOWN VOLTAGE

The cell break down voltage for the two technologies have been determined. Through this analysis the electrochemical terms responsible for the
voltage gap between the two technologies are put under focus. The electrochemical terms taken into consideration for making this analysis are listed
in Table 1.

The determination of the various break down voltage contributions
i.e.:

- anode and cathode potentials (reversible + overpotential)

- electrode structure ohmic drops

- electrolytes resistivities

- membrane inner resistivity and Donnan potential have been made by
  using well known traditional techniques involving:

  - Luggin probes
  - electronic voltmeters at high inner resistance and, finally,
  - in the case of the membrane, a multicompartment lab. cell has also
    been used.

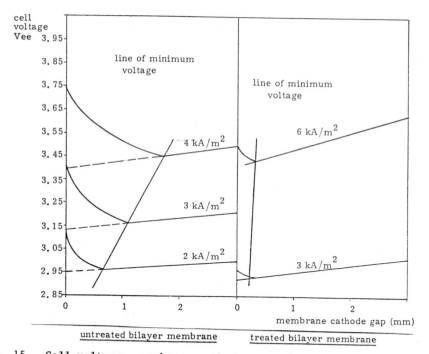

Fig. 15. Cell voltage – membrane cathode gap relationships vs. membrane
type and current density. Working conditions:

	untreated	treated
Anolyte conc.	175 g/l	175 g/l
pH	3.5 – 4.0	3.5 – 4.0
Oper. time	100 hrs	100 hrs
Temperature	90°C	90°C
Catholyte strength	32%	32%
Membrane resistivity	$\approx$ 500 mV/3kA/m^2	$\approx$ 300 mV/3kA/m^2

Concerning the membrane bubble effect term (gas sticking to the membrane) we underline that it has been measured by using a particular lab. size cell with a variable 'membrane-electrode' gap. The bubble effect has been determined only at the cathodic side for sulphonated carboxylated membranes which have been used in this type of experiment.

The adopted procedure for making these measurements includes:

i) cell voltage values determination vs. 'membrane-cathode gap' for values approaching a zero gap configuration.

ii) repeat the above measurements vs. current density scale-up caustic strength increase and operation time.

iii) in the absence of gas sticking phenomena straight lines are observed – wherein the cell voltage values gradually decrease vs. cathode-membrane gap decrease according to the ohmic law ruling the catholyte resistivity contributions. (As it is known, membranes having a sulphonated anodic layer, do not show any gas sticking phenomenon).

iv) in the presence of gas sticking phenomena an abnormal increase in the cell voltage values vs. cathode-membrane gap decrease is, conversely, observed.

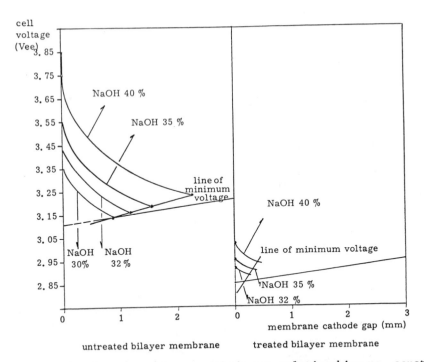

untreated bilayer membrane   treated bilayer membrane

Fig. 16. Cell voltage – membrane cathode gap relationships vs. caustic strength and membrane type. Working conditions:

	untreated	treated
Anolyte conc.	175 g/l	175 g/1
pH	3.0	3.5 – 4.0
Temperature	90°C	90°C
Current density	3 kA/m^2	3 kA/m^2

The graphic representation in this case consists of two parts: one, which is straight line obeying the ohmic drop law ruling the catholyte resistivity, concerns a gap region above a certain critical value depending on working conditions; the second one, which is a curve laying in between zero gap and critical gap, increases more and more vs. membrane cathode gap decrease reaching maximum values at the zero gap position.

Also in this case the values read at zero gap position depend on current density, caustic strength and operation as they increase by increasing these mentioned parameters.

The values of the bubble effect, under certain working conditions, are obtained by making the difference, at zero gap position, between the experimental values laying on the curve and the corresponding theoretical ones laying on the straight line. (All these concepts are briefly summarized and pictured in the next figures showing these relationships (Figures 15, 16, 17, 18, 19 and 20).

By following the classification order written in the Table D-1 graphic relationships for all the involved electrochemical terms - used to determine cell voltage break down - are given.

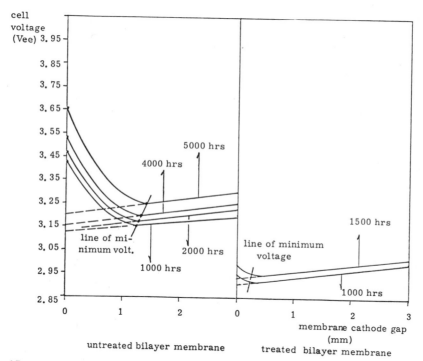

Fig. 17. Cell voltage - membrane cathode gap relationship vs. operation time and membrane type. Working conditions:

	untreated	treated
Anolyte conc.	175 g/l	175 g/l
pH	3.5 - 4.0	3.5 - 4.0
Caustic strength	32%	32%
Temperature	90°C	90°C
Current density	3 kA/m²	3 kA/m²
Operation time (hrs)	250	250
	1000	1000
	2000	1500
	4000	
	5000	

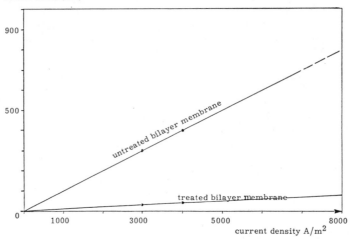

Fig. 18. Membrane cathodic bubble effect – current density relationships
vs. membrane type (250 – 1000 hrs of operation time). Working
conditions: Anolyte conc. 175 g/1 – pH 3.5 – 4.0; temperature
90°C; Caustic strength – 32%; operation time – 250 – 1000 hrs.

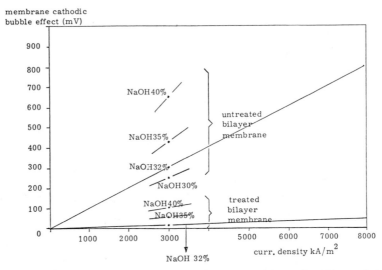

Fig. 19. Membrane cathodic bubble effect – current density relationship
vs. caustic strength (operation time 250–1000 hrs). Working
conditions: Anolyte – conc. 175 g/1 – pH 3.5 – 4.0; temperature
90°C; operation time – 250–1000 hrs.

According to the experimental data detailed and commented above the
cell break down voltage for the two technologies can be redrawn as shown in
Figure 21.

CONCLUSIONS

SPE (M & E) cell technology is an attractive system compared to zero
gap cell technology mainly in the range of high current density (above

296

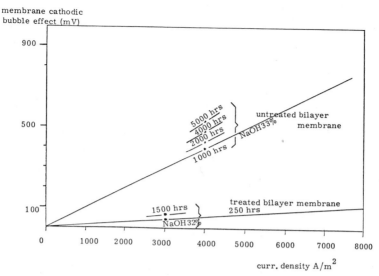

Fig. 20. Membrane cathodic bubble effect – current density relationship
vs. operation time (membrane aging). Working conditions:
Anolyte – conc. 175 g/1; – pH 3.5 – 4.0; temperature 90°;
caustic strength – 32%; cruising current density – 3 kA/m².

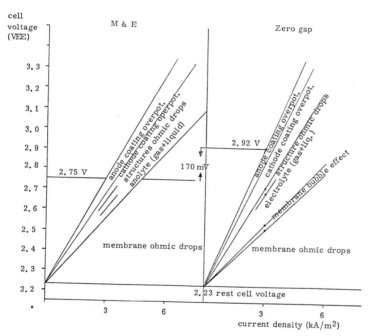

Fig. 21. V/A characteristics (5000 hrs operation time). Working
conditions: Anolyte – conc. 175 g/1; – pH 3.5 – 4.0; caustic
strength – 32%.

3.0 kA/m²) and high caustic soda strength (> 40% w.t.). Moreover the
expected life of the membrane in SPE (M & E) cell technology should be
higher than the one obtained with the zero gap systems under the same
working conditions owing to a better current density distribution which
is one of the most critical parameters for the membrane aging (durability).

FOOTNOTE

For a correct reading and a proper understanding of the V/A characteristics plotted in the previous diagram the following additional notes are
given below.

Electrode Structure - ohmic Drops

The ohmic drop penalties of the electrode structures are determined
for a (100 x 1000) mm active area size (zero gap: $\cong$ 30 mV, M & E: $\cong$ 15 mV).

Electrolyte ohmic Drops

The term electrolyte ohmic drop penalties included in the final diagram consist of both liquid and gas contributions. This contribution has
been determined via the 'indirect way' i.e. by subtracting from the cell
voltage (V.E.E.) all the electrochemical known terms with the exception of
the electrolyte ohmic drop penalties:

$$
\text{Vee} - \left[
\begin{array}{ll}
\text{anode coating potential} & + \\
\text{cathode coating potential} & + \\
\text{membrane resistivity} & + \\
\text{membrane bubble effect} & + \\
\text{structure ohmic drops} &
\end{array}
\right] =
$$

= electrolytes (gas + liquid) ohmic drop contributions

These numbers are in accordance with the measurements obtained by
experimental route of both electrolytes resistivities gas free and the
electrode bubble effects as shown in the following table.

anode side		cathode side	
anolyte ohmic drop (0.5 mm thick)	35 mV	catholyte ohmic drop (0.5 mm thick)	15 mV
anode (chlorine) bubble effect	30 mV	cathode bubble effect	50 mV
overall anode side	65 mV	overall cathode side	65 mV

Through these data the overall anolyte and catholyte (gas liquid)
contribution is 130 mV, which is very close to the 150 mV found via the
indirect way.

REFERENCES

1. R. Jackson, Transport in Porous Catalyst: Chemical Engineering and
   Monograph, pp.25,51,59, Elsevier, New York (1977).
2. P. Gallone, Elettrochimica Industriale, Ed. Tamburini (1975).
3. G. Bianchi, Impianti elettrochimici industriali, Ed. Quadri (1970).
4. R. J. Thomson, Surface tension phenomena, Ed. Paidon, London (1903).
5. J. O'M Bockris and D. M. Drazic, Electrochemical Science, Taylor &
   Francis Ltd., London (1972).
6. M. W. Breiter, Proceedings on the symposium on electrocatalysis,
   pp.1,94,142,156, The Electrochemical Society, Inc., Princeton
   (1974).

ELECTRODIALYSIS IN THE SEPARATION OF CHEMICALS

W. A. McRae

Zurich
Switzerland

ABSTRACT

Modern Electrodialysis (ED) is a simple, convenient, flexible process for rapidly moving low-molecular weight organic and inorganic ions from one solution to another by means of highly conductive, ion-selective membranes and low-voltage direct electric current. Two types of membranes are used simultaneously – one type selective to cations such as sodium, hydrogen, potassium, ammonium, guanidinium, tetra-alkyl ammonium, calcium and magnesium, the other type selective to anions such as chloride, hydroxide, nitrate, perchlorate, acetate, phosphate, citrate, sulfate and bicarbonate. The pressure used is low, only enough to pass the solutions which are to be deionized/respectively ionized over the surfaces of the membranes. Very little water goes through the membranes, essentially only the water of hydration of the ions transferred, although in the case of deionization of concentrated solutions of electrolytes, the transport of such water of hydration can be useful in concentrating non-ionized materials present in the solution deionized. The transport of ions is not generally accompanied by the transport of non-ionized materials of molecular weight greater than about 200 although special membranes are available which permit non-ionized materials of molecular weight up to about 500 to be transported with the water of hydration of the ions.

In contrast to ultrafiltration, diafiltration, dialysis and ion-exchange, ED permits the ions which are removed to be recovered in solution more concentrated than that from which they were removed and free of foreign ions. (Concentrations can be obtained up to about 4N or as may be limited by the solubility of the least soluble electrolyte). Often such concentrations will permit the re-use of the recovered electrolytes (e.g. guanidine hydrochloride). Chemical regenerants are not required as in the case of deionization with granular ion-exchangers. Ultrapure water requirements are minimal in contrast to diafiltration and dialysis. Deionization of 99 percent or more is easily achieved.

Batch-mode laboratory and commercial scale ED can be operated at constant low pressure (about 1.5 atmospheres) and constant low voltage (about 0.5 volts per membrane) and is particularly user-friendly. ED is simply continued until the desired electrolyte level is obtained, as may be measured by electrical conductivity. ED membranes and apparatus can be operated continuously at temperatures up to 60°C and can be sanitized-in-place with strong acids and/or alkalies or dilute solutions of active chlorine. ED apparatus, including the membrane modules, can be easily and rapidly dissassembled for inspection.

Owing to its simple hydrodynamics, ED is relatively tolerant of colloids and particulates. In the event that high molecular weight charged materials are concentrated by electrophoresis against one of the membranes, they can be moved back into solution by a brief reversal of the direction of the electric current.

ED is now being used commercially to reduce the high salt concentration (2N NaCl) present in an alpha-interferon eluate and to recover guanidine hydrochloride in high yield. In the laboratory, ED has been used to precipitate euglobulins from plasma, to salt-out proteins (and subsequently to deionize the supernatant) and to deionize protein and polypeptide solutions (MW as low as 1000).

INTRODUCTION

In 1940, Meyer and Strauss[1] suggested a multi-compartment ED process using ion-selective membranes (see Figure 1). Membranes selective to cations alternated with membranes selective to anions. When a direct current voltage was applied, cations in solutions between the membranes tended to migrate toward the negatively-charged electrode (cathode). They were able to penetrate the cation-selective membranes but not the anion-selective membranes (to the extent that the latter were perfectly selective). Similarly anions, tending to migrate toward the positively-charged electrode (anode), were able to penetrate the anion-selective membranes but not the cation-selective membranes, again to the extent that the latter were perfectly selective. As a result, every other compartment (space defined between any given pair of membranes) became depleted in electrolyte (deionized or partially deionized) and the intervening compartments became enriched in electrolyte. The membranes available to Meyer and Strauss were not commercially practical; those which had high ionselectivity also had high electrical resistance and those which had low electrical resistance also had low ion-selectivity. Further the membranes were not mechanically strong or chemically and biologically stable.

MODERN MEMBRANES

In 1950, Juda and McRae[2] described ion-selective membranes which had high ion-selectivity, low electrical resistance, good mechanical strength and good chemical and biological stability. Chemically these were (and are) ion-exchange resins in sheet form. Modern cation-selective membranes for ED (and related applications) typically consist of cross-linked vinyl

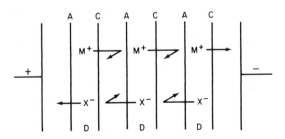

Fig. 1. Schematic diagram of electrodialysis.

polymers having negatively charged sulfonate groups chemically bonded to the polymer. The negative charges of the sulfonate groups are electrically balanced by positively charged cations (counter-ions). The counter-ions are appreciably dissociated from the bound, negatively-charged sulfonate groups into water absorbed by the membrane and are mobile in such water. The high concentration of counter-ions in the sheet-form ion-exchange resin results in low electrical resistance (typically about 1 milli-ohm per square meter). The high concentration of bound, negatively-charged groups tends to exclude negatively-charged ions (co-ions) from solution in contact with the membrane and is responsible for its high ion-selectivity. In general, the ion-selectivity

- increases as the concentration of sulfonate groups on the polymer increases
- decreases as the amount of water in the membrane increases
- decreases as the electrolyte concentration increases in the ambient solution.

Anion-selective membranes generally have positively-charged quaternary ammonium groups chemically bonded to the polymer. In this case, the counter-ions are negatively charged and are the principal carriers of electric current.

Ion-selective membranes for industrial applications of ED are re-inforced with chemically and biologically stable fabrics to improve mechanical properties. Several hundred thousand square meters of ion-selective membranes are now produced annually. The transport properties of most importance are:

- the electrical resistance per unit area
- the fraction of d.c. electric current carried by counter-ions (current-efficiency)
- the quantities of water and of electrically neutral solutes (of low-to moderate-molecular weight) accompanying the electrical transport of counter-ions through the membranes
- the diffusion of electrolytes and of neutral solutes (of low-to moderate-molecular weight) through the membranes under the influence of concentration gradients for each.

Commercial ion-selective membranes typically have thicknesses in the range of from about 0.15 to about 0.6 mm and electrical resistances in the range of from about 0.3 to about 2 milli-ohms per square meter at room temperature in equilibrium with half-normal sodium chloride solution. Electrical resistances will be somewhat higher in more dilute solutions since the bound-ions in the membrane will more effectively exclude co-ions from such solutions.

The fraction of the electrical current which is transported by counter-ions through commercial membranes is typically in the range of from about 85 to 95 percent when the membranes are in equilibrium with dilute electrolyte solutions.

The electrical transport of water, essentially as water-of-hydration of the ions, is generally in the range of from about 100 to about 200 cm^3 per gram-equivalent of counter-ions transferred, towards the low end of the range for anion-selective membranes, toward the high end for cation-selective. In the case of dilute solutions, such electrical transport of water is generally not of practical importance. In more concentrated solutions, e.g. 1 N, electrical water transport can be a significant fraction of the volume of the solution deionized. In some industrial applications, for example the deionization of protein solutions, such water transfer is a process advantage.

Low-molecular-weight neutral molecules tend to be absorbed into the membranes with the absorbed water. However, the pores in the membranes have diameters about the same size as the diameters of such molecules. As a result, both electrical transport (with the water of hydration) and diffusion of dissolved, neutral organic molecules of molecular weight more than 100 tend to be strongly inhibited. For example, concentrated solutions of disaccharides such as lactose and sucrose can be deionized with very little transport or diffusion of the neutral molecules. Almost no protein or polypeptide will be lost when solutions of either or both are deionized by ED. Special membranes are available which permit non-ionized materials of molecular weight up to about 500 to be transported.

"Back-diffusion" is the transport of electrolyte (that is, equivalent amounts of co-ions and counter-ions) under the influence of concentration gradients developed between enriched and deionized compartments during ED. Such back-diffusion is experienced as a decrease in the current efficiency of the apparatus. It increases with the concentration difference between the deionized and the enriched compartments and decreases as the ion-selectivity of the membranes increases. Back-diffusion between water and 1 N sodium chloride solution is generally less than about 20 microgram equivalents per second per square meter, that is, equivalent to about 2 amperes per square meter.

APPARATUS

In principle, ED apparatus is an array of alternating anion-selective and cation-selective membranes terminated by electrodes (see Figure 1). The membranes are separated from each other by gaskets which form compartments through which fluids can be passed. Compartments having anion-selective membranes on the side facing the positively charged electrode (anode) are electrolyte depleting (deionizing) compartments. Since the anion-selective and cation-selective membranes alternate in the array, the deionizing- and enrichment-compartments also alternate. The array of membranes and separating gaskets is symmetrical, hence the choice of which electrode is the positive pole and which the negative is arbitrary. Always, however, those compartments having anion-selective membranes on the side facing the positively-charged electrode will be deionizing compartments. Referring again to Figure 1, the left-hand electrode is represented as positively-charged, one alternating group of compartments deionizing and the other group enriching. If the right hand electrode had been chosen as the positive electrode, the roles of the two groups of alternating compartments would have been interchanged. Brief periodic reversal of polarity or even regular, symmetric reversal is often used in ED to keep membrane surfaces free of high molecular weight charged colloids.

In an array of ion-selective membranes and separating gaskets, holes in the membranes and gaskets (for example, as in Figure 2) register with each other to provide two pairs of internal manifolds to carry liquid into and out of the compartments. One pair of manifolds communicates with the deionization compartments. A contiguous group of two membranes and the two fluid compartments defined by them is called a "cell-pair". A group of cell-pairs with their end electrodes is called a "pack" or a "stack". Several hundred cell-pairs may be arrayed in a single stack, the number depending on the ED capacity required, the uniformity of flow distribution among compartments of the same type in the stack and the maximum DC voltage which is acceptable. One or more stacks are contained in a press which compresses the membranes and gaskets against the pressure of the liquid flowing through the compartments, thereby preventing leaks to the outside and internal cross-leaks between adjacent compartments (Figure 3). For small presses, such compression is usually provided by tie-rods; for large presses hydraulic rams are frequently used.

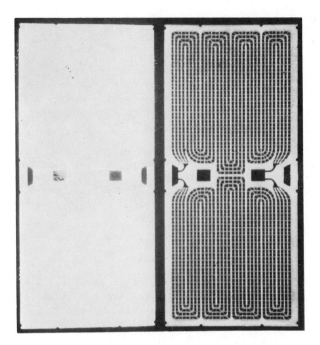

Fig. 2.   457 x 1016 mm anion-selective membrane and an intermembrane
turbulence-promoting gasket.   (Courtesy of Ionics, Inc.).

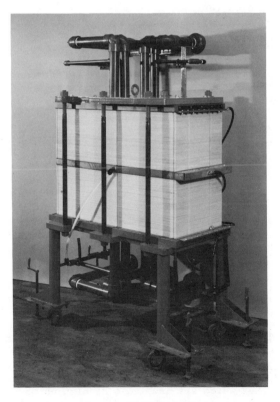

Fig. 3.

As mentioned above, membranes typically have thicknesses in the range of from about 0.15 to about 0.6 mm. Compartments between membranes typically have thicknesses between about 0.5 and about 2 mm. The thickness of a cell-pair is therefore from about 1.3 to about 5.2 mm, typically 3.2 mm. The effective area of a cell-pair for ion transport is generally in the range of from about 0.2 to about 2 m^2, typically about 0.3. (Stacks suitable for laboratory use typically have effective areas in the range 0.0005 to 0.02 m^2, i.e. 5 to 200 cm^2. A complete laboratory apparatus has been recently introduced having a price of less than US\$ 1,000.00.

In most applications of ED, a direct current of 30 amperes will transfer about 1 gram-equivalent of electrolyte per hour for each cell pair. A typical commercial stack operating at 100 amperes and having 300 cell-pairs will transfer about 1000 gram-equivalents per hour. If the electrolyte is sodium chloride this is about 60 kilograms per hour or about 480 tonnes in an operating year of 8000 hours.

LIMITING VOLTAGE

When electrolytes more concentrated than about 0.1 N are deionized, the direct electric current applied to the stack will be limited by engineering and economic considerations to about 300 amperes per square meter (about 100 amperes in a typical commercial stack). In the case of more dilute electrolytes, current will be limited by the ability of ions to diffuse to the membranes from the interior of solution passing through the deionizing compartments. As mentioned above, the current efficiency for transport of counterions through a membrane is generally 85 to 95 percent. However, in the interior of a solution, roughly half the current is carried by anions and half by cations. For example in the case of sodium chloride, only about 60% of the current is carried by chloride (the remainder by sodium). Hence 25 to 35% of the chloride passing through the anion-selective membranes must be transported to the membrane-solution interface by diffusion and convection. Near the interface, convection is ineffective and the difference between the current efficiencies of the membrane and the solution must be provided by diffusion alone. The concentration of electrolyte at the interface becomes reduced from the bulk, interior value to provide that concentration gradient sufficient to give the necessary diffusion. In order to reduce the thickness of the diffusion layer, thereby to increase the rate of diffusion, designers usually include some kind of structure in the ED compartments (for example, as shown in Figure 2) to mix the interfacial region and bring bulk electrolyte as close (and as uniformly close) as possible to the membrane surface by convection.

If the membrane-solution interface becomes depleted of electrolyte almost completely then the electrical resistance of the apparatus will increase significantly, even though the solution in the interior of the deionizing compartments may still have an appreciable concentration of electrolyte. Such increase in electrical resistance is called "concentration-polarization" or more simply "polarization". When the apparatus is polarized, the concentration of electrolyte at the membrane-solution interface will be comparable to the concentration of ions resulting from the normal dissociation of water. Hence, for example, in the case of polarization of an anion-selective membrane, a significant fraction of the current through such membrane will be carried by hydroxide ions into the enrichment compartment and hydrogen ions will be carried into the interior of the solution in the deionizing compartment. Changes in the pH's of the compartments are a second indication of polarization, though such changes can be obscured by the presence of buffers. If the polarization is substantial, electrolytes which are insoluble at high pH may precipitate at the interface between the anion membrane and the solution in the enrichment compartment. Local acidity at the interface between the anion membrane and the solution in the deionizing compartment can result in agglomeration of colloids sensitive to low pH's. Such precipitation in the enrichment compartment or colloid agglomeration in the deionizing compartments is a third indication of polarization. If such precipitates or agglomerates are excessive they can seriously reduce flow through the relevant compartment, significantly increase electrical resistance and, in extreme cases, damage or destroy membranes.

Such polarization effects can generally be avoided by:

- maintaining rapid flows of solutions through each type of compartment (a pressure head of about 1.5 atmospheres to each compartment is generally sufficient)
- maintaining the direct current voltage at not more than about 0.5 volts per membrane (This is an effective rule since in the case of the dilute electrolytic solutions subject to polarization, the principal electrical resistance of the stack will be due to the solution in the deionizing compartments).

Recent evidence[3] indicates that the amount of hydroxide and hydrogen ions generated at an anion-selective membrane during polarization depends strongly on the type of anion-selective membrane and its age and abuse.

APPLICATIONS

The largest application of ED in terms of installed membrane area is probably the concentration of seawater to about 18 to 22% (about 4 N) for the ultimate production of solid salt[4]. This experience is pertinent to the recovery of valuable electrolytes from fermentation liquors or derivatives thereof. Capacity installed or under construction for seawater concentration is equivalent to about 1.6 million metric tonnes of solid salt per year, presenting roughly 1.4 million square meters of installed membrane. This capacity consists of a few large plants.

The next largest application (in terms of installed membrane area) is probably the deionization of saline water for potable, process or pure water use. In this case the desired product is the deionized water. More than 1200 ED plants have been installed for this purpose, most of them pre-assembled, modular, standard designs of the Symmetrical Reversing Type for desalting brackish water[5]. Total capacity is about 300,000 cubic meters/day, representing roughly about 750,000 square meters of installed membrane. Symmetrically Reversing ED (EDA) has been developed to overcome membrane scaling and fouling problems often encountered during deionization of naturally occurring saline waters[6], without the addition of acids or other chemicals and without extensive pretreatment of the liquid to be processed. This process typically reverses the electrical polarity of the ED stacks 2 to 4 times per hour (though reversal frequencies as long as once a day have been reported). At the same time, automatic valves (preferably located immediately adjacent to the stacks) interchange stream flows so that the deionization and enrichment flow paths are interchanged. Generally any deposits on the membranes accumulated in one half-cycle are substantially completely removed during the next half-cycle.

Symmetrical reversal has generally eliminated or substantially reduced the need for addition of acid or other chemicals to the feed, deionizing or enriching streams during ED of saline waters and many industrial process streams. Reversing the direction of deionization is unique to a symmetrical ED stack and is not possible with any other desalting process. It has permitted, for example, the preparation of enrichment streams more than

400 percent saturated in calcium sulfate. It has been permitted steady operation of ED on secondary sewage effluents with pretreatment limited to particulate removal.

The third largest application of ED is the deionization of cheese-whey. Total installed capacity is equivalent to more than 100,000 tonnes per year of 90 + % demineralized whey powder[7], representing roughly 25,000 square meters of installed membrane and 1.6 million tonnes of raw whey. Most such deionized whey powder is used as the base for formulating human mother's milk replacement. The apparatus is much the same as that used for deionizing saline water or concentrating seawater except much greater attention must be paid to sanitary design, materials and procedures. Piping, pumps and heat-exchangers are dairy-quality stainless steel and all other components in contact with the whey are dairy quality. Generally the apparatus is cleaned and sterilized with warm, dilute caustic and hydrochloric acid at least once a day. Care must be taken in design and operation to minimize losses of organics. This has led to some disagreement among designers about the efficacy of short-cycle Symmetrically Reversing ED in this application. About one-third of the whey deionizing ED plants are of this reversing type. Most of the whey ED plants operate in a batch mode. Concentrated whey of about 20 to 25 percent total solids is recirculated from a holding tank through one or more ED stacks until the desired electrolyte content is obtained, as measured by an electrical conductivity meter. Such mode is particularly userfriendly and forgiving. Regardless of variations in electrolyte composition, ED is simply continued until the desired electrical conductivity is achieved.

Successful application of ED to cheese whey has led to substantial interest in deionizing other solutions of biological origin. For example, two ED plants (Figure 4) have been constructed to reduce the high salt concentration (2 N NaCl) present in an alpha-interferon eluate[8]. Another has been constructed to recover guanidine hydrochloride in high yield[8].

The following examples from laboratory apparatus illustrate the scope of ED:

- Bovine Serum Albumin (BSA, MW 67,000): One percent BSA solution containing 0.2 N NaCl was 95 percent deionized in about 4 hours.

Fig. 4.

The loss of BSA was about 1.4 percent and no significant pH change occurred[9].

- Low Molecular Weight Polypeptide: The polypeptide 1 - leucine - tryptophan - methionine - arginine - phenylalanine - acetate (MW 830) containing 0.2 N NaCl was 95 percent deionized in about 4 hours. There was a gradual increase in pH of the deionizing stream. Polypeptide lost to the enrichment stream was below 1%[9].

- Converting Enzyme Inhibitor (CEI, MW 1117): 15.7 micromolar CEI containing 1 N sodium acetate was about 75 percent deionized in about 3 hours without measurable loss of CEI. Deionizing further to 98 percent during another hour resulted in adsorption of peptide on the membranes. The final yield of peptide was 38 percent[9].

- Vasopressin (MW about 1084): A vasopressin solution containing 0.2N NaCl and KCL was 99 percent deionized in about 1.5 hours. There was no significant pH change and no vasopressin was found in the enrichment stream. There was some adsorption of vasopressin on the membranes[9].

- Angiotensin I-125 (MW about 1172): A solution containing about 1 nanogram per ml of I-125 labeled angiotensin and 0.15 N NaCl was 99 percent deionized in about 1 hour, without significant change in pH. The yield of angiotensin I-125 was 80 to 85 percent. Loss of angiotensin I-125 to the enrichment stream was about 2 percent. The yield of angiotensin I-125 could be improved by rinsing the deionizing compartments with ammonium acetate solution. The ammonium hydroxide or acetate were later removed by lyophilization. Such technique appears to be useful whenever proteins or peptides are absorbed on the membranes as a result of high degrees of deionization[9]. Such absorbed materials can also be removed and recovered by briefly reversing the direction of the electric current while the deionizing stream is flowing or while the deionizing compartments are being rinsed.

- Fractionation of Plasma Proteins: As the electrolyte content of plasma is reduced by ED, most of the non-albumin proteins precipitate out and the supernatant is enriched in albumin and globulins such as IgA. By continuing the ED, it is possible to obtain an IgA-rich precipitate and an albumin rich supernatant[10].

- Salting-Out of Proteins from Plasma: 80 percent of IgA can be removed from plasma by adding sodium sulfate to the plasma by ED to a level of about 1.1 N. In this case the plasma is the enriching stream. The supernatant can be deionized by ED, the sodium sulfate being transferred to plasma in the enriching stream[10].

REFERENCES

1. K. H. Meyer and W. Strauss, Permeability of membranes: VI. Passage of current through selective membranes, Helv.Chim.Acta, 23:795-800 (1940).
2. W. Juda and W. A. McRae, Coherent ion-exchange gels and membranes, J.Am.Chem.Soc., 72:1044 (1950).
3. R. Simons, Nature, 280:30 (1979); Desalination, 28:41 (1979) and O. Kedem, Weizmann Institute of Science, Rehovot, Personal Communication.
4. H. Miyauchi, Asahi Chemical Industry Co. Ltd., Tokyo, Personal Communication.
5. W. A. McRae, European Society of Membrane Science and Technology, Symposium on Synthetic Membranes in Science and Industry, Tubingen, 6-9 Sept. 1983.
6. W. Juda and W. A. McRae, U.S. Patent 2,863,813.
7. W. A. McRae, Recent developments in electrodialysis with ion exchange membranes, Society of Chemical Industry Symposium: "Ion Exchange

Membranes," 12-13 April 1983, Runcorn, Cheshire, U.K.

8. S. M. Jain and P. B. Reed, Electrodialysis, in: "Comprehensive Bio-technology and Bioengineering – Principles, Methods and Applic-ations," M. Moo-Young, ed., Pergamon Press, London (1983).

9. S. M. Jain, Ionics, Inc., Watertown, MA 02172, Private Communication.

10. S. M. Jain, "Electric Membrane Processes for Protein Recovery," Proceedings Biochemical Engineering Conference III, Santa Barbara, CA, Sept. 1982, (New York Academy of Sciences).

CARRIER FACILITATED TRANSPORT AND EXTRACTION THROUGH ION-EXCHANGE MEMBRANES: ILLUSTRATED WITH AMMONIA, ACETIC AND BORIC ACIDS

D. Langevin, M. Metayer, M. Labbe, M. Hankaoui and B. Pollet

ERA 471 CNRS
University of Rouen
76130 Mont Saint Aignan, France

ABSTRACT

Extraction and transport of a non-ionic substrate S can be facilitated through a charged membrane when a counter-ion T^z is a moving carrier for S. Coupling of transport and fast reaction is considered, with various values of stability constant. Consecutive reactions are investigated. Illustration, with S=NH$_3$(in solution) + T^z=H$^+$ or Ag$^+$ and S=Acetic or Boric acid + T^z=OH$^-$, is given. When polarization phenomena controls the transport, a critical flux J_{cri} is reached. J_{cri} is 80 to 200 times higher than passive flux.

INTRODUCTION

Transport can be facilitated in a system which is able to maintain a high concentration gradient. Due to their permselectivity and high capacity, charged membranes have this ability when they have a counter ion T^z as a moving carrier for a permeating S. Here one reaction (1) or two consecutive reactions (1) and (2) are considered.

$$S + T^z \rightleftharpoons ST^z \tag{1}$$

$$S + ST^z \rightleftharpoons S_2T^z \tag{2}$$

with the stability constants of the complexes (or products):

$$ST^z : \quad K_1 = C_1'/C_oC_1 \tag{3}$$

and S_2T^z: $\quad K_2 = C_1''/C_oC_1 \tag{4}$

(C_o, C_1, C_1' or C_1'' = concentrations of the species S, T^z, ST^z or S_2T^z)

The following assumptions are made:

a) Reactions are fast enough to consider Equation (3) or (4)

b) Co-ions are excluded from the membrane by the Donnan effect. Thus the total concentration C of counter-ions is related to the concentration

X of fixed ionic groups in the membrane by:

$$C_T = \omega X/Z \tag{5}$$

where $\omega = +1$ or -1 (cation or anion exchanger)

c) Flux J_2 of the co-ion is negligible, owing to the membrane perm-selectivity.

d) The steady state is reached[1,2].

COUPLING OF TRANSPORT AND REACTION (1)

From Equations (3) and (5) (where $C_T = C_1 + C_1'$) carrier concentration C_1 and product concentration C_1' are related to permeating concentration C_o;

$$C_1 = \omega X/Z(1+K_1C_o); \quad C_1' = K_1C_o\omega X/Z(1+K_1C_o) \tag{6a; 6b}$$

and thus, indirectly related to overall permeant concentration C_R (in its two forms, S and ST^Z).

$$C_R = C_o + C_1' \tag{6c}$$

By using the Nernst-Planck equation:

$$J_i = -D_i \left(\frac{dC_i}{dx} + Z_iC_i \frac{F}{RT}\frac{d\phi}{dx}\right) = -D_iC_i\left(\frac{d\ln C_i}{dx} + Z_i\frac{F}{RT}\frac{d\phi}{dx}\right) \tag{7}$$

we obtain the ratios J_i/D_iC_i for S ($C_i=C_o$ and $Z_i=0$),T^Z and ST^Z ($C_i=C_1$ and C_1', $Z_i=Z$) and the following equations:

$$J_o/D_oC_o+J_1/D_1C_1 - J_1'/D_1'C_1' = -\frac{d}{dx}(\ln C_o+\ln C_1-\ln C_1') = \frac{d}{dx}\ln k_1 = 0$$

or

$$J_1'/D_1'C_1' = J_1/D_1C_1 + J_o/D_oC_o \tag{8}$$

Moreover for the permeant in the S and ST^Z form, the overall flux is:

$$J_o + J_1' = J_R, \text{ const. (extraction flux)} \tag{9}$$

And in the absence of an electric current ($\Sigma Z_iJ_i=0$), negligible diffusion of the co-ions ($J_2=0$) gives:

$$J_1 + J_1' = 0 \tag{10}$$

As a result the reagent fluxes J_o and J_1 and product flux J_1' are related to extraction flux J_R and concentrations C_o, C_1, C_1':

$$J_1' = J_R/(1+D_oC_o/D_1C_1+D_oC_o/D_1'C_1') = -J_1 \text{ and } J_o=J_R-J_1' \tag{11}$$

and thus indirectly related only to the parameters J_o and C_o. As all the parameters of the system are now related to the constant extraction flux J_o and the local concentration C_o, the whole system will be defined, i.e. the parameter profiles will be known, when the C_o profile is determined.

From Equations (5) (7) and (9) we can relate J_R and dC_R/dx as follows:

$$J_R = D_o\frac{dC_o}{dx} + D_1'\left(\frac{dC_1'}{dx} + \frac{C_1'}{\omega X}B\right) = \left(D_1'\frac{dC_1'}{dC_R} + D_o\frac{dC_o}{dC_R}\right)\frac{dC_R}{dx} + \frac{D_1'C_1'}{\omega X}B \tag{12}$$

310

As the electric potential is related to fluxes J_1 and J_1' by the relation:

$$B = \Sigma - z_i J_i / D_i = \frac{d}{dx} \Sigma z_i C_i + \frac{F}{RT} \frac{d\phi}{dx} \Sigma z_i^2 C_i = Z\omega X \frac{F}{RT} \frac{d\phi}{dx} \tag{13}$$

The dC_R/dx gradient is only related to J_R and C_R and thus the C_R profile is accessible. The particular case $D_1 = D_1' = D_o (=D)$ is interesting as it gives the following simplifications:

$B = 0$ (see Equations (10) and (12)

and $J_R = D \frac{d}{dx} (C_o + C_1') = D \frac{d}{dx} C_R$ (see Equations (6) and (13)) (14)

Here C_R is a linear function of x and the shape of parameter profiles is given in Figure 1, between two appropriate concentrations C_R.

According to the value of the constant K_1, reaction (1) is localized in a more or less extended volume, delimited by two planes, parallel with the membrane surfaces and corresponding to $J_R/J_o = \epsilon$ and $1 - \epsilon$. When K_1 is very large ($>10^5$), only one plane, the reaction plane P_R, can be considered. On the other hand, when K_1 becomes small ($<10^{-2}$), species ST^z is not produced and transport of permeant S becomes passive diffusion. When the diffusion coefficients D_o, D_1 and D_1' are not equal, but in the same order of magnitude, these conclusions can be extended.

COUPLING OF TRANSPORT AND TWO CONSECUTIVE REACTIONS (1) AND (2)

From Equations (3), (4) and (5) (where $C = C_1 + C_1' + C_1''$) the concentrations are related to C_o (reagent S) and thus to C_R (permeant is in its three forms):

$$C_1 = \omega X / Z (1 + K_1 C_o + K_1 K_2 C_o^2); \quad C_1' = K_1 C_o;$$
$$C_1'' = K_1 K_2 C_o^2 \text{ and } C_R = C_o + C_1' + 2C_1'' \tag{6'}$$

As previously the ratios $J_i / D_i C_i$ can be used to give Equation (8)

and $J_1''/D_1''C_1'' = J_1'/D_1'C_1' + J_o/D_o C_o$ (8')

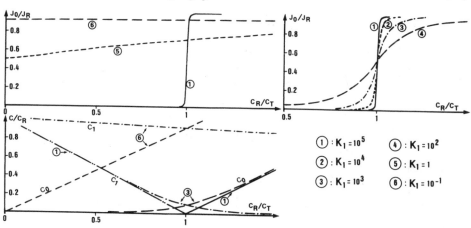

Fig. 1. Dependence of flux J_o (S) and of concentration C_o or C_1 (reagent S or T^z) and C_1' (product ST^z) on the overall permeant concentration C_R. In the particular case where $D_o = D_1 = D_1'$. Various values of stability constant K_1 are considered.

For the permeant $(S+ST^Z+S_2T^Z)$, the overall flux is:

$$J_0 + J_1' + 2J_1'' = J_R, \text{ const. (extraction flux)} \qquad (9')$$

The membrane permselectivity leads to:

$$J_1 + J_1' + J_1'' = 0 \qquad (10')$$

From Equations (8), (8'), (9') and (10') all the fluxes are expressed by:

$$J_1'/J_R = c(d-f)/(af+bd); \quad J_0/J_R = g \; (1+aJ_1'/cJ_R)/d \qquad (11'a)$$
$$J_1/J_R = -(1+J_1'/J_R-J_0/J_R)/2; \quad J_1''/J_R = (1-J_0/J_R-J_1'/J_R)/2$$

where

$$a = D_1'C_1' + 2D_1C_1; \quad b = D_1'C_1' + 2D_1''C_1''; \quad c = D_1'C_1'; \qquad (11'b)$$
$$d = D_0C_0 + 2D_1C_1; \quad f = D_0C_0 + 2D_1''C_1''; \quad g = D_0C_0$$

and are thus related to J_R and C_R. In the particular case $D_1 = D_1' = D_1'' = D_0$ (=D) C_R is again a linear function of x and variations of fluxes and concentrations are shown in Figure 2.

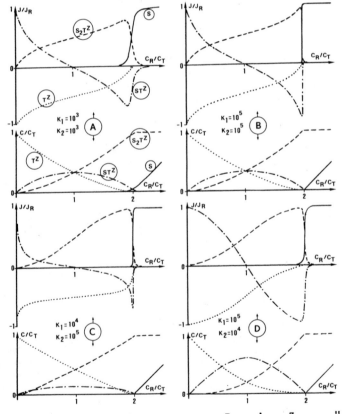

Fig. 2. Dependence of fluxes J_0 (S), J_1 (T^Z), J_1' (ST^Z) or J_1'' (S_2T^Z) and of the concentrations C_0, C_1, C_1' or C_1'' on the overall permeant concentration C_R. Various values of consecutive constants are considered.

From the four cases examined it can be noted:

a) In a large range ($C_R < 2C$ and $J_o \ll J_R$), only the equilibrium:

$$S_2T^z + T^z \rightleftarrows 2ST^z, \quad K_3 = C_1'^2/C_1''C_1 = K_1/K_2 \tag{15}$$

can be considered and the shapes of profiles depend on K_3 alone. Thus the curves of Figures 2A and 2B are identical ($K_3=1$). Ions S_2T^z and T^z are interdiffusing and act as substrate and carrier respectively. ST^z is a product, the greater part of which diffuses in the opposite direction to either T^z when $C_R < C_T$ (or $J_1'/J_R > 0$) or S_2T^z, when $C_R > C_T$.

b) When J_o is no longer negligible ($C_R \cong 2C_T$), the complex ST^z acts as an intermediate carrier. According to the value of K_2, the reaction (2) is localized in a more or less limited volume (reaction volume or plane). This explains the difference between the curves of Figures 2A and 2B in the corresponding range.

c) When J_o is close to J_R ($C_R > 2C_T$), transport becomes passive diffusion of permeant S.

d) Usually, as its stability increases, the contribution of intermediate complex ST^z increases with K_3.

Two boundary cases can be examined when K_2 is large enough:

a) When $K_3 \to 0$ (see Figure 3A), complex ST^z can be disregarded and only one reaction is considered.

$$2S + T^z \rightleftarrows S_2T^z \tag{16}$$

and is localized in plane P_R^o where $C_R = 2C_T$ ($=2C_1''$). T^z and S_2T^z interdiffuse in the layer where $C_R < 2C_T$.

b) When $K_3 \to \infty$ (see Figure 3B), the complex ST^z plays a major part in the transport. Two reaction planes P_{R1}^∞ and P_{R2}^∞ appear, for $C_R = C_T$ and $C_R = 2C_T$ respectively. On P_{R1}^∞, the reaction (16) takes place by collision of carrier T^z and substrate S_2T^z, these species diffusing in opposite directions. Half of product ST^z is drained towards the side of the extractor from which T^z comes. The other part of ST^z diffuses in the opposite direction and acts as a carrier for S, as is shown by reaction (2) localized on P_{R2}^∞. This theoretical approach can be extended to multiple consecutive reactions.

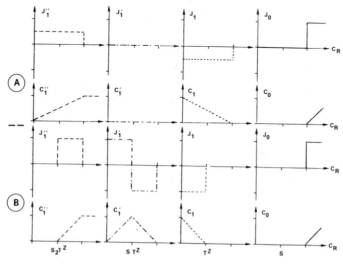

Fig. 3.   Boundary cases of Figure 2C and 2D when K_2 is large.
(A) $K_3 \to 0$; (B) $K_3 \to \infty$.

313

POLARIZATION PHENOMENA

In our measurements the membrane is between two solutions containing substrate S in solution I (concentration C_0^I) and carrier T^z in solution II (concentration C_1^{II}). Two polarization layers (') and (") build up on both sides of the membrane must now be considered. Figure 4A shows the case of one reaction localized on intramembrane plane P , at a distance e' from interface ('). Distance e' depends on C_0^I and C_0^{II}. When C_0 is constant and C_1^{II} increases, e' decreases and becomes equal to 0. A "critical flux" $J_{cri} = C_0^I D_0 / \delta'$ is reached[1] (Figure 4B). As long as the solution I is free from electrolyte and the flux of the co-ion negligible, J cannot exceed J_{cri} and the plane P cannot enter the polarization layer. The curve $J_R = f(C_1^{II})$ exhibits a plateau. For sub-critical conditions, the membrane interface (') is completely in ST^z form. Beyond, it becomes partially in T^z form. The validity of this model has been verified experimentally by the extraction of NH_3 (S) with carrier H^+ (T^z), in a system previously characterized by ion exchange isotherms, diffusion coefficients, etc...[1]. The notion of "critical flux" can also be extended to multiple successive reaction systems.

SOME EXPERIMENTAL SYSTEMS

Measurements were made in the steady state. Here, the concentrations of the bulk solutions on both sides of the membrane are equal to the outlet concentrations C^{out} (see Figure 5). We studied the extraction of NH_3 by H^+ through a cation exchange membrane and the extraction of CH_3COOH by OH^- through an anion exchange membrane, in order to illustrate the coupling between the transport and reaction (') when this reaction is localized in reaction plane P_R.

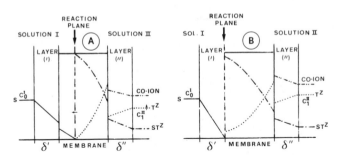

Fig. 4. Concentration profiles in membrane and polarization layers. Steady state when the reaction is localized in plane P_R. (A) Sub-critical conditions ($J_R < J_{cri}$); (B) Over-critical conditions ($J_R = J_{cri}$).

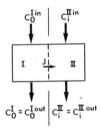

Fig. 5. Scheme of experimental device for measurements in the steady state. i = 1 (extraction) or 0 (other cases).

The facilitated extraction of NH_3 through a Nafion membrane (see Figure 6C) is very similar to that previously obtained through a poly-ethylene sulfonic membrane (see Figure 6A and[1]).

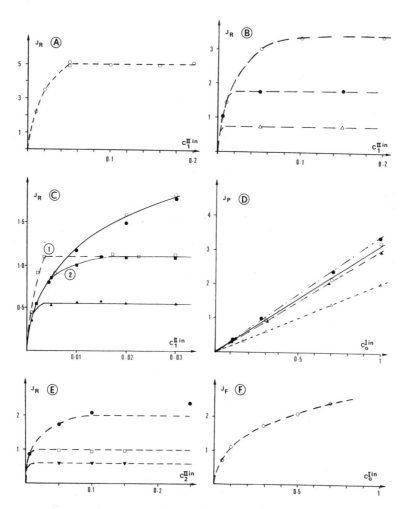

Fig. 6. **Experimental fluxes** (J, 10^{-5} mmol $cm^{-2}s^{-1}$). 1) J_R vs. C_1^{IIin}: **Extraction** of S [c_o^{Iin} = 0.025 (0), 0.01 (●), 0.005 (□ or ■), 0.0043 (△), 0.003 (▲) or 0.0025 (▼)] by T^z (c_1^{IIin}) in solution II; [S/memb./T^z (sol.II)] (fg.) = [NH_3/PES/H^+ (HCl)] Ⓐ, [CH_3COOH/RPA/OH^- (NaOH)] Ⓑ, [NH_3/Nafion/H^+ (HNO_3 -□-) or Ag^+ (1/2 Ag NO_3 + 1/2 HNO_3) Ⓒ, [Boric Ac./RPA/OH^- (NaOH)] Ⓔ. 2) J_p vs. c_o^{Iin}: **Passive diffusion** of S (c_o^{Iin}) through memb. in Y form; [S/mb- Y ( )] = [NH_3/PES-NH_4^+ (-.- ●)], [NH_3/Nafion -NH_4^+ (—— o) or 1/2 Ag^+ + 1/2 NH_4^+ (- - - ▲)], [CH_3COOH/RPA - CH_3COO^- (--- △)] Ⓓ. 3) J_F vs. c_o^{Iin}: **Facilitated transport** of Boric Ac. (c_o^{Iin}) through RPA in (Poly) borate(s) form Ⓕ. (See Table 1; C: molar scale).

315

The favorable factors for obtaining the critical conditions are:

a) A high membrane capacity and a high carrier mobility, increasing critical fluxes.
b) A high film thickness δ' and low substrate mobility, reducing J_{cri}.

The proportionality of J_{cri} to C_0^{Iout} is a criterion to verify that the critical conditions are actually reached. For a given flow q' of the ammonia solution, the critical flux is also proportional to C_0^{Iin}, because of the relation:

$$J_{cri} = (C_0^{Iin} - C_0^{Iout}) \, q'/A \qquad (17)$$

where A is the area of the membrane. This proportionality is shown by curves of the Figure 6B, 6C and 6E.

The previous explanation clearly show why the highest critical fluxes are obtained with the NH_3/H^+ system. In the NH_3/Ag^+ system there is coupling between the transport and the two consecutive reactions. But here it is necessary to use a Ag^+/NH_4^+ mixture in order to prevent precipitation of silver hydroxyde. As previously there are critical conditions which only appear for lower substrate concentrations and higher carrier concentrations (see Figure 6C and especially curves 1 and 2). The transport of two NH_3 molecules by one Ag^+ ion leads to an enhancement of the membrane capacity and thus of the extraction. Nevertheless this advantage is balanced by the presence of NH_4^+ ions which fill an appreciable part of the sulfonic groups of the membrane. In addition Ag^+ mobility is lower than that of H^+.

For the critical conditions, reaction (2) $NH_3 + Ag\ (NH_3)^+ \rightleftarrows Ag\ (NH_3)_2^+$ is localized at interface (') as constant K_2 is close to its solution value[3], i.e. 10^4. These conditions are those of Figure 2D when $C_R \cong 2C_T$.

However, the reaction (15):

$$Ag\ (NH_3)_2^+ + Ag^+ \rightleftarrows 2Ag\ (NH_3)^+$$

extends through the whole thickness of the membrane because constant K_3 is close to $10^{0.5}$, corresponding to an intermediate case between those described in Figures 2A, 2B and 2C.

In all the extractions, carried out under identical hydrodynamic conditions, polarization film thickness δ' had the average value $5 \cdot 10^{-3}$ cm (see Table 1) independently of the membrane nature and of the transport-reaction system. Nevertheless, the extraction flux is much higher in this case than in passive diffusion systems, although it is controlled by the polarization. Experiments in passive diffusion conditions were carried out with a membrane, in ST^z form (NH_4^+, CH_3COO^-, $Ag(NH_3)_2^+$...), dividing two substrate solutions, one at concentration C_0^I and the other infinitely diluted. Results show that passive flux J_p is like J_{cri}, proportional to C_0^I, but depends on the membrane used (see Figure 6D). For a given C_0^I value, J_{cri} is 80 to 200 times higher than Jp (see Table 1). That illustrates the spectacular effect of the diffusion-reaction coupling very well. We checked also that in all these systems, the back-diffusion was negligible, the permselectivity of the membrane showing a "valve" effect.

To illustrate the more general case of consecutive reactions, the investigation was extended to the boric acid (LH) extraction, by ion OH^-, through anion-exchange membrane (see Figure 6E). The experimental conditions were as above. The critical flux can lead to the following values of parameters: Substrate diffusion coefficient $D_0 = 1.4 \times 10^{-5}$ cm²s⁻¹ and polarization layer thickness δ' = 4.9×10^{-3} cm. When the membrane, in the

Table 1. Critical and Passive Fluxes

Membrane*	P.E.S.	Nafion		R P A
Carrier	H^+	H^+	Ag^+	OH^-
Substrate	NH_3	NH_3	NH_3	CH_3COOH
δ'(cm)	0.006	0.0046	0.0046	0.0049
J_{cri}/JP	200	150	150	80

*P.E.S.: Sulfonic membrane. Polyethylene matrix [see reference 1].
NAFION: Nafion 120 - Sulfonic membrane - PTFE matrix.
R P A : Strong base anion exchange membrane.
Flow of the solutions: $q' = 80$ ml h^{-1}.

borate form, divides two solutions of substrate HL, one at concentration C_o, the other infinitely diluted ($C \ll C_o$), the flux is no longer proportional to C_o and in the low concentration range, exceeds the passive diffusion flux (see Figure 6F).

We can assume that a facilitated transport appears, with formation (in the concentrated side), interdiffusion (inside the membrane) and dissociation (in the diluted side) of polyborates since the formation of polyborates ions has already been shown in boric acid absorption on strong base-anion exchange resin[4,5].

CONCLUSION

In this paper, the ability of ion exchange membranes to facilitate the extraction or the transport of a non ionic substrate is shown. Rare cases of such facilitated transport are found in the literature[6], but many examples of convenient reactions on ion exchange resins are known[7]. There is here a wide scope for studying such couplings between transports and reactions where systems could be selected according to their theoretical or practical interest.

REFERENCES

1. D. Langevin, Thesis, Rouen, France (1982).
   M. Metayer, D. Langevin, M. Labbe, and E. Selegny, Extraction facilitée de l'ammoniac en solution, in: "Filtra 82," Société Francaise de Filtration, ed., Paris (1982).
   M. Metayer, D. Langevin, M. Labbe, and E. Selegny, Couplage de diffusions et de réaction au travers d'une membrane chargée, in: "Physical Chemistry of Transmembrane ion motions," G. Spach, ed., Elsevier, Amsterdam (1983).
   D. Langevin, M. Metayer, and M. Labbe, Diffusion-reaction through charged membranes, in: "Charge and Field Effects in Biosystems," M. J. Allen and P. N. R. Usherwood, eds., Abacus Press (1984).
2. E. Selegny, Transport in reactive membranes and asymmetry rules, Europe-Japan Congress on Membranes and Membrane Processes, Stresa, Italy (1984).
3. L. G. Sillen and A. E. Martell, "Stability Constants of Metal Ion Complexes," p.153, Burlington House, London (1964).
4. R. Rosset, H. Fould, M. Chemla, H. Labrousse, J. Hure et B. Tremillon, Separation des isotopes du Bore à l'aide de résines échangeuses d'anions, Bull.Soc.Chim.France, 110:607 (1964).

5. T. Tomizawa, Studies of the absorption of boric acid on anion exchange resin. I, <u>Denki Kagaku</u>, 47(10):602 (1979).
6. O. H. Leblanc, W. J. Ward, S. L. Matson and S. G. Kimura, Facilitated transport in ion exchange membranes, <u>J.Memb.Sci.</u>, 6(3):339 (1980).
7. B. Tremillon, "Les Séparations par les Résines Echangeuses d'Ions," Gauthier Villars, Paris (1965).

A THIN POROUS POLYANTIMONIC ACID BASED MEMBRANE

AS A SEPARATOR IN ALKALINE WATER ELECTROLYSIS

R. Leysen, W. Doyen, R. Proost and H. Vandenborre

Studiecentrum voor Kernenergie, S.C.K./C.E.N.
Department of Electrochemistry
Boeretang 200, B-2400 MOL, Belgium

ABSTRACT

Polyantimonic acid based membranes have been evaluated as a separator in alkaline water electrolysis.

Therefore, thin sheets of polyantimonic acid–polysulfone in different weight ratios have been prepared using a film casting technique. According to differences in the preparation method, the mechanical as well as the electrochemical properties of these membranes may be varied. Since hydroxyl ion exclusion is not a requirement in the use of the membrane as a separator in the alkaline water electrolysis, we have been preparing membranes having an optimum porous structure in order to obtain a very low resistance to ionic migration.

For a 100% polymer film it is known that different types of pores and pore size distributions can be obtained by changing the preparation parameters. Scanning Electron Micrographs of these films reveal the existence of a "skin", which contains very fine pores, and underneath there are large fingerlike pores. Measurements of the ionic conduction of these films in alkaline medium were performed and the results of these measurements show a large scattering.

In contrast to these results for polysulfone sheets, the membrane resistance data for 80 wt % polyantimonic acid – 20 wt % polysulfone membranes show a very good reproducibility. Mean values in the range of 0.22–0.30 Ωcm^2 could be found at room temperature, when at 90°C the resistance equals 0.14–0.16 Ωcm^2.

Since the thickness of these membranes equals 0.04 cm the membrane resistance at room temperature is about a factor of 2–3 higher than that of the free electrolyte.

1. INTRODUCTION

In the present alkaline electrolysis technology, the asbestos diaphragm limits the working temperature of the cell up to 90°C. Without any special precautions taken, as for instance, the addition of silicates into the electrolyte[1], the asbestos diaphragm is going to corrode in alkaline medium at temperatures above 100°C. This increase in temperature will be necessary in order to make hydrogen production by water electrolysis competitive with conventional methods. By raising the operating temperature of the cells well over 100°C, it should be possible to lower the cell voltage which results in an energy efficiency of more than 90%.

In the framework of the Indirect Energy Program of the Commission of the European Communities (C.E.C.), we have been developing a substitute for chrysotile asbestos as a separator in the alkaline water electrolysis. This kind of separator is based on a combination of an inorganic "filler" material and an organic binder. As the organic binder, polysulfone has been chosen for its excellent stability at temperatures up to 120°C in the alkaline medium.

Most of the separators which have been made in our laboratory up till now use polyantimonic acid as the inorganic filler material. Polyantimonic acid which is a hydrous oxide of antimony shows a good thermal and chemical stability in sodium hydroxide electrolytes.

In section 2 of this paper we describe the preparation of these membranes and also indicate some differences between the "loaded" membranes and the 100% polymer films. In section 3 we discuss the membrane resistance measurements as well as the performance in real electrolysis conditions. Finally, in section 4 some major conclusions will be drawn from the foregoing results.

2. EXPERIMENTAL

2.1. Membrane Preparation

Thin sheets of 100% polysulfone (PSF) and of polyantimonic acid (PAM)-polysulfone in different weight ratios have been obtained using a film casting technique. The membranes have been prepared by a combination of the normal phase inversion and thermally induced phase inversion technique. This technique has already been used extensively in the case of homogeneous polymer films, e.g. for preparing membranes for reverse osmosis[2].

The dope is cast with a wet thickness of about 750 microns onto a glass plate. Immediately after casting, the glass plates are transferred into an oven at 70°C. After 35 minutes of evaporation in the oven, the glass plate is immersed in dionized water. After a period of 20 minutes in the water bath, the films are peeled off the glass plate and kept under water for another 15 to 30 minutes in order to have a complete precipitation. The exact procedure of the preparation of the polyantimonic acid based membranes has been reported before and also been patented[3-4]

2.2. Membrane Characterization

The porous structure and homogeneity of the film casted membranes have been investigated using such techniques as Scanning Electron Microscopy (SEM) and Energy Dispersive Analysis of X-rays (EDAX). A detailed analysis of the structure of the polyantimonic acid based membranes has been made by R. Leysen et al.[5].

In view of the envisaged application of these membranes in water electrolysis cells, the resistance towards ionic migration has been measured, using a simple DC-method[6]. The areal membrane resistance (R, Ωcm^2) can easily be deduced from the measured iR-drop across the membrane by using the equation, $R = A.\Delta V/i$, where i is the current, A is the exposed area of the membrane and ΔV is the measured voltage across the reference electrodes upon passage of the current.

3. EXPERIMENTAL RESULTS AND DISCUSSION

3.1. SEM and EDAX-results

The different membranes structure for the polyantimonic acid based membranes as well as for the 100% polysulfone membranes have been studied using Scanning Electron Microscopy.

Figure 1 shows SEM-photographs of the cross-sections of a 100% poly-sulfone membrane and a membrane composed of 80 wt% polyantimonic acid - 20 wt% polysulfone. The SEM-picture of the 100% polysulfone membrane (Figure 1a) reveals the "classical" structure of a skin which contains very fine pores, and underneath there are large fingerlike pores. In contrast to this photograph, Figure 1b shows the cross-section of a polyantimonic acid-polysulfone membrane. The latter is anisotropic too, but consists of both fingerlike cavities, that are much shorter and a sponge-like structure underneath.

The following important remark has to be made here: according to differences in the preparation method, the mechanical as well as the structural properties of the polyantimonic acid based membranes may be varied[5]. However, the type of membrane we are describing here is the one best suited for use in an alkaline water electrolysis cell.

The distribution of the inorganic compound, i.e. polyantimonic acid, in the polyantimonic acid based membranes has been measured by EDAX-analysis. Indeed, polyantimonic acid, which in its simplest form may be represented as Sb_2O_5, can easily be identified by the large Sb-emission peak at 3.604 keV. Figure 2 shows a typical example of such analysis. Apart from the different Sb-emissions, other emissions were found: S and

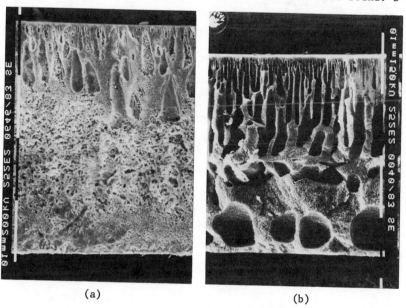

(a)   (b)

Fig. 1.   SEM-photographs of the cross-sections of a 100% polysulfone membrane (1a) and a 80 wt % polyantimonic acid - 20 wt % polysulfone membrane (1b).

Cl. The chlorine peak can be attributed to the polyantimonic acid since chlorine is left over as an impurity from the preparation of the poly-antimonic acid. The sulfur peak has to be ascribed to the polysulfone binding material.

The homogeneous distribution of the polyantimonic acid in the membrane has been confirmed by several "point analysis" over the entire cross-section of the membrane.

3.2. Membrane Resistance

We will now discuss the important question of the membrane resist-ance, since the latter partially determines the overall performance of the electrolysis cell. Two kinds of experiments will be discussed here: first, a comparison is made between the 100% polysulfone membranes and the poly-antimonic acid based membranes, whereby the membrane resistance has been measured in a small (4 cm²) laboratory experimental set-up in moderate conditions of electrolyte concentration and temperature. Secondly, the performances of the polyantimonic acid based membranes under real elec-

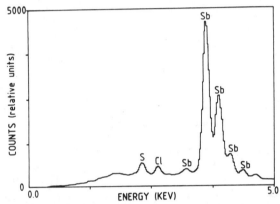

Fig. 2. EDAX-analysis of the cross-section of a polyantimonic acid based membrane.

Table 1. Membrane Resistances for 100% Polysulfone Membranes in 1M NaOH, 30°C

"dope" concentrations	batch No.	Range of Membrane Resistance (Ωcm^2)
15 % PSF in casting solution	1	1.7 - 3.4
	2	3.4 - 5.1
20 % PSF in casting solution	1	35 - 190
	2	27 - 88
	3	20 - 114

322

trolysis conditions (high current densities and temperatures, concentrated alkaline solution) have been evaluated.

Table 1 indicates the membrane resistances for 100% polysulfone membranes in 1 M NaOH at a temperature of 30°C for two different casting solutions. Different membrane samples of 4 cm^2 geometric area were cut out of several batches. These results clearly indicate a large scattering in resistance values not only between different batches but also within the same batch. The membrane resistances range from 2-200 Ωcm^2.

In contrast to the results obtained for 100% polysulfone membranes, the resistance values for several membrane samples cut out of a large sheet (60 x 60 cm^2) of a polyantimonic acid-polysulfone membrane range from 0.55 to 0.78 Ωcm^2. It can therefore be stated that first, the absolute values of the membrane resistances are much lower than in the case of the polysulfone membranes and secondly, there is less scattering in the membrane resistances.

The resistance of the polyantimonic acid based membranes in 15 wt% NaOH (4,3 M), which is the electrolyte currently used in alkaline water electrolysis, has also been measured. Mean values in the range of 0.22 - 0.30 Ωcm^2 could be found at 30°C, when at 90°C (operating temperature for a classical alkaline electrolyzer), the resistance equals 0.14 - 0.16 Ωcm^2.

The significance of the presence of polyantimonic acid inside these membranes on the ionic migration characteristics has been explained by H. Vandenborre and co-workers[7]. These authors stated that a delicate balance exists between the amount of polyantimonic acid and its hydrophilicity on the one hand, and the polymer structure of the polysulfone, which is hydrophobic in nature on the other hand.

3.3. Water Electrolysis Performances

As already indicated in the introduction, the main goal of this study is the application of the polyantimonic acid based membrane into alkaline water electrolysis cells. We have therefore measured membrane resistances and cell performances in real electrolysis conditions: 15 wt% NaOH, high current densities (up to 1000 mA cm^{-2}) and high temperatures (up to 120°C).

Figure 3 gives a schematic view of the electrolysis cell under study. The electrolysis cell is composed of a polyantimonic acid-polysulfone membrane having catalytically active electrodes attached to both sides of the membrane. As electrocatalysts, nickel sulphide for the cathode and NiCo$_2$O$_4$ or Co$_3$O$_4$ for the anode, both on a perforated Ni-support have been used. The geometrical surface area of the membrane and electrode equals 40 cm^2.

The experimental results as far as cell voltage and cell resistance is concerned, as a function of current density and temperature are summarized in Figure 4 and 5. Figure 4 shows the cell resistance (ohmic drop) as a function of current density and temperature. The cell resistance has been measured as the voltage drop between two Hg/HgO reference electrodes. One can notice the very low resistance at 120°C, which gives a penalty of about 180 mV at 1000 mA cm^{-2}. At 90°C, the cell resistance equals 0.24 Ωcm^2, which is higher than the value of the membrane resistance alone (0.14 - 0.16 Ωcm^2) as mentioned earlier in the same electrolyte and temperature. The fact that higher values of cell resistances are found than those obtained on the basis of the membrane resistance alone can be explained by the additional resistances caused by the perforated nickel plates.

323

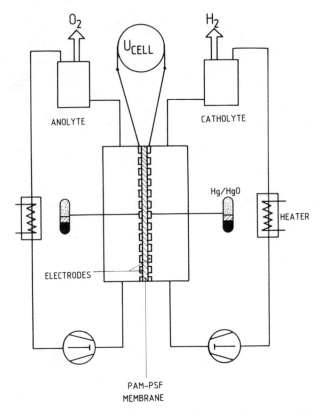

Fig. 3. Schematic view of the electrolysis experimental set up.

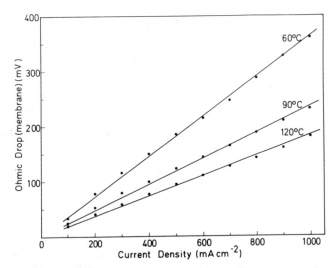

Fig. 4. Membrane resistance as a function of a current density and temperature.

Figure 5 gives the results of the cell voltages as a function of temperature and current density. Cell voltages of about two volts have been measured at 60°C, decreasing towards 1.7 volt at 120°C. Endurance tests have shown the long term stability of this cell combination for up to 5000 hours. The upscaling of the components, membrane and electrodes towards 2000 cm², which is industrially attractive has already been accomplished.

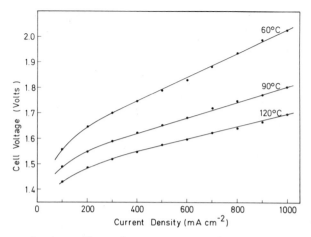

Fig. 5. Electrolysis cell performances as a function of current density
and temperature.

4. CONCLUSIONS

This polyantimonic acid-polysulfone membranes as well as 100% polysulfone membranes have been prepared by film casting. SEM and EDAX-
analysis reveal the differences in structure between the two kinds of
membranes as well as the homogeneous distribution of the polyantimonic acid
inside the "loaded" membranes.

The electrochemical characterization of the polyantimonic acid based
membranes has made it clear that these membranes offer a good possibility
of being used in advanced alkaline water electrolysis. Very low membrane
resistances are being found as opposed to the 100% polysulfone membranes.

Cell voltages of about 1.7 V at 1 A cm^{-2} and 120°C can be obtained,
which is a net gain of several hundred millivolts as compared to
"classical" water electrolysis systems.

Acknowledgements

The authors are much indebted to Dr P. Dejonghe and Dr G. Spaepen for
their continuous interest in this research. The SEM and EDAX-analysis were
performed at the Metallurgical Department by L. Knaepen, and his contribution is hereby gratefully acknowledged. The authors would also like to
thank J. Stroobants for the preparation of the membranes.

Furthermore, the authors wish to return thanks to the European
communities for the financial support of this research under contract
EHB-015-011-BG.

REFERENCES

1. C. Bailleux, Advanced water alkaline electrolysis: A two-year running
    of a test plant, Int.J.of Hydrogen Energy, 6(5):461 (1981).
2. H. Strathmann, P. Scheible, and R. W. Baker, A rationale for the
    preparation of Loeb-Sourirajan cellulose acetate membranes,
    J.Appl.Pol.Sci., 15:811 (1971).
3. R. Leysen and H. Vandenborre, Polyantimonic acid-polysulfone membranes
    as a separator in advanced water electrolysis, 33rd Meeting of the
    I.S.E., Lyon, 1:422 (1982).

4. R. Leysen, Ph. Vermeiren, L. Baetslé, G. Spaepen, and J. B. Vandenborre, Method of preparing a membrane consisting of polyantimonic acid powder and an organic binder, U.S. Patent 4, 253:926 (1981).

5. R. Leysen, W. Doyen, and H. Vandenborre, On the structure of polyantimonic acid - polysulfone membranes, Submitted to the J.of Membr.Sci., (1984).

6. J. M. Breen, ed., The Hydrogen-Halogen Energy Storage System, BNL-report No.50924, 1 (1978).

7. H. Vandenborre, R. Leysen, T. Nenner, and M. Roux, Membranes and diaphragms for advanced alkaline water electrolysis based on inorganic ion exchangers, Proceedings of the 3rd International Seminar on "Hydrogen as an Energy Carrier," Lyon, 25-27 May (1983).

POROUS MEMBRANES IN GAS SEPARATION TECHNOLOGY

U. Eickmann and U. Werner

University of Dortmund
Germany

The increasing use of membranes in gas separation technology is limited on a technical scale to solubility-membrane, in which the transport-mechanisms through the solid phase are realized by adsorption, diffusion and desorption. Nevertheless, studies can still be found in the literature in which the use of porous membranes for the separation of gases is proposed as viable. Indeed, the development of new membranes and modul arrangements appears to have lent new, interesting aspects to this classical procedure[1,2,3]. The separation-effect of porous systems is based on pore-diffusion which, according to Knudsen[4], is inversely proportional to the molecular weight of the diffusing gases. If gases are able to diffuse through the pores without mutual influencing of the various gas-molecules, then – with a binary gas mixture with both light and heavy molecules – the following relationship obtains for the permeating strength of molecular flow and also for the molefractions on the low-pressure-side:

$$\frac{y}{1-y} = \alpha_{max} \cdot \frac{x}{1-x} \tag{1}$$

where $\alpha_{max} = [M_s/1]^{1/2}$.

The Knudsen separation-factor α_{max} can take on considerable value, depending on the gas mixture. In a porous membrane Knudsen-diffusion will be evident when the mean free path of the molecules is far greater than the diameter of pores. Should this condition be not fulfilled then other mass-flow-mechanisms e.g. viscous flow and Fick-diffusion will blur the separation-effect.

Solubility-membranes separate gases by means of differing solubility and diffusion patterns in the dense layer. Their selectivity is generally higher than that of porous membranes – however, their permeability is several degrees slighter. Since solubility-membranes are nearly always made out of polymeric materials they are only deployable at temperatures of under 100°C, whereas porous membranes made out of inorganic materials such as ceramic, metal and glass can be exposed to several hundred degrees centigrade. Obviously porous membranes can not compete with solubility-membranes because when choosing the type of membrane for gas separation purposes higher selectivity is preferable to higher permeability by the same degree. However, the use of porous membranes would be conceivable where solubility-membranes are impractical because of high temperatures and where the positive separation properties of the polymers are no longer exploitable. At the same time, it is also worth considering that the development of porous membranes towards smaller pore-diameter as well as the development of new types of processes (e.g. membrane-rectification) might well open up new areas for the deployment of porous membranes.

Mass-Flow Through Porous Membranes

For the examination of the mass-flow of gases through porous systems two pore-models usually come under scrutiny: the "pseudo-capillary-model" and the "dusty-gas-model[5,6]

Figure 1 shows the specifying equation of a capillary model in which the porous matter is viewed as a bundle of parallel long capillary tubes with circular cross-section. By means of experimental parameters this model is adjusted to the actual behavior of the porous solid matter.

With the dusty gas model the solid matter is seen as supplementary gas-components with very large, permeant and homogeneously distributed molecules. Thus the binary gas-diffusion in the porous matter is treated as a ternary system.

Figure 2 shows the transport-equations which similarly have to be adjusted by means of experiments.

Influence of Different Process-Parameters on the Separation-Factor of Porous Membranes

Using permeation and counter diffusion methods, porous systems made of metal and polymer were examined and experimental adjustment-parameters were obtained for the capillary model[7]. In this way it was possible to specify separation behavior in relation to the process-parameters. The separation-factor α is defined as the ratio between the component-flows behind and in front of the membrane, and describes the enrichment of a gas-component. For any given gas mixture it is influenced by membrane-specific and process-specific parameters.

Figure 3 shows the influence of the mean pore-radius r_K and of the pressure-ratio p_R on the separation-factor of a polymer-membrane whereby the basic gas mixture N_2/CO_2 possesses a maximum separation-factor

$$\dot{n}_1 = - \frac{8 n_k r_k^3 \beta_k}{3} \left[\frac{\pi}{2kTm_2} \right]^{\frac{1}{2}}$$

$$\left[\frac{\alpha_{max}}{1 + B p \beta_k} \left\{ x \cdot \frac{dp}{dz} + p \frac{dx}{dz} \right\} + \frac{f_0 B x p \beta_k}{1 + B p \beta_k} \cdot \frac{dp}{dz} + \frac{A}{\beta_k} \cdot x \cdot p \cdot \frac{dp}{dz} \right]$$

$$\dot{n}_2 = - \frac{8 n_k r_k^3 \beta_k}{3} \left[\frac{\pi}{2kTm_2} \right]^{\frac{1}{2}}$$

$$\left[\frac{1}{1 + B p \beta_k} \left\{ (1-x) \frac{dp}{dz} - p \frac{dx}{dz} \right\} + \frac{f_0 B (1-x) p \beta_k}{1 + B p \beta_k} \cdot \frac{dp}{dz} + \frac{A}{\beta_k} (1-x) p \frac{dp}{dz} \right]$$

| separative Knudsen flow | diffusion and momentum losses | nonseparative Poiseuille flow |

experimental factors : r_k = radius of pores (average)

n_k = apparent number of parallel capillaries

β_k = proportionality - factor

Fig. 1. Mass-flow equation for a capillary model.

$$\dot{n}_i = - [C] \cdot K_1 \underbrace{\frac{d_K^2}{32\eta} \cdot \frac{dp}{dz}}_{\text{Poiseuille flow}} - \underbrace{K_2 \cdot [F]^{-1} \left[\frac{dc}{dz}\right]}_{\substack{\text{diffusion and} \\ \text{momentum losses}}} - \underbrace{[D_S]\left[\frac{dc}{dz}\right]}_{\substack{\text{surface} \\ \text{diffusion}}}$$

$$\left. \begin{aligned} F_{ij} &= -\frac{x_i}{D_{ij}} \\[2mm] F_{ii} &= \frac{1}{D_{ki}} + \sum_{\substack{h=1 \\ h \neq i}}^{n} \frac{x_h}{D_{ih}} \end{aligned} \right\} \quad \text{elements of matrix } [F]$$

experimental factors : K_1 = dimensionless factor
K_2 = dimensionless factor
D_S = surface diffusion coefficient

Fig. 2. Mass-flow equation of the dusty gas model.

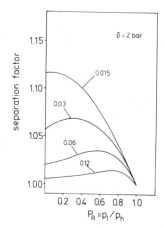

Fig. 3.   Influence of the pressure-ratio P_R on the separation-factor
(parameter: pore-radius r_K [µm], $\bar{p}$ = 2 bar).

α_{max} = 1.253, which results in the limiting case of vanishing pressure
level and pressure ratio. With significant pressure behind the membrane
there occurs a back diffusion of the enriched components which in the event
of $p_h = p_l$ ruins the whole separation effect. α takes on for $\bar{p} \neq 0$ a
maximum which, in the absolute height and position relating to P_R, is
dependent on the pore-radius r_K. The smaller the pores are the less the
viscous part of the flow is able to cover the separative Knudsen-diffusion
and, therefore, the greater α is.

Figure 4 shows the influence of the mean pressure $\bar{p}$ on the membrane
with variable pore-size. With increasing $\bar{p}$ the viscous part of the flow
increases and the separation-factor decreases. The picture on the left
corresponds to commercial pore-size, whilst the pore-size on the right
corresponds to new membrane developments, e.g. from γ –Al_2O_3[4]. The
sensitivity of the separation-factor decreases sharply from $\bar{p}$, whilst α
remains significantly under the value maximum of 1.253. In particular,
the separation-factor remains dependent on the pressure-ratio P_R.

A temperature increase enlarges the mean free path of the gas-
molecules and along with it the Knudsen-flow. The separation-factor rises
as the result of increased Knudsen-diffusion.

Porous Membranes in Modul Arrangements

In this present report the shown relationship of separation behavior in porous membranes holds for a single separating stage with an ideal mixture on both sides. Since with only one stage only a limited enrichment can be obtained, different modul arrangements[5] are used. Figure 5 shows the processes of a recycling cascade and the membrane-rectification. With the recycling cascade the permeate from one stage reaches the next as feed by means of a compressor whilst the retained is led back to the previous stage. In other words, a feed-compressor is required for each stage. Membrane-rectification needs both a feed-and a circuit-compressor in order to completely separate a gas-mixture down into its basic components. Membrane-rectification is based on the principle of rigorously applied countercurrent-flow. With fitting-parameters obtained from experiments[7] some modul-processes were calculated for a normal commercial porous membrane with a pore-radius of 150 Å.

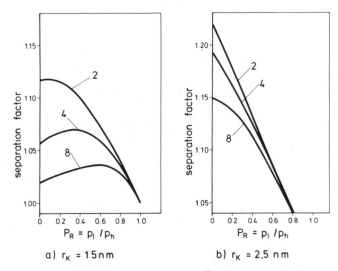

a) $r_K = 15\,nm$ b) $r_K = 2{,}5\,nm$

Fig. 4.   Influence of the pressure-level $\bar{p}$ and the pore-radius r_K on the separation-factor (parameter: $\bar{p}$ [bar]).

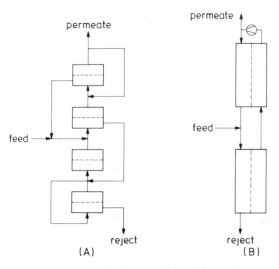

Fig. 5.   Membrane-processes: (A) cascade; (B) membrane-rectification.

The gas mixture H_2/CH_4 with a maximum separation-factor $\alpha = 2.83$ and a feed-concentration $x_{H2} = 0.625$ is separated in a depleted stream with $x_R = 0.40$ and enriched stream with $y_{H2} = 0.80$. The feed is assumed at 1 mol/s and the permeability of the membrane at about 10^{-4} [cm^3_N · cm/s · cm^2 · cm Hg].

An analysis of the gas stream to be compressed and of the membrane-areas which exert a decisive influence on the costs of separation revealed that for the given separation-project the cascade-process chosen (with five stages) demanded the smallest outlay. It should be immediately seen that the required separation can only be achieved by an excessive outlay in compression whilst the outlay in membrane-area is negligible due to the high permeability.

By lowering the pressure ratio the mass-flow can be diminished as the separation-factor increases (Figure 7). However, it still amounts to 38 times that of the feed-flow. The through exploitation of concentration gradients in countercurrent flow modules (e.g. membrane rectification) can also bring down outlay in compression.

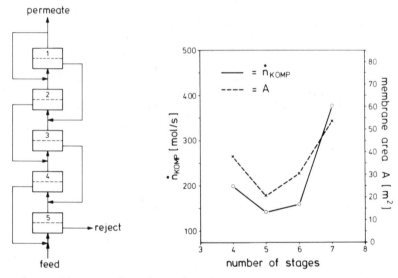

Fig. 6. Influence of number of stages on separation-processes

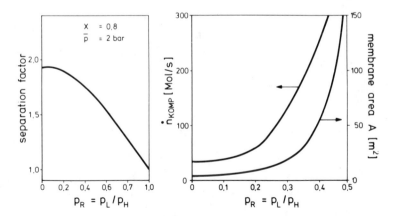

Fig. 7. Influence of the pressure-ratio p_R on separation-factor and cascade-size.

331

Figure 8 shows the influence of the pressure-ratio P_R and the mean pressure $\bar{p}$ on the gas stream to be compressed in the circuit compressor and on the necessary membrane area. They fall exponentially as P_R and $\bar{p}$ decreases - here due to the increase in the separation-factor. For a mean pressure $\bar{p}$ = 2 bar and a pressure ratio P_R = 0.33 the circuit stream still amounts to 8.5 times that of the feed.

The influence of a rise in temperature on the gas-separation also has a positive effect with membrane-rectification. With a process-temperature of 300°C the flow decreases to 5.4 mol/s since the separation-factor increases with the temperature.

CONCLUSION

For a porous membrane with a mean pore-diameter of 150 Å it has been demonstrated that the cost of the compression of the gas streams is considerably more expensive than the cost of membrane-area since highly permeable membranes only require small areas. It has been known for a long time that the deployment of porous membranes for the separation of gas mixtures leads to a large number of cascade stages. Membrane-rectification effectively reduces the number of stages to two but it may not be over-looked that the shown dependencies of the separation behavior on pressure-level, pressure-ratio and temperature always lead to a separation-factor that lies under the ideal Knudsen-value and therefore leads to large gas streams. Deployment of porous membranes is therefore only conceivable if the separation-process operates at a low pressure-level, at a low pressure-ratio across the membrane and at high temperatures. These physical depend-

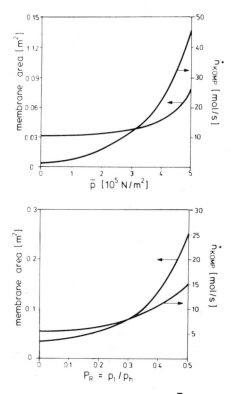

Fig. 8. Influence of the pressure-level $\bar{p}$ and pressure-ratio P_R on membrane-rectification

encies also remain if porous membranes with increasingly smaller pores are used. Admittedly, smaller pores reduce the dependence of the separation-factor on the pressure-level but the influence of the pressure-ratio P is not diminished. A technical application of porous membranes appears to be conceivable only when pressure-less process gases accrue, gases which also possess greatly different molecular weight. A case in point would be the large amount of gases from coke-ovens or the product of brown-coal low-pressure gasification where, however, the problem of cleaning the gases beforehand at such high temperatures would have to be solved.

Polymeric membranes would not be suitable with the high temperatures predominating there and it would be necessary to fall back on microporous ceramics, for example. A further development of pore-structures towards pore-diameters that either lead to steric hindrance of the flow or to capillary condensation of a component could increase the separation-factor, reduce the pressure-relationship shown and last but not least make the use of porous membranes in many areas economically much more interesting.

NOMENCLATURE

$[C]$	Concentration matrix	$[mol/m^3]$
D_{ij}	Diffusion coefficient	$[m^2/s]$
D_{Ki}	Diffusion coefficient of Knudsen flow	$[m^2/s]$
$[D_s]$	Matrix of surface diffusion coefficients	$[m^2/s]$
d_K	Pore diameter	$[m]$
K_1	Dimensionless factor	$[1]$
K_2	Dimensionless factor	$[1]$
k	Boltzmann constant	$[J/K]$
M_1	Molecular weight (light component)	$[kg/mol]$
M_s	Molecular weight (heavy component)	(kg/mol)
m	Mass of a molecule	$[kg]$
$\dot{n}_i$	Permeate flow	$[mol/m^2/s]$
n_K	Apparent number of parallel capillaries	$[1/m^2]$
$\dot{n}_{Komp}$	Total compressorload	$[mol/s]$
P_R	Pressure ratio $P_R = (P_1/P_h)$	$[1]$
P_h	High pressure	$[Pa]$
P_1	Low pressure	$[Pa]$
r_K	Pore radius	$[m]$
T	Temperature	$[K]$
x	Mole fraction (high pressure side)	$[1]$
y	Mole fraction (low pressure side)	$[1]$
z	Length of a pore	$[m]$
α_{max}	Ideal separation factor of Knudsen flow	$[1]$
β_k	Proportionality factor	$[1]$
η	Viscosity	$[Pa\ s]$

REFERENCES

1. J. Dworschak, Trennung von Gasgemischen mittels Membranen, Verfahrenstechnik, 9(3):103–105 (1975).
2. J. W. Thorman, Engineering Aspects of Capillary Gas Permeators and the Continuous Column, Ph.D. Thesis, Iowa (1979).
3. A. F. M. Leenaars, K. Keizer, and A. J. Burgraaf, Thin Supported Layers with very small Pores, Solid State Chemistry 1982, Proceedings of the Second European Conference, Veldhoven, Netherlands, 7.-9.6. 1982, in: "Studies in Inorganic Chemistry," Vol. 3, R. Metselaar, H. J. M. Heijligers, and J. Schoonman, eds., Elsevier Scientific Publishing Co., Amsterdam, 401:404 (1983).
4. M. Knudsen, Die Gesetze der Molekularströmung und der inneren Reibungsströmung durch Röhren, Ann.Physik, 28:75–130 (1909).
5. D. Massignon, "Gaseous diffusion in uranium enrichment," S. Vallani, ed., Springer (1979).
6. E. A. Mason, A. P. Malinauskas, and R. B. Evans III, Flow and diffusion of gases in porous media, J.Chem.Phys., 46(8):3199–3216 (1967).
7. U. Eickmann and U. Werner, Gastrennung mit Porenmembranen, Chem.Ing. Techn., 56(9):720–721 (1984).

SEPARATION OF HYDROCARBON MIXTURES BY

PERVAPORATION THROUGH RUBBERS

J-P. Brun, C. Larchet and B. Auclair

Université Paris - Val de Marne
Avenue du Général De Gaulle
94 010 Creteil, Cedex, France

SUMMARY

The pervaporation of binary mixtures through moderately swollen membranes is analyzed as a 'Solution-Diffusion' process and on the assumption that the diffusivity of each permeant is an exponential function of both concentrations. A model is derived, in which changes in selectivity and fluxes are related to major external conditions: the upstream mole fraction in the feed and the downstream total pressure of the pervaporate. In order to test the validity of this model, it has been applied to a set of experimental data for the pervaporation of hydrocarbon binary mixtures through rubber membranes.

INTRODUCTION

Through a number of published works on the pervaporation of hydrocarbon mixtures through different polymers[1,2,3,4,5,6], it appears that the composition of the upstream feed has a drastic effect on the selectivity. The order of magnitude of the selectivity is in the same time far lower than the ratio of the fluxes of pure permeants. This behavior is usually interpreted as a plasticizing effect of permeants on the membrane. Although it is well admitted that both kinetic and equilibrium properties of permeants are equally involved in the pervaporation process, diffusivities are most of the time more sensitive to permeant concentrations than solubilities[2,4,6,7], even if the penetrant/polymer system is diluted enough to remain within Henry's law limit. A linear dependance of permeants diffusivity is not convenient to describe large plasticizing effects. These effects may be explained in terms of diffusivities varying exponentially with concentrations.

THEORY

The basic idea for the following development was suggested by a number of experimental data on permeation and pervaporation: If logarithm of the diffusivity D_i of a pure permeant i is plotted as a function of its own concentration $\bar{C}_i$ in a membrane, a quasilinear relationship is generally found over a more or less extended area[4,6,8,9], leading for diffusivities to 'LONG-model' expressions:

$$D_i = D_i^o \exp(A_i \bar{C}_i) \qquad \text{(pure permeant i)} \tag{1}$$

$$D_j = D_j^o \exp(A_j \bar{C}_j) \qquad \text{(pure permeant j)} \tag{2}$$

where A_i and A_j are specific arguments for i and j.

For a mixture, similar expressions may be derived, in which the exponential depends separately on the concentrations:

$$D_i = D_i^o \exp(A_{ii} \bar{C}_i + A_{ij} \bar{C}_j) \tag{3}$$

$$D_j = D_j^o \exp(A_{ji} \bar{C}_i + A_{jj} \bar{C}_j) \tag{4}$$

Let us introduce activity coefficients of the mixture γ_i and γ_j, and 'internal activity coefficients' of i and j in the membrane $\bar{\gamma}_i$ and $\bar{\gamma}_j$, defined by:

$$\gamma_i x_i = \bar{\gamma}_i \bar{C}_i \quad \text{and} \quad \gamma_j x_j = \bar{\gamma}_j \bar{C}_j \tag{5}$$

These coefficients are assumed to remain constant across the membrane (validity of Henry's law); this excludes many reacting or strongly interactive systems[2,10], but includes most of the moderately swelling ones.

Let us define the pervaporation selectivity (separation factor) at the steady state:

$$\alpha = \frac{J_i}{J_j} \frac{x_{j1}}{x_{i1}} = \frac{y_{i2}}{y_{j2}} \frac{x_{j1}}{x_{i1}} \tag{6}$$

where J_i, x_{i1} and y_{i2} are respectively the molar flux, the upstream ("1") mole fraction and the downstream ("2") mole fraction of permeant i. The molar flux ratio J_i/J_j, which is a constant across the membrane, may be written:

$$\frac{J_i}{J_j} = \frac{D_i^o \exp(K_i \bar{C}_i) \, d\bar{C}_i}{D_j^o \exp(K_j \bar{C}_j) \, d\bar{C}_j} \tag{7}$$

where

$$K_i = A_{ii} - A_{ji} \quad \text{and} \quad K_j = A_{jj} - A_{ij} \tag{8}$$

α is derived from Equation (7) by integrating separately 'i' and 'j' differential terms from the upstream to the downstream face of the membrane:

$$J_j \int_{\bar{C}_{i2}}^{\bar{C}_{i1}} D_i^o \exp(K_i \bar{C}_i) \, d\bar{C}_i = J_i \int_{\bar{C}_{j2}}^{\bar{C}_{j1}} D_j^o \exp(K_j \bar{C}_j) \, d\bar{C}_j \tag{9}$$

One obtains:

$$\alpha = \frac{D_i^o K_j [\exp(K_i \bar{C}_{i1}) - \exp(K_i \bar{C}_{i2})] x_{j1}}{D_j^o K_i [\exp(K_j \bar{C}_{j1}) - \exp(K_j \bar{C}_{j2})] x_{i1}} \tag{10}$$

336

where:

$$\bar{C}_{i2} = \frac{y_{i2} \, P_2}{\bar{\gamma}_i \, P_i^{sat}} = \frac{\alpha \, x_{i1} \, \rho}{\bar{\gamma}_i \, [1 + (\alpha-1) \, x_{i1}]} \tag{11}$$

and

$$\bar{C}_{j2} = \frac{y_{j2} \, P_2}{\bar{\gamma}_j \, P_j^{sat}} = \frac{x_{i1} \, P_i^{sat} \, \rho}{\bar{\gamma}_j \, [1 - (\alpha-1)x_{j1}] \, P_j^{sat}} \tag{12}$$

where p_i^{sat} and P_j^{sat} are the saturated vapor pressures of permeants i and j at the pervaporation temperature, and ρ the ratio of total downstream pressure P_2 to P_i^{sat}. At zero downstream pressure ($\rho = 0$) selectivity has two limit values:

$$\alpha_1 = \underset{x_{i1} \to 0}{\text{Lim}} = \frac{D_i^o \, K_j \, \gamma_{i1}}{D_j^o \bar{\gamma}_i \, [\exp(K_j/\bar{\gamma}_j) - 1]} \tag{13}$$

and

$$\alpha_2 = \underset{x_{i1} \to 1}{\text{Lim}} = \frac{D_i^o \, \bar{\gamma}_j \, [\exp(K_i/\bar{\gamma}_i) - 1]}{D_j^o \, K_i \, \gamma_{j1}} \tag{14}$$

To express molar fluxes, let us now integrate the Equation (7) from the upstream face to abscissa 'z':

$$\frac{J_i}{J_j} \quad \frac{D_i^o \, K_j \, [\exp(K_i \bar{C}_{i1}) - \exp(K_i \bar{C}_i(z))]}{D_j^o \, K_i \, [\exp(K_j \bar{C}_{j1}) - \exp(K_j \bar{C}_j(z))]} \tag{15}$$

$\exp(K_j \bar{C}_j(z))$ may be expressed as a function of $\bar{C}_i(z)$. Defining:

$$E = \frac{J_j \, D_i^o \, K_j}{J_i \, D_j^o \, K_i} \tag{16}$$

$$\Omega = \exp(K_j \bar{C}_{j1}) - E \exp(K_i \bar{C}_{i1}) \tag{17}$$

and

$$\psi(z) = \exp(K_j \bar{C}_j(z)) = \Omega + E \exp(K_i \bar{C}_i(z)) \tag{18}$$

$$\psi(z) = \exp K_i \bar{C}_i(z) = \frac{\exp K_j \bar{C}_j(z) - \Omega}{E} \tag{19}$$

337

Combining Equations (15) to (19), we derive the reduced molar fluxes of permeants (products $J_i \cdot \ell$ and $J_j \cdot \ell$, where ℓ is the thickness of the membrane) by integrating their differential expressions:

$$J_i \cdot \ell = D_i^o \int_{\bar{C}_{i2}}^{\bar{C}_{i1}} [\psi (z)]^{A_{ij}/K_j} \cdot \exp(A_{ii}\bar{C}_i) \, d\bar{C}_i \qquad (20)$$

$$J_j \cdot \ell = D_j^o \int_{\bar{C}_{j2}}^{\bar{C}_{j1}} [\Psi (z)]^{A_{ji}/K_i} \cdot \exp(A_{jj}\bar{C}_j) \, d\bar{C}_j \qquad (21)$$

Most of the classical variations of the pervaporation selectivity and fluxes may be described with this semi-empirical model[11,12], provided a realistic assumption, i.e. the absence of large variations of the partition coefficients throughout the membrane.

RESULTS AND MODELING

Pervaporation of Benzene/n-heptane Mixtures Through N.B.R. Membranes

Recent work on the pervaporation of benzene/n-heptane mixtures through a polybutadiene acrylonotrile rubber (NBR) membrane has been published by Larchet[6]. Data have been identified with the model. Ten 'well-constrained' parameters are useful for computations (molar volumes, saturated vapor pressures, activity coefficients of benzene and n-heptane in the mixture, internal activity coefficients of benzene and heptane in the membrane, reduced molar flux of pure permeant i, downstream pressure of the pervaporate). Activity coefficients are computed on the basis of the modified theory of regular solutions[13] and are plotted as a function of the composition of the mixture in Figure 1. Other parameters are issued from the literature[14] and from Larchet's data[6].

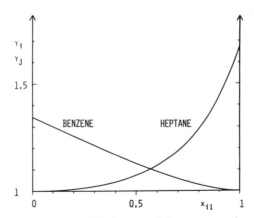

Fig. 1. Computed activity coefficients of benzene and n-heptane in their mixtures at 60°C, according to the modified theory of regular solutions[13].

Larchet's experiments were always performed under a downstream pressure of 2670 Pa (20 torrs, ρ = 0.05). Selectivity measurements[6] are plotted in Figure 2, and compared with the model. Between x_{i1} = 0.3 and 0.85, the change in selectivity displays a quasilinear variation (the slope and initial starting point respectively equal to 8.447 and 11.001). The coefficient of correlation is excellent, reaching 0.979. The decrease in selectivity in the high dilution area may be related to a non vanishing value of the downstream pressure. Unfortunately, this cannot be verified, since experiments have not been performed for mole fractions values below 0.26. Flux measurements are also in good agreement with the model (Figure 2). Best fitting values are listed in Table 1.

Pervaporation of 1,3-butadiene/isobutene Mixtures Through N.B.R. Membranes

The pervaporation of 1,3-butadiene/isobutene mixtures through the same type of membranes has been studied by Brun[3] and Vasse[4]. The observed ideal behavior of the mixture at every composition[13], implies that the activity coefficients are equal to 1. The 'well-constrained' parameters are issued from the literature[14,15] and from Vasse's data (4).

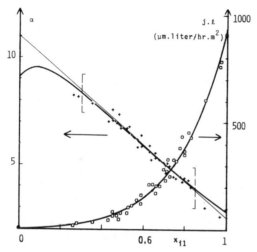

Fig. 2. Variation of selectivity and reduced volumic flux (j·ℓ) with mole fraction benzene x_{i1} in the pervaporation of benzene/n-heptane mixtures through a butadiene acrylonitrile rubber membrane at 60°C and under a downstream pressure of 2670 Pa. The present model (heavy lines) is compared with the experimental selectivity (crosses) and reduced volumic flux (squares) obtained by Larchet[6]. Also shown is a linear fit to the data (thin line).

Table 1. Computed Coefficients of the Model

Coefficients	Benzene/heptane	Butadiene/isobutene
D_i° (m²/s)	2.65 X 10⁻¹²	5.56 X 10⁻¹²
D_j° (m²/s)	6.17 X 10⁻¹²	1.94 X 10⁻¹²
A_{ii} (m³/mole)	900 X 10⁻⁶	1060 X 10⁻⁶
A_{ij} (m³/mole)	2200 X 10⁻⁶	106 X 10⁻⁶
A_{ji} (m³/mole)	988 X 10⁻⁶	1574 X 10⁻⁶
A_{jj} (m³/mole)	689 X 10⁻⁶	530 X 10⁻⁶

The first group of data (Figure 3) corresponds to the variation of selectivity and total mass flux with upstream composition, at a constant downstream pressure close to 10700 Pa (80 torrs, $\rho = 0.04$). Figures 3 shows an excellent agreement between data and the model. Again the values of parameters for the best fitting model are listed in Table 1.

The second group of data (Figure 4) relates to the variation of selectivity and total mass flux with downstream pressure, at a constant upstream composition close to equimolarity ($x_{i1} = 0.51$). It can be shown that the model accounts satisfactorily for the mass flux, but not for the selectivity. Significant variations of the partition coefficients from the upstream to the downstream face are probably occuring, since preferential solvation phenomena have been observed by Vasse[4] during sorption experiments. These variations may be attributed to the possible occurence in the system of strong interactions between 1,3-butadiene and nitrile groups of the polymer[3,4].

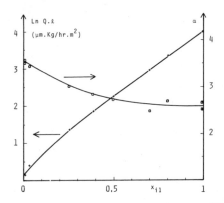

Fig. 3. Variation of selectivity and reduced mass flux ($Q \cdot \ell$) with mole fraction butadiene x_{i1} in the pervaporation of 1,3-butadiene/isobutene mixtures through a butadiene acrylonitrile rubber membrane at 22°C and under a downstream pressure of 10 700 Pa. The present model (heavy lines) is compared with the experimental selectivity (squares) and reduced mass flux (crosses) obtained by Vasse[4].

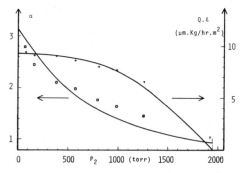

Fig. 4. Variation of selectivity and reduced mass flux ($Q \cdot \ell$) with downstream pressure P_2 in the pervaporation of 1,3-butadiene/isobutene mixtures through a butadiene acrylonitrile rubber membrane at 22°C and for an upstream mole fraction in butadiene equal to 0.51. The present model (heavy lines) is compared with the experimental selectivity (squares) and reduced mass flux (crosses) obtained by Vasse[4].

CONCLUSION

The introduction of diffusivity of permeants exponentially related to their concentrations in a solution-diffusion model allows to extend the LONG-model to the case of binary mixtures. The model is suitable to describe the pervaporation of binary mixtures, provided a moderate swelling and in absence of strong physico-chemical interactions between permeants and the membrane.

REFERENCES

1. R. Y. M. Huang and V. J. C. Lin, Separation of liquid binary mixtures by using polymer membranes. Permeation of binary organic liquid mixtures through polyethylene, J.Appl.Polym.Sci., 12:2615-2631 (1968).
2. P. Aptel, Pervaporation à travers des membranes de polytétrafluoroéthylene modifiées par greffage de monomères polaires, Thèse de Doctorat es-Sciences Physiques, Université de Nancy - 1 (1972).
3. J-P. Brun, G. Bulvestre, A. Kergreis, and M. Guillou, Hydrocarbons separations with polymer membranes. I. Butadiene-isobutene separation with nitrile rubber membranes, J.Appl.Polym.Sci., 18:1663-1683 (1974).
4. F. Vasse, Contribution à l'étude de l'extraction du 1,3-butadiene des coupes petrolières par pervaporation à travers des membranes permselectives, Thèse de Docteur-Ingénieur, Université de Paris - 6 (1974).
5. M Fels and R. Y. M. Huang, Theoretical interpretation of the effect of mixture composition on separation of liquids in polymers, J.Macromol.Sci-Phys., B5(1):89-109 (March 1971).
6. C. Larchet, J-P. Brun, and M. Guillou, Separation of benzene-heptane mixtures by pervaporation with elastomeric membranes. I. Performance of membranes, J.Membrane Sci., 15:81-96 (1983).
7. R. A. Shelden and E. V. Thompson, Dependence of diffusive permeation rate on upstream and downstream pressures. III. Membrane selectivity and implications for separation processes, J.Membrane Sci., 4:115-127 (1978).
8. I. Cabasso and H. Leon, Pervaporation of organic liquid mixtures through hollow fiber membranes, Preprints A.I.Ch.E. 80th Nat. Meeting, Boston, Mass. (1975).
9. D. R. Paul, The solution-diffusion model for swollen membranes, Sep.and Purif.Methods, 5(1):33-50 (1976).
10. J. G. Sikonia and F. P. McCandless, Separation of isomeric xylenes by permeation through modified plastic films, J.Membrane Sci., 4:229-241 (1978).
11. J-P. Brun, Etude thermodynamique du transfert selectif par pervaporation à travers des membranes élastomères d'espèces organiques dissoutes en milieu aqueux, Thèse de doctorat es-Sciences Physiques, Université Paris-12 (1981).
12. J-P. Brun, C. Larchet, R. Melet, and G. Bulvestre, Modeling of the selective pervaporation through moderately swollen, non-reactive membranes (submitted to the J.Membrane Sci., 1983).
13. G. Maffiolo, J. Vidal, and A. Asselineau, Coefficients d'équilibre liquide-vapeur dans les mélanges d'hydrocarbures à pression modérée, Chem.Eng.Sci., 30:625-630 (1975).
14. R. H. Perry and C. H. Chilton, Chemical Engineers' Handbook, fifth edn., McGraw-Hill, New York (1973).
15. R. W. Gallant, Physical properties of hydrocarbons, Hydroc.Process., 44, 8, 127-156 (1965).

AN ENERGY-EFFICIENT MEMBRANE DISTILLATION PROCESS

T. J. van Gassel and K. Schneider

Enka AG, P.O. Box 200916
5600, Wuppertal 2
FDR

Although the principle of the process has been known for many years, an increasing interest in membrane distillation has been observed recently. This undoubtedly is because membranes for the process have been available only in recent years.

Characteristics of membrane distillation, among other membrane processes, are that it is a thermally driven process, and so the driving force is a temperature difference. Whereas in most other processes the driving force is a pressure difference or a concentration difference. The process will be called TMD (Trans Membrane Distillation), as proposed in [3].

PRINCIPLE

A hydrophobic, or a water repellant, microporous membrane separates two liquids with the microporous volume as a gas or vapor space in between (Figure 1).

If there is a temperature difference between the liquids (e.g. a salt solution and a pure water distillate), then a vapor pressure difference will exist and water-vapor will be transported from the warm to the cold side of the membrane.

Between two different situations can be distinguished:

a) When non-condensable gases (e.g. air) are present in the membrane pores, the vapor must <u>diffuse</u> through the gas: the transport mechanism is a diffusion mechanism.
b) When no non-condensable gases are present (so a vacuum), we can assume a <u>convective</u> transport through the pores of the membrane.

The latter situation will result in the highest vapor fluxes.

MEMBRANES

The most important membrane requirement is the necessary non-wettability. This means not only that hydrophobic polymers have to be applied, like PTFE, PVDF or PP, but also the maximum pore sizes should not

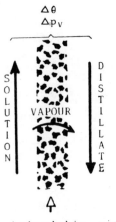

hydrophobic, micro-
porous membrane

Fig. 1. Principle of the TMD-Process

exceed certain limits; because in fact the pressure at which water penetrates the pores is inversely proportional to the pore diameter, and a
certain difference in absolute pressure between the two liquids, of course,
should be allowed in practical applications.

Suitable membranes for membrane distillation should therefore be:

- hydrophobic
- highly porous (e.g. porosity of 70%)
- thermally and chemically resistant

and should have a narrow pore distribution.

These conditions are fulfilled by example PP and PVDF membranes,
produced by the "Accurel[R] -Process" which is a thermal phase-separation
process[1]. Also microporous membranes produced by a stretching process
are being used for membrane distillation[2].

INFLUENCE PARAMETERS

Figure 2 shows the flux vs. the temperature difference between the
solution and the distillate for different membrane thickness. Fluxes
increase with increasing temperature differences and this increase is
exponential, as can be expected, because vapor pressure increases exponentially with the temperature. So a temperature difference of e.g. 5°C on
a level of 90°C is much more effective than the same 5°C at a level of
50°C. Further one can see that fluxes increase with decreasing wall-
thickness of the membrane.

The influence of non-condensable gases in the membrane can be seen in
Figure 3. By application of an under-pressure on one of the two liquids
(here on the solution side) one removes part of the non-condensable gases
and increases the flux because the diffusion resistance in the membrane is
being used, as indicated before. It has been possible to develop an
equation which relates the flux to the vapor pressure difference over the
membrane for a given membrane and which takes into account the influence
of under-pressure in the membrane.

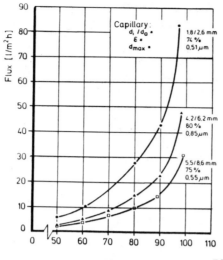

Fig. 2. Specific flux vs. temperature solution for different capillaries $\theta_2 = 18 - 38°C$, P_{v1}, $P_{v2} = 1$ bar abs.

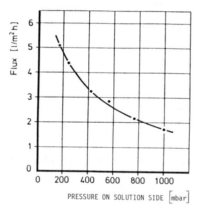

Fig. 3. Specific flux vs. absolute pressure on the solution $\theta_1 = 40°C$, $\theta_2 = 20°C$, $P_{v2} = 1,100$ mbar, membrane 4.2/6.1.

The membrane or module specific flux rate can be determined experimentally and then fluxes can be calculated with a fair agreement with experimental results.

REALIZATION OF THE PROCESS

Membrane distillation, operated in the most simple way, involves a unit, a module, with two chambers, containing liquids of different temperatures, separated by a non-wettable membrane.

In fact, the TMD modules we propose will consist of a bundle of hollow fiber membranes with a diameter of e.g. 1-2 mm. The construction of such a module is much like a shell and tube heatexchanger.

Since heat is required for evaporation and is being released at condensation, the solution will cool down when flowing through the module

345

and the distillate will heat up. In doing so, it is necessary to supply the heat of evaporation (apart from some losses) to the solution, and this heat will be transferred to the distillate together with the produced distillate. Compared with a normal single stage evaporation, the process will then only have the advantage that the distillate is extremely pure, since no entrapment can take place. However, it is possible to recover a considerable part of the energy put in.

If we use countercurrent flows, as in the example in Figure 4, it will be clear that a heatexchange between the heated up distillate and the cooled down solution can be realized, thus transferring back part of the energy to the solution. This process scheme is being considered as a "typical" TMD process unit. A TMD module is combined with a heatexchanger or a HX module which can be a polymer hollow fiber module with a similar construction as the TMD module.

Two circulation flows are created: the warm solution flow and the less warm distillate flow. In this example the cooled down solution from 90°C to 50°C is preheated in a recovery HX module up to 80°C and a final heating from 80°C up to 90°C is required by an external heat source. This means that by adding 10°C to the solution flow, 40°C are available for evaporation; or in other words, ∿4 kg distillate can be produced by 1 kg of vapor. The unit acts - in this example - as a multi-effect evaporator with a performance factor of 4. This phenomena is being considered as one of the biggest advantages of the TMD process; that one unit can act as a multi-effect evaporator with the resulting low energy consumption. By increasing the highest temperature of the solution (the 90°C in this example) and by decreasing the temperature difference over the membrane and the HX module (in this example: 5°C in both cases) higher performance ratios are possible (Figure 5).

Lower temperature differences, however, mean larger membrane areas and higher investments, so an optimization between energy costs and investment, costs will be the way to the lowest costs.

It is possible to combine the membrane as well as the heatexchange area in one device. Such a device has been proposed by others, and it is conceived as a spiral wound module for a flat membrane, respectively a flat heatexchange foil. A drawback of this system is that the distillate cannot be moved by a pump circulation system, thus resulting in lower heat

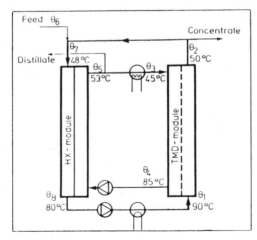

Fig. 4. TMD-process with heatrecovery (temperatures are examples).

transfer coefficients and the relation between membrane and heatexchange
area always has to be one.

APPLICATIONS

Applications of the process are being seen in two directions:

1. The production of ultra pure, desalted or demineralized water from
   various sources, ranging from seawater to waste water.
2. Concentration of aqueous solutions of salts, inorganic acids or lyes
   up to high concentrations.

Of course, a combination of the two possibilities or a situation in
which waste energy can be used will be most attractive.

Figure 6 shows in one example both, the ability of the process to
produce a highly pure, fully demineralized water and the ability to concentrate to very high concentrations (much higher than RO could do it). At
higher concentrations the flux will slightly decrease, as expected, because

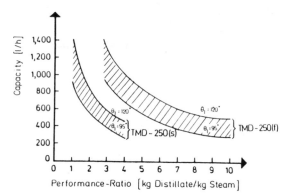

Fig. 5. Capacity (f) and (s) type TMD-250 modules vs. performance-ratio
(situation without boiling point increase).

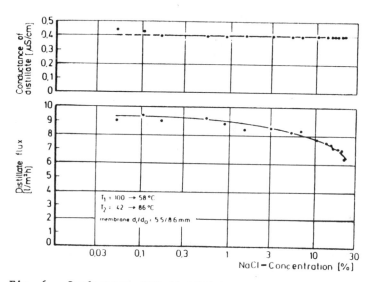

Fig. 6. Conductance TMD distillate vs. concentrate solution.

347

of the vapor pressure reduction with increasing salt concentrations. The quality of the distillate remains unchanged, even up to high concentrations. That the conductance is not even lower than this already quite nice figure of 0.4 μS/cm, has mainly to do with the CO_2 content of the solution. CO_2, so as any other gases, can go freely through the membrane; if disturbed they have to be removed separately.

Apart from a low conductance, the produced water is also particle and bacteria/pyrogene-free, thus making the process also suitable for medical or pharmaceutical purposes. An important advantage of the process might be that it seems to be sensibly low for membrane fouling (a phenomena which in most other membrane processes is a serious problem). A test unit, running on 5μ filtered river water for several months, did not show a fouling effect.

More pilot units, based on 50 - 100 m^2 modules, for several applications are under construction.

REFERENCES

1. K. Schneider, Kunststoffe, 71:183-184 (1981).
2. D. W. Gore, Proc. 10th Ann. Conf. Water Suppl. 25-29 July, Honolulu (1982).
3. K. Schneider and T. J. van Gassel, Chemie.Ing.Techn., 56,7:514-521 (1984).

USE OF HYDROPHOBIC MEMBRANES IN THERMAL SEPARATION

OF LIQUID MIXTURES: THEORY AND EXPERIMENTS

G. C. Sarti and C. Gostoli

Istituto di Impianti Chimici, Facoltà di Ingegneria
Università di Bologna, Viale Risorgimento 2
40136 Bologna, Italy

ABSTRACT

The thermodialysis process separating solutes from aqueous liquid mixtures was studied by using highly porous PTFE membranes. An extremely high rejection was observed for heavy solutes as inorganic salts. The separation was found to be due not to a Soret effect, but rather to an evaporation-diffusion-condensation process enhanced by capillary forces. A crucial role is played by the hydrophobicity of the membrane, which prevents the liquid phases from entering into the pores. The resulting mathematical model is entirely predictive (no adjustable parameters) and proves to be in satisfactorily good agreement with experimental data for different solutes.

INTRODUCTION

In analyzing thermally induced separation through membranes it is convenient to consider two different processes:

  i) thermoosmosis[1-10] occurring across dense membranes, usually showing relatively small fluxes of the transported species, and
 ii) thermodialysis[11-16] occurring across grossly porous partitions which do not show selection properties "per se".

In the latter case the observed fluxes are usually much higher than for the thermoosmotic case.

Typically the theoretical framework used to analyze and correlate the data is found in the thermodynamics of irreversible processes[17-19], which first introduced a general tool for the description of coupled transport phenomena. In the case of thermodialysis, another theoretical formulation was found to be useful, based on the radiation pressure idea[11-20].

In the present work we will focus our attention on the thermodialysis process obtained by using PTFE porous membranes and water solutions as liquid phases. The process is first investigated experimentally and the conclusion is drawn that a sort of low temperature distillation, enhanced by capillarity, is taking place. A theoretical model is then developed based on the following characteristic features. Only a gaseous phase is contained within the pores of a hydrophobic membrane when:

a) the membrane was never subject to spontaneous imbibition of a wetting liquid;
b) the membrane was never subject to forced drainage of a non-wetting liquid;
c) the membrane is in contact with non-wetting liquids which are at a pressure smaller than the minimum entry pressure.

Under the above conditions, a liquid-vapor interface is formed on either side of the liquid repulsing membrane; typically evaporation takes place at the interface of the warmer side and, after mass transport through the vapor phase within the pores, condensation takes place at the interface of the colder side. An essential prerequisite of the process is found in the capillary forces which prevent the liquid phase from entering into the pores.

The above process, which will be called <u>capillary distillation</u>, is based on the use under non-isothermal conditions of the "gaseous membrane" idea used by Imai et al.[21] in isothermal process.

Of course capillary distillation must not be confused with another membrane distillation as pervaporation[22]; in the latter case, separation is obtained through the partition between a dense membrane and an external vapor phase.

EXPERIMENTAL

Commercial polytetrafluoroethylene (PTFE) membranes manufactured by Gelman Instrument Co. as TF 1000 were used. It is a composite membrane formed by an actual porous PTFE layer (the thickness of which was measured in δ = 60 µm), supported by a polypropylene mesh; the nominal overall thickness is 175 µm. The nominal pore size is 1 µm. By using mercury porosimetry the void fraction ε = 0.60 and the pore size distribution were measured.

The experimental apparatus is schematically shown in Figure 1. Mass transfer took place across a membrane area A = 37.4 cm^2. The two halfcells WS and CS are filled with two liquids; the liquid depths, from the halfcell bottoms to the membrane, are 0.8 cm. The two semicell bottoms are kept at two different temperatures, recorded by two thermocouples (Th), by circulating thermostated fluids. For clarity sake a Soret configuration was used (horizontal membrane and higher temperatures in the upper semicell); in that case the heat transfer from the upper to the lower semicell takes place simply by pure conduction.

A continuous flow of bidistilled water, NaCl or NH$_4$Cl solutions was maintained through the warm semicell; the cold semicell was initially

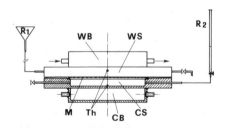

Fig. 1.   Experimental apparatus.

filled either by water or by salt solutions. In order to avoid bubble formation, all liquids were suitably degassed before use, at least to the equilibrium conditions with pure air at 1 atm. and at the highest temperature used in the run. The total flux exchanged was measured by reading the liquid level in the burette R2. The solute flux was calculated through the salt concentration measured in the cold halfcell at the end of each run.

For the cases in which salt solutions flow through the warm semicell and pure water is held in the cold semicell, the water flux is reported in Figures 2 and 3 as a function of the temperature differences between the semicells bottoms, ΔT_b; the different curves refer to different concentrations of NaCl and NH$_4$Cl, ranging from .1 M to .9 M. Two different average temperatures are shown in Figures 2 and 3. Remarkably, the curves relative to the two different salts are identical for the same ionic concentration.

In the case of pure water on both sides, see Figure 4, the water flux is always directed from the warm to the cold semicell and increases with

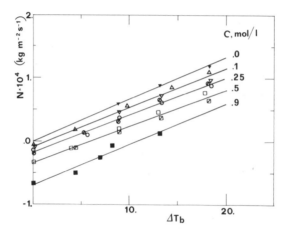

Fig. 2. Water flux vs. ΔT_b (°C) from different NaCl and NH$_4$Cl solutions at T_m = 30°C; NaCl symbols: △ ○ □ ■. NH$_4$Cl symbols: ▽ ⊘ ⊠. TF 1000.

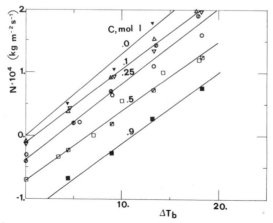

Fig. 3. Water flux vs. ΔT_b (°C) from different NaCl and NH$_4$Cl solutions at T_m = 50°C; NaCl symbols: △ ○ □ ■. NH$_4$Cl symbols ▽ ⊘ ⊠. TF 1000.

increasing the average temperature T_m. In the case of solutions, on the contrary, water flux can have either directions. At lower temperature differences water flows towards the solution side, viz. the osmotic effect prevails; at higher temperature differences, on the contrary, water flows from the warm solution towards the cooler side containing pure water. Consistently with different data reported in Reference [16] a temperature ΔT_b° exists at which no water flux is observed, which typically is of few degrees Celsius for osmotic pressures of several mega Pascal. The water flux vs ΔT_b slope is nearly independent of solute concentration at the values inspected, while it increases by increasing the average temperature.

In Figure 5 we report the purely osmotically driven water flux obtained at $\Delta T_b = 0$, for the case of NaCl solutions; clearly that flux is linear in the ionic mole fraction; just as it is usually observed for an osmotic flux.

For the case in which both warm and cold semicells initially contain the same salt solutions, the water flux observed is reported in Figure 6; different initial salt concentrations were inspected. In such cases, the water flux is always directed towards the lower temperature side; its value slightly decreases with increasing salt concentration.

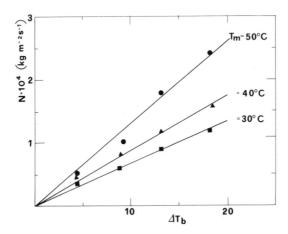

Fig. 4. Water flux vs. ΔT_b (°C) for the pure water case. TF 1000.

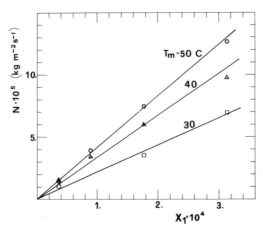

Fig. 5. Water flux vs. ion mol fraction, under isothermal conditions, at different average temperatures T_m. TF 1000.

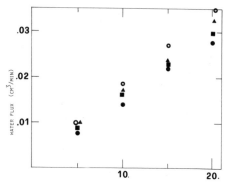

Fig. 6. Water flux vs. the temperature difference between the baths ΔT_B (°C), for the case of NaCl solutions of equal concentrations on both sides. o pure water; ▲ .1 M; ■ .25 M; ● .5 M. TF 1000.

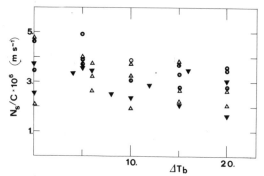

Fig. 7. Salt flux per unit concentration vs. T_b from NaCl solutions; o .1 M; △ .25 M; ▽ .5 M. TF 1000.

In Figure 7 the solute flux N_s is reported as a function of salt concentration and of the temperature difference. The salt flux has definitely low values, thus showing the very high selectivity of the process. As a consequence experimental data are rather scattered, although they clearly show that a) solute flux is independent of the temperature gradient and b) the solute flux is roughly proportional to the concentration gradient.

We also attempted to perform analogous experiments by using water-ethyl alcohol solutions. It was observed that for alcohol concentrations higher than 8% by wt the membrane was no longer a barrier for fluid flow and spontaneous imbibition took place.

However, under our experimental conditions, when the membrane is in direct contact with salt water solutions, the pores inside the membrane contain only a vapor phase. A clearcut evidence of that was obtained by measuring the electric resistance offered by the membrane imbedded into two liquid solutions. In case (a) the membrane was surrounded by a water solution of NaCl, 0.2 M, in case (b) the membrane was surrounded by a .17 M solution of NaCl in a water-ethyl alcohol mixture 50% by wt; the latter solution shows spontaneous imbibition into the membrane. The electrical resistance was then measured between two electrodes, 4.0 cm distant from each other, located on the opposite sides of the membrane. The ratio of the resistivities of the two solutions is $\rho_a/\rho_b = 0.167$. On the contrary, the ratio of the resistances measured across the membranes is $R_a/R_b = 845$, thus indicating that in case (a) the liquid cannot be present inside the pores.

The Theory of Capillary Distillation

A solid smooth surface in contact with a vapor and with a non-wetting liquid shows a contact angle ν, measured through the liquid phase, which is larger than $\pi/2$ [23]. By neglecting all the uncertainties outlined in Reference [24], the equilibrium contact angle requirement imposes a curvature radius on smooth surfaces, either inside or outside the pores. The same is not true when the three-phase line is at a solid edge, such as the edge at the pore entrance. It is well known, after Gibbs[25], that at a solid edge the meniscus can assume a variety of stable equilibrium configurations which range between the two extreme situations in which an angle equal to the contact angle is formed respectively with the two surfaces converging into the edge.

In the case of capillary distillation, a liquid vapor meniscus is formed at the pore entrances; the mean curvature radius of the meniscus, ρ, is not imposed by the contact angle requirement but is given by the Laplace equation. When the center of curvature belongs to the liquid phase, we have:

$$P_L - P_V = \frac{2\sigma}{\rho} \tag{1}$$

where P_L and P_V indicate the pressure values kept on the liquid and on the vapor phases respectively.

For any capillary internal radius r, the maximum curvature admissible, $1/\rho_M$, is obtained when

$$\rho_M = \frac{r}{|\cos \nu|} \tag{2}$$

The pressure difference corresponding to ρ_M in view of Equation (1), is the capillary pressure for the pore, above which the liquid flows through the pore. For lower pressure values on the liquid phase, the meniscus is kept at the pore entrance. Therefore a liquid-vapor equilibrium is maintained at the pore inlet. By equating the liquid and vapor fugacities of all components, and following the same technique used for flat interfaces, the liquid vapor equilibrium is given as:

$$P_i = P_i^* x_i \gamma_i \exp\left[\frac{P_L P_i^*}{RT} \tilde{V}_{iL}\right] \tag{3}$$

We restrict our attention to the case of water-salt solutions.

For P_L values in the order of the atmospheric pressure and for temperatures close to the ambient value, the exponential factor appearing in Equation (3) is actually unity to within a very good approximation. In addition, the activity coefficient γ_w for water is often unity, unless high salt concentrations are considered.

The driving force for water transport is the difference between the water vapor partial pressure at the two pore ends, and is calculated as:

$$P_{w1} - P_{w2} = P_{w1}^* x_{w1} \gamma_{w1} - P_{w2}^* x_{w2} \gamma_{w2} \cong (P_{w1}^* - P_{w2}^*)(1-x_m) - P_m^*(x_1 - x_2) \tag{4}$$

The symbols x_1 and x_2 indicate the total ionic mole fractions.

In the case in which pure water is maintained on both sides of the membrane, Equation (4) predicts that the driving force for water transport is always directed from the warm to the cold side. In the case of salt solutions, the driving force given by Equation (4) can be either positive or negative; according to the relative roles played by temperature and activity differences, the net water flux can be directed either out of, or into the higher temperature semicell.

The temperature difference between the pore ends at which the water flux vanishes ΔT_o, can be easily calculated by using the Clausius-Clapeyron equation; we obtain:

$$\Delta T_o = \frac{\tilde{V} \, T_m}{\tilde{\lambda}_m} \frac{\pi_1 - \pi_2}{1 + \frac{\tilde{V}}{2\lambda_m}(\pi_1 + \pi_2)} \tag{5}$$

For the osmotic pressure π of phases 1 and 2, use has been made of the equation:

$$\pi = \frac{RT}{\tilde{V}} \ln \frac{1}{x_w \, \gamma_w} \tag{6}$$

In view of Equation (5) we observe that a rather small temperature difference is usually needed in order to overcome a rather high osmotic pressure difference. Indeed in the case of water solutions at room temperature, an osmotic pressure difference of 5 MPa is overcome by a temperature difference between the pore ends as small as 0.6°C.

In order to calculate the water flux through the membrane, we will refer to the very usual case in which air is the gas initially contained within the pores and the liquid solutions are nearly saturated with respect to air. Under those conditions diffusion of water vapor through a substantially stagnant film of air takes place. By accounting for the tortuosity factor χ and for the void fraction ε, the water mass flux N can be written as[26].

$$N = \varepsilon M \frac{T_m^{b-1}}{R\chi\delta} \left(\frac{P\mathcal{D}}{T^b}\right)_m \frac{P_{w1} - P_{w2}}{P_{A \, ln}} \tag{7}$$

The diffusion coefficient $\mathcal{D}$ should in general account for both ordinary and Knudsen diffusion when the pore radii are of the order of the molecular mean free path. For the membranes used in our experiments, however, the major contribution was given by ordinary diffusion alone.

After substitution of Equation (4) into Equation (7) we obtain:

$$N = K_D \left[(1-x_m) \frac{\Delta P_w^*}{P_m^*} - (x_1 - x_2) \right] \tag{8}$$

In Equation (8) K_D is a sort of mass transport coefficient and is a function of the average temperature given by:

$$K_D = \varepsilon M \frac{T_m^{b-1}}{R\chi\delta} \left(\frac{P\mathcal{D}}{T^b}\right)_m \frac{P_{wm}^*}{P_{A \, ln}} \tag{9}$$

355

For a given gas pair, $p^{\mathcal{D}}/T^b$ is a constant; in the case of air-water vapor diffusion $b = 2.334$ [26].

By using the Clausius-Clapeyron equation the temperature difference across the membrane, ΔT, can be introduced into Equation (8); in view of the relatively small temperature differences across the membrane; by neglecting only terms of the third order we have:

$$\Delta P_w^* = \left(\frac{P_w^{*\tilde{}}\lambda}{RT^2} \right)_m \Delta T \tag{10}$$

Standard steady-state heat transfer considerations give the temperature differences across the membrane ΔT in terms of the temperature difference ΔT_b between the two semicell bottoms. In view of the phase change occurring at both membrane surfaces we have:

$$\Delta T = \frac{U\delta}{K_p} \left[\Delta T_b - \frac{N\hat{\lambda}}{U_L} \right] \tag{11}$$

The heat transfer coefficients U and U_L are given by:

$$\frac{1}{U} = \frac{1}{h_1} + \frac{1}{h_2} \; ; \; \frac{1}{U} = \frac{1}{U_L} + \frac{\delta}{K_p} \tag{12}$$

From Equations (8), (10) and (11) we finally obtain for the water mass flux:

$$N = \frac{K_D}{1 + K_T K_D \frac{\hat{\lambda}_m}{U_L}} \left[K_T \Delta T_b - (x_1 - x_2) \right] \tag{13}$$

The quantity K_T is a function of both the average temperature T_m and of the average ion mole fraction x_m; one has:

$$K_T = (1 - x_m) \frac{\tilde{\lambda}_m}{RT_m^2} \frac{U\delta}{K_p} \tag{14}$$

It is worth observing that Equation (11) is linear in both temperature and ionic mol fraction differences; that result is consistent with simple prediction based on the thermodynamics of irreversible processes. Our model, however, leads to predictive expressions for the so-called phenomenological coefficients and, in addition, specifies under what simplifying assumptions the linear behavior holds true.

Theoretical Predictions and Experimental Data: a Comparison

In order to compare experimental results and theoretical predictions, the slopes have been calculated of the water flux vs. ΔT_b lines for different average temperatures and different salt concentrations. A least mean square interpolation was used among all data obtained for both NaCl and NH$_4$Cl solutions at the same molar concentrations. The data are reported in Table 1.

According to the theory, the driving force for water flux is simply given by the difference in the water partial pressures existing at the two membrane surface. For the salt concentrations used, the water partial pressure only depends upon the total ion mole fraction in the liquid, independently of their chemical nature. Consistently with that, the data obtained for a given concentration of both NaCl and NH_4Cl solutions are superimposable with each other; the standard deviation shown by the data of both solutions is the same as the standard deviation shown by each solution.

Since the theory is entirely predictive, the theoretical slopes have been calculated as well, based on Equation (13). The calculations have been performed by using the numerical values for the relevant physical and transport properties reported hereafter. The measured void fraction is $\varepsilon = 0.60$, the active PTFE layer is $\delta = 60$ µm thick, while the tortuosity factor is typically $\chi = 2$ [21]. The thermal conductivity of the porous membrane, K was calculated as for a homogeneous porous medium, i.e. $K_p = \varepsilon K_v + (1-\varepsilon) K_s$; K_v and K_s are the thermal conductivities of the vapor phase and of Teflon respectively; we obtained $K_p = 0.036$ W/mK. Since the liquid contained in either semicell is 0.8 cm thick we have $U_L = 35.3$ W/m^{2}K; the diffusivity of water vapor in air, at 0°C and at normal pressure is 0.22 cm^2/s.

The theoretical values of K_D and K_T thus obtained are shown in Table 2; in Table 1 we show the theoretical values of the slope of the water flux vs. ΔT_b lines, at the different concentrations and temperature difference.

The numerical comparison between predicted values and experimental ones, Table 1, shows a rather good agreement.

Another test of the theory can be obtained by calculating the values ΔT_b°, at which water flux vanishes, based on Equation (13) and on the predicted K_T values. The comparison is shown in Table 3 and again the comparison is really satisfactory.

CONCLUSIONS

The capillary distillation process is a thermally driven separation process for liquid mixtures based on the following features. A porous and liquid-repulsing partition separates two liquid phases at different temperatures and composition, the capillary forces prevent the liquids from entering the pores in which only a gaseous phase is present. On both sides of the porous partition, local liquid vapor equilibria hold true which give rise to a mass transfer process across the pores. The theory based on the above points proves to be entirely predictive and in excellent agreement with the experimental results.

<u>Acknowledgements</u>

This work has been supported by the Italian Ministry of Education. Several discussions with Prof. F. Bellucci stimulated our interest in the field. We wish to thank Ing. A. Vivarelli and Ing. S. Matulli for their help in performing the experiments. Finally we wish to thank Prof. F. Sandrolini for helping us in the mercury porosimetry tests.

Table 1. Experimental and Calculated Values of the Slopes of the Water Flux vs. ΔT_b Lines, in $kg/m^2 sK \times 10^6$; Different Average Temperatures (°C) and Molar Concentrations (mol/1).

	T_m \ C	.0	.1	.25	.5	.9	aver.	SD%
Experimental	30	6.72	6.17	5.89	5.79	6.47	6.21	6.3
	40	8.72	9.46	9.01	8.02	8.51	8.74	6.2
	50	13.00	11.53	11.99	10.90	11.30	11.75	6.8
Theoretical	30	7.00	6.99	6.97	6.94	6.89		
	40	8.92	8.91	8.89	8.86	8.81		
	50	10.57	10.56	10.55	10.52	10.47		

Table 2. Calculated Values of K_D and K_T for the TF 1000 Membrane.

T_m	K_D $(kg/m^2 s)$	$K_T/(1-x_m)$ (K^{-1})
30	4.25×10^{-3}	3.18×10^{-3}
40	7.72 "	2.95 "
50	13.46 "	2.74 "

Table 3. Experimental and Calculated Values of ΔT_b° as a Function of Salt Concentration (mol/1) and Average Temperature (°C).

	T_m \ C	.1	.25	.5	.9
Experimental	30	1.34	2.95	6.00	10.77
	40	1.55	3.32	6.23	11.45
	50	.98	3.19	6.40	11.18
Theoretical	30	1.13	2.85	5.75	10.39
	40	1.22	3.07	6.20	11.19
	50	1.31	3.30	6.67	12.04

NOMENCLATURE

Notation

$\mathscr{D}$ diffusion coefficient

h_1, h_2 heat transfer coefficient for a single liquid phase

K_p heat conduction of the membrane

K_D defined by Equation (9)

K_T defined by Equation (14)

M   molecular weight

N   mass flux

P   pressure

P*   vapor pressure of pure water

r   pore radius

x   mole fraction

T   temperature

U   heat transfer coefficient

Greek letters

δ membrane thickness

γ activity coefficient

ε void fraction

$\hat{\gamma}$ specific heat of vaporization

$\tilde{\gamma}$ molar heat of vaporization

ρ curvature radius

χ tortuosity factor

σ surface tension

π osmotic pressure

Subscripts

b   bulk

m   average

s   solute

w   water

1   warm side

2   cold side

L   liquid

V   vapor

ln   log mean

REFERENCES

1. C. W. Carr and K. Sollner, New experiments on thermoosmosis, J.Electrochem.Soc., 109:616 (1962).
2. R. P. Rastogi, R. L. Blokhra, and R. K. Agarwal, Cross phenomenological coefficients. Part I. Study on thermoosmosis, Trans.Farad.Soc., 60:1386 (1964).
3. R. Haase, H. J. de Greiff, Thermoosmose in Flüssigkeiten, Z.Physik. Chem.,N.F., 44:S-301 (1965).
4. R. P. Rastogi and K. Singh, Cross phenomenological coefficients. Part 5. Thermoosmosis of liquids through cellophane membranes, Trans.Faraday Soc., 62:1754 (1966).
5. R. Haase, Zur Theorie der Permeation und Thermoosmose, Z.Physik.Chem. N.F., 51:S-315 (1966).
6. R. Haase, H. J. de Greiff, and H. J. Buchner, Thermoosmose in Flüssigkeiten. IV. Z.Naturf., 25a:1080 (1970).
7. M. S. Dariel and O. Kedem, Thermoosmosis in semipermeable membranes, J.Phys.Chem., 79:1773 (1971).
8. W. E. Goldstein and F. H. Verhoff, An investigation of anomalous osmosis and thermoosmosis, A.I.Ch.E.J., 21:229 (1975).
9. H. Vink and S. A. A. Chisti, Thermal osmosis in liquids, J.Membrane Sci., 1:149 (1976).
10. J. I. Mengual, J. Aguilar, and C. Fernandez-Pineda, Thermoosmosis of water through cellulose acetate membranes, J.Membr.Sci., 4:209 (1978).
11. F. S. Gaeta and D. G. Mita, Non-isothermal mass transport in porous media, J.Membr.Sci., 3:191 (1978).
12. F. Bellucci, M. Bobik, E. Drioli, F. S. Gaeta, D. G. Mita, and G. Orlando, Separation by thermodialysis of acetic acid aqueous solutions, Canad.J.Chem.Eng., 56:698 (1978).
14. F. S. Gaeta and D. G. Mita, Thermal diffusion across porous partitions: the process of thermodialysis, J.Phys.Chem., 83:2276 (1979).
15. F. Bellucci, E. Drioli, F. S. Gaeta, D. G. Mita, N. Pagliuca, and D. Tomadacis, Temperature gradient affecting mass transport in synthetic membranes, J.Membr.Sci., 7:169 (1980).
16. D. G. Mita, F. Bellucci, M. G. Cutuli, and F. S. Gaeta, Non-isothermal mass transport in NaCl and KCl aqueous solutions, J.Phys.Chem., 86:2975 (1982).

17. I. Prigogine, Thermodynamics of Irreversible Processes, Interscience, New York (1954).
18. S. R. De Groot and P. Mazur, Non-equilibrium Thermodynamics, North Holland, Amsterdam (1962).
19. R. Haase, Thermodynamics of Irreversible Processes, Addison-Wesley Publ., Reading, Mass. (1969).
20. F. S. Gaeta, Radiation pressure theory of thermal diffusion in liquids, Phys.Rev., 132:289 (1969).
21. M. Imai, S. Furusaki, and T. Miyauchi, Separation of volatile materials by gas membranes, I.E.C.Proc.Des.Dev., 21:421 (1982).
22. A. S. Micha, Progress in Separation and Purification, E. S. Perry, ed., Wiley-Interscience, New York (1968).
23. J. T. Davies and E. K. Rideal, Interfacial Phenomena, Academic Press, New York (1963).
24. E. B. Dussan V., On the spreading of liquids on solid surfaces, Am.Rev.Fluid Mech., 11:371 (1979).
25. J. W. Gibbs, The Collected Works, p.326, Vol.I, Longmans Green and Co. New York (1931).
26. R. B. Bird, W. E. Steward, and E. N. Lightfoot, Transport Phenomena, J. Wiley, New York (1960).

THE WATER STRUCTURE IN MEMBRANE MODELS STUDIED BY

NUCLEAR MAGNETIC RESONANCE AND INFRARED SPECTROSCOPIES

C. A. Boicelli*, M. Giomini** and A. M. Giuliani***

 * Istituto dei Composti del Carbonio contenenti Eteroatomi e
   loro Applicazioni - CNR - Via Tolara di Sotto
   89 - 40064 Ozzano Emilia (Bologna), Italy
** Dipartimento di Chimica, Università degli Studi
   Pizzale Aldo Moro, 5 - 00185, Roma, Italy
***CNR - ITSE - Area della Ricerca di Roma - C.P. 10 - 00016
   Monterotondo Stazione (Roma), Italy.

INTRODUCTION

The arrangement of lipids, proteins and water in a membrane is intrinsically dynamic, changing to suit its instantaneous functional needs: therefore some insight on the control mechanisms of membrane functions can be obtained from the comprehension of the dynamic properties of its components.

In particular, water plays a central role in all functions of biomembranes and the knowledge of how different factors can modulate the water organization and mobility is a prerequisite for the understanding of membrane properties.

Infrared (IR) and Nuclear Magnetic Resonance (NMR) spectroscopies are ideal tools for approaching the water problem in membranes and in their models since they can give the required information without the need of extraneous probes and since they are sensitive to phenomena characterized by different life-times (less than 10^{-10} seconds for IR, from that value up to seconds for NMR)[1].

THE TECHNIQUES

IR investigations mainly consist in the observation of the absorptions due to vibro-rotational transitions. When the molar extinction coefficients ε are known, quantitative evaluations are in principle possible. Since every chemical group is associated with a specific set of transitions, the identification of the molecular species and the determination of their relative amounts is therefore possible[2].

The NMR parameters, on the other hand, are of different nature. Some are steady state parameters (the chemical shift δ and the scalar J, dipolar D and quadrupolar Q coupling constants) related to the molecular structure of the system under study; they give information about the chemical environment of the observed nucleus and about its stereochemical relationships with other nuclei close in the space. Other parameters (the long-itudinal or spin-lattice relaxation time T_1 and the transverse or spin-spin relaxation time T_2) are connected with the motional characteristics of the nuclear ensemble. Since all nuclei with a non-zero spin are NMR active, useful informations can be obtained from ^1H, ^2H, ^{13}C, ^{15}N, ^{17}O and ^{31}P, i.e. from all the positions of the molecule of the membrane components[3].

MEMBRANE SPECTROSCOPIC STUDIES

The large majority of IR and NMR studies on membranes and membrane models deals with the structure of the lipidic matrix[2,3], while consider-ably fewer investigations have been devoted to the water[4]. This occurs mainly because the permeability of the membrane allows free exchange between inner and outer compartments, yielding only "averaged" information on the water and making the water outside the cavity delimited by the membrane indistinguishable from the water secluded in it.

The problem of the water must therefore be approached using suitable models. Single bilayer and multilamellar vesicles (SUV and MLV) both retain the same complication of inside-outside exchange; on the other hand, reverse micelles, where water is present only inside the lipidic monolayer, allow the study of the inner water, without interference from the outer compartment[5].

MATERIALS AND METHODS

Reverse micelles have been prepared with egg phosphatidylcholine (EPC; Sigma Che. Co., type VII-E) in an ice-bath and under nitrogen atmosphere to prevent oxidative phenomena. The EPC concentration was 50 mg mL^{-1} and the micelles were obtained by adding the required amount of water or aqueous solution to the EPC solution in hexadeuterobenzene and sonicating in a bath-type sonifier.

NMR high-resolution experiments were obtained with a variety of spectrometers (Varian XL100-A12, Bruker WH-90 and Bruker WP-200), while the proton longitudinal relaxation times were measured with a low-resolution Bruker Minispec p20 operating at 20 MHz. The usual inversion-recovery pulse sequence was used for the determination of the ^1H and ^{31}P long-itudinal relaxation times and the T_1 values were obtained from the slope of the semilogarithmic plots of the magnetization decay as a function of the interpulse time.

IR spectra were recorded with a double-beam Perkin-Elmer 580B instru-ment, using 0.1 mm NaCl cells; the reference was selected for each set of experiments to yield the required information. The OH stretching region (4000 – 3000 cm^{-1}) was analyzed with a DuPont 310 curve resolver.

THE WATER IN REVERSE MICELLES

The main components of reverse micelles are phospholipids, which form the micellar wall, and water, which occupies the internal cavity of the vesicles. The water molecules are distributed in hydration layers around the polar heads of phospholipids and they influence the organization and the microdynamics of the whole system. The micellar water is highly struc-tured and several populations in slow interchange have been evidenced by low-resolution proton NMR and by IR spectroscopies. The motional charac-teristics of these water populations and their relative amounts can be modified in several ways.

362

The broad OH stretching band in the region 4000 - 3000 cm^{-1} can be instrumentally deconvoluted in three Gaussian components, which are assigned to water populations with different degrees of organization, "polymeric" or highly structured, "oligomeric" i.e. of intermediate degree of organization, and "monomeric" i.e. the more free water. Preparation of reverse micelles with increasing amounts of water has allowed to define the spectral characteristics of these Gaussian components (Table 1)[6]. The distribution of the water molecules among these three populations depends critically upon the number of water molecules added per polar head (Figure 1).

The IR and NMR measurements on the water inside a micellar[5] or vesicular[7] cavity indicate that the water pool is organized in layers characterized by individual correlation times (τ_c), i.e. motional properties. The τ_c values depend on the geometrical constraints to which the molecules are exposed and on the intermolecular interactions (water - water, water - polar heads, water - other chemical species). While a large range of different motional characteristics can in principle exist, only a limited number of populations can be experimentally observed, since fast exchange may average the individual dynamic properties. A definite water population can only be identified when the exchange with other groups of molecules associated with different "classes" or correlation times is slow on the time scale of the used technique.

THE PHOSPHOLIPID POLAR HEADS IN REVERSE MICELLES

In reverse micelles, the glycerophosphorylcholine moiety of phospholipids is found at the water interface, like in SUV, while the fatty acid chains extend into the organic phase and behave as in solution. The experimental support to this statement comes from high-resolution ^{1}H-T$_1$ measurements[8]. The ^{31}P longitudinal relaxation time of the phosphate groups is sensitive to the number of water molecules per polar head (Figure 2). The T_1 value, moreover, changes with the pH of the inner water pool producing a titration curve, from which an apparent pK value of 6.8 ± 0.1 can be obtained. However, when phosphate buffers are secluded inside the micelles, only one ^{31}P resonance is observed, whose chemical shift is independent of the pH and also the micellar water has a constant ^{1}H chemical shift value in the whole pH range considered[9]. It may be concluded that the extended network of hydrogen bonds within the micellar cavity prevents the change of the ionization state of the orthophosphate and that the ^{31}P-T$_1$ value is the only parameter which can be used to monitor the modifications induced in the micelle interior by the buffers.

Table 1   Infrared Characteristics of the Water Populations in Reverse Micelles of EPC in Benzene. (Reference: EPC in Benzene 50 mg mL^{-1}; Experimental Values ± Half Dispersion).

water population	ν (cm^{-1})	$\Delta\nu_{1/2}$ (cm^{-1})	$\varepsilon \cdot 10^3$ (cm^2 mol^{-1})
POLYMERIC	3250 ± 20	400 ± 20	57 ± 5
OLIGOMERIC	3450 ± 20	300 ± 20	83 ± 5
MONOMERIC	3590 ± 20	180 ± 20	32 ± 2

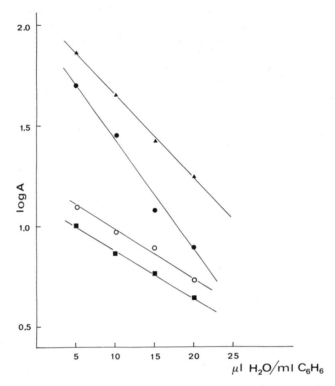

Fig. 1. Effect of the amount of water on the areas (A, area in sq. cm) of the OH stretching band of its Gaussian components for reverse micelles of EPC in benzene (EPC concentration: 50 mg mL^{-1}; the reference was prepared with 30 μl of water per mL of benzene solution). Water populations: ▲ total; ● polymeric; ■ oligomeric; o monomeric.

EFFECT OF WATER-SOLUBLE CHEMICAL SPECIES ON REVERSE MICELLES PROPERTIES

The internal micellar organization may be perturbed by addition of water-soluble species, like electrolytes or aminoacids and small peptides.

The aminoacids and the peptides compete with the polar heads for the water molecules, since they build up their own hydration shell. The effect of the presence of these molecules is clearly observed in the IR spectrum; not only the fractional populations of water are perturbed, but in certain cases one of the populations disappears (Figure 3). Large effects of these perturbing species are observed also on the longitudinal relaxation times of the water protons; two values of T_1 can be calculated in each case, corresponding to two water types in slow exchange with each other on the NMR time scale. A linear relationship

$$T_1^A = 532 - 6 \cdot 3 \, X_A \qquad (1)$$

has been found between the fractional population X_A of the more structured water and its longitudinal relaxation time T_1^A. The maximum allowed value for X_A, obtained for T_1^A equal to zero, is 85 ± 1; this corresponds to ca. 25 molecules of tightly bound water per polar head, since the reverse micelles were prepared with ca. 30 water molecules per phospholipid molecule. The value is in excellent agreement with literature data obtained for SUV[7].

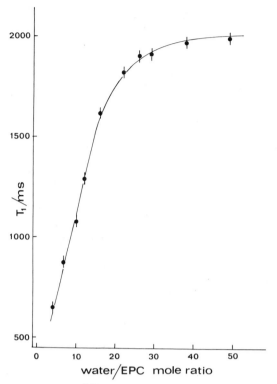

Fig. 2. Dependence of the ^{31}P longitudinal relaxation time on the water/phospholipid mole ratio in reverse micelles ($B_o = 2.35$ T; T = 303 ± 1 K).

While aminoacids and peptides influence essentially the water organization, electrolytes may modify also the polar heads characteristics, because of specific binding and screening phenomena[10-12]. The effect of ionic concentration on the IR and NMR parameters of reverse micelles has been studied for a series of cations having Cl^- as the common anion, and for a number of sodium salts; several nitrates have also been secluded in the micelles. Marked differences are observed in the case of preparation of reverse micelles in the presence of electrolytes: cations strongly interacting with phosphate groups (like Ca^{2+}, Mg^{2+}, etc.) generally require long times of sonic irradiation to yield clear micellar suspensions, while solutions of weakly interacting ions, like Na^+ and K^+, are easily dispersed in the benzene phase. It has been found impossible to prepare reverse micelles containing Ba^{2+} ions in the aqueous pool: the ionic dimensions does not seem to be critical in this respect, since the hydration radius of Ba^{2+} (3.92 Å) is smaller than that of Mg^{2+} (4.65 Å)[13].

The effect of some selected cations, in the form of chlorides on the proton T_1 of the more organized water population is reported in Figure 4.

Significant changes in the ^{31}P longitudinal relaxation time (Figure 5) are caused by divalent cation chlorides, when the concentration of the electrolyte solution exceeds 10^{-2}M, i.e. when the cation/EPC mole ratio is larger than ca. 0.005. The effect is much less substantial for monovalent cations.

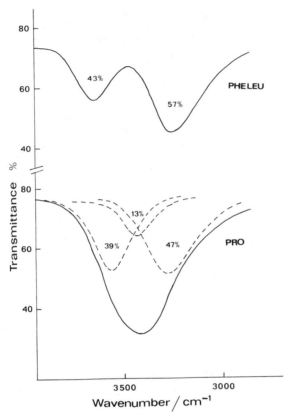

Fig. 3. Infrared spectral pattern of reverse micelles secluding different hydrophilic solutes (EPC concentration: 50 mg mL^{-1}; the reference was prepared with 30 μL of water per mL of benzene solution).

Similarly, for concentrations of the electrolyte solutions larger than 10^{-2} - 10^{-1}M, marked high-field shifts are observed for the ^{31}P resonance in the presence of the divalent cations; practically no effect is induced by the monovalent ions.

Significant differences are observed also when the anion of the electrolyte is changed. While the ^{31}P chemical shift is practically unaffected, as expected, the dynamic parameter T_1 increases gradually (from ca. 1680 to ca. 2170 ms) in the series of Na$^+$ salts (Cl$^-$ < NO$_3$$^-$ < formate < acetate < propionate) at the same electrolyte concentration 10^{-4} M (i.e. Anion/EPC mole ratio = ca. $5 \cdot 10^{-5}$). Moreover, while the concentration of NaCl has negligible influence on the ^{31}P-T_1 value, increasing above ca. 10^{-2} M the concentration of the other electrolytes a marked decrease of the longitudinal relaxation time is induced.

The IR spectra for two different Na$^+$ salts (chloride and nitrate) for concentrations changing from 10^{-5} to 1 M are reported in Figure 6: the marked effect of the ionic concentration and of the change of the anion on the relative amounts of the three water populations identified by IR is immediately evident.

Small shifts have also been measured for the N$^+$(CH$_3$)$_3$ resonance in the proton high-resolution spectra at concentrations of the electrolyte solutions larger than 10^{-2} - 10^{-1}M.

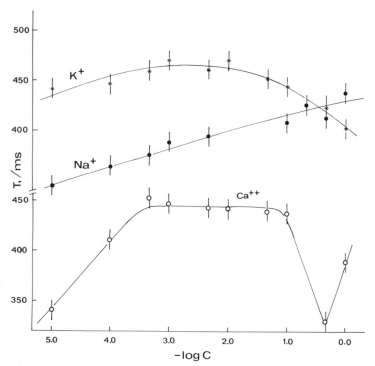

Fig. 4. Effect of cation concentration on the proton longitudinal relaxation time of the tightly bound water in reverse micelles (B_o = 0.47 T; T = 313 ± 1 K; anion Cl^-).

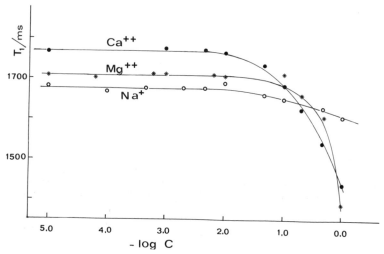

Fig. 5. Effect of cation concentration on the ^{31}P longitudinal relaxation time of reverse micelles (B_o = 4.7 T; T = 303 ± 1 K; anion Cl^-).

The reported experimental findings may reflect several phenomena: electrostatic screening, specific binding of the ions of the electrolyte to the charged groups on the micellar wall, a modification of the packing of the phospholipid polar heads or even a different spatial arrangement of the phospholipids. All these effects have been observed for phospholipid

367

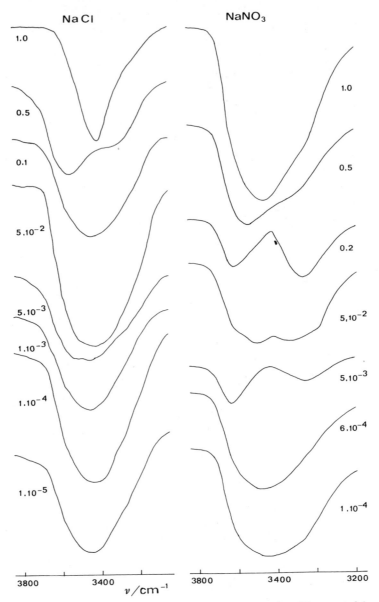

Fig. 6. Influence of the cation concentration on the OH stretching band of reverse micelles secluding two different Na^+ salts (reference: reverse micelles with no electrolytes; number on curves are molar concentrations).

vesicles in the presence of ions: screening has been observed by several authors[10,11] for different vesicular systems, binding of cations and anions to neutral phospholipids has been measured[12,14-16] a conformational change of the polar heads and a squeezing out of water molecules from their hydration layer has been deducted by means of ^7Li, ^{23}Na and ^2H NMR quadrupole splitting measurements[17]. Moreover, a change in vesicle size has been associated to the shift of the trimethylammonium resonance[18].

All these factors play probably a role in our micellar systems and the experimental findings are the result of their relative importance in the different conditions. At the present stage, it is not possible a quantitative evaluation of the individual influences of the various factors.

CONCLUSIONS

The description of the structural, dynamic and functional properties of membranes and membrane-mimetic microenvironments is to date only approximate and can be considered largely tentative. Indeed, very many factors in these systems play fundamental rôles, some of which may be underestimated or still unknown, and the approach to the problem through models is often oversimplified.

Since the information obtained often raises more problems than it solves, the physico-chemical investigation of membrane systems, even by means of such powerful techniques as IR and NMR, inevitably proceeds by slow steps. Thus, an approach to the subject, based on experimental observations more than on theories, seems at present the most appropriate and the most promising for the exploitation of natural or artificial membranes in biomedicine and technology.

REFERENCES

1. G. A. Marshall, "Biophysical Chemistry," J. Wiley & Sons, New York p.384 (1978).
2. U. P. Fringely, and Hs. H. Günthard, Infrared membrane spectroscopy, in: "Membrane Spectroscopy," E. Grell, ed., Springer-Verlag, Berlin (1981).
3. S. I. Chan, D. F. Bocian, and N. O. Petersen, Nuclear magnetic resonance studies of the phospholipid bilayer membrane, in: "Membrane Spectroscopy," E. Grell, ed., Springer-Verlag, Berlin (1981).
4. E. G. Finer and A. Darke, Phospholipid hydration studied by deuteron magnetic resonance spectroscopy, Chem.Phys.Lipids, 12:1 (1974).
5. C. A. Boicelli, F. Conti, M. Giomini, and A. M. Giuliani, Water organization in reversed micelles, in: "Physical Methods on Biological Membranes and their Models," F. Conti, ed., NATO-ASI, A Series, Plenum Publ.Corp., New York (1984), in the press.
6. C. A. Boicelli, M. Giomini, and A. M. Giuliani, Infrared characterization of different water types inside reverse micelles, Appl.Spectry, 38:537 (1984).
7. E. G. Finer, Interpretation of deuteron magnetic resonance spectroscopic studies of the hydration of macromolecules, J.Chem.Soc. Faraday Trans.II, 69:1590 (1973).
8. C. A. Boicelli, F. Conti, M. Giomini, and A. M. Giuliani, Interactions of small molecules with phospholipids in inverted micelles, Chem.Phys.Lett., 89:490 (1982).
9. C. A. Boicelli, F. Conti, M. Giomini, and A. M. Giuliani, The influence of phosphate buffers on the ^{31}P longitudinal relaxation time in inverted micelles, Spectrochim.Acta, 38A:299 (1982).
10. S. G. A. McLaughlin, G. Szabo, and G. Eisenman, Divalent ions and the surface potential of charged phospholipid membranes, J.Gen. Physiol., 58:667 (1971).
11. S. McLaughlin and H. Harary, The Hydrophobic adsorption of charged molecules to bilayer membranes; a test for the applicability of Stern equation, Biochem., 15:1941 (1976).

12. A. McLaughlin, C. Grathwohl, and S. McLaughlin, The adsorption of divalent cations to phosphatidylcholine bilayer membranes, Biochim.Biophys.Acta, 513:338 (1978).
13. K. H. Stern and E. S. Amis, Ionic Size, Chem.Revs., 59:1 (1959).
14. J. G. Stollery and W. J. Vail, Interactions of Divalent cations or basic proteins with phosphatidylethanolamine vesicles, Biochem. Biophys.Acta, 471:372 (1977).
15. H. Akutsu and J. Seelig, Interaction of metal ions with phosphatidyl-choline bilayer membranes, Biochem., 20:7366 (1981).
16. H. F. Hahn, J. M. Collins, and L. J. Lis, Anion influence on the binding of divalent cations to phosphatidylcholine, Biochim. Biophys.Acta, 736:235 (1983).
17. O. Söderman, G. Arvidson, G. Lindblom, and K. Fontell, The inter-actions between Monovalent ions and phosphatidylcholines in aqueous bilayers, Eur.J.Biochem., 134:309 (1983).
18. C. G. Brouillette, J. P. Segrest, T. C. Ng, and J. L. Jones, Minimal size phosphatidylcholine vesicles: effects of radius of curvature on head group packing and conformation, Biochem., 21:4569 (1982).

PREPARATION AND IN VITRO DEGRADATION

PROPERTIES OF POLYLACTIDE MICROCAPSULES

K. Makino, M. Arakawa, and T. Kondo

Faculty of Pharmaceutical Sciences
Science University of Tokyo
Shinjuku-ku, Tokyo, Japan 162

INTRODUCTION

In recent years, the use of polylactide (PLA), a synthetic biodegradable polymer, as drug-loaded matrix has been examined in the hope that the matrix will decay and decompose after releasing the drug in a sustained manner over a long period of time in the human body[1-8]. However, no detailed work has been done on the degradation rate of PLA even under in vitro conditions. The present paper deals with the rate of in vitro degradation of PLA microcapsules in various environmental conditions.

EXPERIMENTAL

Preparation of PLA Microcapsules

Poly(D,L-lactide) and poly(L-lactide) were used to prepare PLA microcapsules. The preparation of PLA microcapsules was made as follows[9]. Fifty ml of n-hexane was added to 150 ml of 1.5% (w/v) aqueous Pluronic F68 solution and the mixture was stirred for 5 min to yield an O/W emulsion. To this emulsion was added by drop 20 ml of 1.5%(w/v) poly(D,L-lactide) (Polyscience, U.S.A.) or poly(L-lactide) (Mitsui Toatsu, Japan) solution in dichloromethane, and the system was stirred under reduced pressure for 2 hr to allow the organic solvent to evaporate completely. Poly(D,L-lactide) or poly(L-lactide) microcapsules thus prepared were centrifuged at 2000 rpm for 10 min and washed three times with water. The average capsule diameter was determined to be about 1.5 μm from electron micrographs of the microcapsules.

Estimation of Degradation

The degree of degradation of PLA microcapsules was estimated from the rate of decrease in the weight-average molecular weight $\bar{M}_w$, of PLA evaluated by gel permeation chromatography (GPC), and from the amount of lactic acid generated as the final product in bulk solution, determined at different degradation periods according to the Barker-Summerson method[10]. In GPC analysis, the eluent was chloroform and the columns were Toso 2000H6 (Toyosoda, Japan) and Shodex AC804 (Shoko-tsusho, Japan). A certain sample suspension was withdrawn from the test suspension at suitable intervals and centrifuged at 3000 rpm for 10 min, and the settled microcapsules were extracted into chloroform. The extract was used as the sample for GPC analysis.

Degradation of PLA Microcapsules

In the degradation experiments, PLA microcapsules were dispersed in various solution media at the final concentration of 0.15% (v/v).

Five buffer solutions, HCl/trisodium citrate, trisodium citrate/ Na_2HPO_4, KH_2PO_4/Na_2HPO_4, KH_2PO_4/Na_2HPO_4, and $NaHCO_3$/Na_2CO_3, were used to observe the effect of pH of the medium on the degradation rate of PLA microcapsules. The pH values of the buffers were 1.6, 3.0, 5.0, 7.4, and 9.6, respectively. PLA microcapsules dispersed in these media were kept at 37°C.

The degradation rate of PLA microcapsules was measured at 21, 37, and 45°C in a phosphate-buffered physiological saline solution. The effect of ionic strength of the medium on the degradation rate was studied at pH 9.6 and 37°C. The ionic strength was adjusted to 0.07, 0.15, 0.30 and 0.45 by adding NaCl to the solution.

The effects of neutral salts and urea on the degradation rate of PLA microcapsules were examined at pH 3.0 and 37°C. The salts used were $AlCl_3$, $CaCl_2$, KCl, NaCl, and NaSCN. The concentrations of the salts and urea were 0.0001M and 0.1M, respectively.

The degradation rate of PLA microcapsules was also studied in the presence of carboxylic esterase (EC 3.1.1.1). PLA microcapsules were suspended in a phosphate buffer solution (pH 7.6) containing the enzyme and kept at 37°C. Determination of enzyme activity was carried out by the Kobayashi method[11], using 0.9 mM p-nitrophenylbutylate as the substrate. Since the enzyme activity was proportional to the concentration of p-nitrophenol converted from the substrate, the rate of increase in the absorbance at 420 nm during the reaction was measured spectrophotometrically. The initial enzyme activities in solution were 1.45, 2.22, and 3.96 U/ml.

RESULTS AND DISCUSSION

Effect of pH

The decrease of weight-average molecular weight $\bar{M}_w$ of PLA in the degradation of poly(D,L-lactide) microcapsules dispersed in each of the buffer solutions is shown in Figure 1 as a function of immersion time. The observed decrease of $\bar{M}_w$ indicated that a number of the ester bounds in the polymer are cleaved and that poly(D,L-lactide) microcapsules are degraded significantly faster in the highly alkaline buffer solution (pH 9.6) than in the highly acidic and slightly alkaline ones. However, no appreciable change of $\bar{M}_w$ was observed during the immersion time in the slightly acidic buffer solutions. These findings are in accordance with the features of the general acid-base catalysis. Since the cleavage of ester bond takes place according to the different mechanisms depending on the pH of the medium (acid and base catalyzed reactions), and the microcapsule membranes are considered to be composed of both crystalline and noncrystalline zones, the observed difference in the degradation behavior of the microcapsules at different pH should be interpreted in these terms.

In the highly alkaline solution, the molecular weight distribution changed with progressive degradation as shown in Figure 1. The molecular weight distribution for PLA of poly(D,L-lactide) microcapsules was re-

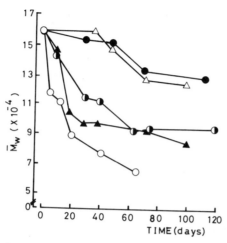

Fig. 1. Effect of pH of medium on the decrease of weight-average
molecular weight of PLA in the degradation of poly(D,L-lactide)
microcapsules. pH: ▲, 1.6; △, 3.0; ●, 5.0; ◐, 7.4; ○, 9.6.

latively narrow and the modal molecular weight was about 60000 at 0 day.
After 6 days, corresponding to the onset of a significant weight loss, the
molecular weight distribution became trimodal with two shoulders developed
at the low molecular weight tail of gel permeation chromatogram and the
modal molecular weights of the two fractions were about 12000 and 2500,
respectively. On further degradation, the molecular weight distribution
still continued to change. That is, the fraction of highest molecular
weight decreased while the fraction of lowest molecular weight increased
and the fraction of middle molecular weight diminished. Similar degradation behavior was also observed in the other buffer solutions in terms of
molecular distribution after much longer immersion times.

These experimental findings indicate that large quantities of low
molecular weight PLA are formed by the de-esterification reaction via
production of polymers of intermediate molecular weights. The depoly-
merization of PLA molecules constituting the microcapsules may have been
accelerated by the production of the intermediate molecular weight
fractions because the decrease of $\bar{M}_w$ was not so rapid until these fractions
appeared in the molecular weight distribution. It is suggested that water
molecules penetrate first into the amorphous zone of the semicrystalline
structure of the material and hydrolysis starts in this zone and then moves
gradually into the crystalline zone. In view of this, the fraction with a
molecular weight of about 60000 is supposed to be a stable unit of the
crystalline zone of PLA microcapsules as suggested in Figure 2. Consequently, the hydrolytic cleavage of polymer chains of PLA is considered to
proceed by unzipping process and not randomly.

Figure 3 shows the weight loss of poly(D,L-lactide) microcapsules
during the period of degradation. The weight loss was evaluated from gel
permeation chromatographic analysis of the undegraded microcapsules.
Poly(D,L-lactide) microcapsules in the strongly alkaline medium lost their
weight very remarkably. This suggests that a fraction of low molecular
weight species which is soluble in the aqueous medium was generated from
the polymer. The fraction is expected to be of lactic acid as the final
product of the degradation.

Poly(L-lactide) microcapsules were degraded less rapidly than
poly(D,L-lactide) microcapsules as shown in Figure 4 where the amount of

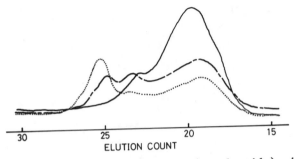

Fig. 2. Gel permeation chromatogram of poly(D,L-lactide) microcapsules dispersed in a buffer solution (pH 9.6) as a function of immersion time: ———, 0 day; - —, 6 days; ...., 20 days.

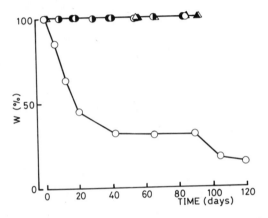

Fig. 3. Effect of pH of medium on weight loss of poly(D,L-lactide) microcapsules. pH: ▲, 1.6; △, 3.0; ●, 5.0; ◑, 7.4; ○, 9.6.

lactic acid in the bulk solution is given as a function of degradation period. The dependency on pH of the degradation rate was almost the same as that for poly(D,L-lactide) microcapsules. The slower degradation rate for poly(l-lactide) microcapsules would be due to its higher crystallinity.

Effect of Temperature

The degradation rate was affected by temperature. The Arrhenius plots for the degradation of poly(D,L-lactide) and poly(L-lactide) microcapsules gave the activation energies of 19.9 kcal.mol for the former and 20.0 Kcal/mol for the latter, which are comparable to the values for the hydrolysis of alkylacetates[12]. Consequently, the degradation of the polymer can be safely said accessible to hydrolysis.

Effect of Ionic Strength

Figure 5 shows the effect of ionic strength on the rate of decrease in $\bar{M}_w$ of PLA in the degradation of poly(D,L-lactide) microcapsules in the buffer solution of pH 9.6. No significant dependence of the rate on the ionic strength of the bulk solution was detected. However, as indicated in Figure 6, the amount of lactic acid generated from the polymer increased with time, and the generation rate increased as the ionic strength of the medium rose. This would be brought about by a decrease in the thickness of electric double layer around the microcapsules which are negatively charged

374

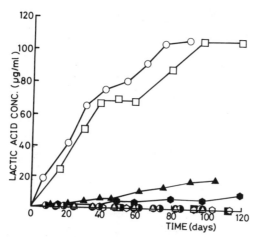

Fig. 4. Variations of the amount of lactic acid generated from poly(D,L-lactide) and poly(L-lactide) microcapsules at different pH. poly(D,L-lactide) microcapsules, pH: ▲, 1.6; △, 3.0; ●, 5.0; ◑, 7.4; ○, 9.6, poly(L-lactide) microcapsules, pH:⬡, 1.6;◯, 3.0; ■, 5.0; ◨, 7.4; □, 9.6.

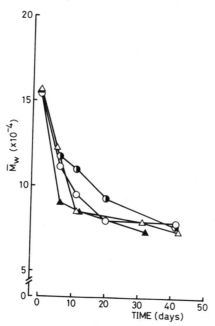

Fig. 5. Effect of ionic strength of medium on the decrease of weight-average molecular weight of PLA in the degradation of poly(D,L-lactide) microcapsules dispersed in a buffer solution (pH 9.6). Ionic strength: ○, 0.07; ◑, 0.15; ▲, 0.30; △, 0.45.

in alkaline solutions. The microcapsules are more accessible to OH⁻ attack in the medium of higher ionic strength.

Effects of Neutral Salts, Urea, and Enzyme

The average molecular weight of PLA was reduced most remarkably when poly(D,L-lactide) or poly(L-lactide) microcapsules were dispersed in the

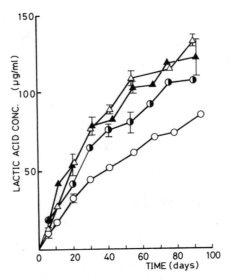

Fig. 6. Variations of the amount of lactic acid generated from poly(D,L-lactide) microcapsules with ionic strength of medium. Ionic strength: ○, 0.07; ◑, 0.15; ▲, 0.30; △, 0.45.

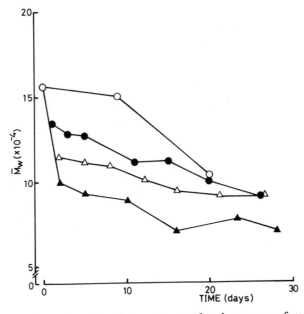

Fig. 7. Effect of carboxylic esterase on the decrease of weight-average molecular weight of PLA in the degradation of poly(D,L-lactide) microcapsules. Enzyme unit (U/ml): ●, 1.45; △, 2.22; ▲, 3.96; ○, control.

medium containing NaSCN. Urea had a similar effect on the degradation of the microcapsules. It is suggested, therefore, that destruction of the water structure on the surface of the microcapsules promotes the hydrolytic degradation of the polymer.

Figure 7 shows the decrease with time in $\bar{M}_w$ of PLA in the degradation of poly(D,L-lactide) microcapsules suspended in the phosphate buffer solution (pH 7.6, 0.1M) containing carboxylic esterase. The rate of $\bar{M}_w$ decrease increased with the increase in the carboxylic esterase concentration. As $\bar{M}_w$ decreased, the corresponding gel permeation chromatogram became less sharp to make the molecular weight distribution of PLA of poly(D,L-lactide) microcapsules broader at the lower molecular weight tail.

Such transformation of gel permeation chromatogram with time was not seen in the simple hydrolytic degradation in the absence of the enzyme. Hence, carboxylic esterase is supposed to cleave the ester bonds in the polymer unzippingly and randomly at any region, probably after being adsorbed on the surface of the microcapsules. Carboxylic esterase has a molecular weight of about 96000, and it is impossible for the enzyme to penetrate into the matrix of PLA. This means that the enzyme can hydrolyze only the ester bonds located on the microcapsule surface.

REFERENCES

1. R. J. Kostelnik, ed., "Polymer Delivery Systems," Gordon and Breach Science Publishers, (1978).
2. D. L. Wise, J. D. Greser, and G. J. McCormick, J.Pharm.Pharmacol., 31:201 (1979).
3. D. L. Wise, H. Rosenkrantz, J. B. Gregory, and H. J. Esber, J.Pharm.Pharmacol., 32:399 (1980).
4. S. Yolles and J. F. Morton, Acta Pharm.Suec., 15:382 (1978).
5. C. G. Pitt, M. M. Gratzl, A. R. Jeffcoat, R. Zweidinger, and A. Schindler, J.Pharm.Sci., 68:1534 (1979).
6. D. A. Wood, Int.J.Pharm., 2:1 (1980).
7. N. Wakiyama, K. Juni, and M. Nakano, Chem.Pharm.Bull., 29:3363 (1981).
8. N. Wakiyama, K. Juni, and M. Nakano, Chem.Pharm.Bull., 30:2621 (1982).
9. K. Uno, Y. Ohara, M. Arakawa, and T. Kondo, J.Microencapsulation, 1:3 (1984).
10. S. B. Barker and W. H. Summerson, J.Biol.Chem., 138:535 (1941).
11. Y. Kobayashi, Seikagaku, 36:355 (1964).
12. R. W. A. Jones and J. D. R. Thomas, J.Chem.Soc., (London) Part 1B:661 (1966).

ELECTRONMICROSCOPIC STUDY OF ULTRATHIN SOLUTE BARRIER
LAYER OF COMPOSITE MEMBRANES AND THEIR SOLUTE TRANSPORT
PHENOMENA BY THE ADDITION OF ALKALI METAL SALTS

T. Uemura and T. Inoue

Membranes Research Laboratory, Pioneering R & D
Laboratories Toray Industries, Inc., 3-Chome
Sonoyama, Otsu, Shiga 520 Japan

INTRODUCTION

The morphological structure of the membrane has a large influence
on membrane performances[1]. The membrane structures have been designed
to suit each purpose for which membranes are used. Many kinds of
membrane processes have been proposed theoretically, and some of them are
now becoming commercially feasible processes, such as reverse osmosis,
ultrafiltration, dialysis, ion exchange, gas separation and so on[2].
Among them the reverse osmosis membrane has been investigated intensively
in the United States and Japan, since the discovery of the asymmetric
membrane by Loeb and Sourirajan. Consequently a new class of membrane
materials and it's fabrication methods has emerged, so-called 'thin film
composite' membranes[3]. They are made by depositing a very thin layer
of solute-rejecting barrier onto the surface of a suitable finely porous
substrate and each of those components can be designed separately to have
excellent performance. Some of the thin film composite membranes have
reached a commercial stage, such membranes being tradenamed as 'PA-300'[4],
'FT-30'[5] and 'PEC-1000'[6,7]. They exhibit high performance, which
made it possible to produce a potable water by single stage seawater
desalination, which cannot be accomplished by the asymmetric flat sheet
membranes. At the same time a few of them are known to have a potentiality
for the separation of low molecular weight water-soluble organic compounds
such as ethanol, which is expected as a new energy resource from biomass,
and acetic acid and so on[6,7]. Although the thin film composite membranes have been designed, their fine structural feature has remained
unclear.

The aims of this paper are to describe the micro structure of the
ultrathin solute barrier layers of some representative composite reverse
osmosis membranes by transmission electronmicroscopy and consider the
effect of micro structure on the membrane performance.

Other objectives are to evaluate and to compare the membrane performance for the ethanol-water and acetic acid-water systems. During this
study, an interesting transport phenomenon was found when alkali metal
salts were added to the solution. Ethanol or acetic acid were concentrated
in the permeate in some cases.

Membrane Fabrications

Seven different kinds of composite reverse osmosis membranes have been prepared. The four kinds of membranes 'NS-100'[8], 'PA-100'[9], 'PA-300'[10] and 'RC-100'[10] are made by the same fabrication method. Their ultrathin solute barrier layers are deposited by the crosslinking of water soluble polymers with two functional reagents on the micropous substrate by interfacial method.

'NS-200'[11] and 'PEC-1000'[12] are also made by the same fabrication method. The water soluble monomers are polymerized and crosslinked on the microporous substrate with an acid catalyst to form the ultrathin solute barrier layer. 'FT-30'[13] is made by a different technique. The ultrathin solute barrier layer is deposited by reacting interfacially m-phenylenediamine with trimesoylchloride on the microporous substrate.

Membrane Structure

'NS-100' was selected as the representative sample of the type that was made by crosslinking of water soluble polymers. The transmission electronmicrograph and its optical diffraction patterns are shown in Figure 1. The microporous polysulfone substrate, intermediate transport layer, which is made from heat crosslinked polyethyleneimine, and ultrathin solute barrier layer, which is made from polyethyleneimine and tolylenediisocyanate, are clearly visible. The ultrathin solute barrier layer can be estimated about 100 Å in thickness and this thickness is different from the thickness of about 250 to 300 Å reported in the literature[4,14]. The latter was estimated from a not very detailed photograph taken by the scanning electronmicroscopy.

'PEC-1000' is selected as the representative composite membrane made from the water soluble monomers. From the transmission electronmicrophotograph (Figure 2), it is also observed clearly that the ultrathin solute barrier layer is uniformly formed on the microporous polysulfone substrate about 300 Å in thickness and some part of the ultrathin layer penetrates into the substrate forming a support layer, which is almost similar to the conclusion by Kurihara et al.[6,15].

'FT-30' is also investigated by the transmission electronmicroscopy (Figure 3). The ultrathin solute barrier layer can be seen clearly like

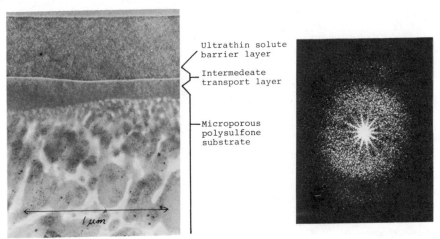

Ultrathin solute barrier layer

Intermedeate transport layer

Microporous polysulfone substrate

Fig. 1. Transmission electronmicrograph of the cross-section of NS-100 and its optical diffraction pattern.

Fig. 2. Transmission electronmicrograph of the cross-section of PEC-1000.

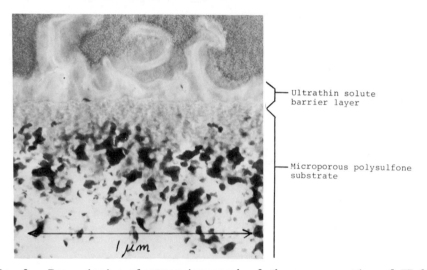

Fig. 3. Transmission electronmicrograph of the cross-section of FT-30.

the corona of the sun and the thickness is not uniform and estimated about a few hundred Å to a few thousand Å. This fact differs from that of published data by Cadotte et al.[3].

Membrane Performance for the Separation of Aqueous Ethanol

The membrane performances of seven kinds of the composite membranes and asymmetric cellulose acetate membrane (SC-5000 type made by Toray Ind. Inc.: herein after CA) were determined for aqueous sodium chloride solution and aqueous ethanol solution in reverse osmosis (Table 1). The aqueous sodium chloride solution was used for checking the membrane performances whether the membranes were fabricated successfully and to see if there is any relationship between salt rejections and ethanol rejections. In the case when the feed solution contained 3.5% sodium chloride and 0.1% ethanol, the rejection performance for sodium chloride was not affected by the addition of ethanol. Only 'PEC-1000' showed high ethanol rejection, 82.7%, while the other composite membranes showed 40% to 60% rejection and

Table 1. Membrane Performance for Ethanol-water and Sodium Chloride-water Systems

Run	Membrane Feed Solution	NS-100	NS-200	PA-100	PA-300	RC-100	FT-30	PEC-1000	CA
1	3.5% NaCl	99.36-297	99.54-261	96.4-443	97.4-262	99.02-363	99.00-453	99.967-372	98.81-321
2	3.5% NaCl	99.39 -283	99.55 -240	96.6 -418	97.6 -243	99.05 -345	99.00 -421	99.967 -342	98.78 -313
	0.1% EtOH	56.3	59.6	41.8	53.8	54.8	46.2	82.7	- 1
3	0.1% EtOH	80.6 -513	78.2 -437	65.3-916	78.6-451	76.7 -704	72.8 -870	91.8 -695	26.2 -703
4	1 % EtOH	81.9 -419	80.4 -342	64.7-799	77.2-392	76.0 -605	72.5 -710	92.7 -532	24.0 -644
5	5 % ETOH	77.5 -218	76.3 -167	58.2-486	65.1-245	69.1 -355	65.1 -368	88.4 -230	20.3 -512
6	10 % EtOH	66.2 -109	66.7 - 77	49.0-300	48.5-171	57.8 -215	52.9 -205	79.4 - 74	16.5 -443
7	3.5% NaCl	99.23-237	99.55-178	96.2-372	97.0-187	98.91-335	98.95-312	99.947-314	98.33-295

Operating Conditions: 56 kg/cm^2, 25°C
Rej (%)-Flux (kg/m^2·d) in each column.
Figures above: NaCl-Rej(%), figures below: EtOH Rej(%) in Run 2.

CA shows no rejection toward ethanol. After this, the feed solution is changed for the aqueous solution containing 0.1%, 1%, 5%, 10% ethanol. As the concentration of ethanol was raised, the ethanol rejection and water flux decreased for every kind of membrane due to the declining of the effective pressure difference across the membrane. Among them 'PEC-1000' showed excellent ethanol rejection performance as is described by Ohya et al.[16] and Kurihara et al.[7,17]. At the end of the test, the membrane performance for 3.5% sodium chloride solution was examined and it was confirmed that all the membranes were not degradated throughout the test. These results showed the curious phenomenon that the ethanol rejection is lowered when sodium chloride is added. In the case of CA, ethanol was concentrated in the permeate. So the effect of adding alkali metal salts to the ethanol solution was studied.

The Membrane Performance for the Ethanol/Alkali Metal Salts/Water Systems

At first the ethanol concentration was fixed at 0.1%, and the sodium chloride concentration was changed from 0.25% to 6.9% (Table 2). As the concentration of sodium chloride increased, both water flux and ethanol rejection for every kind of membranes were reduced. Negative ethanol rejections were observed for 'NS-100', 'FT-30' and CA.

In the next test, the sodium chloride concentration was fixed at 5.0% and the ethanol concentration was changed from 0.1% to 4.3% (Table 3). As the concentration of ethanol increased, the water flux and ethanol rejection were again reduced.

It can be presumed that the alkali metal salts behave as the carrier for the permeation of ethanol, so the addition of other alkali metal salts such as, sodium bicarbonate and potassium chloride were tested. The same phenomena were observed in both cases (Table 4 and 5). Ethanol rejections were reduced in almost the same manner by the addition of the same molar concentration of salts. Therefore it can be assumed that the alkali metal salts do not act as carriers but as reducers of water activity.

Table 2. Membrane Performance for Ethanol–Sodium Chloride–Water Systems-(1)

Run	Feed Solution	NS-100	PA-100	RC-100	FT-30	PEC-1000	CA
1	0.25% NaCl	99.81-642	98.2 -855	99.65-800	99.67-737	99.95-669	98.94-686
2	0.25% NaCl 0.1 % EtOH	99.78 -628 72.6	98.5 -863 60.8	99.65 -790 70.3	99.64 -710 68.4	99.94 -661 90.6	99.10 -688 20.8
3	3.5 % NaCl 0.1 % EtOH	99.32 -291 54.0	95.8 -420 38.1	98.7 -361 49.5	99.00 -348 45.5	99.93 -320 82.2	98.4 -324 -4.0
4	5.0 % NaCl 0.1 % EtOH	98.9 -160 34.0	98.5 -243 18.6	97.9 -204 29.4	98.1 -194 21.6	99.86 -170 69.6	97.3 -182 -24.0
5	6.9 % NaCl 0.1% EtOH	92.1 - 33 -6.6	82.5 - 80 3.0	89.9 - 50 7.1	91.2 - 47 -10.1	99.00 - 13 46.0	90.5 - 50 -31.3

Operating Conditions: 56 kg/cm^2, 25°C
Rej (%)–Flux (kg/m^2·d) in each column.
Figures above: NaCl–Rej (%), figures below: EtOH Rej (%).

Table 3. Membrane Performance for Ethanol–Sodium Chloride–Water Systems-(2)

Run	Feed Solution	NS-100	PA-100	RC-100	FT-30	PEC-1000	CA
1	0.1 % EtOH	98.9 -160 34.0	98.5 -243 18.6	97.9 -204 29.4	98.1 -194 21.6	99.86 -170 69.6	97.3 -182 -24.0
2	1.0 % EtOH	98.8 -122 43.5	93.1 -187 31.4	97.7 -159 40.1	98.1 -139 33.8	99.80 -104 73.9	97.4 -162 -12.1
3	4.3 % EtOH	95.9 - 36 17.2	86.2 - 87 13.6	93.4 - 58 18.4	95.0 - 46 8.1	98.6 - 12 60.9	96.2 - 88 -21.8

Operating Conditions: 5% sodium chloride, 56 kg/cm^2, 25°C
Rej (%)–Flux (kg/m^2·d) in each column
Figures above: NaCl–Rej (%), figures below: EtOH–Rej (%).

Membrane Performance for the Separation of Aqueous Acetic Acid Solution

The separation characteristics for aqueous acetic acid of 7 kinds of composite membranes and CA were measured by the same procedure as in the case of aqueous ethanol (Table 6). The acetic acid rejections were not good enough for the industrial application except for 'PEC-1000', which had been reported by Toray Ind. Inc.[17], for the recovery of acetic acid. The same permeation phenomenon by the addition of sodium chloride was observed in this case as was found in the case of ethanol separation. The concentration of all the ions in permeate solution was shown in Table 7.

Table 4. Membrane Performance for Ethanol-NaHCO$_3$-Water Systems

Run	Feed Solution	NS-100	PA-100	RC-100	FT-30	PEC-1000	CA
1	0.35% NaHCO$_3$	99.78-541	97.5 -838	98.9 -688	99.61-720	99.76-736	99.53-657
2	0.35% NaHCO$_3$ 1.0 % EtOH	99.74 -464 81.2	97.8 -749 57.7	99.19 -587 65.3	99.62 -640 61.5	99.78 -612 91.5	99.63 -619 21.6
3	4.8 % NaHCO$_3$ 1.0 % EtOH	99.38 -235 58.7	97.0 -386 39.6	98.6 -295 52.1	99.66 -338 42.1	99.89 -340 84.8	99.38 -340 -0.5

Operating Conditions: 56 kg/cm^2, 25°C
Rej (%)-Flux (kg/m$^2 \cdot$d) in each column
Figures above: NaHCO$_3$-Rej (%), figures below: EtOH-Rej (%).

Table 5. Membrane Performance for Ethanol-KCl-Water Systems

Run	Feed Solution	NS-100	PA-100	RC-100	FT-30	PEC-1000	CA
1	4.6% KCl	99.00-271	94.3 -394	98.0 -342	97.7 -354	99.73-447	97.3 -311
2	4.6% KCl 1.0% EtOH	98.8 -214 48.0	93.5 -325 29.6	97.7 -282 44.4	97.4 281 35.2	99.56 -324 75.5	97.3 -281 -8.2

Operating Conditions: 56 kg/cm^2, 25°C
Rej (%)-Flux (kg/m$^2 \cdot$d) in each column
Figures above: KCl-Rej (%), figures below: EtOH-Rej (%) in Run 2.

Table 6. Membrane Performance for Acetic Acid-Water, Acetic Acid-NaCl-Water Systems

Run	Feed Solution	NS-100	NS-200	PA-100	PA-300	RC-100	FT-30	PEC-1000	CA
1	3.5% NaCl	99.54-174	99.54-239	99.05-433	97.8-408	99.00-159	98.7 -554	99.96-308	98.7-384
2	3.5% NaCl 0.1% AcOH	98.2 -294 57.6	98.7 -270 45.2	96.8 -575 39.4	96.2 -464 32.1	97.7 -230 56.1	97.5 -435 59.7	99.63 -273 84.3	96.8 -397 -24.0
3	0.1% AcOH	79.2 -374	78.9 -570	49.7-1164	48.2-959	70.5 -427	70.8-1109	94.5 -656	20.4-819
4	1 % AcOH	83.3 -301	77.8 -465	54.1-1058	51.2-843	75.5 -334	75.2 -871	93.5 -570	19.1-726
5	5 % AcOH	79.5 -180	71.7 -284	49.5 -725	38.9-678	68.0 -208	70.7 -468	90.0 -315	13.5-618
6	10% AcOH	68.5 - 95	64.0 -152	46.2 -462	32.4-548	58.8 -126	64.1 -214	84.0 -141	6.24-565
7	3.5% NaCl	99.20-255	99.54-205	98.1 -439	95.6-384	99.06 -171	99.10-495	99.86-315	96.6-398

Operating Conditions: 56 kg/cm^2, 25°C
Rej (%)-Flux (kg/m$^2 \cdot$d) in each column
Figures above: NaCl Rej, figures below: AcOH Rej (TOC).

Table 7. Permeate Ionic Analysis for Acetic Acid–NaCl–Water System
(x 10^{-3} mol/ℓ)

Membrane Ion	Feed Solution	NS-100	NS-200	PA-100	PA-300	RC-100	FT-30	PEC-1000	CA
CH_3COO^-	16.65	8.54	11.04	12.07	13.5	8.74	8.13	3.13	25.0
H^+	16.65	9.06	1.42	14.2	14.4	9.38	10.0	0.047	19.15
Cl^-	598	8.18	3.27	12.4	18.9	9.30	9.59	0.705	22.57
Na^+	598	5.87	12.8	10.0	16.96	7.61	5.57	4.91	20.0

Operating Conditions: 56 kg/cm², 25°C
CH_3COO^-, H^+, Cl^-, Na^+ were measured by TOC, titration, ion
electrode and atomic absorption, respectively.

In case of 'PEC-1000' and 'NS-200', sodium acetate permeates selectively.
While in the case of the other membranes acetic acid and sodium chloride
have similar permeability. This phenomenon can be explained due to the
strong anionic charge of the two membranes, 'PEC-1000' and 'NS-200', and
the other membranes are not so strongly charged. The acetic acid was
concentrated in the permeate in the case of CA.

CONCLUSION

1. The three different kinds of composite reverse osmosis membranes,
    'NS-100', 'FT-30' and 'PEC-1000' were chosen as the representative
    membranes, and the micro structures of their ultrathin solute barrier
    layers were clarified by the transmission electronmicroscopy.
2. The characteristic structures in their ultrathin solute barrier layers
    are related to their fabrication methods. The effective membrane
    thickness of 'NS-100', 'FT-30' and 'PEC-1000' were found to be about
    100 Å, 2500 Å and 300 Å, respectively.
3. The membrane separation characteristics for the ethanol–water, acetic
    acid–water were evaluated for 8 kinds of reverse osmosis membranes.
    Among them 'PEC-1000' was found to have the highest performance for
    both ethanol and acetic acid separation.
4. The addition of alkali metal salts to the solution was found to cause
    an interesting transport phenomenon. In some cases ethanol or acetic
    acid were concentrated in the permeates. The alkali metal salts were
    presumed to play a role as the reducer of water activity.
5. In the case of 'PEC-1000' and 'NS-200', sodium ion and acetate anion
    permeate more selectively than hydronium ion and chloride anion when
    the aqueous solution contains acetic acid and sodium chloride, and
    this phenomenon can be explained to be due to the anionic charge of
    these membranes.

Acknowledgement

This work has been carried out under the management of the Research
Association for Basic Polymer Technology for synthetic membranes for new
separation technology as a part of the project on Basic Technology for
Future Industries funded by the Agency of Industrial Science and
Technology, Ministry of International Trade and Industry.

The authors are indebted to Dr Y. Murata and Mr K. Yoshimura, Toray
Research Center Inc., for the transmission electronmicrographs.

REFERENCES

1. T. Matuura, "Goseimakunokiso" section 3 Kitamishobo Japan (1981).
2. T. Nakagawa et al., "Kinouseimaku no saisentangijutsu," R & D Report No.18, CMC Japan (1981).
3. J. E. Cadotte and R. J. Petersen, Synthetic membranes I, ACS Symposium Ser., 153:305 (1980).
4. R. L. Riley, P. A. Case, A. L. Lloyd, C. E. Milstead, and M. Tagami, Desalination, 36:207 (1981).
5. J. E. Cadotte, R. J. Petersen, R. E. Larson, and E. E. Erickson, Desalination, 32:25 (1980).
6. M. Kurihara, N. Kanamaru, N. Harumiya, K. Yoshimura, and S. Hagiwara, Desalination, 32:13 (1980).
7. M. Kurihara, N. Harumiya, N. Kanamaru, T. Tonomura, and M. Nakasatomi, Desalination, 38:449 (1981).
8. J. E. Cadotte, USP 4,039,440 (1977).
9. W. J. Wrasidlo, USP 3,926,798 (1976).
10. W. J. Wrasidlo, USP 3,951,815 (1977).
11. J. E. Cadotte, USP 3,926,798 (1975).
12. M. Kurihara, T. Watanabe, and T. Inoue, USP 4,366,062 (1982).
13. J. E. Cadotte, USP 4,277,344 (1981).
14. W. Pusch and A. Walch, Desalination, 35:5 (1980).
15. J. Y. Chen, M. Kurihara, and W. Pusch, Desalination, 46:379 (1983).
16. H. Ohya, E. Kazama, Y. Negishi, Y. Urayama, and M. Kitazato, Kagakukogakuronbunshu, 7:372 (1981).
17. M. Kurihara, Chemical Engineering, 28(13):34 (1983).

ACTIVE TRANSPORT IN ARTIFICIAL MEMBRANE SYSTEMS

Y. Kobatake, N. Kamo and T. Shinbo

Faculty of Pharmaceutical Sciences
Hokkaido University, Sapporo and
National Chemical Laboratory for Industry
Tsukuba, Japan

SYNOPSIS

Active transports in biological membranes are characterized by the
following conditions: (1) Transport of a chemical species against its
electrochemical potential gradient (uphill transport) driven by vectorial
metabolic reactions. (2) Tight coupling between reaction and transport
with the aid of mediator in the membrane. (3) Asymmetry of the membrane
system. When these conditions are satisfied, active transport may be
realized even in an artificial model system. Any chemical reactions such
as redox, hydrolysis or neutralization reaction can be used for energy
supply for uphill transport across the membrane. Moreover, if the above
conditions are fulfilled, the reactions against their free energy difference are driven reversibly by downhill transport across the membrane as is
observed in a living membrane.

The present article shows the active transport of ions occurs across
liposomal and/or liquid membranes by redox or photoredox reactions combined
with appropriate mediators. Inversely, the rate of the chemical reaction
was controlled by downhill transport of ions across the membrane showing an
intimate coupling between the reaction and the transport. The relationship
between the reaction and the transport was estimated theoretically at the
steady state, and the equations obtained agreed well with experimental
results.

INTRODUCTION

One of the important functions of a living membrane is the active
transport, in which specific chemical species necessary to living cells are
transported against gradients not only of concentration but of electrochemical potential[1-4]. The characteristics of the active transport are
its high selectivity of the chemical species to be transported and high
efficiency of the energy transduction for performing the uphill transport.
The uphill transport must naturally be supported by chemical energy caused
by a series of chemical reactions, called as the "vectorial metabolism",
which must be coupled tightly with transport of the chemical species
together with the aid of mediators or carriers in the membrane. One more
important thing to be considered in the active transport in a living
membrane is the asymmetry of the membrane system. The active transport can

be realized even in an artificial model system when the characteristics described above are satisfied[5-8]. The chemical reaction for supplying the energy is not specified in the metabolic reactions in the living system, but any chemical reactions such as neutralization, redox reaction as well as hydrolysis may be used. If the coupling between reaction and transport is close enough, the reaction across the membrane are driven reversibly by the downhill transport of some chemicals through the membrane as is observed in living systems.[9-10].

In the present article, we discuss the active transport of ions, mainly liposoluble ions occurring across liposomal and/or liquid membranes by redox and photoredox reactions in conjunction with appropriate mediators. The coupling between reaction and transport of ions in a liquid membrane is slightly different from that in a liposomal membrane. Therefore, we will treat these two membrane systems separately. When the coupling between reaction and transport is close enough, the rate of the chemical reaction is controlled by downhill transport of ions through the membrane. Relationships between the reaction and transport are derived theoretically in several typical cases, which are shown to agree with experiments.

REDOX REACTION AND TRANSPORT IN LIQUID MEMBRANE

The system to be considered here consists of tow aqueous solutions containing redox agents which are separated by a liquid membrane (dichloroethane) including an appropriate mediator. The oxidizing agent used is ferricyanide, while the reducing agent is either ferrocyanide or ascorbate. EDTA is also used if necessary. The mediator is ferrocene (Fc), its derivatives (i.e. dibutyl or amylferrocene), phenazine methosulfate (PMS) or tetramethyl-p-phenylenediamine (TMPD). For convenience in subsequent discussion, the bulk phases containing reducing and oxidizing agents are named hereafter phases I and II, and the liquid membrane as phase III, respectively. The ions to be transported are lipophilic anions (e.g. picrate, ClO_4^-, SCN^-, etc.) and lipophilic cations (e.g. K^+ with crown compound).

Mediator in the membrane, e.g. ferrocene can be oxidized by ferricyanide to form a positively charged ion (Fc^+) which can be reduced to a neutral molecule (Fc) by ascorbate (pH 8.0) or ferrocyanide as follows:

$$Fc + Fe(CN)_6^{3-} \longrightarrow Fc^+ + Fe(CN)_6^{4-}$$

$$Fc^+ + (1/2) \text{ ascorbate} \longrightarrow Fc + (1/2) \text{ dehydroascorbate} + H^+.$$

Anion Transport Driven by Transmembrane Redox Reaction[11]

Figure 1 shows a typical example, where the concentrations of picrate in two aqueous phases are plotted as a function of time. The initial condition of the system was: aqueous phase I contained 10 mM Na-ascorbate with 10^{-4} M K-picrate, while phase II consisted of 10 mM K-ferricyanide with 10^{-4} M K-picrate. The pH of both solutions was adjusted to 8.5 by 50 mM Na-borate buffer. The concentrations of picrate in phases I and II increased and decreased, and them approached respective plateau levels. This result shows that picrate was transported from oxidizing side (phase II) to the reducing side (phase I) against its concentration difference. The overall reaction is as follows:

$$Fe(CN)_6^{3-}(II) + (1/2) \text{ ascorbate}(I) + X^-(II)$$
$$Fe(CN)_6^{4-}(II) + (1/2) \text{ dehydroascorbate}(I) + X^-(I) + H^+(I),$$

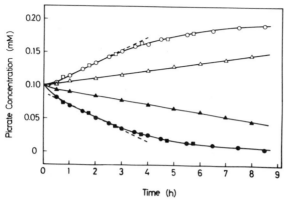

Fig. 1.   Picrate concentrations as a function of time in liquid membrane.
Redox pair: ascorbate/ferricyanide. o in phase I; ● in phase II
(mediator 1 mM but1Fc); △ in phase I; ▲ in phase II (mediator
1 mM Fc). Data obtained with voltage clamp at 0 mV are also shown
□ in phase I; ■ in phase II (mediator 1 mM but1Fc).

where X^- stands for picrate. To examine the role of the transmembrane
redox reaction in picrate transport, we measured both the amount of ferrocyanide produced and the amount of picrate removed in phase II, simultaneously. As shown in Figure 2, these quantities were approximately equal
throughout the time examined. The membrane potential of the present system
was about 140 mV at the steady state with positive polarity in phase I with
respect to phase II. The potential acted in the same direction as the
picrate flux, and hence it might contribute to a part of the anion flux.
However, the rate of picrate transport was not affected at all by clamping
the membrane potential at 0 mV as was shown in Figure 1. In other words,
the membrane potential was independent of the transport of picrate.
Application of uncoupler CCCP also had no effect on the transport of
picrate. These results indicate that picrate was transported against its
electrochemical potential difference, and hence the transport in the
present system is an active transport by definition.

Possible reactions and transports in the system are schematically
illustrated in Figure 3, where Q^+ and R stand for the oxidizing and reducing agents, respectively. The reactions at the membrane surfaces and the
transport of ferrocene are as follows:

$$FcX(I) + R \longrightarrow Fc(I) + R^+ + X^-(I)$$

$$Fc(II) + Q^+ + X^-(II) \longrightarrow FcX(II) + Q$$

$$J_{Fc} = P[Fc(I) - Fc(II)]$$

$$J_{FcX} = P [FcX(II) - FcX(I)]$$

Here P is the permeability coefficient of Fc or FcX, which is assumed to be
equal. The time derivatives of Fc(I), Fc(II), FcX(I) and FcX(II) must be
zero at the steady state. The above equations together with the steady
state conditions of Fc and FcX lead to:

$$\frac{1}{J_X} = \left\{ \frac{1}{P} + \frac{1}{K_I[R]} + \frac{1}{k_2[Q^+]C_x} \right\} \frac{1}{S_t}, \tag{1}$$

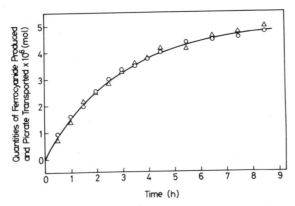

Fig. 2. Simultaneous measurements of the rates of picrate transport and redox reaction. Redox pair: ascorbate/ferricyanide. ○ decrement of picrate in phase II; △ the quantity of ferrocyanide produced in phase II. The electron mediator was 1 mM butlFc.

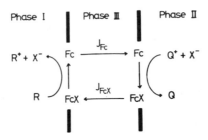

Fig. 3. Schematic representation of the mechanism of the coupling between redox reaction and ion transport.

where S_t stands for the total concentration of the mediator in the membrane, i.e. S_t = Fc(I) + FcX(I) = Fc(II) + FcX(II). Detailed derivation of Equation (1) will be shown later in connection with Equation (2). Equation (1) implies that the reciprocal of the transport rate of X^-, $1/J_X$, is proportional to the inverse concentration of reducing agent [R] or oxidizing agent [Q^+], under a given concentration of mediator S_t. This relation is shown in Figure 4. Also, $1/J_X$ is proportional to $1/C_X$ when [R], [Q^+] and S_t are kept constant. The similar relation holds for other liposoluble anions, e.g. ClO_4^-.

Transmembrane Redox Reaction driven by Diffusion of Ions[12]

The next problem to be considered is the reverse energy transduction, i.e. the transmembrane redox reaction driven by diffusion of ions. The two aqueous phases I and II contained initially 0.1 mM ferricyanide and 0.1 mM ferrocyanide, and were buffered with 25 mM Na-borate (pH 8.5). $NaClO_4$ (1.0 M) and Na-sulfate (0.5 M) were added to phases II and I, respectively, in order to build up a concentration gradient of anion within the membrane. Figure 5 shows the typical data, where the concentrations of ferricyanide and ferrocyanide in both aqueous phases are plotted against time. The ferricyanide concentration in phase II decreased while that in phase I increased with time, and the sum of concentrations of ferri- and ferrocyanides was not changed. When the mediator, dibutylferrocene (butlFc) was

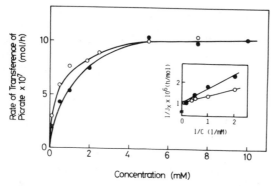

Fig. 4. Transport rate of picrate (J_X) under various conditions. ○ J_X vs. [ascorbate] with fixed [ferricyanide] (10 mM); ● J_X vs. [ferricyanide] with fixed [ascorbate] (10 mM). Inset: ○ $1/J_X$ vs. $1/$[ascorbate]; ● $1/J_X$ vs. $1/$[ferricyanide]. The electron mediator was 1 mM but1Fc.

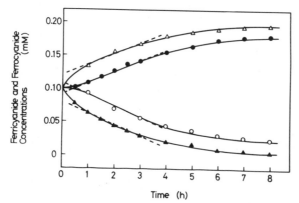

Fig. 5. Plots of the concentrations of ferricyanide and ferrocyanide against time in the presence of perchlorate with a mediator dibutylferocene or ferocene. ● [ferricyanide] in phase II; ○ [ferrocyanide] in phase I with 2 mM Fc as mediator. ▲ [ferricyanide] in phase II; △ [ferrocyanide] in phase I with 2 mM but1Fc as mediator. Without Fc no redox reaction was proceeded.

not present in the membrane, no concentration changes of ferri- and ferrocyanides were observed. Because of low lipophilicity of ferri- and ferrocyanides the changes in the concentration of these chemicals are not attributable to the transport of ferri-and ferro-cyanides themselves, but caused by the transmembrane redox reaction via but1Fc mediated by anion transport. A similar observation was made for picrate instead of perchlorate anion. Even in this case, no redox reaction across the liquid membrane occurred without Fc or but1Fc in the membrane. The rate of production of ferricyanide in phase I was measured when perchlorate anion in phase II was replaced by various anions. As seen in Figure 6, the rate of production of ferricyanide decreased in the following order:

$$ClO_4^- > SCN^- > NO_3^- > Br^- > Cl^-.$$

This series is often found in the sequence of selectivity coefficient in a liquid membrane electrode sensitive to anions. It has been shown that the selectivity coefficient is mainly determined by the difference of the standard chemical potentials of anions between aqueous and oil phases, i.e., by the difference in lipophilicity. The results show that lipophilic anions such as ClO_4^- or SCN^- are potent in inducing the redox reaction suggesting that the permeation of anion is necessary. As shown in Figure 7, the quantity of perchlorate transported from phase II to I agreed well with that of ferricyanide produced in phase 1. This relation holds for all other anions examined in Figure 6.

The membrane potentials for combination of sulfate (phase 1) and perchlorate, thiocyanate, nitrate, bromide and chloride anions (phase II) were 190, 160, 100, 65 and 40 mV at the steady state, respectively, and were negative in phase I with respect to phase II. Thus the potentials acted in the same direction as the resultant electron flow caused by the transmembrane redox reaction.

The reaction which brings about the production of ferricyanide in phase I and the disappearance of ferricyanide in phase II is equivalent to an electron flow across the membrane from phase I to II. The rate of redox reaction was not affected by the clamping of the transmembrane potential at 0, 500 or -500 mV (the value is expressed with respect to phase I).

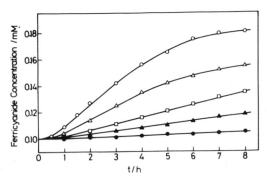

Fig. 6. The dependence of the rate of redox reaction on various anions. o NaClO₄; △ NaSCN; □ NaNO₃; ▲ NaBr; ● NaCl.

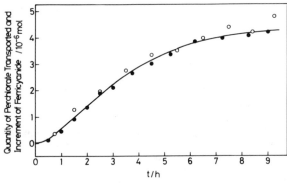

Fig. 7. Simultaneous measurements of the rates of redox reaction and perchlorate transport. o the quantity of ClO_4^- transported to phase I; ● the increment of ferricyanide in phase I. Mediator, but1Fc.

Before addition of a lipophilic anion into phase II, Fc in phase III is not virtually oxidized by ferricyanide in phase II, because oxidized ferrocene has a positive charge and the production of Fc^+ breaks the electroneutrality condition in the membrane phase. Note that both ferri- and ferro-cyanides are not able to be counter-ions of Fc^+ due to their low lipophilicity. When X^- is added to phase II the redox reaction at the interface between phases II and III begins to proceed owing to the presence of extractable negative charge of X^-. In phase III, an ion pair between Fc^+ and X^- diffuses towards phase I. The increase of the concentration of Fc^+ and X^- at the interface of the membrane facing phase I drives the redox reaction, i.e.,

$$Fc^+ + Fe(CN)_6^{4-} \longrightarrow Fc + Fc(CN)_6^{3-}.$$

The inverse reaction occurs at the opposite interface of the membrane facing phase II. X^- is liberated from phase III to phase I due to the production of neutral Fc. The reduced Fc migrates toward phase II and is oxidized there. These processes are repeated. As a result of this cyclic process, X^- is transported from phase II to I, and the transmembrane redox reaction takes place via Fc. The following reaction scheme in the present system may be considered:

$$d[Fc(I)]/dt = K_1 [FcX(I)] C_r - J_{Fc}$$

$$d[Fc(II)]/dt = J_{Fc} - K_2 [Fc(II)] C_r C_x + k_{-2} [FcX(II)] C_r$$

$$d[FcX(I)]/dt = J_{FcX} - K_1 [FcX(1)] C_r$$

$$d[FcX(II)]/dt = k_2 (Fc(II)] C_r C_x - k_{-2}[FcX(I)] C_r - J_{FcX}$$

In the above equations J stands for the flux across the membrane, C_x is the concentration of perchlorate in phase II, C_r stand for the concentrations of redox agents in phases I and II, and they are assumed to be equal each other, i.e. $C_r = [R] = [Q^+]$. At the steady state, all of the left hand sides of the above equations should vanish, leading to the following equation for the rate of the redox reaction V_r:

$$\frac{1}{V_r} = \left\{ \frac{P(k_1 + k_{-2}) + k_1 k_{-2} C_r}{P k_1 k_2 C_r C_x} + \frac{P + k_1 C_r}{P k_1 C_r} \right\} \frac{1}{S_t} \tag{2}$$

Here, S_t and P have the same meanings as before. The right hand side of this equation agrees with that in eq.(1) when either the oxidizing and reducing agents in phases I and II are diminished, or K_{-2} becomes zero. Equation (2) implies that the inverse of the redox reaction rate, $1/V_r$, given by Equation (2) as well as $1/J_x$ given by Equation (1) are proportional to $1/C_x$ when the concentrations of mediator in the membrane S_t and of redox reagents in the bulk solutions C_r are kept constant. This relation is shown in Figure 8, where the rate of redox reaction is plotted as a function of $1/[ClO_4^-]$. As seen in Figure 2 as well as in Equations (1) and (2), the rate of transport J_x agrees with that of redox reaction V_r.

Mediator and Carrier

The mediators used in the present system were neutral in the reduced form, and became cations in the oxidized form which acted as anion carriers

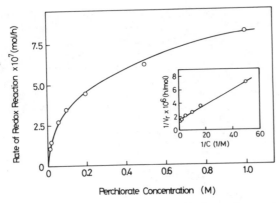

Fig. 8. Rate of redox reaction under condition of varying perchlorate concentration. The mediator was butlFc. The inset shows the plot of the reciprocal of the rate of redox reaction (V_r) vs. the inverse of perchlorate concentration.

themselves. Thus, they were convenient for transporting lipophilic anions such as SCN^-, ClO_4^- or picrate. On the contrary, if we can use an anion type mediator having no charge (neutral) in the oxidized form and negative charge in the reduced form, lipophilic cations are able to be transported across the membrane with coupling to the transmembrane redox reaction. But, even if a cation type mediator is used, the transport of lipophilic cations could be realized in the presence of a lipophilic anion as a co-carrier. Here, we will show that the active transport of K^+ is succeeded when a lipophilic anion, e.g. tetraphenyl borate (TPB^-) is incorporated with a cation type mediator as Fc or phenazine methosulphate (PMS) in the membrane. In this case, we need to use an ion carrier, e.g. crown compound as dicyclohexyl-18-crown-6 (DC-18-C-6) [13].

Figure 9 shows the cases where K^+ is transported actively by coupling to a redox pair of Na-ferricyanide and Na-ascorbate across a liquid membrane. The membrane contained PMS together with DC-18-C-6 and TPB^-. The initial concentrations of K^+ in the two bulk solutions were 1 mM. In the figure, the variations of $[K^+]$ in phases I and II are plotted as a function of time. K^+ was transported toward oxidizing side, i.e. phase II. The transport rate of K^+ was not affected by clamping of the transmembrane potential difference at 0 mV as shown in the same figure. Thus, K^+ was transferred against its electrochemical potential difference across the membrane. The active transport of K^+ in the present system is explained as follows: The oxidized mediator S^+ diffuses with TPB^- from phase II to phase I, and is reduced at the interface of phase I.. The reduced mediator S (neutral) diffuses back with no accompanying ions. While TPB^- at the interface of phase I diffuse back with crown-K^+ toward phase II. K^+ ions reached to phase II interface are released in the bulk solution II. The neutrality condition in the membrane is thus maintained.

In Figure 10, a comparison is made between K^+ transported and $Fe(CN)_6^{4-}$ produced as a function of time. As seen in the figure, the rate of the redox reaction does not agree well with the K^+-transport rate, implying no close coupling in this system as in the case of picrate anion shown in Figure 2. The difference between the redox reaction and the transport of K^+ becomes more appreciable when another carrier as dibenzo-24-crown-8 (DB-24-C-8) is used as shown in Figure 10.

394

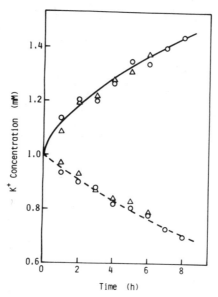

Fig. 9. Active transport of K^+ in a redox system (ferricyanide/ferro-cyanides) PMS + TPB⁻ + DC-18-C-6 dissolved in dichloroethane. With (△) and without (○) voltage clamp at 0 mV.

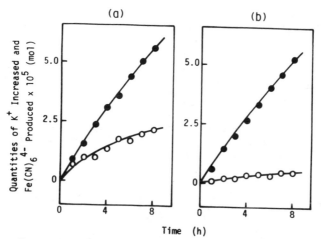

Fig. 10. Comparison between the rates of redox reaction and K^+ transport. The system is the same as in Figure 9 except the species of ion carriers in the liquid membrane. (a) DC-18-C-6; (b) DB-24-C-8; ● $Fe(CN)_6^{4-}$; ○ K^+ transported.

Photoredox Reaction and Transport of Ions in Liquid Membrane[13]

The system to be considered next is a light coupled transport system where perchlorate was transported with coupling to a photoredox reaction. The oxidizing agent Q^+ in phase II was ferricyanide and the reducing agent in phase I was EDTA + proflavine (PF) + methylviologen (MV^{2+}). PF was used as the photosensitizer and MV^{2+} was an electron acceptor. The lipophilic anion to be tested in this experiment was ClO_4^-. The mediator added in the liquid membrane was amylferrocene (amylFc).

On illumination, the transport of perchlorate across the membrane occurred as illustrated in Figure 11. When the rate of the redox reaction occurring under illumination was measured with ion transport simultaneously, both rates agreed with each other as shown in Figure 12. Neither the transport of ClO_4^- nor the redox reaction proceeded without illumination, the presence of photosensitizer PF, reducing agent EDTA or of mediator amylFc in the membrane. In this system, no effect on the transport of perchlorate was observed by voltage clamp at 0 mV. Thus the transport of perchlorate with the redox reaction sensitized by illumination in the present system is regarded as an active transport.

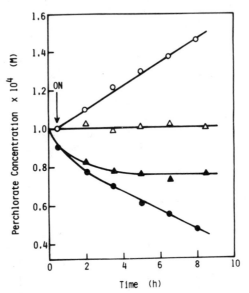

Fig. 11. Transport of ClO_4^- caused by photoredox reaction. $\triangle$ $[ClO_4^-]$ in phase I in the dark; $\circ$ $[ClO_4^-]$ in phase I under illumination; $\bullet$ $[ClO_4^-]$ in phase II under illumination.

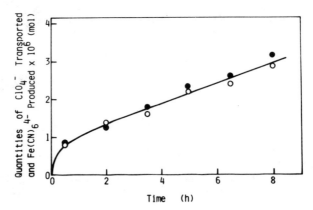

Fig. 12. Simultaneous measurements of the rates of ClO_4^- transport and of photoredox reaction. $\circ$: ClO_4^-; $\bullet$: $Fe(CN)_6^{4-}$ in phase II. No effect was observed by application of uncoupler CCCP or voltage clamp at 0 mV.

PHOTOREDOX REACTIONS IN CONJUNCTION WITH TRANSPORT OF IONS IN LIPOSOME
SYSTEM[14]

Figure 13 shows the potential deflection of an electrode sensitive to
tetraphenylphosphonium cation (TPP$^+$), one of the membrane potential indicators, in response to illumination of the system containing a redox pair
of ascorbate and ferricyanide separated by the liposomal membrane. The
membrane contains phenosafranine (PhS) as a photosensitizer. It is known
that TPP$^+$ distribute passively when the membrane potential is generated by
some means in living cells. The upward deflection of the TPP$^+$ electrode
potential indicates a decrease of the TPP$^+$ concentration in the medium. As
seen in the figure, TPP$^+$ was accumulated rapidly in the liposome with light
illumination and released with light-off. Addition of an uncoupler, e.g.
CCCP (2 μM), also led to immediate release of accumulated TPP$^+$. Since
illumination in the presence of a photosensitizer generates the membrane
potential across the liposomal membrane, and without light or addition of
CCCP eliminates it, this result indicates that TPP$^+$ is transported across
the membrane in response to the membrane potential produced. Thus the
transport of TPP$^+$ in this liposome system is not active. As will be shown
later, this is also true for ordinary redox reaction in liposome system.
The membrane potential and TPP$^+$ uptake (U_r^e) can be calculated by the
deflection of the electrode potential as shown later. The membrane
potential and hence the uptake of the TPP$^+$ as well as the rate of redox
reaction across the membrane strongly depend on the light intensity and
on the wavelength illuminated. In Figure 14A and B, the rates of the
redox reaction and of TPP$^+$-uptake are shown as a function of wavelength
irradiated. For the sake of comparison, the absorption spectrum of
phenosafranine is also shown in the figure. It is seen that the action
spectrum of both redox reaction and the change in the membrane potential
agree well with that of absorption spectrum of PhS. The similar results
were obtained in the case where ascorbate in the liposome was replaced by
EDTA. It is also noted that the magnitude of the membrane potential or
the uptake of TPP$^+$ was increased with increase of the light intensity
irradiated, I. Figure 15 shows the relation between the membrane potential, $\Delta\psi$, and the light intensity irradiated, log 1.

A schematic diagram of possible photoredox reaction which proceeded in
the present system is shown in Figure 16. The redox pair in the present
system is ascorbate/ferricyanide, and the mediator in phase III is pheno-

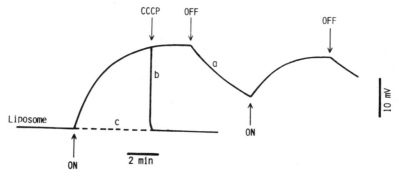

Fig. 13. TPP$^+$ uptake due to the light-induced transmembrane redox
reaction. The experiment was performed in an atmosphere of N_2.
a: TPP$^+$ release on cessation of irradiation; b: TPP$^+$ release
due to addition of 2×10^{-6} mol/dm^3 CCCP; c: TPP$^+$ uptake in the
presence of CCCP or O_2. The concentrations of TPP$^+$, TPB$^-$, PhS,
ascorbate and liposomes were 10^{-4}, 10^{-5}, 5×10^{-5}, 10^{-2} mol/dm^3
and 5.6 mg of lipid/ml.

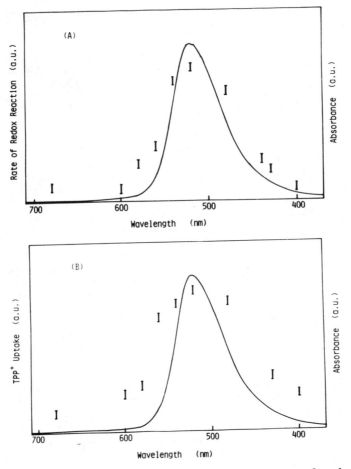

Fig. 14. Comparison of the action spectra of the rates of redox reaction (A) and TPP$^+$ uptake (B) with the absorption spectrum of phenosafranine (PhS).

safranine (PhS). The oxidized phenosafranine PhS$^+$ is transported from phase II to phase I, and activated by light irradiated at the interface between phases I and III. The activated PhS^{+*} is deactivated by colliding with reducing agent R and transformed either oxidized (PhS$^+$) or reduced form (PhS). The reduced mediator PhS is transported across the membrane and oxidized at the interface between phases III and II, and this cyclic reaction is repeated. If we consider only the initial rate of the redox reaction, the time derivatives of PhS and PhS$^+$ in phases I and II are represented as follows:

$$d[PhS^{+*}(I)]/dt = aI_{ab} - k_1[PhS^{+*}(I)] - (k_2 + k_3)[PhS^{+*}(I)][R]$$

$$d[PhS(I)]/dt = k_3[PhS^{+*}(I)][R] - \alpha_I J_{PhS}$$

$$d[PhS(II)]/dt = \alpha_{II} J_{PhS} - k_4[PhS(II)][Q^+]$$

$$d[PhS^+(I)]/dt = - aI_{ab} k_1[PhS^{+*}(I)] + k_2[PhS^{+*}(I)][R] + \alpha_I J_{PhS^+}$$

$$d[PhS^+(II)/dt = k_4[PhS(II)][Q^+] - \alpha_{II} J_{PhS^+}$$

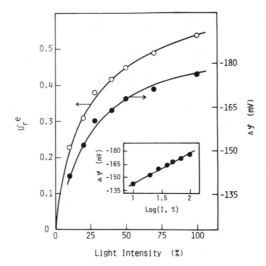

Fig. 15.   TPP$^+$ uptake and the membrane potential as a function of light
intensity irradiated.   O : TPP$^+$ uptake U_r^e ; ● : membrane
potential.   The inset shows the plot of $\Delta\phi$ vs. log I.   The
experimental conditions were the same as those in Figure 13.

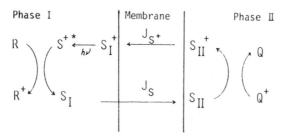

Fig. 16.   Photoredox reaction and transport of anion in liquid membrane.
R: ascorbate, Q$^+$: ferricyanide, S$^+$: phenosafranine (PhS).

Here, α is a proportionality constant relating to the area and volume, and
subscripts I and II stand for phases I (outside) and II (inside) of the
liposomes, respectively, and J's are fluxes of oxidized and reduced phenosafranine.   here, I_{ab} represents the quantity of light absorbed per unit
time.

By using above equations, the flux of reduced mediator J_{PhS} and the
initial rate of the photoredox reaction V_o are represented as follows at
the steady state:

$$V_o = \frac{\gamma\, a\, k_3\, I_{ab}\, [R_o]}{K_1 + (k_2 + k_3)[R_o]} \tag{3}$$

where γ is a constant relating to the geometrical size of the system, $[R_o]$
represents the initial concentration of the reducing agent in phase I,
i.e. outside of liposomes.   The above equation implies that the initial
rate of the photoredox reaction V_o is proportional to the absorbed light

quantity, I_{ab}, and is independent of the concentration of oxidizer in phase II, i.e. internal solution of liposomes. These are confirmed in Figures 17 and 18. As expected, these relationships hold for the present liposome system. The above equations also predict that inverse of reduction rate of ferricyanide in phase II is proportional to inverse of ascorbate concentration in phase I at a given light intensity.

This relation is in line with the experimental data as shown in Figure 19.

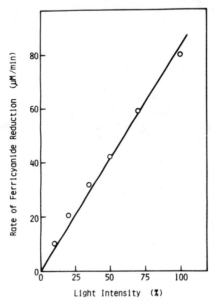

Fig. 17. The rate of photoredox reaction as a function of light intensity irradiated to liposomes. Redox pair: ferricyanide(inside)/ ascorbate (outside) with phenosafranine (PhS) as a photosensitizer. pH was adjusted at 8.0 with tris-Cl both inside and outside of the liposomes.

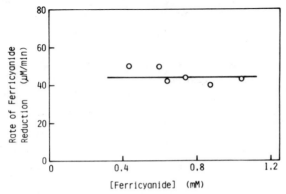

Fig. 18. The dependency of the photoredox reaction rate on the concentration of oxidizing agent in phase II (inside of liposomes).

Figure 20 shows the time course of the potential deflection of a picrate sensitive electrode when a PC-liposome containing 0.25 M K^+- ascorbate is subjected to addition of various chemicals. The external solution consists of 0.4M Na_2SO_4 and 2×10^{-4} M Na-picrate initially. The pH of both inside and outside of the liposome are adjusted to 8.0 by tris-SO_4. The oxidizing agent (20 mM ferricyanide), a mediator (0.1 mM amylFc) and an uncoupler (4 µM CCCP) are added in the medium successively at the points indicated by arrows. The picrate selective electrode is made of PVC-based membrane incorporated with trioctylmethyl-ammonium (TOMA$^+$) as an ion exchanger. The working concentration range of the electrode was between 1.0 M and 2×10^{-6} M with Nernstian response with high selectivity. The upward deflection of the electrode potential corresponds to a decrease of picrate anion in the medium.

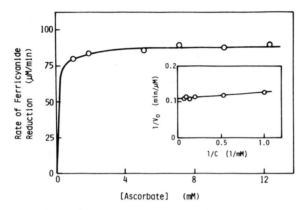

Fig. 19.   The rate of photoredox reaction as a function of reducing agent concentration in phase I, and the inverse relation between redox reaction rate and ascorbate concentration. The experimental conditions were the same as in Figure 17.

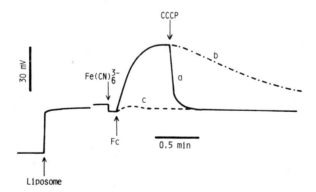

Fig. 20.   Uptake of picrate anion in liposomes due to redox reaction. Redox pair: ascorbate/ferricyanide with amylferocene 0.25 M ascorbate (inside), 20 mM $Fe(CN)_6^{3-}$ (outside) pH of both solutions were adjusted at 8.0 with tris-SO_4. a: release of picrate due to application of uncoupler CCCP; b: natural release of picrate; c: uptake of picrate under the presence of 4 uM CCCP.

As seen in Figure 20, an introduction of the liposome in the medium leads to a rapid decrease of picrate concentration, which stems from an adsorption of picrate on the membrane surface, and follows the permeation of picrate across the liposome membrane. Slight decrease of picrate concentration by addition of ferricyanide solution corresponds to a dilution in the medium. The next decrease of picrate concentration by addition of Fc (amylFc) corresponds to redistribution of the picrate anion in response to development of the redox potential across the membrane (inside positive), implying that the redox potential is developed only when a mediator is present in the membrane. The picrate anions trapped in the liposome are released rapidly with diminution of the membrane potential by addition of an uncoupler, CCCP.

The Nernst equation relates to the membrane potential, $\Delta\psi$, with the concentrations of membrane permeable ion (e.g. picrate) inside and outside the liposome;

$$\Delta\psi = (RT/zF) \ln [C_{out}]/[C_{in}], \tag{4}$$

where z is the valence of the ion, and R, T and F have their usual thermodynamic meanings. The mass conservation of the ion leads to the following equation;

$$[C_{out}] V + [C_{in}] v = [C_o] (V + v),$$

where V and v are the volumes of outside and inside the liposome, respectively, and $[C_o]$ stands for the concentration of the ion at the steady state after addition of liposome (see Figure 20). In the above equation, the adsorption of the ion on the membrane surface is assumed to be constant independent of the membrane potential developed.

The change of concentration of the ion in the medium from $[C_o]$ to $[C_{out}]$ is observed by the potential deflection of the ionsensitive electrode as;

$$\Delta E = (RT/zF) \ln [C_{out}]/[C_o].$$

Elimination of $[C_o]$ from the above two equations leads to

$$\frac{[C_{in}]}{[C_{out}]} = \frac{1 + (V/v)[1 - \exp (zF\Delta E/RT)]}{\exp (zF\Delta E/RT)} \tag{5}$$

Introducing Equation (5) into Equation (4), we have

$$\Delta\psi = (RT/zF) \left\{ \ln (v/V) - (RT/zF) \ln [\exp(zF\Delta E/RT) - 1] \right\} \tag{6}$$

While relative quantity of the probe ion accumulated in the liposome, U_r^e, is represented by;

$$U_r^e, = [1 - \exp (zF\Delta E/RT)] \tag{7}$$

Equations (5),(6) and (7) imply that the membrane potential as well as the concentration of the probe ion accumulated in the liposome can be evaluated in terms of the electrode potential deflection ΔE. Under the condition presented in Figure 20, the steady concentration $[C_{out}]$ was 5.5×10^{-5} M, while $[C_{in}]$ was 6.1×10^{-2} M, implying that the picrate anion was accumulated in the liposome more than 10^3 times higher than that in the outside medium.

However, the accumulated picrate was released by application of CCCP and the concentrations of picrate in two sides of the membrane became the same. Picrate is accumulated against the concentration gradient by the redox reaction, but it is not an active transport as in the case of liquid membrane system.

In conclusion, it is not very difficult to realize the active transport in an artificial membrane if the characteristics of the active transport in a living membrane described above are fulfilled in the model system. Any chemical reactions may be used for energy supply for the uphill transport provided that an appropriate mediator and carrier are incorporated. However, in the case of liposomes a difficulty arises for constructing the active transport system because the leakage or incomplete coupling is involved in a thin lipid layer membrane.

REFERENCES

1. T. Rosenberg, Acta Chem.Scand., 2:14 (1948).
2. H. H. Ussing, Acta Physiol.Scand., 19:43 (1949).
   H. H. Ussing and K. Zerahn, Acta Physiol.Scand., 23:110 (1951).
3. P. Mitchell, Biol.Rev., 41:445 (1966).
4. O. Kedem, in: "Membrane Transport and Metabolism," A. Kleinzeller and A. Kotyk, eds., Acad. Press, New York, p.87 (1961).
5. P. Hinkle, Biophys.Res.Commun., 41:1375 (1970).
6. S. S. Anderson, I. G. Lyle and R. Paterson, Nature, 259:147 (1976).
7. J. J. Grimaldi, S. Boileau and J. M. Lehn, Nature, 265:229 (1977).
8. T. Shinbo, K. Kurihara, Y. Kobatake and N. Kamo, Nature, 270, 277 (1977).
9. B. Chance and G. Hollunger, Nature, 185:666 (1966).
10. J. Garrahan and I. M. Glynn, J.Physiol. (London) 192:237 (1967).
11. T. Shinbo, M. Sugiura, N. Kamo and Y. Kobatake, J.Membr.Sci., 9:1 (1981).
12. T. Shinbo, M. Sugiura, N. Kamo and Y. Kobatake, Chem.Lett., 1979, 1177 (1979).
13. T. Shinbo, PhD Thesis (1982), Hokkaido University.
14. T. Shinbo, M. Sugiura, N. Kamo and Y. Kobatake, J.Chem.Soc.Japan (in Japanese), 1983, 917 (1983).
15. T. Shinbo, N. Kamo, K. Kurihara and Y. Kobatake, Arch.Biochem. Biophys., 187, 414 (1978).

ELECTROPHYSIOLOGICAL ASPECTS OF Na^+-COUPLED COTRANSPORT

OF ORGANIC SOLUTES AND Cl^- IN EPITHELIAL CELL MEMBRANES

Takeshi Hoshi

Department of Physiology, Faculty of Medicine
University of Tokyo
Tokyo 113, Japan

INTRODUCTION

The biological membranes possess various types of transporters as functional elements. The transporters can be classified into three major groups; pumps, carriers and channels. The carriers are further classified into 3 different types, the uniporter, cotransporter and antiporter. The uniporter is a membrane element which facilitates equalizing transport of a specific substrate. Both the cotransporter and the antiporter are characterized as osmo-osmotic coupling agencies which transfer one of the conjugated substrates in an uphill manner in the presence of an electro-chemical potential gradient of the other.

The Na^+-linked cotransport mechanism has been shown to be a general principle in active transport of organic solutes and anions across animal cell membranes[1,2]. This mechanism is also playing an essential role in transcellular active transport of various nutrients and inorganic anions across certain types of transporting epithelia, such as the small intestine and renal proximal tubule. In such epithelia, the apical membrane of cells possesses multiple Na^+-coupled cotransport carriers. These Na^+-linked cotransport systems transfer respective substrates in an uphill manner toward inside the cells since an inwardly directed Na^+ gradient is normally maintained by the activity of the Na^+-K^+ pump locating in the basolateral membrane. When uniport carriers for a cotransported solute are present on the other side of membranes, the substrate is effectively transferred transcellularly in one direction by virtue of cooperation of three different types of transporters. Owing to the uphill uptake by the cotransporter, the substrate is accumulated within the epithelial cells during the transcellular transport. Thermodynamically, the maximum concentration ratio of the substrate achieved by the cotransport mechanism, $[S]i/[S]_o$, is given by the following general equation:

$$\frac{[S]i}{[S]_o} = (\frac{[Na^+]o}{[Na^+]_i})^n \; exp[-(n+z)E_m F/RT] \; ...$$

(1)

where n is the coupling ratio of Na^+ flow to substrate flow (J_{Na}/J_s), z the valency of the substrate, E_m the transmembrane potential difference (F, R and T are of the usual meanings).

Substantial knowledge of kinetic properties of various cotransport systems have been accumulated during past years. However, exact mechanism of coupled translocation of the solutes within the carrier molecule are still not fully understood. Also, our understanding of molecular properties of carrier proteins and the nature of chemical groups involved in translocation is still not mature. To clarify the cotransport mechanism in detail, we need further information from both biochemical and biophysical studies. Electrophysiological studies can yield information mainly concerning the mode or mechanism of charge transfer. In this communication, some basic electrophysiological properties of representative Na^+-linked cotransporters will be outlined.

Electrophysiological Properties of Na^+-linked Cotransport of Electrically Neutral Organic Solutes

Electrophysiological observations made on sugars and neutral amino acids in intact renal and intestinal cells suggest the existence of a gate-opening-like mechanism for Na^+[5]. This mechanism is closely linked to sterospecific binding of the substrate to the carrier site located at the outer portion of the carrier. Such a gate-opening mechanism is manifested by substrate-induced depolarization of the membrane, increase in membrane conductance and flow of Na^+ depending on the existing driving force for Na^+. These phenomena occur immediately after the addition of the substrate to the medium bathing the outer surface of the membrane. The resultant Na^+ flow seems to couple to the substrate flow in the subsequent step. D-Glucose[3] and neutral amino acids[4] are known to be transferred by the respective carriers even in the absence of Na^+ though the transport rate is low. A great stimulation of substrate flow by the induced Na^+-flow appears to be dependent on this coupling mechanism.

It is well established that the cotransport of Na^+ and electrically neutral organic solutes is electrogenic. This has been shown by demonstrating the membrane depolarization in both intact cells[6,7] and isolated membrane vesicles[8,9] and enhancement of vesicular uptake by valinomycin-induced K^+ diffusion potentials[10]. In the intestinal and renal epithelia, the depolarizing response of the cells can be ascribed to change in the electromotive force of the luminal membrane or, more simply to generation of a new inwardly directed electromotive force at the luminal membrane in association with the cotransport[11,12,13]. These epithelia have low-resistance paracellular shunt pathways which permit the current to flow across the basolateral membrane to circulate through the paracellular pathway when such a new electromotive force is generated at the luminal membrane. Consequently, both the luminal and basolateral membranes depolarize and the luminal side becomes more negative.

Saturable nature: One of the important characteristics of the cotransport-associated depolarization is the saturable nature of its amplitude. A Michaelis-Menten type relation is generally seen for sugars and amino acids. Moreover, the half saturation concentration of the substrate determined from the electrical potential changes (ΔPD's) agree well with that for the substrate determined by influx measurements. The difference in K_m value among substrates sharing a common carrier is properly reflected in the kinetic data of ΔPD's.

Voltage dependence: Another important property is that the membrane depolarization is dependent on the initial level of the transmembrane potential[6,14]. When the membrane is depolarized by passing current intracellularly, the size of the depolarizing response becomes smaller and smaller as the initial level is depolarized. At a certain level of the transmembrane potential, the depolarization response is completely suppressed. When the initial level of membrane potential is further depolarized exceeding the nil-response level, the substrate added to the outer medium elicits a hyperpolarization response as shown in Figure 1. This seems to indicate that the membrane potential shift toward the Na^+ equilibrium potential because of an increase in Na^+ conductance, although true Na^+ equilibrium potential across the brush border membrane has not yet been defined clearly in intact cells. Under such conditions, Na^+ flows in the reverse direction as compared with the normal situation. Similar hyperpolarization responses can be observed when intraluminal Na^+ concentration is lowered below a certain concentration or made Na^+-free[15] as shown in Figure 2. When Na^+ is replaced with Li^+, a substrate, e.g. L-alanine, elicits hyperpolarization responses initially, but the responses are gradually changed to depolarization when addition and washout of the substrate are repeated[15]. Li^+ is known to be the only cation which can substitute the role of Na^+ in the Na^+-linked cotransport system though its affinity is much lower than N^+. These findings are easily explained if we assume that the cotransported substrate opens a Na^+ channel when added to the external medium, and that Na^+ flows in accord with the existing driving force for Na^+.

Conductance change: The conductance increase of the apical membrane has been shown in both the small intestine[16,17] and renal proximal tubule[7,18]. In the small intestine, an initial rapid increase in the apical membrane conductance is followed by a subsequent gradual increase in the basolateral membrane conductance, which is dependent on the metabolism[16]. However, in amphibian renal proximal tubule, such an increase in the basolateral membrane conductance is not distinct and only the increase in the apical membrane conductance is dominant. In the case of renal tubule, conductance change can be measured by injecting current pulses of a constant strength into the tubular lumen using a microelectrode, and recording the resultant voltage deflections from both the lumen and inside the cells along the tubular axis. The spatial decay of the voltage deflections can be simulated by using equations obtained from mathematical solution of the equivalent electric circuit of the tubule. By this method, change in each resistance parameter comprising the circuit can be determined quantitatively by simulating the altered distribution of

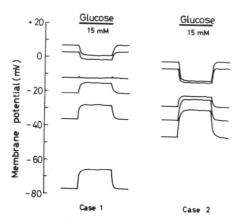

Fig. 1. Effects of the membrane potential level on D-glucose-induced change in cell membrane potential. Triturus renal proximal tubule. The membrane potential level was varied by injecting long-lasting DC current into a cell in the vicinity of the cell impaled with a potential measuring microelectrode. D-glucose (15 mM) was injected into the lumen through a glass-micropipette inserted into the Bowman's capsule during the period indicated by a horizontal heavy line. (From Hoshi and Himukai[5]).

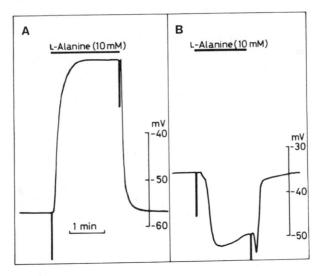

Fig. 2. Responses of the cell membrane potential of Triturus proximal tubule to L-alanine introduced into the lumen. A) A typical response in the presence of Na^+ (normal Ringer solution) on both the luminal and peritubular sides. B) A hyperpolarizing response seen in a Na^+-free medium (D-mannitol-substituted Ringer solution in this particular case). (Modified from Hoshi et al.[15]).

the voltage deflections[18]. The increased apical membrane conductance, as well as the depolarized cell membrane potential, immediately returns to the original state when the added solute is rapidly removed from the outer medium. Prolongation of exposure to the substrate, which brings about a larger intracellular accumulation of the substrate, does not alter such a pattern of rapid restoration, as shown in Figure 3. This suggests that the substrate has no 'gate-opening' effect from inside of the membrane in intact cells. There may be an argument that low intracellular Na^+ concentration can cause such an asymmetric behavior. However, this seems unlikely, since the externally added substrate causes an hyperpolarizing response in a low Na^+ or Na^+-free external medium as described above. There is a possibility that the asymmetric property of the mechanism is impaired or lost by metabolic inhibition in the cells or in isolated membrane vesicles. In the inhibited cells, rapid removal of the substrate frequently causes a transient hyperpolarization response[19]. Also, mirrored flux conditions are shown to cause a similar hyperpolarization response in brush border membrane vesicles[20].

Relation to stereospecific binding of the substrate: The hypothetical gate-opening process appears to be closely linked to the process of stereospecific binding of the substrate to the carrier site. The known specificity of the cotransport system obtained from substrate-flux measurements is properly reflected in the electrical phenomena. For example, intestinal and renal active sugar transport systems are known to require the D-pyranose ring structure, C_2-OH of the same orientation as in D-glucose and C_6 at minimum. When compared with various D-glucose-related compounds or chemically modified derivatives, an affinity scale is seen for these compounds[21], and a sharp discrimination of actively and non-actively transported sugars is difficult as far as the sugars have the D-pyranose ring structure. When the affinity was compared by recording the sugar-evoked potential changes, a very similar pattern was obtained not only in guinea pig intestine (mammal) but also in intestines of other vertebrates, such as bird, reptile, amphibia and fresh water fish. Namely, the sugars

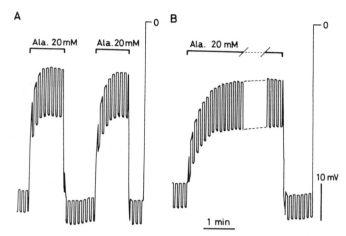

Fig. 3. Changes in the cell membrane potential and voltage divider ratio
induced by intraluminal injection of L-alanine in Triturus
proximal tubule. The membrane potential difference was recorded
across the peritubular (basolateral) membrane and current pulses
of a constant strength (10^{-7} A) were injected into the tubular
lumen at 75 μm apart from the tip of the current electrode. A)
Effects of repeated injections of alanine for a short period of
time (1 min). B) Effects of a prolonged infusion of alanine for
10 min and a rapid washout with alanine-free control solution
(normal Ringer solution). (unpublished data).

without forming the D-pyranose ring,such as L-glucose, D-fluctose and
D-mannitol, did not induce any detectable potential change. Elimination of
C_2-OH or its different orientation and elimination of C_6 greatly reduce the
ability to induce Na^+ current flow in all animal species examined. Modification of hydroxyl groups of C_3, C_4 and C_6 also reduce the ability, but the
degree of reduction of affinity differed among animal species. These
findings suggest that hydroxyl groups of C_3, C_4 and C_6 are also contributing to binding process in some degrees as discussed by Silverman[22], but
their relative importance is different among animal species.

Flow coupling in translocation step: The coupling of flows of conjugated substrates in the cotransport process has been discussed thus far
mainly based on kinetic models which involve a mobile carrier having
separate but closely located binding sites for respective cosubstrates.
Different coupling ratios have been explained in terms of the difference
in number of binding sites for a cosubstrate, particularly for Na^+, on a
single carrier and in the mobility of partially loaded and full-loaded carriers. In intact intestinal epithelium, Na^+/sugar influx ratio was shown
to be unity, and a kinetic model in which the sugar is assumed to be transferred only by forming a full-loaded ternary complex (XNaS) has been proposed[23]. For neutral amino acids, transfer by both a binary and ternary
complex is assumed to take place since the coupling ratio is dependent on
the external Na^+ concentration and a significant carrier-mediated influx is
seen even in a Na^+-free media[4,24]. Recently, however, 1:1 coupling in
Na^+/sugar transport has been questioned from measurements of very short
influxes in membrane vesicles[25,26] which report the ratios ranging 2-3.

Electrically, the sugar-induced increase in Na^+ current flow can be
measured under short-circuited conditions and the increase in short circuit
current can be compared with simultaneously determined sugar influxes
during 1 min. Such measurements yielded 1:1 coupling irrespective of Na^+
concentration[27], supporting the results of Goldner et al.[23]. The

discrepancy between these results obtained in intact cells and membrane vesicles has not been solved at the present time, but these data seem to suggest that the ratio may be variable depending on the experimental conditions.

With regard to this problem, the dipeptide carrier of the intestine seems of particular interest, since carrier-mediated transport of glycylglycine does not show any Na^+-dependence but induces saturable increase in Na^+ current, suggesting the presence of the gate-opening mechanism for Na^+ but the absence of the flow coupling[28]. Figure 4 shows the time course of uptake of glycine and glycylglycine by intestinal brush border membrane vesicles in the presence and absence of a Na^+-gradient. Glycine uptake is stimulated by an imposed Na^+ gradient and exhibited an overshoot uptake, whereas in the case of glycylglycine, there was no such stimulation by a Na^+ gradient. However, both glycine and glycylglycine induce saturable potential changes which obey Michaelis-Menten kinetics. K_m values obtained by the electrical measurements agreed with those obtained by flux measurements for both substances. Glycylproline and carnosine transport across the intestinal and renal brush border membrane has been shown to be independent of Na^+ but pH-dependent[29]. Glycylglycine transport across the same membranes is pH-dependent and cotransported with H^+ (unpublished data). It is speculated that protonation of the carboxyl group of glycylglycine during translocation may be responsible for the absence of Na^+/substrate flow coupling during translocation.

Molecular Properties of the Cotransporters and Putative Models

Recently, a new attractive model of the Na^+-D-glucose cotransporter has been proposed by Semenza[30] and his group[25], which is called 'gated channel (or pore) model'. The model is based on accumulated biochemical information including asymmetric location of the phlorizin-binding sites and asymmetric kinetic behaviors of the transport. The principal features of the model are: (1) the carrier protein which has a relatively large molecular size is embedded within the membrane without diffusive or rotational mobility, (2) the carrier has a channel (or pore) structure within the molecule as pathways for Na^+ and glucose, (3) both Na^+ and sugar binding sites are located in the vicinity of the internal surface of the membrane, and (4) the Na^+ binding site having a carboxyl group can reorient depending on the transmembrane potential.

The molecular size of the Na^+-D-glucose cotransporter has currently been studied by a variety of methods, including reconstitution with purified or partially purified proteins[31,32,33], photoaffinity labeling using a photosensitive phlorizin derivative[34] affinity chromatography on a phlorizin polymer[35], differential labeling of thiol or amino groups with group-reactive reagents[36,37,38], and radiation inactivation of phlorizin binding[39]. At the present time, the reported values of molecular size vary greatly, ranging from 30,000 to 165,000 daltons, a definitive conclusion has not yet been obtained. Our reconstitution study was carried out with soybean phospholipids and membrane proteins extracted from negatively purified brush border membrane vesicles from guinea pig intestine[32]. The protein fractions separated by gel-filtration were finally purified by chromatofocusing using the Polybuffer exchanger (Pharmacia Fine Chemicals). Transport assay revealed that only the protein eluted at pH range of 5.0-5.5 (Fraction D in Figure 5) had the Na^+-dependent cotransport function. This protein could be seen on SDS-polyacrylamide gel electrophoresis as a single band of glycoprotein of 160K daltons in molecular weight. This finding is in accord with that of Malathi and Preiser[31] who demonstrated by reconstitution studies that Na^+-D-glucose cotransporters isolated from small intestines and kidneys of several different animals had a molecular weight in the range of 160K-165K daltons. However, these

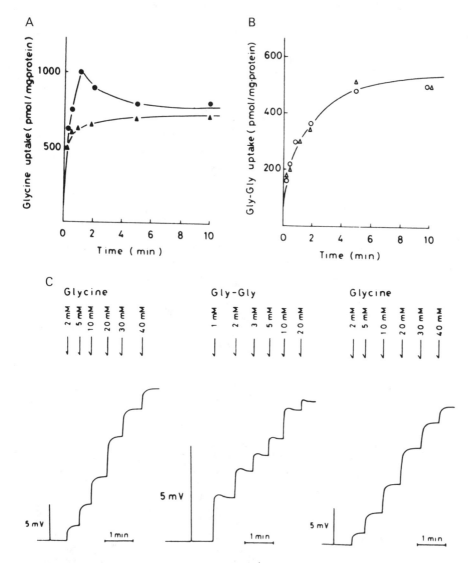

Fig. 4. Upper panel: A comparison of Na$^+$ gradient-dependence of uptake of glycine (A) and glycylglycine (B) by the brush border membrane vesicles from guinea pig small intestine. Lower panel (C): concentration-dependent increments of the transmural potential difference of everted small intestine of guinea pig caused by glycine and glycylglycine. Both substances were added to the solution bathing the mucosal side and the effects of the cumulative concentration were shown. (Modified from Himukai and Hoshi[28]).

values are about 2-fold greater than those of identified phlorizin binding proteins and further greater than those of proteins identified by thiol or amino group modification methods. There is no explanation for these different values obtained by different methods although some investigators are considering a dimer structure[31] or some other subunit structures[32].

   In spite of uncertainty of the exact molecular size, the reported molecular weight of the Na$^+$-D-glucose cotransporters may rule out the possibility of classical diffusive or rotational carrier models. It is

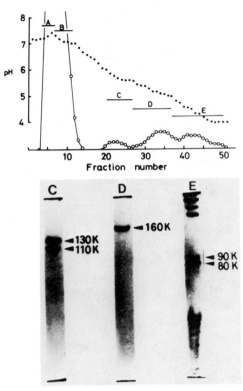

Fig. 5. Chromatofocusing of protein extracts from the brush border
membrane vesicles isolated from guinea pig small intestine and
SDS-polyacrylamide gel electrophoresis patterns for separated
small fractions (C.D.E). The brush border membrane vesicles
prepared by Ca^{++}-precipitation method were first treated with
papain and deoxycholate (negative purification). The proteins
were solubilized by treatment with Triton X-100, and then
roughly separated by gel-filtration in order to remove aggregated
proteins. Thereafter, the sample was subjected to chromatofocusing. The eluted proteins were divided into 5 fractions
according to pH range; A: 7.4-7.3, B: 7.3-7.0, C: 6.5-5.5,
D: 5.5-5.0, E: 5.0-4.0. Reconstitution studies revealed that
only sample D had a Na^+-dependent overshoot uptake activity.
(Modified from Kano-Kameyama and Hoshi[32]).

known that about 3000 daltons polypeptide chain forming an α-helix is
enough to span whole membranes[40]. Also, recent knowledge of vectorial
insertion of membrane proteins during biosynthesis[41] support an asymmetric structure of the carrier proteins. Therefore, the presence of a
channel (or pore) structure is considered to be very probable.

    Based on the electrophysiological findings, an asymmetric carrier
model which has a channel structure is considered as a possible model. In
addition, the existence of a gate-opening mechanism is strongly suggested
at the outer part of the carrier as illustrated in Figure 6. In intact
cells, only the substrate added to the outside medium appears to have the
gate-opening effect. The hyperpolarizing responses to an externally added
substrate observed in artificially depolarized cells or cells facing a
Na^+-free medium are easily explained on the basis of such a model but seem
to be difficult to explain on the basis of 'gated channel model'.

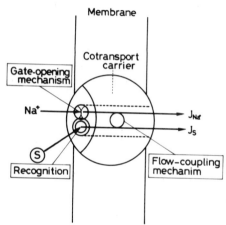

Fig. 6. Mechanism of charge transfer by the Na^+/neutral organic solute cotransporter deduced from electrophysiological findings. Recognition process is a stereospecific binding of the carrier with the substrate. The gate-opening mechanism is closely linked to the recognition process and allow Na^+ to flow into a hypothetical channel penetrating the carrier molecule upon the substrate binding. This mechanism is considered to be located at the outer part of the carrier because of an asymmetric behavior of the mechanism in intact cells. Within the channel, the flow of Na^+ driven by the driving force for Na^+ across the membrane ($E_{Na} - E_m$) is considered to couple to the substrate flow. Because of the substrate concentration-dependence of the changes in membrane potential, membrane conductance and Na^+ current, the Na^+ conductance (g_{Na}) of the membrane is considered to be proportional to the number of the carrier loaded with the substrate.

Electrophysiological Characterization of $(Na^+ + K^+)$-coupled Cl^- Transport

Recently, a $Na^+/K^+/Cl^-$ cotransport has been shown to occur in a variety of cell types, e.g. Ehrlich ascites cells[42], some of cultured cell lines[43], erythrocytes[44], diluting segments of nephron[45,46], shark rectal gland[47] and marine teleost intestine[48]. Both flux measurements and electrophysiological observation in some of these cells have disclosed that the cotransport has a $2Cl^-/Na^+/K^+$ stoichiometry[42,45, 46], thus the cotransport itself is nonelectrogenic. However, the epithelia having such a mechanism show a distinct net charge transfer across the cell which directly depends on the rate of Cl^- entry mediated by the cotransport mechanism. Electrophysiological studies in such epithelia are forwarded mainly to delineation of mechanism of charge transfer by the cells.

Parts of electrophysiological observations made in the diluting segment of Triturus kidneys are shown in Figure 7. This segment normally exhibits a considerably high lumen positive transtubular potential (V_t), which is an indication of transfer of net negative charge in the outward direction. The generation of such a lumen-positive V is completely suppressed by elimination (or replacement) of one of Na^+, K^+ and Cl^- of the luminal fluid. This indicates that the Cl^- entry and resultant net charge transfer require the simultaneous presence of Na^+, K^+ and Cl^- in the luminal fluid. Furosemide which is known to be a common specific inhibitor of $(Na^+ + K^+)$-coupled Cl^- transport also completely suppresses the generation of V_t when added to the luminal fluid. The rate of net charge transfer across the tubule can be measured as the equivalent short-

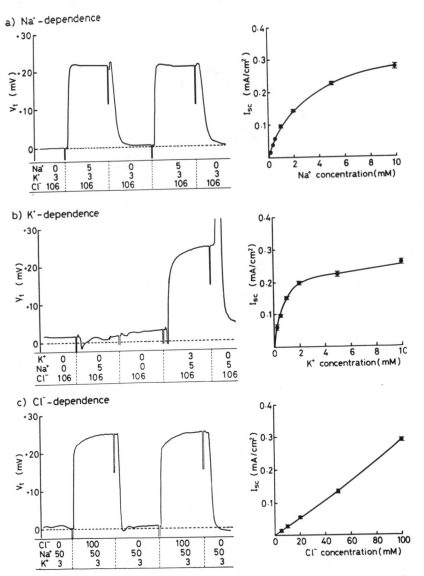

Fig. 7. Dependence of the lumen-positive V_t and net charge transfer (I_{sc}) on Na^+, K^+ and Cl^- concentrations in the luminal fluid in Triturus early distal tubule. a) Na^+-dependence. Left: The effect of removal of Na^+ and addition of 5 mM Na^+ on V_t, Right: Relationship between I_{sc} and Na^+ concentration in the presence of 106 mM Cl^- and 3 mM K^+. b) K^+-dependence. Left: The effect of a Na^+, Cl^--containing solution in the absence and presence of 3 mM K^+. Right: Relationship between I_{sc} and K^+ concentration in the presence of 106 mM Cl^-, 5 mM Na^+ and 2 mM Ba^{++}. c) Cl^--dependence. Left: The effect of replacement of Cl^- with NO_3^- on V_t. Right: Relationship between I_{sc} and luminal Cl^- concentration in the presence of 50 mM Na^+ and 3 mM K^+.

circuit current (I_{sc}) which is given by dividing V_t by the specific trans-tubular resistance measured in the same nephron segment under the same conditions as those of V_t measurement[49]. The I_{sc} increases as the concentration of one of these three ions is increased without changing the concentration of the other two. The relationship between I_{sc} and Na^+ or K^+ concentration is simple Michaelis-Menten type, while the relation of I_{sc} to Cl^- concentration is almost linear, suggesting that one Na^+ and one K^+ are cotransported with multiple Cl^-. The measurements of the cell potential in the presence and absence of furosemide revealed that the Cl^- entry was non-electrogenic, namely the drug caused no depolarization which is expected from excess Cl^- entry but a slight hyperpolarization of the cell potential. These findings support an electrically neutral entry of Cl^- by forming a $Na^+/K^+/2Cl^-$ carrier complex ($XNaKCl_2$).

A further analysis of kinetic data on Cl-concentration-dependence of both K_t for Na^+ and $I_{sc\ max}$ obtained by increasing Na^+ concentration also supported the same mode of cotransport. Namely, the patterns of Cl^- concentration-dependence of K_t for Na^+ and $I_{sc\ max}$ were well explained on the basis of kinetic model in which $XNaKCl_2$ is assumed to be formed in the following binding order and the I_{sc} is directly proportional to the number of $XNaKCl_2$ complexes formed in the luminal membrane: $X+Na(or\ K) \rightleftharpoons XNa(or\ XK)$, $XNa(XK)+ Cl \rightleftharpoons XNaCl(XKCl)$, $XNaCl\ (XKCl)+K(Na) \rightleftharpoons XNaKCl$, $XNaKCl+ Cl$ $XNaKCl_2$[49].

An important problem relating to the cotransport is a much lower concentration of K^+ in the lumen as compared to Na^+ and Cl^- in this segment. There must be a sufficient supply of K^+ into the lumen, otherwise the cotransport will rapidly cease because of K^+ depletion. Some authors[45,46] maintain that K^+ is recirculating across the luminal membrane because the membrane has a high Ba^{++}-sensitive K^+ permeability. However, the supply or recirculation of K^+ was found to be very sensitive to suppression of cation permeability of the paracellular shunt pathways. This is shown by the following experiments. Intraluminal perfusion with a K^+-free solution at a high flow rate suppresses the lumen-positive V_t almost completely as described above. When the perfusion is stopped, a lumen-positive V_t reappears with a relatively rapid time course. This reappearance of V_t is considered to be due to operation of the cotransport due to a supply of K^+ since this restoration of V_t is sensitive to furosemide. Triaminopyrimidine[50] and kinetin[51] are known to reduce cation permeability of the paracellular shunt pathways of some epithelia. These drugs as well as Ba^{++} very strongly suppress the restoration of V_t during the stopped-flow period. These agents, particularly Ba^{++} also suppress the diffusion potentials created by a small change in K^+ concentration. Observations of diffusion potentials created by changes in intraluminal K^+ or Na^+ concentration in the presence of furosemide or in Cl^--free perfusion media indicate that the paracellular pathway is specially permeable to K^+ and this selective permeability is very sensitive to Ba^{++} (unpublished observations). These observations indicate that the paracellular pathways are predominantly important as the recirculation pathway of K^+ at least in Triturus early distal tubule.

Ion transport mechanisms and pathways of ion movements in the early distal tubule thus can be summarized as illustrated in Figure 8. Chloride ions are taken up by the cells by a $2Cl^-/Na^+/K^+$ cotransport mechanism and K^+ recirculates mainly across the basolateral membrane and the paracellular pathways to maintain the cotransport. Chloride ions are accumulated within the cells because of the presence of an inwardly directed driving force for Na^+ which is maintained by the Na^+-K^+ pump locating in the basolateral membrane. Both the recirculating K^+ flow and the passive flow of Cl^- across the basolateral membrane are considered to be responsible for net charge transfer (I_{sc}). However, the nature of the Cl^- pathways of the basolateral membrane has not yet been well understood. Further studies are needed to characterize the Cl^- flow across the basolateral membrane.

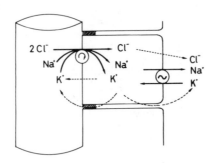

Fig. 8. Coupled transport of Na^+, K^+ and Cl^- across the luminal membrane and related ionic movements across the cell membranes and the paracellular pathways in the early distal tubule of Triturus kidney.

REFERENCES

1. S. G. Schultz and P. F. Curran, Coupled transport of sodium and organic solutes, Physiol.Rev., 50:637 (1970).

2. S. G. Schultz, Ion-coupled transport across biological membrane, in: "The Physiological Basis for Disorders of Biomembranes," T. Andreoli, J. F. Hoffman, and D. D. Fanestil, eds., Plenum Medical Book Co., New York - London, p.273 (1978).

3. S. Hilden and B. Sacktor, Potential-dependent D-glucose uptake by renal brush border membrane vesicles in the absence of sodium, Am.J.Physiol., 242:F340 (1982).

4. P. F. Curran, S. G. Schultz, R. A. Chez, and R. E. Fuisz, Kinetic relations of the Na^+-amino acid interaction at the mucosal side, J.Gen.Physiol., 50:1261 (1967).

5. T. Hoshi and M. Himukai, Na^+-coupled transport of organic solutes in animal cells, in: "Transport and Bioenergetics in Biomembranes," R. Sato and Y. Kagawa, eds., Japan Scientific Societies Press, Tokyo, Plenum Press, New York - London, p.111 (1982).

6. T. Hoshi and Y. Kikuta, Effects of organic solute-sodium cotransport on the transmembrane potentials and resistance parameters of the proximal tubule of Triturus kidney, in: "Electrophysiology of the Nephron," T. Anagnostopoulus, ed., INSERM, Paris, p.136 (1977).

7. E. Frömter, Electrophysiological analysis of rat renal sugar and amino acid transport. I. Basic Phenomena.Pflügers Arch., 393:179 (1982).

8. J. C. Beck and B. Sacktor, Membrane potential-sensitive fluorescence changes during Na^+-dependent D-glucose transport in renal brush border membrane vesicles, J.Biol.Chem., 253:7158 (1978).

9. R. E. Schell, B. R. Stevens, and E. M. Wright, Kinetic of sodium-dependent solute transport by rabbit renal and jejunal brush border vesicles using a fluorescent dye, J.Physiol.(London), 355:307 (1983).

10. H. Murer and U. Hopfer, Demonstration of electrogenic Na^+-dependent D-glucose transport in intestinal brush border membranes, Proc. Nat.Acad.Sci.USA, 71:484 (1974).

11. R. C. Rose and S. G. Schultz, Studies on the electrical potential profile across rabbit ileum. Effects of sugars and amino acids on transmural and transmucosal electrical potential differences, J.Gen.Physiol., 57:639 (1971).

12. J. F. White and W. McD. Armstrong, Effect of transported solutes on membrane potentials in fullfrog small intestine, Am.J.Physiol., 221:194 (1971).

13. T. Maruyama and T. Hoshi, The effect of D-glucose on the electrical potential profile across the proximal tubule of newt kidney, Biochim.Biophys.Acta, 282:214 (1972).

14. N. Iwatsuki and O. H. Petersen, Amino acids evoke short-latency membrane conductance increase in pancreatic acinar cells, Nature, 283:492 (1980).

15. T. Hoshi, K. Sudo, and Y. Suzuki, Characteristics of changes in intracellular potential associated with transport of neutral, dibasic and acidic amino acids in Triturus proximal tubule, Biochim.Biophys.Acta, 448:492 (1976).

16. P. G. Ganter-Smith, E. Grasset, and S. G. Schultz, Sodium-coupled amino acid and sugar transport by Necturus small intestine. An equivalent electrical circuit analysis of a rheogenic cotransport system, J.Membrane Biol., 66:25 (1982).

17. C. A. R. Boyd and M. R. Ward, A microelectrode study of oligopeptide absorption by the small intestinal epithelium of Necturus maculosus, J.Physiol.(London), 324:411 (1982).

18. T. Hoshi, K. Kawahara, R. Yokoyama, and K. Suenaga, Changes in membrane resistance of renal proximal tubule induced by cotransport of sodium and organic solutes, Adv.Physiol.Sci., Vo.11 (Kidney and Body Fluids, L. Takàcs, ed.), p.403 (1981).

19. T. Hoshi and Y. Komatsu, Effects of anoxia and metabolic inhibitors on the sugar-evoked potential and demonstration of sugar-outflow potential in small intestine, Tohoku J.Exp.Med., 100:47 (1970).

20. B. Stieger, G. Burchhardt, and H. Murer, Demonstration of sodium-dependent, electrogenic substrate transport in rat small intestinal brush border membrane vesicles by a cyanine dye, Pflügers Arch., 400:178 (1984).

21. E. Brot-Laroche and F. Alvarado, Mechanisms of sugar transport across the intestinal brush border membrane, in: "Intestinal Transport," M. Gilles-Baillien and R. Gills, eds., Springer-Verlag, Berlin - Heidelberg, p.147 (1983).

22. M. Silverman, Glucose transport in the kidney, Biochim.Biophys.Acta, 457:303 (1976).

23. A. M. Goldner, S. G. Schultz, and P. F. Curran, Sodium and sugar fluxes across the mucosal border of rabbit ileum, J.Gen.Physiol., 53:362 (1969).

24. F. V. Sepùlveda and J. W. L. Robinson, Kinetics of the cotransport of phenylalanine and sodium ions in the guinea pig small intestine: I. Phenylalanine fluxes, J.Physiol.(Paris), 74:569 (1978).

25. J. D. Kaunitz, R. Gunther, and E. M. Wright, Involvement of multiple sodium ions in intestinal D-glucose transport, Proc.Intl.Acad.Sci.USA, 79:2315 (1982).

26. M. Kessler and G. Semenza, The small-intestinal Na^+, D-glucose cotransporter: An asymmetric gated channel (or pore) responsive to $\Delta\phi$, J.Membrane Biol., 76:27 (1983).

27. T. Hoshi, Y. Suzuki, T. Kusachi, and Y. Igarashi, Interrelationship between sugar-evoked increases in transmural potential difference and sugar influxes across the mucosal border of the small intestine, Tohoku J.Exp.Med., 119:201 (1976).

28. M. Himukai and T. Hoshi, Interaction of glycylglycine and Na^+ at the mucosal border of guinea pig small intestine. A non-mutual stimulation of transport, Biochim.Biophys.Acta, 732:659 (1983).

29. V. Ganapathy and F. H. Leibach, Role of pH gradient and membrane potential in dipeptide transport in intestinal and renal brush border membrane vesicles from the rabbit. Studies with L-carnosine and glycyl-L-proline, J.Biol.Chem., 258:14189 (1983).

30. G. Semenza, The small-intestinal Na^+/D-glucose carrier is inserted asymmetrically with respect to the plane of the brush border membrane. A model, in: "Membrane and Transport," A. Martonosi, ed., Plenum Press, New York, Vol.2, p.175 (1982).

31. P. Malathi and H. Preiser, Isolation of the sodium-dependent D-glucose transport protein from brush border membranes, Biochim.Biophys. Acta, 735:314 (1983).

32. A. Kano-Kameyama and T. Hoshi, Purification and reconstitution of Na$^+$/D-glucose cotransport carriers from guinea pig small intestine, Jpn.J.Physiol., 33:955 (1983).

33. H. Koepsell, H. Menuhr, I. Ducis, and T. E. Wissmüller, Partial purification and reconstitution of Na$^+$-D-glucose cotransport protein from pig renal proximal tubules, J.Biol.Chem., 258:1884 (1983).

34. M. Hosang, E. Michael-Gibbs, D. F. Diedrich, and G. Semenza, Photo-affinity-labeling and identification of (a component of) the small intestinal Na$^+$, D-glucose transporter using 4-azidophlorizin, FEBS Lett., 130:244 (1981).

35. J. T. Dacruz, M.E.M., S. Riedel, and R. Kinne, Partial purification of hog kidney sodium-D-glucose cotransport system by affinity chroma-tography on a phlorizin-polymer, Biochim.Biophys.Acta, 640:43 (1981).

36. L. Thomas, Isolation of N-ethylmaleimide-labeled phlorizin sensitive D-glucose binding protein of brush border membrane from rat kidney cortex, Biochim.Biophys.Acta, 291:454 (1973).

37. A. Klip, S. Grinstein, J. Biber, and G. Semenza, Interaction of the sugar carrier of intestinal brush border membranes with HgCl$_2$, Biochim.Biophys.Acta, 598:100 (1980).

38. B. E. Peerce and E. M. Wright, Conformational changes in the intestinal brush border sodium-glucose cotransporter labeled with fluorescein isothiocyanate, Proc.Natl.Acad.Sci.USA, 81 (1984) in press.

39. R. J. Turner and E. S. Kempton, Radiation inactivation studies of the renal brush border membrane phlorizin-binding protein, J.Biol. Chem., 257:10794 (1982).

40. S. Maroux, H. Feracci, J. P. Gorvel, and A. Benajiba, Aminopeptidases and proteolipids of intestinal brush border, in: "Brush Border Membranes," Ciba Foundation Symposium 95, Pitman, London, p.34 (1983).

41. H. Hauri, Biosynthesis and transport of plasma membrane glycoproteins in the rat intestinal cell: Studies with sucrase-isomaltase, in: "Brush Border Membranes," Ciba Foundation Symposium 95, Pitman, London, p.132 (1983).

42. P. Geck, C. Pietrzyk, B. C. Burchhardt, B. Pfeiffer, and E. Heinz, Electrically silent cotransport of Na$^+$, K$^+$ and Cl$^-$ in Ehrlich cells, Biochim.Biophys.Acta, 600:432 (1980).

43. J. F. Aiton, A. R. Chipperfield, J. E. Lamb, P. Odgen, and N. L. Simmons, Occurrence of passive furosemide-sensitive transmembrane transport in cultured cells, Biochim.Biophys.Acta, 646:389 (1981).

44. A. R. Chipperfield, Chloride dependence of furosemide- and phlorizin-sensitive passive sodium and potassium fluxes in human red cells, J.Physiol., 312:435 (1981).

45. R. Greger, E. Schlatter, and F. Lang, Evidence for electroneutral sodium and chloride cotransport in the thick ascending limb of Henle's loop of rabbit kidney, Pflügers Arch., 396:308 (1983).

46. H. Oberleithner, W. Guggino, and G. Giebisch, The effect of furosemide on luminal sodium, chloride and potassium transport in the early distal tubule of Amphiuma kidney, Pflügers Arch., 396:27 (1983).

47. J. Hannafin, E. Kinne-Saffran, D. Friedman, and R. Kinne, Presence of a sodium-potassium chloride cotransport system in the rectal gland of Squalus acanthias, J.Membrane Biol., 75:73 (1983).

48. M. W. Mush, S. A. Orellana, L. S. Kimberg, M. Field, D. R. Halm, E. J. Krasny, and R. A. Frizzell, Na$^+$-K$^+$-Cl$^-$ cotransport in the intestine of a marine teleost, Nature, 300:351 (1982).

49. T. Hoshi, G. Kuramochi, and H. Sakamoto, Kinetic properties of active Cl$^-$ transport across amphibian early distal tubule, in: "Coupled Transport in Nephron-Mechanisms and Pathophysiology," Miura Medical Foundation, Tokyo, in press (1984).

418

50. J. H. Moreno, Blackage of gallbladder tight junction cation-selective channels by 2,4,6-triaminopyrimidinium, J.Gen.Physiol., 66:97 (1975).

51. C. J. Bentzel, B. Hainau, A. Edelman, T. Anagnostopoulos, and E. L. Benedetti, Effect of plant cytokinins on microfilaments and tight junction permeability, Nature, 264:666 (1976).

DISINFECTION OF ESCHERICHIA COLI BY USING WATER

DISSOCIATION EFFECT ON ION-EXCHANGE MEMBRANES

T. Tanaka*, T. Sato* and T. Suzuki**
* Showa Pharmaceutical College, Tsurumaki, Setagaya
  Tokyo 154, Japan
**Department of Applied Chemistry, Yamanashi University
  Takeda, Kofu 400, Japan

ABSTRACT

Escherichia coli (E.coli K-12 W3110) suspended in various electrolyte solutions (10^8 cells/cm^3) were passed through the desalting chamber in the electrodialysis system using the cation and the anion-exchange membranes at 20°C. Factors varied in the tests included current densities, pH values of the solutions, etc. Viable counts of the E.coli cells in effluent were determined by colony formation on nutrient at 37°C.

In general, disinfection efficiency was increased as the current density was increased and pH values of the solutions were decreased.

Remarkable germicidal efficiency was found in the region of current densities greater than the limiting current density (0.8 A/dm^2) where "neutral disturbance" occurs, i.e. water in the vicinity of the two ion-exchange membranes in the desalting chamber is dissociated to H^+ and OH^- ions.

The germicidal efficiency is considered to be due to a synergistic effect of both the H^+ and OH^- ions produced, as in the case of the mixed bed of H^+ and OH^- form ion-exchange resins shown by us[1].

The electrodialysis system has the feasibility of a new and handy "disinfectant" for water treatment instead of chlorine which has recently become known to react with a variety of organic impurities in water and convert them to carcinogens.

INTRODUCTION

Electrodialysis methods using ion-exchange membranes have been generally used for sea water desalting, concentration of electrolyte solutions, etc. During the investigation of the membrane characteristics of this method, we have discovered a bactericidal effect[1,2] in the desalting chamber surrounded by the cation and anion-exchange membranes. We have therefore begun to examine the feasibility of employing the electrodialysis system as a "new disinfectant" for water treatment instead of chlorine which has recently been found[3,4] to react with a variety of organic impurities in water and convert them to carcinogens.

Experimental

The electrodialysis apparatus consisted of five chambers as shown in Figure 1. Both cation and anion-exchange membranes were set alternatively as the diaphragm of each chamber. The area of the membranes (CMV and AMV, Asahi Glass Co.) was 18.4 cm^2 and the distance between the membranes was 1 cm. 0.1N-NaCl solutions were passed through both side polar chambers (I and V chambers) from lower to upper at a flow rate of 25 cm^3/min, and 0.01N-NaCl solutions were fixed in the concentrating chambers (II and IV chambers).

Escherichia coli (E.coli K-12 W3110) suspended in various electrolyte solutions (10^8 cells/cm^3) were passed through the central desalting chamber (III chamber) from lower to upper at 20°C by changing factors such as current densities, pH values of the solutions, etc. and viable counts of the E. coli cells in effluent were determined by colony formation on nutrient agar at 37°C.

Results and Discussion

Table 1 shows viability changes (%) of E.coli cells and pH changes of the effluent physiological saline solutions under the conditions of various current densities (0.13 - 1.63 A/dm^2) and time intervals (7 - 60 min). From the table, it was found that in the region of low current densities (0.13 - 0.54 A/dm^2) there were no pH changes and no viability changes, but the changes emerged at the current density of 0.81 A/dm^2, which was the limiting current density of the system determined by measuring the current-voltage curves. The changes were increased with an increase in the current density, especially at the current density of 1.63 A/dm^2, where the E. coli was perfectly devitalized (0%) and pH value of the solution was lowered to 4.62 for only 7 minutes after beginning the electrodialysis. In this manner, remarkably a bactericidal effect was found in the region of current

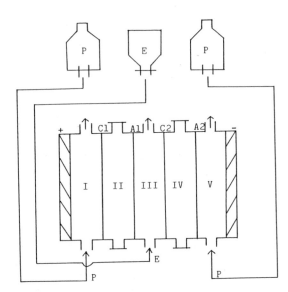

Fig. 1. Schematic diagram of electrodialytic disinfection system. C1, C2 cation-exchange membranes (Selemion CMV-10). A1, A2 anion-exchange membranes (Selemion AMT-20). E: E.coli suspended in various electrolyte solutions (E.coli K-12 W3110 10^8 cells/ml). P: polar solution (0.1 N-NaCl). Distance between membranes: 1 cm. Area of each membrane: 18.4 cm^2. II, IV: fixed solution (0.01 N-NaCl).

Table 1. Viability Changes of Cells (%) and pH Changes in Effluent Solutions Under the Conditions of Various Current Densities and Time Intervals

min CD	7		15		30		45		60	
	pH	%	pH	%	pH	%	pH	%	pH	%
0.13	6.97	87.7	6.95	86.1	6.92	85.5	6.87	86.3	6.85	85.7
0.27	6.95	82.0	6.90	83.9	6.85	80.0	6.78	79.3	6.74	80.6
0.54	6.90	77.4	6.81	78.1	6.70	79.0	6.63	77.6	6.51	78.0
0.81	6.70	20.0	6.57	21.6	6.21	18.9	5.34	20.4	4.34	18.1
1.08	6.32	7.5	5.61	7.8	4.82	6.3	4.10	7.2	3.31	6.0
1.35	5.50	0	4.01	0	3.24	0	2.90	0	2.82	0
1.63	4.62	0	3.20	0	2.63	0	2.59	0	2.54	0

original cell concentration of the physiological saline solution (0.14M NaCl aqueous solution): 10^8 cells/cm^3.

flow rate of the solution: 3 cm^3/min.

temperature of the solution: 20° C.

densities over the limiting current density. This result suggests that the effect is deeply concerned with the "neutrality disturbance phenomenon"[5], i.e. water dissociating to H^+ and OH^- ions observed over the region of the limiting current density in the vicinity of the anion and cation-exchange membranes of the desalting chamber (III).

Figure 2 shows the effect of various anionic species in solution on the disinfection of E. coli at the current density of 1.63 A/dm^2 under the same condition as Table 1. As clearly shown in the figure, SO_4^{2-} and HPO_4^{2-} ions were not effective at all but the NO_3^- ion was found to have the same strong effect as that of the Cl^- ion.

The disinfection effect of the same anions at the current density of 1.08 A/dm^2 is shown in Figure 3. The tendency of the efficiency became clearer than that of Figure 2, that is, the order of the efficiency was $NO_3^- > Cl^- \cong I^- > SO_4^{2-} \cong HPO_4^{2-}$.

Table 2 presents pH changes of the effluent solutions shown in Figure 2 with the elapse of time. From the table, the solutions containing NO_3^- and SO_4^{2-} ions were found to indicate low pH values of 3.60 and 3.15, respectively, for 60 minutes after beginning the electrodialysis and the pH values were almost equal to that of Cl^- system shown in Table 1, while the effluent solution with HPO_4^{2-} ion was indicated by the high pH values of 9.25 after 60 minutes.

The result means that pH changes do not parallel the disinfection effect of the solution, i.e. the pH value itself does not affect the devitalization of E. coli in a striking manner.

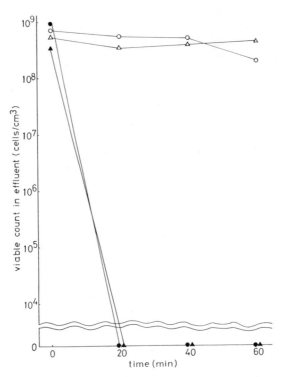

Fig. 2. Disinfection effect of anionic species in solution at the current density of 1.63 A/dm^2. ○ 0.1M Na$_2$SO$_4$; △ 0.1M Na$_2$HPO$_4$; ● 0.1M NaCl; ▲ 0.1M NaNO$_3$.

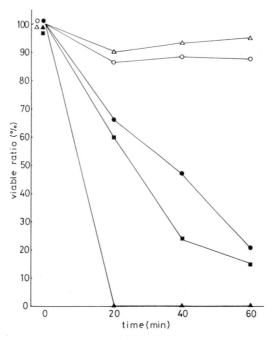

Fig. 3. Germicidal efficiency (%) of anionic species at the current of 1.08 A/dm^2. △ 0.1M Na$_2$HPO$_4$; ○ 0.1M Na$_2$SO$_4$; ● 0.1M NaCl; ■ 0.1M NaI; ▲ 0.1M NaNO$_3$.

Table 2. pH Changes of the Solutions Containing NO_3^-, SO_4^{2-} and HPO_4^{2-} Ions with the Elapse of Time at the Current Density of 1.63 A/dm².

time(min)	0.1M-NaNO₃	0.1M-Na₂SO₄	0.1M-Na₂HPO₄
0	4.55	5.00	8.85
5	5.30	5.30	8.95
10	4.00	3.70	9.25
15	3.85	3.35	9.35
20	3.85	3.20	9.25
25	3.85	3.15	9.25
30	3.80	3.15	9.25
35	3.75	3.15	9.25
40	3.75	3.15	9.25
45	3.75	3.15	9.25
50	3.70	3.15	9.25
55	3.70	3.15	9.25
60	3.60	3.15	9.25

The disinfection effect of cationic species was also investigated under the same conditions of Table 1 and shown in Figure 4.

From the figure, it was found that the devitalization efficiency order of the species is $K^+ \cong Na^+ \gg Ca^{2+} \cong Mg^{2+}$ and the tendency of pH changes in the solutions is almost equal to the anionic system shown in Table 2, that is, pH changes in the solutions are not linear with the disinfection effects of the solution.

From the results of Figures 2, 3 and 4, the solution containing K^+ and NO_3^- ions was found to be most effective for the devitalization of E. coli cells and this is considered to be mainly concerned with the largest permeability of the ions through the ion-exchange membranes.

The disinfection effect of current density is represented in Figure 5. Notice that there is no devitalization effect in the system without ion-exchange membranes even at the high current density of 1.63 A/dm².

425

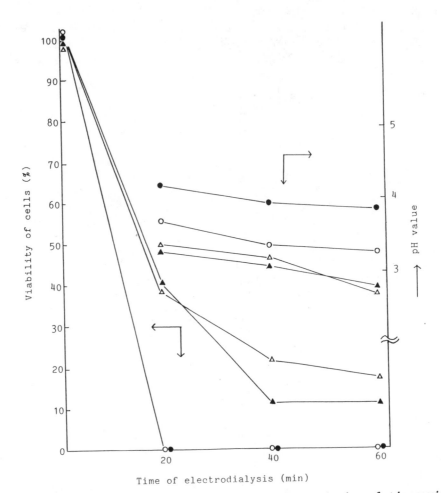

Fig. 4. Viability changes and pH changes of the cationic solutions with the elapse of time at the current density of 1.63 A/dm^2. $\triangle$ 0.1M MgCl$_2$; $\blacktriangle$ 0.1M CaCl$_2$; $\bullet$ 0.1M NaCl; $\circ$ 0.1M KCl.

Taking the results obtained into account, we can conclude that the remarkable disinfection effect of the electrodialysis system is due to a synergistic effect of current density, pH change and surface characteristics of ion-exchange membranes. As shown in Figure 6, the germicidal effect is caused by both accelerated H$^+$ and OH$^-$ ions which are produced on the anion-exchange membrane and the cation-exchange membrane of the desalting chamber, respectively, in the region of the current densities greater than limiting current density where "neutrality disturbance phenomenon" occurs and water in the vicinity of the two ion-exchange membranes is dissociated to H$^+$ and OH$^-$ ions.

There was no degradation of the ion-exchange membranes even after 10 h from the beginning of the electrodialysis at the current density of 1.63 A/dm^2.

The electrodialysis system has the feasibility of a new and handy "disinfectant" for water treatment instead of chlorine.

The details of economic aspects of the system will be discussed in a following paper.

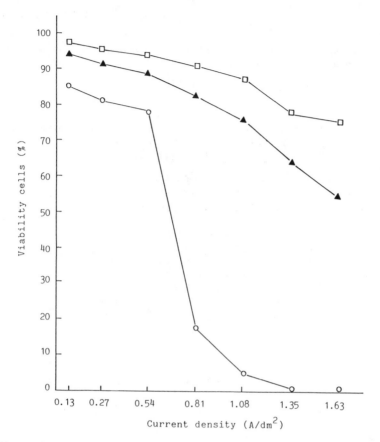

Fig. 5.   Relationship between viability cells (%) and current density in
the system of 0.1M-NaCl solution.   1) Using ion-exchange
membranes - ○; 2) Using Cellophane membranes - ▲; 3) Current
only - □.

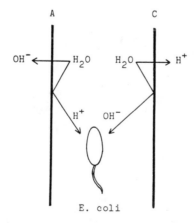

E. coli

Fig. 6.   Disinfection mechanism of E.coli with both H^+ and OH^- ions
produced by the ion-exchange membranes.   A: anion-exchange
membrane; C: cation-exchange membrane.

Acknowledgement

This work was supported by Grant-in-Aid from the Japanese Ministry of Education (Project Number 57550497).

REFERENCES

1. T. Suzuki, Y. Iwata, S. Goto, T. Sato, and T. Tanaka, Disinfection of E. coli by using surface characteristics of ion-exchange resins, Absts. 36th Symposium of Coll.& Surface Chem.Japan, 3C 19:296 (1983).
2. Y. Hiruma, E. Watanabe, T. Sato, T. Tanaka, and T. Suzuki, Disinfection of E. coli by electrodialysis using ion-exchange membranes, Absts.49th Annual Meeting of Chem.Soc.Japan, 3R 33:698 (1984).
3. Y. Baba, Comparison of drinking water disinfected by Cl_2 and O_3, Kagaku to Kogyo (Chemistry & Chemical Industry), 31:492 (1978).
4. T. Suzuki and L. T. Fan, A disinfection mechanism of triiodide type anion-exchanger, J.Fermentation Technol., 57:578 (1979).
5. Y. Tanaka and M. Seno, Concentration polarization and dissociation of water in ion exchange membrane electrodialysis. IV. Water dissociation layer, Denki Kagaku (J.Electrochem.Soc.Japan), 50:821 (1982).

THE CHARACTERIZATION OF POLYMER MEMBRANES

AS PRODUCED BY PHASE INVERSION

C. T. Badenhop and A. L. Bourgignon

Seitz-Filter-Werke, Membrane Development
Bosenheimer Str. 225
6550 Bad Kreuznach, Germany

INTRODUCTION

The development of microfilters for special applications requires the use of polymers that have the chemical, thermal and mechanical properties needed to fit these applications. Many of these polymers are not particularly soluble, making it difficult to produce membranes by the common phase inversion methods used in the manufacture of cellulosic microfilters. It is often necessary to form the membranes by precipitating the polymer solution in a bath in much the same way as the Loeb-Sourirajan procedure for the formation of UF and RO membranes[1,2,3]. Since it is desired to produce microporous filters with flow and retention characteristics similar to the existing cellulose ester membranes, it is necessary to adjust the precipitation conditions in order to control the development of the pores.

The methods used to control the pore formation of polymer membranes are illustrated by two specific examples, namely, polysulfone and a polyamide. The mechanisms for the control of the pore development for each of these membranes are considerably different from one another. The effects of the changes in processing conditions as well as the changes in the formulations of the casting solutions can be confirmed by means of scanning electron microscope photography[4,5].

Polysulfone Membrane

This polymer is soluble in a variety of solvents, the most effective of which are the chlorinated aliphatic compounds such as chloroform. To avoid the use of chlorinated chemicals it was chosen to use the strong organic solvents such as DMF, n-MP, DMAc etc, and since these solvents are not volatile, the solvent removal would be through washing rather than evaporation.

Precipitating the simple solvent solution of the polysulfone in water results in an ultrafilter or filter with a skin that is very tight and with a porous under structure. Therefore a third component was added to the solution to act as a pore-former. This solvent must be relatively compatible with the polymer so that precipitation of the film would occur when the solution is submerged in water and at the same time not precipitate the polymer from the casting solution. In addition the pore-former must be totally soluble in water. The pore-formers which were evaluated are the ethylene glycol ether series (EG to PEG 400).

The following figures illustrate the type of membrane formed when the casting solution containing the pore-former is quenched in water. Figure 1 shows a membrane cross-section when a solution is cast in water at 12°C. This causes a strong asymmetry in the membrane. Figure 2 shows the cross-section of a membrane cast in 30 degree water with the result that although the membrane is still asymmetric, as can be seen in Figures 3 and 4 and macropores are visible on the bottom surface, the membrane is beginning to attain the desired microporous structure. Cast at 50°C (Figure 5), the membrane attains the final symmetrical structure (see Figures 6 and 7). Figure 8 shows the relationship between the temperature of the casting bath and the water flow rate through the filter formed at this temperature[6]. The two mixtures differ only in the percentage of pore-former, with the first mixture containing the greater amount.

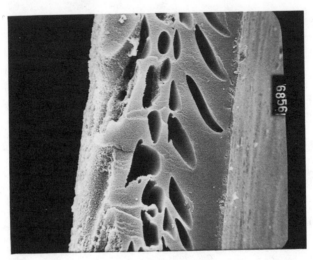

Fig. 1.   Cross-section of a PSU membrane cast in water at 12°C, magnification 479.6 x (12.1% PSU, 25.0% n-Methylpyrrolidone, 62.9% Polyethyleneglycol 400).

Fig. 2.   Cross-section of a PSU membrane cast in water at 30°C, magnification 750 x (composition as in Figure 1).

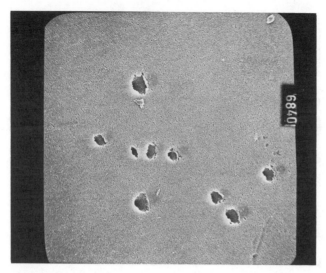

Fig. 3. PSU membrane cast at 30°C, bottom (side in contact with a
suitable carrier as a continuous belt), magnification
120 x (composition as in Figure 1).

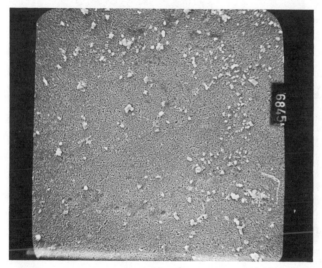

Fig. 4. PSU membrane cast at 30°C, top, magnification 750.3 x
(composition as in Figure 1).

A range of structures were possible to form by adjusting the percent
pore-former and temperature of the gelation bath. The precipitation rate
is almost instantaneous being somewhat slower when the bath is cold. The
precipitation rate was fast enough to eliminate any appreciable transport
of pore-former or solvent from the casting mixture before the membrane had
precipitated.

These membranes were all cast under laboratory controlled conditions.
With membranes cast continuously, as is the case in production, the same
general tendencies are observed. The additional variable of the residence
time of the casting solution in air, however, also influences the results.

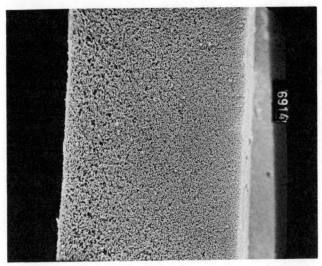

Fig. 5.   Cross-section of a PSU membrane cast in water at 50°C, magnification 470 x (composition as in Figure 1).

Fig. 6.   PSU membrane cast at 50°C, bottom, magnification 750.75 x (composition as in Figure 1).

Polyamide Membranes

The solution of the polyamide is much more difficult to obtain in that the plastic is not soluble in the solvents without the presence of inorganic salts to increase the polarity of the solvent. The pore-former necessary to structure the membrane in a symmetrical form also was chosen from the ethylene glycol series.

The initial attempts to influence the structure of the membrane with change in gelation temperature proved to be unsuccessful. All membranes produced were strongly asymmetric. After changing the gelation bath from pure water to water containing some of the components already in the casting solution two effects occurred. The first is to reduce the rate with which the precipitant, water, diffuses into the casting solution, the second, to retard the transfer of the components from the casting solution into the gelation bath[7].

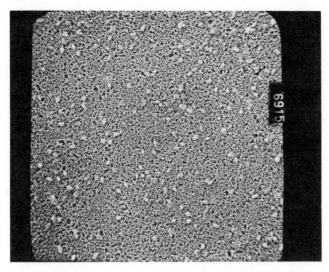

Fig. 7. PSU membrane cast at 50°C, top, magnification 1100 x
(composition as in Figure 1).

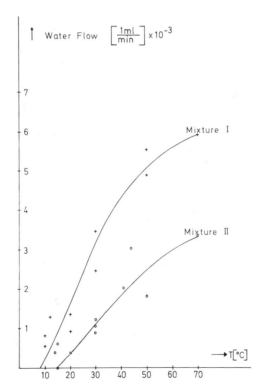

Fig. 8. Diagram of the relationship between casting bath temperature
and flow rate through the filter (filtration area = 12.5 cm^2).
Mixture I = 12.1% PSU, 25.0% n-Methylpyrrolidone, 62.9%
Polyethyleneglycol 400. Mixture II = 15.3% PSU, 24.0%
n-Methylpyrrolidone, 60.7% Polyethyleneglycol 400.

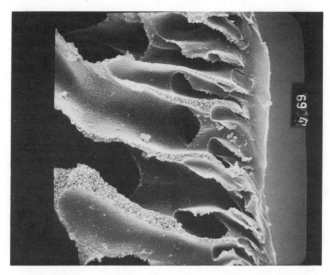

Fig. 9. Cross-section of a PA membrane cast in water at 30°C,
magnification 3296.4 x (11.0% Nomex®, 4.3% LiCl, 35.9%.
Dimethylformamide, 38.6% Dimethylacetamide, 3.0% dist.
water, 7.2% Ethyleneglycol).

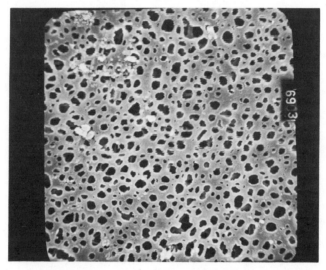

Fig. 10. PA membrane cast at 30°C, bottom, magnification 7504 x
(composition as in Figure 9).

The following figures illustrate the effect upon the structure of the
membranes with the various changes in the gelation bath. Figure 9 shows a
cross-section of the PA membrane gelled in 30 degree water. Figures 10 and
11 confirm that the membrane is strongly asymmetric with the top side of
the membrane showing almost no porosity even at high magnifications.

The next figures show the improvement of the membrane structure
through the alteration of the gel bath. Figure 12 shows the cross-section
of a PA membrane that has been gelled in a bath containing the same ratio
of pore-former to solvent as is present in the casting solution. These
figures (Figure 12-14) should be compared with the following figures

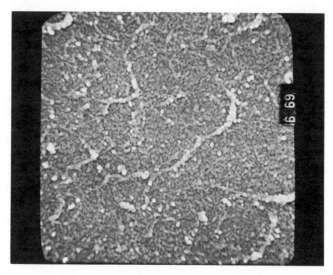

Fig. 11.   PA membrane cast at 30°C, top, magnification 14873 x
(composition as in Figure 9).

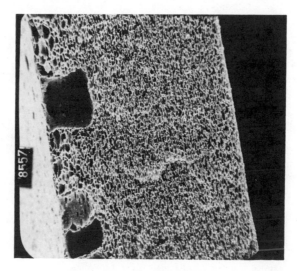

Fig. 12.   Cross-section of a PA membrane cast in a bath with the same
ratio of pore-former to solvent as in the casting solution-
water content 30%, magnification 1100 x.

(Figures 15-17) of a membrane cast in a bath with less precipitant, water,
as in the first. The second membrane appears to be symmetric in construction, however, the top surface of the membrane is not as porous as the
bottom surface as is seen in the Figures 16 and 17.

When the membrane is precipitated in a bath that contains no pore-
former, the pore-former being replaced with solvent and the concentration
of the water in the bath remaining the same, the membrane formed is quite
asymmetric in structure (see Figure 18). The rate of precipitation of this
membrane was significantly slower than the rate of precipitation of the
membrane when the bath contained the pore-former. In fact, the rate of
precipitation could not be used as a measure of the eventual structure of
the membranes.

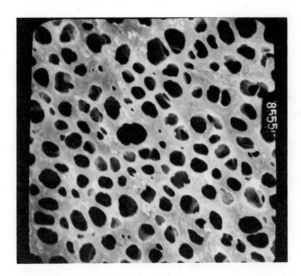

Fig. 13. PA membrane as Figure 12, bottom, magnification 7500 x.

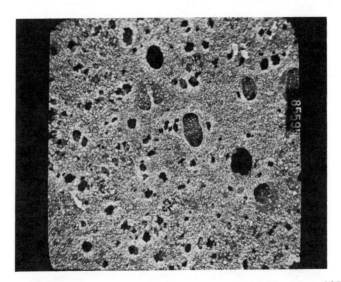

Fig. 14. PA membrane as Figure 12, top, magnification 2050 x.

The PA membranes precipitate very slowly in relation to the polysulfone, with the amount of water being required to precipitate the resin
being significantly greater than with the polysulfone. To prevent a
considerable change in the composition of the casting mixture during
precipitation the bath must contain the solvent and pore-former in sufficient concentration to inhibit diffusion of these components. The final
composition of the bath is adjusted to reduce the formation of skin on the
membrane surface.

It was found that the chemical structure of the pore-former was of
great significance. Figures 19-21 show that the structure of the membranes
varied with the specific pore-former used and that the best precipitant in
this series was PEG 200. This is not to say that with another solvent
system this precipitant would necessarily be ideal.

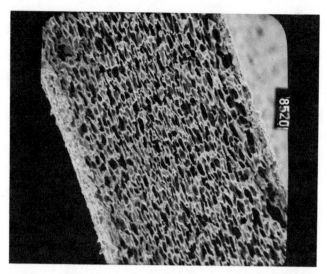

Fig. 15. Cross-section of a PA membrane cast in a bath with the same ratio of pore-former to solvent as in the casting solution—water content 25%, magnification 1728 x.

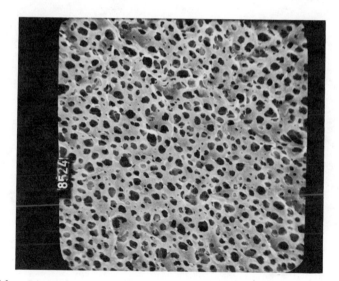

Fig. 16. PA membrane as Figure 15, bottom, magnification 3200 x.

In the case of the polysulfone membrane the temperature of the gelation bath was the overriding controlling factor in establishing the structure, symmetry or asymmetry, of the membrane. The polyamide membrane was more influenced by the composition of the casting solution and the gelation bath than the temperature of the bath. The temperature of the bath did play a role in the refinement of the structure with an ideal temperature being dependent upon the composition of the bath and mix.

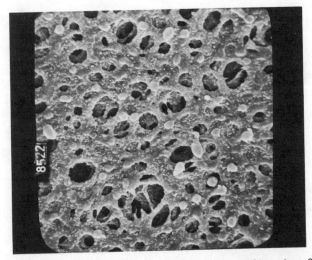

Fig. 17. PA membrane as Figure 15, top, magnification 3120 x.

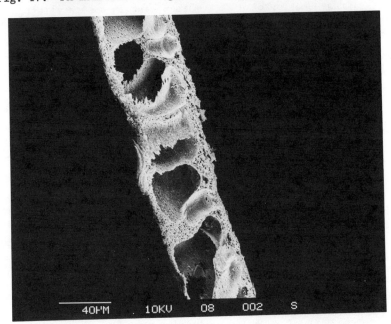

Fig. 18. Cross-section of a PA membrane (composition 56.5% Dimethylacetamide, 10% Nomex®, 3.9% LiCl, 29.6% Polyethyleneglycol 400)
cast in a bath containing only solvent and water, no pore-former
(30% water, 70% Dimethylacetamide – Temp: 24°C).

Pore Distribution of the Membrane

The structure of the membranes produced were illustrated here by
scanning electron microscope photographs. This method of evaluation is
time consuming and cumbersome especially when a large number of castings
are being made in the laboratory. The following test method has been shown
effective in not only characterizing the pore size of the membranes being
produced, but is effective in demonstrating the degree of anisotropisity
present in the structure of the membrane. The method is limited to
membranes with pore sizes equal to or greater than 0.1 um mean flow
pore[8,9,10].

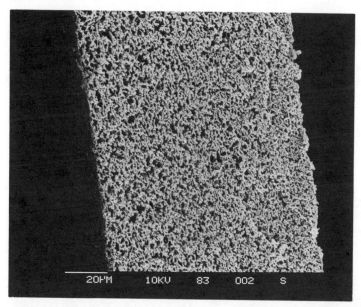

Fig. 19. Cross-section of a PA membrane (composition 72.5% Dimethylformamide, 8% Nomex®, 3.2% LiCl, 14.1% Triethyleneglycol, 2.2% dist. water) cast in a bath containing TEG as poreformer (48% Dimethylformamide, 25% Triethyleneglycol, 27% water — Temp: 24°C).

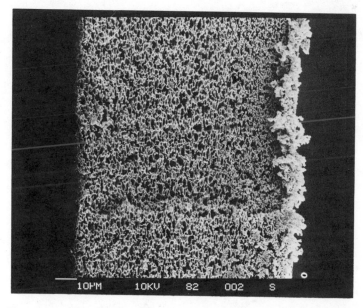

Fig. 20. Cross-section of a PA membrane (65.6% Dimethylformamide, 10% Nomex®, 3.9% LiCl, 17.6% Polyethyleneglycol 200, 2.9% dist. water) cast in a bath containing PEG 200 as poreformer (44% Dimethylformamide, 29% Polyethyleneglycol 200, 27% water — Temp: 24°C).

The principle of the test method is the same as that used in the ASTM test method[11] for the mean-pore-size of membranes; in that a wetting liquid is displaced from the pores of the membrane by air by increasing the transmembrane pressure in steps until the membrane is completely free of entrained liquid. The gas flow through the membrane as a function of pressure can be used to calculate the pore distribution. The relationship between this flow data and the pore distribution is an integral equation. This integral equation describing the flow distribution has been solved numerically through the use of a computer program.

In Figure 22 the first equation is the flow of a gas through a porous structure[12]. At low Reynolds numbers entrance and exit effects can be

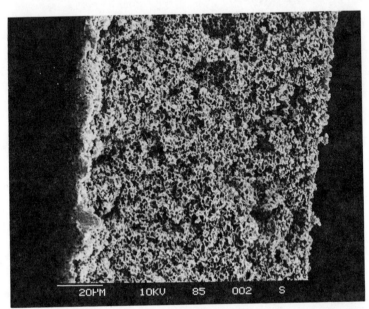

Fig. 21.   Cross-section of a PA membrane (65.6% Dimethylformamide, 10% Nomex®, 3.9% LiCl, 17.6% Polyethyleneglycol 300, 2.9% dist. water) cast in a bath containing PEG 300 as pore-former (44% Dimethylformamide, 29% Polyethyleneglycol 300, 27% water – Temp: 24°C).

$$p \overset{\circ}{V} = \frac{X \, \overline{p} \, \Delta p \, D^2}{32 \, n \, l} \left\{ 1 + \frac{4 \, Kn}{\overline{p} \, D} \right\}$$

$$p_{\circ} V_{wet_i} = U_i \int_{D_i}^{D_{BP}} W(D) \, G(D) \, dD$$

Fig. 22.   The two basic equations of the gas flow through a dry and a wetted membrane. D: Diameter of pore or capillary (cm); Kn: Knudsen number; N: Number of pores; U_i: Integration coefficient (sec^{-1}); V: Velocity of volume flow per unit area $(cm^3 * cm^{-2} * sec^{-1})$; V_{wet}: Wet flow velocity $(cm^3 * cm^{-2} * sec^{-1})$; x: Void volume (porosity) (%); l: Capillary length (cm); n: Viscosity of gases $(g * cm^{-1} * sec^{-1})$; p: Absolute pressure $(dyn * cm^{-2})$; p: Pressure difference $(dyn * cm^{-2})$; p_0: Standard absolute pressure $(dyn * cm^{-2})$.

neglected[13]. The membranes can be characterized as if it was a bundle of capillaries of various diameters.

This equation predicts that the function pV/Δp will be linear when plotted against the average pressure within the membrane.

Figure 23 shows a plot of the pV/Δp function vs. the average pressure for the 0.6 um Nuclepore membrane. The line through the origin is the Hagen-Poiseuille (viscous flow prediction); the displacement of the measured line from the H-P line is the effect of gas-slip through the small membrane pores. It is possible to differentiate between number, length and diameter of the pores even with simple gas flow measurements.

In Figure 22 the second equation is the integral equation relating the flow of a gas through a wetted membrane and the pore distribution of the membrane[14]. This equation is solved by generating the necessary flow

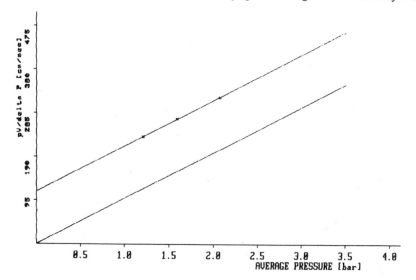

Fig. 23.   Plot of the pV/Δp function vs. the average pressure for a
           0.6 um Nuclepore membrane.

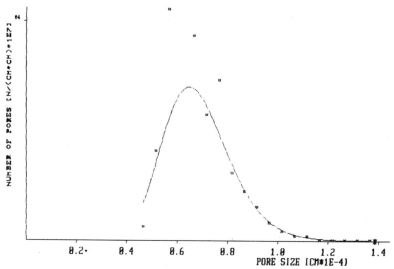

Fig. 24.   Pore distribution of a 0.6 um Nuclepore membrane.

data through a membrane that has been wetted with a suitable wetting fluid such as a low viscosity silicone oil. This data is fed into a computer programmed to solve the equation.

With the aid of computer graphics, the solution to the integral equation can be displayed and printed in graphic form. The next figures show the pore distribution of several membranes to illustrate the effectiveness of the method.

Figure 24 shows the pore distribution of a Nuclepore membrane[15], with a pore size listed as 0.6 um, as measured using this method. There is a very good agreement between the given pore size and the results of the pore distribution measurements.

From the pore distribution, it is possible to calculate the flow distribution both for water and for air. The water flow distribution for this membrane is shown in Figure 25.

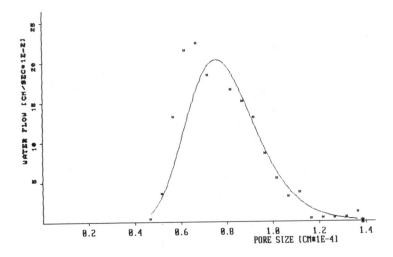

Fig. 25. Water flow distribution of a 0.6 um Nuclepore membrane.

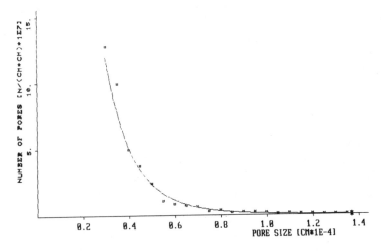

Fig. 26. Pore distribution of a SEITZ 0.45 um non-woven reinforced membrane, bottom or reinforcement side.

The non-woven reinforced membranes produced by Seitz show a degree of anisotropisity illustrated by Figures 26-29. The numerical pore distribution of this membrane is typical for mixed ester of cellulose. The maximum is seldom reached due to the either excess pressure required for the measurement, in excess of 8 bar, or air-flow rates well above 150 liter per minute, exceeding the limits of the test equipment or the filter holder. The flow distributions clearly indicate the asymmetric nature of the membranes.

Figures 30-33 show the asymmetric nature of experimental polysulfone membranes. Here the degree of asymmetry is large and can be seen in the numerical distribution as well as in the flow distribution.

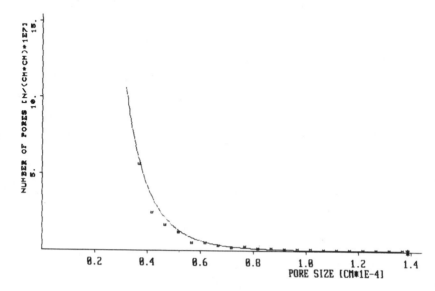

Fig. 27. Pore distribution of a SEITZ 0.45 um non-woven reinforced membrane, top or membrane side.

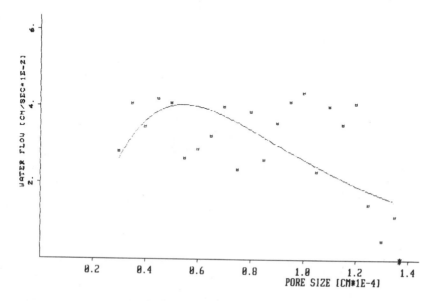

Fig. 28. Water flow distribution of a SEITZ 0.45 um non-woven reinforced membrane, bottom.

443

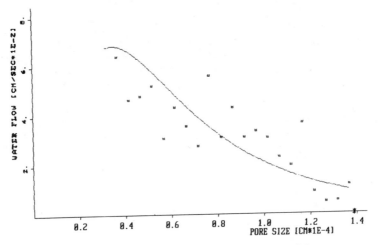

Fig. 29. Water flow distribution of a SEITZ 0.45 um non-woven reinforced membrane, top.

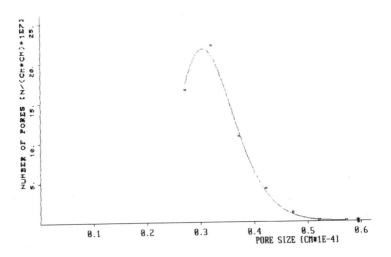

Fig. 30. Pore distribution of a SEITZ 0.45 um PSU membrane, bottom.

CONCLUSIONS

It has been demonstrated that the pore size and shape of a membrane precipitated in a casting bath can be controlled. The asymmetry of the membrane is a function of several parameters. Of major importance is the polymer type being cast. In the case of the polysulfone the temperature of the casting bath controls the degree of asymmetry. With the polyamide the composition of the precipitating bath controls this factor.

These results are only empirical. It appears that there are more than one mechanism affecting the pore structure of the membranes. The composition of the membrane casting solution determines the pore size of the membrane where the rate of precipitation, temperature of the precipitation bath, and composition of the bath effects the asymmetry more than the pore size of the membranes.

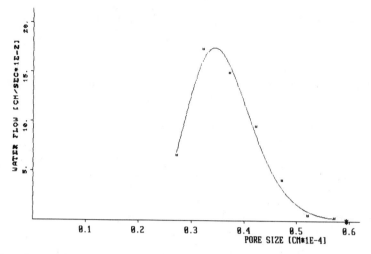

Fig. 31. Water flow distribution of a SEITZ 0.45 um PSU membrane, bottom.

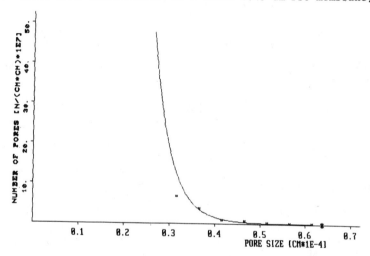

Fig. 32. Pore distribution of a SEITZ 0.45 um PSU membrane, top.

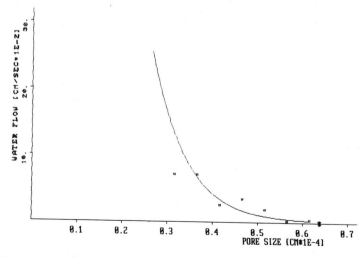

Fig. 33. Water flow distribution of a SEITZ 0.45 um PSU membrane, top.

In addition a test method was described for measuring the pore distribution and the asymmetry of the new polymer membranes. This test method was validated by use of Nuclepore track edged polycarbonate membranes, which have capillaries with defined pore size.

These membranes yield narrow pore distribution curves corresponding to the actual nature of the membrane.

REFERENCES

1. S. Loeb and S. Sourirajan, Advan.Chem.Ser., 38:117 (1962).
2. S. Loeb and S. Sourirajan, Advan.Chem.Ser., 33:117 (1963).
3. S. Loeb and S. Sourirajan, U.S. Patent 3,775,308.
4. G. Moll, Elektronenmikroskopische Demonstration des Filtrations-Mechanismus von Membranfiltern, Kolloid-Zeitschrift, 203 1:20-27 (1965).
5. H. J. Preußer, Elektronenmikroskopische Untersuchungen an Oberflächen von Membranfiltern, Kolloid-Zeitschrift, 218:129-136 (1967).
6. Seitz-Filter-Werke, Bad Kreuznach, Fed. Rep. of Germany, German Patent Application P 33 42 824.7, 26.11.1983.
7. Seitz-Filter-Werke, Bad Kreuznach, Fed. Rep. of Germany, German Patent Application P 33 42 823.9, 26.11.1983.
8. C. T. Badenhop, The determination of the pore distribution and the methods leading to the prediction of retention characteristics of membrane filters, Dissertation, University of Dortmund, Department of Chemical Engineering, Fed. Rep. of Germany.
9. E. Honold and W. Bachmann, Application of mercury-intrusion method for determination of pore-size distribution to membrane filters, Science, 120 11:805-806 (1954).
10. K. Erbe, Die Bestimmung der Porenverteilung nach ihrer Größe in Filtern und Ultrafiltern, Kolloid-Zeitschrift, 63 3:277-285 (1933).
11. ASTM Standard F 316-70, Pore size characteristics of membrane filters for use with aerospace fluids.
12. P. C. Carman, Flow of Gases Through Porous Media, Butterworths Scientific Publications, London, pp.62-78 (1956).
13. J. Machacova, J. Hrbek, V. Hampl, and K. Spurny, Analytical methods for determination of aerosols by means of membrane ultrafilters. XV. Hydrodynamical properties of membrane and nuclear pore filters, Collection Czechoslov.Chem.Commun., 35:2087-2099 (1970).
14. M. M. Agrest, M. L. Aleksandrov, and L. S. Reifman, Structural characteristics of porous membranes, Russian J.Phys.Chem., 50 9:541-545 (1976).
15. Nuclepore Corp. 7035 Commerce Circle, Pleasanton, CA 94566, U.S.A., Literature, private communication.

TRANSPORT EQUATIONS AND COEFFICIENTS OF REVERSE

OSMOSIS AND ULTRAFILTRATION MEMBRANES

S. Kimura

Institute of Industrial Science, University of Tokyo
7-22-1 Roppongi, Minatoku
Tokyo 106, Japan

INTRODUCTION

Qualification of membrane natures based on the appropriate transport equations and coefficients are necessary for understanding transport mechanisms and developing better membranes. The following is what we have done along this line for the unified treatment of both reverse osmosis and ultrafiltration membranes.

TRANSPORT EQUATIONS

Concentration Polarization

The rejection characteristics of a membrane are usually described by the observed rejection R_{obs}, defined as

$$R_{obs} = (C_b - C_p)/C_b \qquad (1)$$

where C_b and C_p denote bulk and permeate concentration, respectively. But a concentration at the membrane surface, C_m, is always larger than C_b because of the concentration polarization phenomenon. The real rejection R is defined as

$$R = (C_m - C_p)/C_m \qquad (2)$$

C_m is given by the following equation as

$$(C_m - C_p)/(C_b - C_p) = \exp(J_v/K). \qquad (3)$$

where K is a mass transfer coefficient in a boundary layer next to the membrane.

Transport Equations Inside of Membrane

The following equations are used throughout our work to obtain a reflection coefficient σ and a solute permeability P.

$$R = \sigma (1 - F) / (1 - \sigma F) \qquad (4)$$

$$F = \exp[\ -(1 - \sigma)\ J_v/P)] \qquad (5)$$

which were derived by Spiegler and Kedem[1]. As for the volume flux J_v, the original equation developed by Kedem and Katchalsky[2], is used, which is given as

$$J_v = L_p(\Delta P - \sigma \cdot \Delta \pi) \qquad (6)$$

where Lp is a hydraulic permeability. Among the three transport coefficients used above, Lp was obtained from pure water permeation experiments using Equation (5), and σ and P were obtained by the curve-fitting method by plotting R vs. $1/J_v$. An example of such plots[3] is shown in Figure 1, where it is seen that R approaches σ, when $1/J_v \to 0$ as suggested by Equations (4) and (5). The extrapolation is not done by a straight line, except for large and small R values, and this brings some errors to estimate values.

TRANSPORT MECHANISM OF ULTRAFILTRATION MEMBRANES

Pore Flow Theory

Transport mechanism of an ultrafiltration membrane is often assumed as a solute particle flow in a cylindrical pore in a membrane, which is called "pore flow theory". According to the most recent theory[4], L_p, σ and P are given as follows

$$\sigma = 1 - g(q) \cdot S_F \qquad (7)$$

$$P = D\ f(q) \cdot S_D \cdot (A_k/\Delta x) \qquad (8)$$

$$L_p = (r_p^2/8\mu) \cdot (A_k/\Delta x) \qquad (9)$$

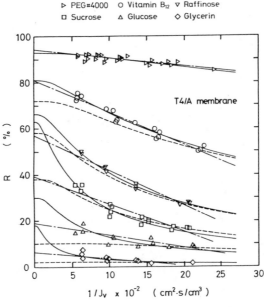

Fig. 1. Plot of R vs. $1/J_v$ for ultrafiltration membrane. ——: determined by the method of curve fitting; ---: determined by using the logarithmic mean concentration; —·—: extrapolation of $1/J_v \to 0$ for the determination of σ.

where

$$f(q)=(1-2.105q+2.0865q^3-1.7068q^5+0.72603q^6)~/(1-0.75857q^5) \qquad (10)$$

$$g(q)=(1-(2/3)q^2-0.20217q^5)/(1-0.75857q^5) \qquad (11)$$

$$S_D=(1-q)^2 \qquad (12)$$

$$S_F=2(1-q)^2-(1-q)^4 \qquad (13)$$

and $A_k/\Delta x$ is a fractional pore area divided by a pore length. By the careful comparison between the pore flow theory and the frictional interpretation of transport parameters, we came to the new relation of σ, which is given as[5].

$$\sigma=1-S_F[g(q)+(16/9)q^2f(q)] \qquad (14)$$

Experimental Proof of the Theory

From the experimentally determined values of σ using various molecular weight solutes, a pore radius of membrane can be estimated by using Equation (7) or Equation (14). Using this pore radius $A_k/\Delta x$ is calculated by using both Equations (8) and (9). When these two $A_k/\Delta x$ values coincide, the theory and the pore radius can be considered correct. Our experimental results of cellulose acetate membranes showed[5], that this coincidence was not sufficient and was improved by eliminating wall correction factors f and g. This led us to propose our "steric hindrance-pore model", which is given as

$$\sigma=1~-S_F[1+(16/9)q^2] \qquad (15)$$

$$P=D\cdot S_D\cdot(A_k/\Delta x) \qquad (16)$$

An example of our result is shown in Figure 2. Research works extending this analysis to various other membranes made of different materials are now in progress.

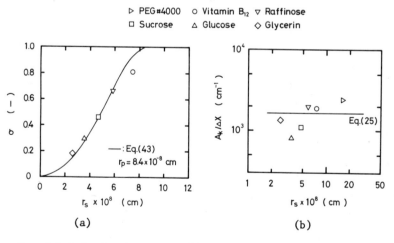

▷ PEG#4000   ○ Vitamin B₁₂   ▽ Raffinose
□ Sucrose   △ Glucose   ◇ Glycerin

(a)

(b)

Fig. 2.   (a) Relation between reflection coefficient and Stokes radius of solute. (b) Values of $A_k/\Delta x$ calculated by Equation (8).

449

Transport Parameter Estimation

Estimation of transport parameters can be done by the same procedures explained above. Rejections of inorganic solutes by cellulose acetate membranes are shown in Figure 3, where it is seen that the general trend is quite different from Figure 1, and σ is very close to 1, even if the solute rejection is low. The reason for this will be explained later, but it can be concluded that solute distribution in the membrane is very small, while the solute permeability is large.

Estimation of transport parameters for solutes, that show negative rejection, can also be done by the same procedure[6]. An example is shown in Figure 4.

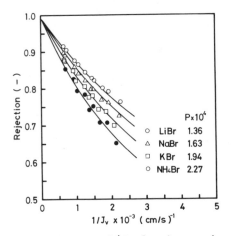

Fig. 3. Plot of R vs. $1/J_V$ for inorganic solutes.

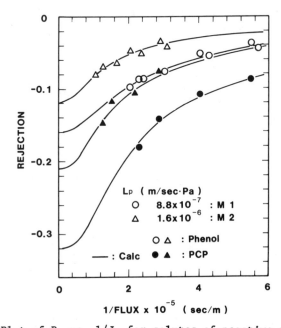

Fig. 4. Plot of R. vs. $1/J_V$ for solutes of negative rejection.

Nature of Dense Membranes

From transport parameters determined above, we want to proceed further to clarify the transport mechanism. But the active layer of a asymmetric membrane is so thin that it is difficult to measure various natures of membrane, such as, distribution coefficient and water content. So we cast symmetric dense membranes, whose thickness is about 10μm[6]. Transport parameters of these membranes were determined by the reverse osmosis experiment. Flux of dense membranes was so small and it was difficult to obtain data in a wide range. An example is shown in Figure 5.

After the reverse osmosis experiment the same membrane was used to determine the diffusion coefficient, the distribution coefficient of the same solute and finally the water content. Solute used were the same as those used in the asymmetric membrane experiment, so that comparison of both asymmetric and symmetric data could become possible.

In Figure 6 distribution coefficient K is plotted against water contents of membrane, ϕ_w. The lower solid line corresponds to the case of no preferential sorption of solute to membrane, and k is proportional to ϕ_w. The upper line corresponds to the case of no solute rejection. Solutes under this line show the preferential sorption to the membrane, but their rejection is normal. Solutes under the lower line are not preferentially sorbed and are mostly inorganic salts.

In Figure 7 diffusion coefficients D_{AM} are plotted against ϕ_w. First it is seen that D_{AM} greatly changes by a small change of ϕ_w, that may show that a decrease of ϕ_w brings a decrease of pore radius and this results in a great decrease of D_{AM}. Also it is seen that solutes which have a large K value, have a small D_{AM} value, which may show that preferential sorption force between solutes and membrane is strongly acting to reduce D_{AM}.

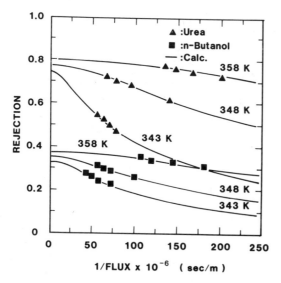

Fig. 5. Plot of R vs. $1/J_v$ for dense membrane.

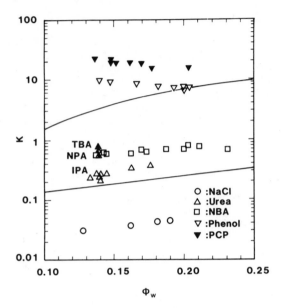

Fig. 6. Relation between K and ϕ_w.

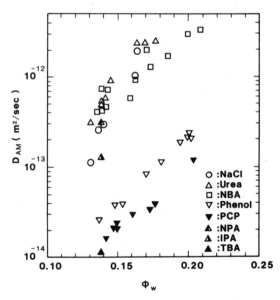

Fig. 7. Relation between D_{AM} and ϕ_w.

Correlation Between Asymmetric and Dense Membrane

Solute and water permeabilities are inversely proportional to the membrane thickness δ, while σ is independent of δ. When we select data of both membranes which have the same σ values, we can estimate the membrane thickness of asymmetric membrane δ_A, by comparing both L_P values. If δ_A is correct also P values should coincide by this correction. This is shown in Figure 8, and δ_A is estimated to be 0.1 μm, which can be considered reasonable comparing reported values[7,8].

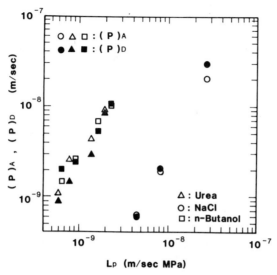

Fig. 8. Correlation between symmetric and asymmetric membranes.

Table 1. Values of Friction Coefficients

No.	Solute	$L_p \times 10^9$ [m/(s·MPa)]	$P \times 10^9$ [m/s]	σ [-]	$f_{wm} \times 10^{-14}$ [$\frac{N \cdot s}{mol \cdot m}$]	$f_{sw} \times 10^{-14}$ [$\frac{N \cdot s}{mol \cdot m}$]	$f_{sm} \times 10^{-14}$ [$\frac{N \cdot s}{mol \cdot m}$]	$\frac{K}{\phi_v}$	$\dfrac{f_{sw}+f_{wm}\cdot(\frac{\bar{V}_s}{\bar{V}_w})}{f_{sw}+f_{sm}}$
1	NaCl	2.7	30	0.88	0.78	0.51	2	0.30	4.0×10^{-1}
2		0.82	2.1	0.97	2.1	1.3	25	0.23	1.3×10^{-1}
3		0.45	0.6	0.99	4.3	1.5	100	0.17	5.9×10^{-1}
4	Urea	2.3	18	0.74	0.39	0.83	11	1.7	1.5×10^{-1}
5		1.7	5.3	0.79	0.51	3.4	29	1.4	1.5×10^{-1}
6		0.93	3.2	0.80	0.87	4.8	47	1.5	1.3×10^{-1}
7		0.70	3.0	0.80	1.2	4.7	52	1.3	1.3×10^{-1}
8	NBA	2.2	16	0.33	0.27	2.0	13	3.2	2.1×10^{-1}
9		1.4	7.2	0.35	0.40	5.0	30	3.3	2.0×10^{-1}
10		0.84	4.1	0.37	0.65	8.7	50	3.2	2.0×10^{-1}
11		0.60	3.3	0.40	0.98	10	59	2.8	2.1×10^{-1}
12	Phenol	2.3	(22)	(-0.03)	0.24	2.3	110	33	3.1×10^{-2}
13		1.7	(8.3)	(-0.08)	0.28	7.3	290	37	2.9×10^{-2}
14		0.93	(5.1)	(-0.12)	0.50	12	620	51	2.2×10^{-2}
15		0.70	(4.5)	(-0.16)	0.79	16	900	55	2.1×10^{-2}
16	PCP	2.3	(24)	(-0.12)	0.24	2.3	200	69	1.6×10^{-2}
17		1.7	(8.4)	(-0.21)	0.29	9.6	720	81	1.5×10^{-2}
18		0.70	(4.8)	(-0.32)	0.66	14	1200	99	1.3×10^{-2}

( ): calculated using asymmetric membrane's data

Frictional Interpretation

According to the frictional interpretation[1], a reflection coefficient σ, solute permeability P, and water permeability L_p, can be expressed as

$$\sigma = 1 - (K/\phi_v) \cdot [(f_{sw}+f_{wm}(V_s/V_w))/(f_{sm}+f_{sw})] \qquad (17)$$

$$P = KRT/[\delta(f_{sw}+f_{sm})] \qquad (18)$$

$$L_p = \phi_v \cdot V_w / f_{wm} \cdot \delta \qquad (19)$$

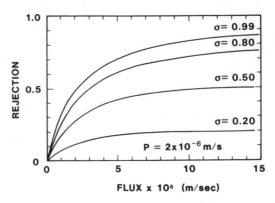

Fig. 9. Effect of reflection coefficient on R.

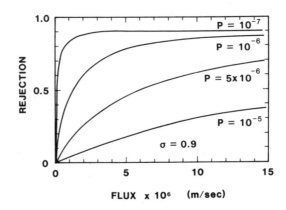

Fig. 10. Effect of solute permeability on R.

Using the above equations it is possible to calculate f_{wm}, f_{sw} and f_{sm}. A reflection coefficient consists of a so-called "exclusion term", which include a distribution coefficient, and a so-called "kinetic term", which includes frictional coefficients. These two terms are compared in Table 1, where it is seen that "exclusion term" has a decisive importance in determining σ, which is a maximum value of a solute rejection. Actual solute rejections are lower than σ and dependent also on P, which is dependent mainly on f_{sm}, which is dependent both on the pore size and the solute-membrane interaction. L_p is dependent on f_{wm}, which is mainly dependent on the pore size of membrane. Effect of σ and P and R is shown in Figures 9 and 10, where behaviors of membranes having different σ and P are shown.

REFERENCES

1. K. S. Spiegler and O. Kedem, Desalination, 1:311 (1966).
2. O. Kedem and A. Katchalsky, J.Gen.Physiol., 45:143 (1967).
3. S. Nakao and S. Kimura, J.Chem.Eng.of Japan, 14:32 (1981).
4. A. Verniory, R. Du Bois, P. Decoodt, J. P. Gassee, and P. P. Lambert, J.Gen.Physiol., 62:483 (1973).
5. S. Nakao and S. Kimura, J.Chem.Eng.of Japan, 15:200 (1982).
6. M. Okazaki and S. Kimura, J.Chem.Eng.of Japan, 17:192 (1984).
7. S. Kimura, Proc. 4th Int. Symp. on Fresh Water From the Sea, 4:197 (1973).
8. E. Glueckauf, Proc. 1st Int. Symp. on Water Desalination, 1:143 (1967).

TRANSFER OF SOLUTES THROUGH COMPOSITE

MEMBRANES CONTAINING PHOSPHOLIPIDS

E. Sada, S. Katoh and M. Terashima

Kyoto University
Kyoto
Japan

SUMMARY

Artificial membranes containing a phospholipid were prepared, and their permeabilities to uncharged and charged solutes were measured around the phase transition temperature of the phospholipid. An abrupt change of permeability was observed at the phase transition temperature. The permeability of the membranes to hydrophobic and uncharged solutes was higher by two orders of magnitude than that to charged solutes.

INTRODUCTION

Biomembranes, consisting of a liquid bilayer backbone into which are incorporated proteins and polysaccharides, act as semipermeable and selective barriers for a variety of solutes, including ions, nutrients and drugs. As experimental models of biomembranes, bilayer lipid membranes and liposomes have been studied with respect to their permeability to these solutes. Recently, stable artificial membranes containing phospholipid or artificial amphiphiles have been prepared, and their similarities to biomembranes in permeability characteristics have been studied[1,2].

Previously we reported that composite membranes containing phospholipids showed similarities to biomembranes in their permeability to solutes of different polarities and in the response of permeability to temperature[3]. In the present work the permeability of several solutes through composite membranes containing phospholipid was measured, and the effect of electric charge on the permeability was studied.

EXPERIMENTAL

A Millipore membrane (VSWP 04700, thickness 117 μm; average pore size 0.025 μm; and void volume 70 %) was soaked for 15 min in a chloroform solution of a known concentration (40 - 100 mg/cm^3) of D,L-α-dipalmitoyl phosphatidylcholine (abbreviated as DPPC), dried at room temperature and stored in water. The amount of DPPC present as bilayer structure in the membranes was estimated from the enthalpy change obtained by differential thermal analysis.

Solutes used were salicylic acid, salicylamide n–butanol, n–propanol, D,L–lysine and L–monosodium glutamate, all of reagent grade.

A permeability test cell similar to that previously reported[3] had two chambers of 12.7 cm³ and a membrane area of 4.85 cm². A phosphate buffer was continuously supplied to one chamber of the cell at a constant flow rate and the buffer containing a solute was circulated through the other chamber by microtube pumps. The concentration of the solute in the effluent solution from the former chamber was measured with a spectrophotometer (Shimadzu UV–100–02, for salicylic acid 298 nm; salicylamide 205 nm), a gas chromatography (Shimadzu GC–4A, for n–butanol) or a fluorometer (Gilson Model 3301, for amino acids).

The specific permeability KD_m is defined as:

$$KD_m = P_m l = C_i q l / A C_o \tag{1}$$

where P_m is the solute permeability, l the membrane thickness, q the flow rate, A the membrane area, C_i the solute concentration in the effluent, and C_o the solute concentration in the circulating solution.

RESULTS AND DISCUSSION

Differential Thermal Analysis

Figure 1 shows the DTA curves for DPPC and the Millipore/DPPC membrane equilibrated with water. The membrane containing DPPC showed the endothermic peak at the same temperature as pure DPPC, which indicates that phospholipid in the membrane underwent a phase transition from the gel state to the liquid–crystalline state.

The amount of the phospholipid present in the membrane, which was estimated from enthalpy change obtained by the DTA measurement, was proportional to the DPPC concentration of the solution used in preparation of the membrane.

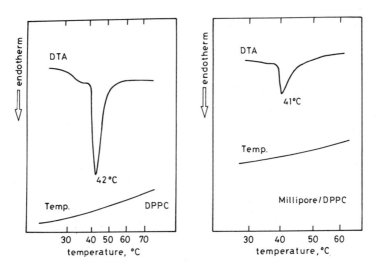

Fig. 1. DTA curves for DPPC and Millipore/DPPC membrane.

Solute Permeability

In Figure 2, the values of the permeability of salicylic acid through the Millipore/DPPC membranes prepared with the DPPC solutions ranging from 40 to 100 mg of DPPC/cm³ are plotted against 1/T. The permeability increased with the temperature and showed an abrupt change around the phase transition temperature of the phospholipid. This change is caused by the fluidity change of the hydrophobic layer of the phospholipid lamellae above and below the phase transition. The proportionality of the amount of the phospholipid present in the membranes and the inverse proportionality of the permeability to the DPPC concentration of the solution used for preparation indicate that the phospholipid layer limits the permeation rate of the solute.

Figure 3 shows the flux of salicylic acid through the Millipore/DPPC membrane (100 mg of DPPC/cm³, 25°C) plotted against the degree of dis-

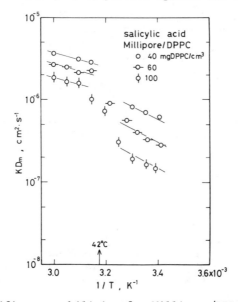

Fig. 2.   Specific permeabilities for Millipore/DPPC membranes.

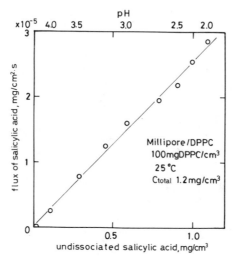

Fig. 3.   Effect of degree of dissociation of salicylic acid on permeability.

sociation calculated from the dissociation constant and the pH value of the salicylic acid solution. The flux is nearly proportional to the concentration of undissociated salicylic acid. This is because the permeability of dissociated (charged) salicylate ion through the phospholipid layer is much smaller than that of the undissociated acid.

Figure 4 shows the permeability of alanine (pI 6.0) through the Millipore/DPPC membrane (100 mg of DPPC/cm^3, 50°C) plotted against pH. Because of the amphoteric nature of amino acids, the permeability of alanine, even at its isoelectric point, where the charges in the molecules were balanced, was lower than that of undissociated salicylic acid. Moreover, it decreased at lower and higher pH values, where alanine had a net positive or negative charge.

In this manner the electric charge of the solute affected its permeability through the Millipore/DPPC membranes. Permeabilities of several solutes are summarized in Figure 5: amino acids showed lower permeabilities by two orders of magnitude than uncharged solutes. In these hydrophilic solutes the permeability changes around the phase transition temperatures were smaller than hydrophobic solutes. The hydrophilic solutes might permeate through the hydrophilic region of the phospholipid layers consisting of the polar head group, where the effect of the fluidity change is small[4]. Like biomembranes, the Millipore/DPPC membrane acts as a semipermeable membrane for hydrophobic and uncharged solutes and as a barrier for charged solutes. Therefore a carrier system is necessary to transport charged solutes, such as ions and amino acids, through the phospholipid-containing membranes. A selective transport system can be constructed by use of this type of membrane containing a carrier, which is reported elsewhere[5].

CONCLUSIONS

Millipore/DPPC membranes are similar to biomembranes in the response of permeability to temperature and in the effect of the electric charge on the permeability of solutes. These membranes are very stable and can be used to construct selective transport systems similar to those of biomembranes.

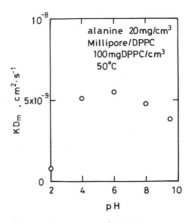

Fig. 4. Specific permeabilities of alanine.

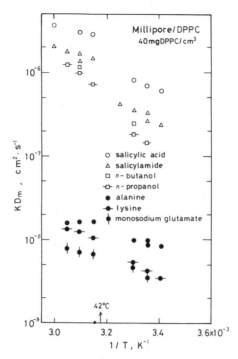

Fig. 5. Specific permeabilities of several solutes.

REFERENCES

1. S. Oda, T. Kajiyama, and M. Takayanagi, Permselectivities of metal ions by polymer/liquid crystal/crown ether composite membrane, Rept.Progr.Polym.Phys.Japan, 24:249 (1981).
2. A. Akimoto, K. Dorn, L. Gros, H. Ringsdorf, and H. Schupp, Polymer model membrane, Angew.Chem., 93:108 (1981).
3. E. Sada, S. Katoh, and M. Terashima, Permeabilities of composite membranes containing phospholipids, Biotech.& Bioeng., 25:317 (1983).
4. E. Sada, S. Katoh, M. Terashima, S. Yamana, and N. Ueyama, Effects of solute hydrophobicity and phospholipid composition on permeability of composite membranes containing phospholipid, Biotech.& Bioeng., under submission.
5. E. Sada, S. Katoh, M. Terashima, and Y. Takada, Carrier-mediated transport across phospholipid-composite membranes containing valinomycin, A.I.Ch.E.J., in press.

NEW INORGANIC AND INORGANIC-ORGANIC ION-EXCHANGE MEMBRANES

G. Alberti, M. Casciola, U. Costantino, and D. Fabiani

Dipartimento di Chimica, Università di Perugia
Via Elce di Sotto, 10
06100 Perugia, Italy

SUMMARY

The great technological success of membranes has in turn stimulated additional interest in new possible applications in processes in which the operating conditions are too drastic for the organic membranes presently available.

Attention has been given to inorganic membranes because an inorganic material is usually more stable than an organic one in drastic conditions (temperature, oxidizing agents, ionizing radiations and so on).

In this paper the inorganic and inorganic-organic ion-exchange membranes have been subdivided in various subclasses and, for each subclass, a brief panoramic review and an account of the results obtained in our laboratory is reported. Among the inorganic ion-exchange membranes, particular attention has been given to the preparation and properties of membranes consisting of fibrous cerium phosphate and of pellicular α-zirconium phosphate. Furthermore the electrochemical properties of some heterogeneous membranes consisting of microcrystals of inorganic ion-exchangers (e.g. α-$[Zr(PO_4)_2]H_2$, $[Zr_2PO_4(SiO_4)_2]Na_3$ and β-alumina) and a suitable organic binder (e.g. polyvinylidene fluoride) are examined and discussed.

The effect of a small concentration of ionogenic groups in the binder (e.g. 5% of -SO_3H groups introduced by sulphonation) is also reported.

INTRODUCTION

A very large number of membranes, in many cases exhibiting high specific selectivities for various solutes, are available today. These membranes have found increasing use in several technological processes and the great success obtained has, in turn, also stimulated additional interest in possible applications in processes in which the operating conditions (temperature, oxidizing agents, ionizing radiations and so on) are too drastic for the organic membranes, available at present. There is no doubt that the technological applications of the membranes could be greatly enlarged if the instability of the organic matrix could be overcome or reduced.

In addition than to fluoropolymeric membranes (e.g. NAFION) attention has been given to inorganic materials because an inorganic structure is usually more stable than an organic one in drastic working conditions[1]. Furthermore, in recent years, there has been an increasing interest in the use of solid inorganic electrolytes as separators in high temperature fuel cells, in the production of hydrogen by the electrolysis of water at high temperature, in solid state rechargeable batteries (e.g. sodium/sulphur at 250°-350°C) or in various electrochemical devices such as sensors for gaseous species, oxygen and hydrogen pumps, ion-selective electrodes in the temperature range 100°-400°C and so on[2-4]. These exciting new applications have also contributed greatly to an increasing interest in charged inorganic membranes.

Unfortunately, only a limited success has been obtained up to now. In fact, if it is true than an inorganic structure is more stable to temperature, to oxidizing solutions, to radiations and to liquid alkali metals, it is also true that thin, large and flexible polymeric membranes are more difficult to obtain with an inorganic than with an organic structure. Furthermore, the presence of water is very harmful (hydrolytic reactions, solubility etc.) for many inorganic compounds carrying fixed charges.

The present report is essentially an account of the results obtained in our laboratory and does not constitute an attempt at a comprehensive review of all the known inorganic membranes; therefore no overall coverage of the present literature in the field will be attempted. However in order to put our results in context, a brief panoramic review on inorganic ion-exchange membranes, will be given.

For such a purpose, it is useful to subdivide the inorganic membranes into various subclasses and then to examine these subclasses separately. Table 1 shows an attempt at such a classification.

Table 1. Classification of Inorganic Ion-exchange Membranes

Membranes with crystalline network	- Single crystal membranes
	- Membranes obtained by sintering of microcrystals
	- Membranes obtained by fibrous microcrystals
	- Membranes obtained by lamellar microcrystals
Membranes with amorphous network	- Membranes obtained by ion-conductive glasses
	- Membranes made of amorphous inorganic polymers
Heterogeneous membranes:	- obtained by crystalline or amorphous powders of inorganic ion-exchangers and suitable inorganic binders. (e.g. silicates, glasses).

Table 2 shows a similar classification for inorganic-organic ion-exchange membranes.

INORGANIC ION-EXCHANGE MEMBRANES

Single Crystal Membranes

In recent years some crystalline compounds with very high ionic conductivity at room temperature (e.g. $RbAg_4I_5$)[5] at medium temperatures (e.g. β'' alumina[6] and NASICON)[7] or at high temperatures (e.g. doped zirconia)[8] have been found.

In the cases where it is possible to obtain large single crystals, thin sheets may be obtained by a proper cutting of the crystals and such sheets could be directly employed as membranes in miniaturized electro-chemical devices where low current intensities are required.

Although a careful examination of the literature has not been made, it is the authors opinion that such a possibility has, until now, scarcely been considered. Nevertheless, membranes made of thin crystals could find interesting practical applications in microbatteries or as sensors for some ionic or gaseous species, especially if selective ionic transport occurs through the crystal. Owing to the high thermal stability of inorganic crystals, the sensors obtained by cutting single crystals could find useful employment even at temperatures where sensors made with liquid or organic membranes cannot be used.

Table 2. Classification of Inorganic-organic Ion-exchange Membranes

Homogeneous membranes (Inorganic-organic ion-exchangers)	- Membranes obtained by single crystals of an inorganic-organic ion-exchanger. - Membranes obtained by fibrous microcrystals of an inorganic-organic ion-exchanger. - Membranes obtained by lamellar microcrystals of an inorganic-organic ion-exchanger. - Membranes obtained by amorphous polymers of an inorganic-organic ion-exchanger
Heterogeneous membranes (Inorganic powders + organic binders)	$\alpha-[Zr(PO_4)_2]H_2 \cdot H_2O$ microcrystals + Kynar (polyvinylidene fluoride). $\alpha-[Zr(PO_4)_2]H_2 \cdot H_2O$ microcrystals + Kynar at low sulphonation degree ($\sim 5\%$). $\alpha-[Zr(PO_4)_2]HNa \cdot 5H_2O$ microcrystals + Kynar at low sulphonation degree ($\sim 5\%$). β''-alumina in Na form microcrystals + Kynar at low sulphonation degree ($\sim 5\%$). NASICON microcrystals + Kynar at low sulphonation degree ($\sim 5\%$).

Because of excellent ionic sieve properties of $LiZr_2(PO_4)_3$ (only H^+, Li^+, Na^+ and Ag^+ have been found to diffuse easily within this skeleton structure[9]) we are presently studying the possibilities of using single crystals of this ion-exchanger as a selective sensor for such cations.

Finally, single crystal membranes could also be very useful in theoretical studies, the study of the transport mechanism being facilitated by an exact knowledge of the crystalline structure.

Membranes Obtained by Sintering of Microcrystals

Large membranes must obviously be employed in processes where high current intensity is required. Since the preparation of large membranes using single crystals is unimaginable, attempts have been made, where possible, to obtain large sheets by sintering microcrystalline powders. Sintered or ceramic membranes of zirconia, β" - alumina and NASICON have already been obtained[10]. Among the various applications of such ceramic membranes one may recall the use of zirconia sheets in high temperature fuel cells[11] and that of β" - alumina in Na/S rechargeable batteries[12], in the conversion of heat into electricity[13], in the electrolysis of water at high temperature[14] and in oxygen pumps[15].

On the other hand, no ceramic membranes with good proton conduction are currently available. In fact, condensation of acid groups usually occurs at temperature lower than that necessary for sintering while the replacement of Na^+ with protons in already sintered membranes usually leads to the breaking of the membrane itself. Simple contact with water may also lead to the breaking of the membrane especially in the cases where a strong volume change of the crystals occurs for the hydration of the cationic sites (e.g. in β" - alumina).

Inorganic Ion-Exchange Membranes Obtained By Fibrous Crystals

Inorganic crystals having a fibrous structure usually tend to mat and porous inorganic sheets similar to cellulose paper can easily be obtained. For example, two inorganic ion exchangers having a fibrous structure, the cerium and the thorium acid phosphates, were synthesized in our laboratory some years ago[16,17]. From these fibrous exchangers inorganic papers useful for chromatographic separation of some cations have been obtained[18].

If these papers are pressed, or if suitable fillers are added, the porosity is reduced and they can be used as ion-exchange membranes. Some electrochemical properties of these membranes have already been reported[19].

A serious problem is swelling in aqueous media which notably decreases the mechanical strength of fibrous membranes. We are now trying to press these membranes at relatively high temperatures in order to create some cross-linking between adjacent fibers.

Inorganic Ion-Exchange Membranes Obtained By Crystals With Lamellar Structure

If the inorganic ion-exchanger possesses a lamellar structure, attempts may be made to obtain a colloidal dispersion of single layers from these crystals. Large and flexible membranes can then be obtained by packing these lamellar colloids. A well known example is the thin films obtained by a swollen dispersion of montmorillonite[20].

A more recent example is the membranes made of lamellar α-zirconium phosphate obtained in 1982 in our laboratory[21].

The $\alpha-[Zr(PO_4)_2]H_2 \cdot H_2O$ is presently a well known inorganic ion-exchanger with a lamellar structure[22]. Cations and protonable molecules such as NH_3 are easily intercalated in the interlayer region. The interlayer distance increases but the strong ionic forces between counterions and adjacent layers do not permit the crystal to swell infinitely. Several amines, too, have a strong tendency to be intercalated in the interlayer region where they are protonated[23].

It was recently found in our laboratory that if the intercalation is carried out with an aqueous solution of propylamine, (up to about 50% of the full intercalation capacity under strong stirring), delamelling of the crystals occurs and a very stable colloidal suspension of delaminated zirconium phosphate is obtained[21]. The colloidal suspension remains stable enough even after regeneration of the exchanger in the H^+-form by a dilute mineral acid. The flat colloidal particles tend to adhere to the surfaces of solid materials (e.g. glass, ceramic, metals, etc.) completely covering the surface itself with a thin film of α-zirconium phosphate. For this reason we have called the exchanger so obtained "pellicular zirconium phosphate" in analogy with pellicular ion-exchange resins. If the colloidal suspension is filtered on a flat plastic filter and air dried, the resulting pellicle can easily be detached from the filter, owing to the weak cohesion forces between the plastic material and the zirconium phosphate. Large, thin and flexible membranes which are mechanically resistent enough in air or in non-aqueous solvent, can be obtained easily,. However, when these membranes are put in contact with liquid water there is a considerable swelling which greatly decreases their mechanical stability.

This fact should not give rise to any inconvenience for some uses in ultrafiltration since the applied pressure makes the membranes very compact over a porous rigid support. For electrochemical uses in an aqueous medium a mechanical stabilization of the membranes is required in order to avoid breakage.

Similarly to what has previously been discussed for membranes consisting of fibrous exchangers, we are currently trying to obtain more stable layered membranes by heating at a high temperature under high pressure. Also mixed membranes made of pellicular zirconium phosphate and fibrous cerium phosphate have been prepared.

Preliminary experiments carried out at 200°C under a pressure of 3-4 ton/cm^2 have shown that it is possible to obtain appreciable improvements of the mechanical resistance to water.

Figure 1 shows the proton permselectivity (calculated from the concentration potential) of a membrane of pellicular $\alpha-[Zr(PO_4)_2]H_2 \cdot H_2O$ as a function of the HCl activity (curve a). The permselectivity of a mixed lamellar fibrous membrane is also reported (curve b). It may be noted that there is already a strong decrease in permselectivity at medium concentration of HCl.

We hope that better results can be obtained by heating at higher temperatures under high pressure. Regarding the thermal stability of α-Zirconium phosphate it must be remembered that the condensation of $\equiv$ P-OH groups to pyrophosphate occurs in the temperature range 450° – 550°C[24]. If the condensation is carried out at temperatures below 700°C a pyrophosphate with lamellar α-structure, schematically shown in Figure 2, is formed[24].

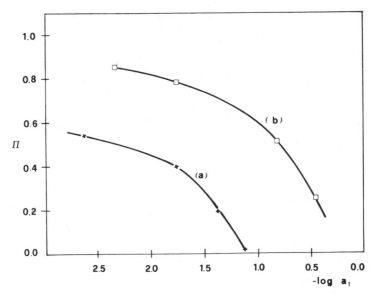

Fig. 1. Protonic permselectivity of membranes separating HCl solutions
(Π) with activity ratio $a_1/a_2 = 2$. (t = 20°C). Curve a):
membrane made of pellicular zirconium phosphate and fibrous
cerium phosphate (ratio 1:2).

Note that a microporous structure is formed in the interlayer region.
In our opinion attention should be given to this or to similar porous
structures for potential uses in the sieve separation of small gaseous
species.

Some other porous structures will be discussed later. It is the
opinion of the authors that considerable improvements on pellicular ion
exchange membranes will be obtained in the near future. For example, a
very thin pellicle of zirconium phosphate (less than 10 µm) can easily be
deposited over various porous supports so that dynamic membranes useful for
ultrafiltration may be obtained. For some electrochemical devices pellicular membranes could be stabilized between two porous supports. If the
porous supports consist of an electronic conducting material (e.g. fritted
nickel) the system could be used as an H_2 sensor or as hydrogen pump.

Membranes With An Amorphous Network

In contrast to crystalline materials, large inorganic ion-exchange
membranes may easily be prepared by amorphous inorganic solid electrolytes.

Oxide glasses are by far the fullest investigated materials for such a
purpose[25]. They associate some amphoteric oxides, such as SiO_2, GeO_2,
Al_2O_3, B_2O_3 and P_2O_5 with some metal oxides such as Li_2O, Na_2O, K_2O, Ag_2O,
CaO, BaO and so on.

The first oxides, called network formers, ensure the rigidity of the
macromolecular structure, while the metal oxides, called network modifiers,
introduce ionic bonds giving rise to cationic conductance. Unfortunately
the ionic conductivity of the presently known oxides glasses is low
($10^{-4} - 10^{-3}\Omega^{-1}cm^{-1}$) and their use is therefore restricted in particular
applications such as glass electrodes for pH-meters or for alkali metal
sensors.

466

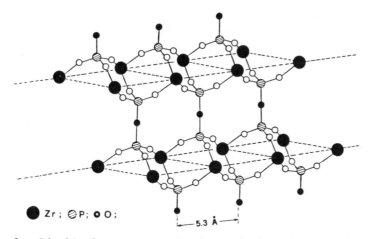

Zr ; ◉ P; ○ O; ├───5.3 Å───┤

Fig. 2. Idealized structure of α-layered zirconium pyrophosphate.

Inorganic membranes with an amorphous network are very attractive and several attempts are in course to improve their ionic conductivity.

INORGANIC–ORGANIC ION-EXCHANGE MEMBRANES

Homogeneous Membranes

The same considerations as made above for completely inorganic membranes can be applied to inorganic-organic ion-exchange membranes. The only differences are in the crystalline compounds used.

Our attention will be limited to just some new inorganic-organic compounds having layered or threedimensional structures. It was found in our laboratory[26] that layered zirconium bis(benzenephosphonate) can easily be prepared if the phosphoric acid is replaced with the benzene phosphonic acid in the synthesis of α-zirconium phosphate.

By using other phosphonic acids, a very large number of organic derivatives of α-zirconium phosphate of general formula $Zr(RPO_3)_2$ (where R is an organic radical) have been obtained in our and in other laboratories[27-30].

A variety of R groups anchored to the layered inorganic matrix is shown schematically in Figure 3. It is expected that several other organic derivatives will be obtained in the near future owing to the present interest in these compounds for their potential uses in catalysis.

If an ionogenic group such as $-SO_3H$, $\equiv P-OH$, $-COOH$ is present in the R organic radical, the compound may be considered to be an inorganic-organic ion-exchanger. For example, Figure 4 shows a schematic picture of the layered structure of the zirconium bis(carboxymethane phosphonate) in which carboxyl groups are present[27]. Dines and collaborators have obtained zirconium bis(3-sulphopropylphosphonate) containing $-SO_3H$ groups, which acts as a strong cationic exchanger[30]. These new inorganic-organic ion-exchangers exhibit good thermal stability up to about 200°C.

Pellicles of these inorganic-organic compounds may be obtained by using the same intercalation-deintercalation procedure used for $\alpha-[Zr(PO_4)_2]H_2 \cdot H_2O$.

467

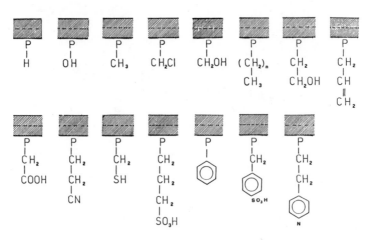

Fig. 3. Schematic representation of some organic radicals anchored to α-layered zirconium phosphate.

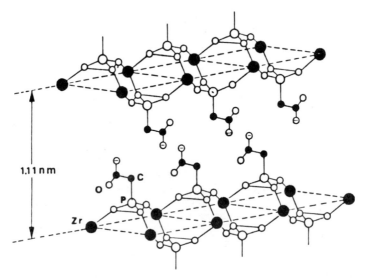

Fig. 4. Idealized structure of α-zirconium bis(carboxymethanephosphonate).

We are now investigating the possibility of preparing dynamic membranes by filtering colloidal dispersions of these compounds on porous supports.

As well as organic derivatives with a lamellar structure, some inorganic-organic compounds with a threedimensional structure have been obtained[29] by using bifunctional phosphonic acids, $R/PO(OH)_2)_2$. Such cross-linked, or pillared zirconium phosphates are of interest for the possibility of obtaining a tailor made porosity in the interlayer region[19]. In other words, the final goal is to obtain molecular or ion-sieves in which the size of the cavities can be varied as desired by changing the length of the R radical and/or by diluting the R radicals with phosphite groups.

Attempts are being made in our and in other laboratories to obtain compounds similar to that schematically shown in Figure 5 in which functional groups are present in large cavities.

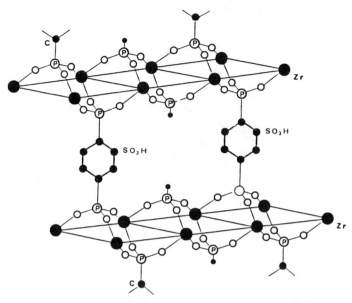

Fig. 5.   Schematic representation of an organic derivative of α-layered
          zirconium phosphate in which a microporosity in the interlayer
          region may be created by replacing some phosphonic groups with
          smaller ones, e.g. phosphites.

The development of such a new class of molecular sieves is important,
not only in the field of the sorbents and catalysts, but also in membrane
technology. Membranes consisting of such compounds could exhibit selective
behavior because of their molecular or ion sieve properties.

Selective separation of mixtures of molecules having different sizes
could be performed and new ion-sieve electrodes could also be obtained.

Heterogeneous Inorganic-Organic Ion-Exchange Membranes

Several investigations have been carried out on heterogeneous membranes obtained by powders of inorganic ion-exchangers and suitable organic
binders. Some general aspects of these membranes, such as their preparation and the choice of the binder in connection with the particular use of
the membrane have already been reported[19].   Most of our previous research has been carried out on membranes made of zirconium phosphate using
polyvinylidene fluoride (Kynar) as binder. With this binder, large and
flexible membranes, with good thermal stability below 180°C, can easily be
obtained. However, a disadvantage of these membranes is their high
electrical resistance due to the presence of the organic binder which
insulates a large part of the microcrystals of the inorganic ion-exchanger.
In order to lower the electrical resistance, we are now using some binders
with a certain degree of ionic conductance.

Encouraging results have already been obtained using Kynar at a low
degree of sulphonation (∿5%). Due to the presence of -SO$_3$H groups the
binder does not completely insulate the inorganic particles. A comparison
between the conductance of zirconium phosphate membranes, when Kynar and
sulphonated Kynar are used as binders respectively, is shown in Figure 6.
Note that the electrical conductance increases by about 10 times when the
sulphonated binder is used.

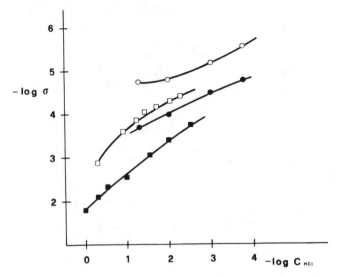

Fig. 6.   Electrical conductivity, σ, of heterogeneous α-zirconium phosphate
membranes against HCl concentration (t = 20°C). □: Membrane pre-
pared by evaporation of a solution of Kynar in N,N'-dimethylaceta-
mide with dispersed α-zirconium phosphate microcrystals (ratio
α-zirconium phosphate/Kynar 1:1)[19]. ■: Membrane obtained as
above but using 5% sulphonated Kynar. ○,●: The same membranes
after heating at 150°C under a pressure of 3t/cm^2.

The permselectivity of these membranes, (calculated from the concentration potential in HCl solution) as a function of the HCl activity is
shown in Figure 7. Results comparable with those obtained with non-
sulphonated Kynar, have been observed.

We have recently enlarged our investigation to membranes obtained
with β"-alumina in Na-form and with NASICON having the composition
$[Zr_2PO_4(SiO_4)_2]Na_3$. These materials have been chosen taking into account
their good ionic conductivity at medium temperatures and their thermal
stability[31,32].

Figure 8 shows the permselectivity of these membranes using Kynar as
binder. At room temperature results comparable with those obtained with
zirconium phosphate were obtained. However it was found that when a sample
of β"-alumina was put in contact with water for 5 days at 90°C its electrical resistance increased about 100 times while no appreciable variations
on its X-ray powder pattern were found. It may be concluded that β"-alum-
ina is not a good material for preparing membranes to be used in an aqueous
medium at high temperature. We are studying the reasons for the increase
in resistance but, for the moment, the research on β"-alumina membranes has
been stopped.

In the case of NASICON membranes, too, our preliminary data seem to
indicate a significant decrease of the conductivity after treatment with
water at 90°C for several days. We are now studying lamellar γ -zirconium
phosphate, γ-$[Zr(PO_4)_2]H_2 \cdot 2H_2O$, which seems to possess better conductance
than α-zirconium phosphate.

We are also planning to study membranes made with α-$[Zr(PO_4)_2]H_2 \cdot 5H_2O$
in which, owing to the high amount of water in the inter-layer region, a
good proton conductivity is to be expected.

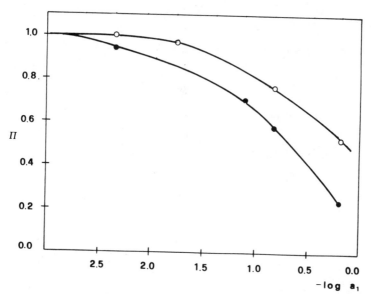

Fig. 7. Proton permselectivity (Π) for membranes made of α-zirconium phosphate microcrystals supported by Kynar (o) and sulphonated Kynar (●), separating HCl solutions with activity ratio $a_1/a_2 = 2$. (t = 20°C). Both the membranes, obtained by evaporation technique [19], were heated at 150°C under a pressure of $3t/cm^2$.

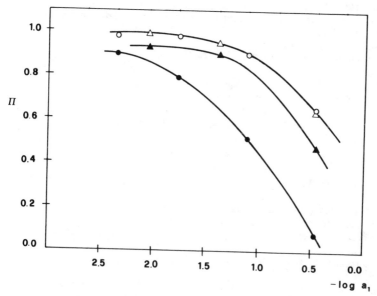

Fig. 8. Proton permselectivity (Π) of NASICON-Kynar (▲) and β"Alumina-Kynar (●) membranes, obtained by evaporation technique[19]. △,o: The same membranes after heating at 150°C under a pressure of $3t/cm^2$. Experimental conditions: activity ratio of the HCl solutions in the two sides of the cell $a_1/a_2 = 2$ and t = 20°C.

471

CONCLUSION

We have seen that inorganic and inorganic-organic ion-exchange membranes are more difficult to obtain than the organic membranes. We have also seen that until now the field of the inorganic membranes has been characterized by increasing interest but not by rapid progress. However, the need for membranes more resistant to drastic operating conditions is continuously increasing and it is likely that it will be even greater in the near future. Now, when a research is characterized by high practical interest the progress is usually fast because an increasing number of researchers are engaged in the field.

Thus, a considerable acceleration in the field of inorganic ion-exchange membranes, also leading to some good application, is to be expected, in our opinion, in the next ten years.

Acknowledgement

The work was supported by Finalized Project on F.C.S.C. of the C.N.R.

REFERENCES

1. G. Alberti, Inorganic ion exchange membranes, in: "Study Week on Biological and Artificial Membranes and Desalination of Water," R. Passino, ed., Pontificia Academia Scientiarum, Città del Vaticano (1976).
2. H. Rickert, "Electrochemistry of Solids," Springer Verlag, Berlin (1982).
3. P. Hagenmuller and W. van Gool, Application prospects of solid electrolytes, in: "Solid Electrolytes," P. Hagenmuller and W. van Gool, eds., Academic Press (1978).
4. K. P. Jagannathan, S. K. Tiku, H. S. Ray, A. Ghosh, and E. C. Subbarao, Technological applications of solid electrolytes, in: "Solid Electrolytes and Their Applications," E. C. Subbarao, ed., Plenum Press (1980).
5. J. N. Bradley and P. D. Greene, Solids with high ionic conductivity in Group 1 halide systems, Trans.Faraday Soc., 63:424 (1967).
6. S. A. Weimer, Ford Motor Co., Research on Electrodes and Electrolytes for the Ford Sodium Sulfur Batteries, Annual Report, Contract NSF-C 805 July, 1975, 1976, 1977.
7. P. Hong, Crystal structures and crystal chemistry in the system $Na_{1+x}Zr_2Si_xP_{3-x}O_{12}$, Mat.Res.Bull., 11:173 (1976).
8. T. H. Etsell and S. N. Flengas, Electrical properties of oxide electrolytes, Chem.Rev., 70:339 (1970).
9. S. Allulli, M. A. Massucci, and N. Tomassini, Procedimento per la preparazione di scambiatori inorganici in forma idrogeno, prodotti ottenuti e loro utilizzazione per la purificazione degli ioni lito e sodio e/o argento, Italian Patent No.48852A/79.
10. See chapter 16 and 23 of reference 3; and chapter 7 of reference 4.
11. See for example F. J. Rohr, chapter 25 of reference 3.
12. See for example B. B. Scholtens and W. von Gool, chapter 27 of reference 3.
13. N. Weber, Application of beta alumina in the energy field, in: "Superionic Conductors," G. D. Mahan and W. L. Roth, eds., Plenum Press, New York (1976).
14. P. Hagenmuller, Hydrogen as an energy carrier. A European perspective of the problem, Adv.Chem.Ser., 163:1 (1977).
15. D. Yuan and F. A. Kröger, Stabilized Zirconia as an oxygen pump, J.Electrochem.Soc., 111:594 (1969).

16. G. Alberti, U. Costantino, F. Di Gregorio, P. Galli, and E. Torracca, Preparation and ion exchange properties of cerium (IV) phosphate of various crystallinities, J.Inorg.Nucl.Chem., 30:295 (1968).

17. G. Alberti and U. Costantino, Fibrous thorium phosphate, a new inorganic ion-exchange material suitable for making (support-free) inorganic sheets, J.Chromatog., 50:482 (1970).

18. G. Alberti, M. A. Massucci, and E. Torracca, Chromatography in inorganic ions on (support-free) cerium (IV) phosphate sheets, J.Chromatog., 30:579 (1967).

19. G. Alberti, M. Casciola and U. Costantino, Inorganic ion-exchange membranes made of acid salts of tetravalent metals. A short review, J.Memb.Sci., 16:137 (1983).

20. H. Van Olphen, "An Introduction to Clay Colloidal Chemistry," John Wiley & Sons, New York (1977).

21. G. Alberti and U. Costantino, Pellicole Inorganiche a Scambio Ionico Costituite da Sali Acidi Insolubili di Metalli Tetravalenti con Struttura a Strati e/o loro Derivati e relativo Procedimento di Preparazione, Italian Patent No.48437 A/82 (1982).

22. A. Clearfield, Zirconium phosphates, in: "Inorganic Ion Exchange Material," A. Clearfield, ed., CRC Press, Boca Raton, Florida (1982).

23. G. Alberti and U. Costantino, Intercalation chemistry of acid salts of tetravalent metals with layered structure, in: "Intercalation Chemistry," M. S. Whittingham and A. J. Jacobson, eds., Academic Press, New York (1982).

24. U. Costantino and A. La Ginestra, On the existence of pyrophosphates of tetravalent metals having a layered structure, Thermochim.Acta, 58:179 (1982).

25. D. Ravaine and J. L. Souquet, Ionic conductive glasses, chapter 17 of reference 3.

26. G. Alberti, U. Costantino, S. Allulli, and N. Tomassini, Crystalline $Zr(R-PO_3)_2$ and $Zr(R-OPO_3)_2$ compounds (R = organic radical). A new class of materials having a layered structure of zirconium phosphate type, J.Inorg.Nucl.Chem., 40:1113 (1978).

27. G. Alberti, U. Costantino, and M. L. Luciani Giovagnotti, Synthesis and ion exchange properties of zirconium bis(carboxymethanephosphonate), a new organic-inorganic ion exchanger, J.Chromatog., 180:45 (1979).

28. L. Maya, Structure and chromatographic applications of crystalline $Zr(OPO_3R)_2$. R = butil, lauryl and octylphenil, Inorg.Nucl.Chem. Letters, 15:207 (1979).

29. M. B. Dines and P. M. Di Giacomo, Derivatized lamellar phosphates and phosphonates of M(IV) Ions, Inorg.Chem., 20:92 (1981).

30. P. M. Di Giacomo and M. B. Dines, Lamellar zirconium phosphonates containing pendant sulphonic acid groups, Polyhedron, 1:61 (1982).

31. R. Collongues, J. Thery, and J. P. Boilot, β-Aluminas, in: reference 3, chapter 16.

32. J. B. Goodenough, Skeleton Structures, in: reference 3, chapter 23.

THE INFLUENCE OF THE SURFACTANTS ON THE PERMEATION

OF HYDROCARBONS THROUGH LIQUID MEMBRANES

P. Pluciński

Institute of Chemical Engineering and Heating Equipment
Technical University of Wrocław
ul. Norwida 4/6, 50-373 Wrocław, Poland

INTRODUCTION

According to the literature[1-5] the permeation process of hydrocarbons consists of the following steps:

- dissolving and diffusion of feed components in the outer layer of
  surfactant molecules adsorbed on both sides of the aqueous membranes,
- dissolving of the latter and their further diffusion in the core of
  membrane.

The permeation rate and selectivity depend on the membrane thickness
and on the surfactant used for the membrane stabilization[1-7]. In the
case of a thick membrane, the permeation of hydrocarbons is governed by
their dissolving and diffusion in the aqueous phase of the membrane. When
the membrane thickness decreases the roles of solubility and diffusion in
surfactant layers become more important and finally these parameters
control the permeation process. In spite of intensive investigations of
the permeation of hydrocarbons through liquid membranes, the mechanism of
this process is still not clear.

The purpose of this work is to explain the hydrocarbons permeation
through liquid membrane. The possibility of using the literature data
concerning surfactants' properties, such as solubility parameter and
Cohesive Energy Balance for description of hydrocarbons permeation have
also been investigated.

The Scatchard-Hildebrand theory of solubility[8,9] may be used while
considering the influence of surfactants on the permeation of hydrocarbons.
This theory takes into account the cohesion energy between the solvent and
dissolved molecules. In the oil-water system, the cohesion forces within
the separation phase makes the mutual solubility difficult, which in turn
is facilitated by these forces acting between the phases. The number of
intermolecular interactions increases by the addition of surface active
agent. Cohesion energies among particles of feed elements and the mixture
of lipo- and hydrophilic surfactant molecules seem to have a considerable
influence on the permeation rate and selectivity. Both energies can be
described as C_{OH} and C_{OL}. For the whole surfactant molecule the cohesion
energy is given by Beerbower and Hill[10] as:

$$C_{OS} = 2 \phi_o V_S \delta_O \delta_S \qquad (1)$$

where: ϕ_O - volume fraction
        V_S - surfactant molar volume
        $\delta_O \delta_S$ - solubility parameters of oil and surfactant respectively

For the given surfactant the cohesion energy is directly proportional to the solubility parameter of the diffusing hydrocarbon molecules – δ_O. In the case of hydrocarbons, mentioned in the literature[1-7], the cohesion energies are arranged in the series given below:

styrene ($\delta_O = 9,3$ cal$^{1/2}$/m$^{3/2}$) > benzene (9,2) > toluene (8,9) > ethylbenzene, o-xylene, m-xylene (8,8) > p-xylene (8,6) > cyclohexane (8,2) > methylcyclohexane (7,8) > n-heptane (7,4) > n-hexane (7,3).

The literature values[8,9] of solubility parameter at 25°C are given in the brackets.

Thus it seems possible that the styrene molecules are the particles closely attached to the molecules of surfactant forming the outer layer, the attachment of n-hexane being the weakest one. The styrene molecules will also be most strongly absorbed into micelles formed in sufficiently thick membranes[11], while the absorption of n-hexane will be the lowest one. In the case of a thin membrane, where the diffusion through palisade layer of surfactant is a decisive step, the value of selectivity coefficient of separation of aromatic form aliphatic hydrocarbons will be the lowest, because of strong attachments of aromatics to surfactant molecules at the interface. For the membrane with significant amount of aqueous phase (where the micelles may be formed), the selectivity coefficient will take a higher value, since cohesion energy of aromatics and surfactant molecules inside the micelles will increase. In this case the aromatics will be absorbed into the micelles and their transport across the membrane resulting from concentration gradient will be faster.

EXPERIMENTS

In order to verify the above proposed mechanism of the permeation and the influence of surfactants properties on permeation process of hydrocarbons we have performed the experiments using the emulsion – treating technique. To this end we have selected three fractions namely a model one: toluene – n-heptane in 1:1 mole ratio, and two industrial ones: "xylene fraction" produced by Mazovian Refining Plant in Pock (Poland) and solvent AR-16, produced by Petroleum Refining Plant in Gdańsk (Poland). The composition of separate fractions are given below:

xylene fractions:	chain hydrocarbons		– 22,0% by weight
	toluene		1,5
	ethylbenzene		9,7
	m+p – xylenes		48,5
	o-xylene		18,3
solvent AR-16	n-paraffins	25,3	56,0
	iso-paraffins	30,7	
	naphtenes		24,0
	aromatics		20,0

The surfactants selected for stabilizing of emulsion are given in Table 1. The contents of hydrocarbons in permeate and in purified solution were determined by means of GLC. In all the experiments the balance error has not exceeded 10%.

Table 1. The Characteristics of the Surfactants Tested

Surfactant	Structure	Purity	Producer
Rokanol K20	$RO\,(CH_2CH_2O)_{20}\,H$ R - alkil moiety from cocount oil	tech.	NZPO "Organika Rokita" Brzeg Dolny, Poland
Rokwinol 60	Tween's type polyoxyethylene ester of fatty acids and sorbitan	tech.	NZPO "Organika Rokita" Brzeg Dolny, Poland
Rokafenol N10	$C_9H_{19} - C_6H_4O\,(CH_2CH_2O)_{10}\,H$	tech.	NZPO "Organika Rokita" Brzeg Dolny, Poland
Rokafenol N8	$C_9H_{19} - C_6H_4O\,(CH_2CH_2O)_8\,H$	tech.	
Sodium laurate	$C_{11}H_{23}COONa$	pure	P.O.Ch. Gliwice, Poland
Potassium palmitate	$C_{15}H_{31}COOK$	pure	P.O.Ch. Gliwice, Poland
Sodium oleate	$C_{17}H_{33}COONa$	pure	P.O.Ch. Gliwice, Poland
Sodium dodecyl sulfate	$C_{12}H_{25}OSO_3Na$	pure	ROTH, West Germany
Sodium dodecyl benzene sulfonate	$C_{12}H_{25}C_6H_4SO_3Na$	tech.	Z.Ch.G. "Pollena" Wrocław, Poland

RESULTS AND DISCUSSION

Selectivity coefficient of separation of aromatic from aliphatic hydrocarbons versus membrane thickness (volume phases ratio in emulsion) is shown in Figure 1. At first the separation coefficient increases with the amount of aqueous phase, and then it reaches a constant value. It seems, that the membrane thickness responsible for the change of permeation mechanism is determined by the value of volume phases ratio, for which the selectivity coefficient becomes constant. This value, below which the permeation process depends significantly on the diffusion through surfactant palisade layer, is called critical value of volume phases ratio $(V_w/V_o)_{cr}$. When this value is exceeded the diffusion of hydrocarbons through aqueous phase of membrane appears to be limiting step. The values of $(V_w/V_o)_{cr}$ were found from selectivity coefficient versus phases ratio in emulsion plots for all surfactants, and are given in Table 2. It appears that the critical volume phase ratio does not depend strictly on the type of surfactant. Using, for instance polyoxyethylene nonylphenols (Rokafenol N8 and N10) we have stated that the values of $(V_w/V_o)_{cr}$ were low. These values were much smaller than those for ionic surfactants. The greatest values of $(V_w/V_o)_{cr}$ were found for nonionic surfactants such as: polyoxyethylene ester of fatty acids and sorbitan (Rokwinol 60) and polyoxyethylene alcohols from coconut oil (Rokanol K20). That means that for these types of surfactants the influence of surface layers on the permeation process is the highest one. On the contrary, for the ionic surfactants with lipophilic straight chains (fatty acid salts) the values of $(V_w/V_o)_{cr}$ were lower and the influence of surface layer were also lower. It appears that the compounds with aromatic ring in lipophilic moiety (polyoxyethylene nonylphenols and sodium dodecyl benzene sulfonate) had the lowest values of $(V_w/V_o)_{cr}$. This fact can be explained by the additivity of the solubility parameter of surfactant molecule. According to Little[12] this parameter can be determined from the equation:

$$\delta_S = \phi_L\,\delta_L + \phi_H\,\delta_H \tag{2}$$

where: L - lipophilic, H - hydrophilic.

Substituting Equation (2) for the Equation (1) we obtain:

$$C_{OS} = 2\phi_o\delta_O\,(V_L\delta_L + V_H\delta_H) \tag{3}$$

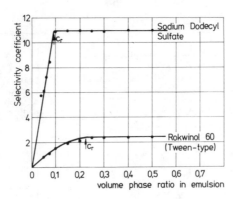

Fig. 1. The selectivity coefficient of the toluene – heptane system versus volume ratio of emulsion phase.

Table 2. The Experimental Data for Separation of Toluene-heptane mixture

Surfactant	$(V_w/V_o)_{cr}$	δ_s $cal^{1/2}/m^{3/2}$	CEB –	α_{max} –
Rokwinol 60	0,200	9,04	177,2	2,42
Rokanol K20	0,150	9,01	276,3	2,95
Sodium oleate	0,130	9,30	29,4	3,65
Potassium palmitate	0,120	9,60	12,96	4,55
Sodium laurate	0,100	9,60	12,96	4,12
Sodium dodecyl sulfate	0,095	14,1	2,247	11,00
Rokafenol N-10	0,095	8,91	$2,93 \cdot 10^4$	3,06
Rokafenol N-8	0,090	8,86	1707	3,23
Sodium dodecyl benzene				
Sulfonate	0,075	13,0	2,517	4,60

Permeate phase: transformer oil

Feed: permeate ratio = 1 : 2 by vol.

Surfactant concentration 0,08 kMOL/m^3

The values of cohesion energies of hydrophilic and lipophilic moieties can be different depending on the structure of surfactant molecule. So far we are not able to explain exactly what kind of cohesion energy is responsible for mass transfer both through surface and aqueous layers of the membrane. Also there are no data which would allow an exact evaluation of the values of particular cohesion energies.

Similar results were also obtained for tetracomponent mixture: solvent AR-16. It has been observed that with the decreasing water content in the emulsion the selectivity coefficient of n-paraffins and naphtenes increased, while the selectivity coefficient for isoparaffins and aromatics decreased (see Table 3).

The results obtained (Figure 1, Tables 2 and 3) correspond well with the proposed mechanism of hydrocarbons'permeation through liquid membranes. That means that the selectivity coefficient depends on cohesion energies

Table 3.   The Influence of Membrane Thickness
on Selectivity Coefficient of Solvent
AR-16 Separation.

α Hydrocarbons	emulsion phase ratio		W : O
	1 : 10	1 : 20	1 : 50
n-paraffin	0,602	1,002	1,023
iso-paraffin	0,852	0,757	0,713
naphtenes	0,559	0,653	0,919
aromatics	2,734	1,917	1,532

Surfactant: sodium dodecyl sulfate- 2 % by weigth.

Permeate phase - transformer oil

Table 4.   The Influence of the Surfactant Kind
on Selectivity Coefficient of Xylene
Fraction Separation.

Surfactant	δ_s	α
Sodium dodecyl sulfate	14,1	15,42
Sodium dodecyl benzene sulfonate	13,0	10,05
Potassium palmitate	9,6	7,66
Sodium laurate	9,6	7,15
Rokafenol N8	8,9	5,30
Rokanol K20	9,01	3,45
Rokwinol 60	9,04	3,38

Permeate phase: paraffinic oil

Surfactant concentration - 0,08 kMol/m^3

Feed: O/W type emulsion 10 : 1 vol : vol

Emulsion: Permeate natio 1 : 1 by vol.

among oil and surfactant molecules both in the adsorbed layers and
in micelles. Cohesion energies depend also on the solubility parameter
of the surfactant used. For the surfactants investigated, the values of
cohesion energies are arranged as follows: sodium dodecyl sulfate (δ_s =
14,1) > sodium dodecyl benzene sulfonate (13,0) > potassium palmitate,
sodium laurate (9,6) > sodium oleate (9,3) > Rokwinol 60 (9,05) > Rokanol
K20 (9,0) > Rokafenol N10 (8,91) > Rokafenol N8 (8,86). From the data on
separation of mixture: toluene – n-heptane (Table 2) and xylene fraction
(Table 4). However, it can be assumed that the selectivity coefficient
depends strongly on the kind of surfactant. The highest value of the
selectivity coefficient was found for sodium dodecyl sulfate while the
lowest one was for Rokwinol 60, regardless of the feed used. The changes
of selectivity coefficient are usually in agreement with the change of
solubility parameter of surfactants (Table 2 and 4). It is worth noting,
however, that better correlation was obtained by comparing selectivity
coefficient and Cohesive Energy Balance (CEB) value (see Table 2),
defined[13] by equation:

$$CEB = \frac{V_{01}(\delta_{01} - \delta_S)^2}{V_{02}(\delta_{02} - \delta_S)^2} \qquad (4)$$

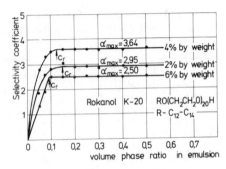

Fig. 2.   The critical value of volume phases ratio versus Rokanol K20
concentration.

This parameter seems suitable for description of some bulk properties of
surfactant in aqueous solution, contrary to Beerbower's[10] Cohesive Energy
Ratio (CER), which describes the energy balance in the interfacial zone of
a well-formulated emulsion. The results obtained for the separation of the
toluene-heptane mixture enable us to find the dependence of the $(V_w/V_o)_{cr}$
or CEB value. Thus the highest $(V_w/V_o)_{cr}$ value is observed for the surfactants for which CEB value is the highest one. This correlation is valid
for both lipophilic groups (alkyl and alkylbenzene).

The influence of Rokanol K20 concentration on the $(V_w V_o)_{cr}$ value was
also investigated (Figure 2). Critical value of volume phases ratio increases with the decreasing surfactant concentration. This fact can be
explained by the solubilization phenomenon. The increase of the surfactant
concentration is accompanied by the increase of the solubility in the
aqueous layer of the membrane. Thus the increasing influence of diffusion
through water layer in membrane is manifested in the decrease of $(V_w/V_o)_{cr}$
value.

It seems that the results presented above confirm the validity of the
proposed mechanism of hydrocarbons transport through liquid membranes.
Based on Scatchard and Hildebrand's theory of solubility we can foresee the
type of surfactant which would give a high separation selectivity. Since
the technological effect of separation depends either on selectivity or on
mass transfer kinetics, the knowledge of volume phase ratio value in
emulsion, represented by the critical volume phase ratio $(V_w/V_o)_{cr}$ (i.e.
the membrane thickness causing a high process rate) not only proper selection of surfactant, is indispensible.

REFERENCES

1. N. N. Li, Permeation through liquid surfactant membranes,
    AIChE Journal, 17:459 (1971).
2. N. N. Li, Separation of hydrocarbons by liquid membrane permeation,
    Ind.Eng.Chem.Process Des.Develop., 10:215 (1971).
3. N. D. Shah and T. C. Owens, Separation of benzene and hexane with
    liquid membrane technique, Ind.Eng.Chem.Prod.Res.Develop., 11:58
    (1972).
4. G. Casamatta, L. Boyadzhiew, and H. Angelino, Chem.Eng.Sci., 29:2005
    (1974).
5. P. Alessi, B. Canepa, and I. Kikic, Influence of operating parameters
    on selectivity in a separation process by liquid membrane
    permeation for the system styrene - ethylbenzene, Can.J.Chem.Eng.,
    57:54 (1979).

6. P. Alessi, I. Kikic, and M. Orlandini-Visalberghi, Liquid membrane permeation for the separation of C_8 hydrocarbons, Chem.Eng.J., 19:221 (1980).
7. P. Alessi, I. Kikic, B. Canepa, and P. Costa, Inversion of selectivity in the liquid membrane permeation process, Sep.Sci.and Technol., 17:613 (1978).
8. J. H. Hildebrand and R. L. Scott, "The Solubility of Non-electrolytes," Reinhold Pub. Corp., New York (1950).
9. J. H. Hildebrand and R. L. Scott, "Regular Solutions," Prentice-Hall Inc., Englewood Cliffs, N.J. (1962).
10. E. Beerbower and M. W. Hill, The cohesive energy ratio of emulsions – a fundamental basis for the HLB concept, in: "McCutcheon's Detergents and Emulsifiers Annual," ed., Allured Publ. Co., Ridgewood, N.J. (1971).
11. P. Pluciński, Stabilitat von flussigen Membranen – Auswahl des Tensids und Einfluss der Solubilisation auf die Permeation von aromatischen Kohlenwasserstoffen, Chem.Techn., 34:543 (1982).
12. R. C. Little, Correlation of surfactant hydrophile – Lipophile balance HLB with solubility parameter, J.Colloid Interface Sci., 65:587 (1978).
13. L. Marszall, Further approach to the energy balance of nonionic surfactants, Colloid and Polymer Sci., 254:440 (1976).

SEPARATION OF ISOMERIC XYLENES BY MEMBRANES

CONTAINING CLATHRATE-FORMING METAL COMPLEXES

S. Yamada and T. Nakagawa*

Materials Department, Industrial Products
Research Institute, 1-1-4, Yatabe-machi, Higashi
Tsukuba-gun, Ibaraki 305
*Department of Industrial Chemistry, Meiji University
1-1-1, Higashi-mita, Tama-ku, Kawasaki 214

SUMMARY

The synthesis and preparation of polymeric membranes including Werner type metal complexes which are known to form selective clathrate compounds with specific xylene isomers were investigated in order to study the feasibility of specific membranes which would separate xylene isomers. A porous PTFE membrane was radiation grafted with 4-vinylpyridine. The grafted membrane was combined with Werner type metal complexes such as $Ni(4-methylpyridine)_4(SCN)_2$, $Co(4-methylpyridine)_4(SCN)_2$ etc. through the exchange of ligands between 4-methylpyridine of metal complex and the pyridyl group of the side chain. It was found that the membrane thus obtained is selective for p-xylene in pervaporation process and the separation factor of p-xylene to m-xylene was 1.26 beginning with a feed containing equal weights of the three xylene isomers. Studies were also conducted on the effects of reaction time, concentration, continuous or intermittent procedure, grafting percent and composition of ligand upon the uptake amount of metal complex in the graft membrane so as to find the best condition of membrane preparation.

INTRODUCTION

Recently much attention has been paid to membrane separation because it is believed to be much less energy consuming process than conventional separation techniques such as distillation. Pervaporation, which is one of membrane separation process, can be generally used to separate organic liquids which are difficult to separate by distillation such as azeotropic mixture and mixtures of close boiling components. The membrane separation of isomeric xylenes, which are quite identical as far as boiling point are concerned, is very attractive because the production of pure p-xylene is of great industrial importance.

Among xylene isomers p-xylene is in strong demand in industry market as a major raw material for the production of dimethyl terephthalate and terephthalic acid, intermediates used in the production of polyester fiber and film. The xylene isomers must be separated in order to be used in their many applications as chemical intermediates.

Concerning the separation of xylene isomers Michaels and coworkers attempted to make a pervaporation experiment for the first time and proposed "Molecular Sieve Model"[1]. No major studies of synthetic polymeric membranes containing highly selective carriers or additives to mediate and separate a specific molecule have been reported until McCandless et al.[2] investigated the separation of the xylene isomers with the modified poly (vinylidene fluoride) membrane including Werner complexes. Specific interactions (molecular recognition) between the permeant and the immobilized carrier in the membrane constitute this transport process just as generally occurs in biological membrane.

Similarly Lee[3] developed a separation membrane by incorporating cyclodextrin additives into hydroxypropylmethyl cellulose and made pervaporation experiments for isomeric xylene mixture with the carrier-containing membrane. The highest separation factor for para/meta isomers of xylenes was 1.5 for the cellulose derivatives membrane containing 25% α-cyclodextrin.

On the other hand it is known that some class of Werner complexes exhibits a sharp selectivity for forming clathrate crystals containing specific isomers of compounds such as xylenes, as first reported by Schaeffer et al.[4].

In Table 1 are summarized pervaporation data of xylene isomers done by previous workers[1,2,3].

The objective of this research work is to demonstrate and establish a possible method for preparing a separation membrane by immobilizing highly selective carriers into porous sublayers so that the function of carriers may be most effective in membrane separation. It is not clear in which part, carrier or matrix of polymeric membrane, selectivity is specifically occurring in the study of McCandless[2] and Lee[3]. However when carrier is immobilized in the micropore of sublayer and impermeable matrix is chosen, selectivity should take place in the carrier immobilized region.

EXPERIMENTAL APPARATUS AND PROCEDURE MATERIALS

Porous polytetrafluoroethylene (PTFE) membrane used in this study was Fluoropore Membrane Filters (Type FP010) with pore size of 0.10 μm obtained from Sumitomo Electric Ind. Ltd. Guaranteed grade chemicals were used in the production of the metal complexes and clathrates with no further purification. Purities of xylenes used were more than 98.0% for ortho, meta and para-isomers.

Metal Complex and Clathrate Preparation

The preparation of metal complexes and clathrates was accomplished by a method proposed by de Radzitzky and Hanotier[5]. Here, for example, p-xylene clathrate of Ni(SCN)$_2$ (4-methylpyridine)$_4$ was obtained by first forming a solution of Ni(SCN)$_2$ by double decomposition of 9.51 g (0.04 mol) of NiCl$_2$6H$_2$O with 7.78 g (0.08 mol) of KSCN dissolved in 100 ml of water. This solution was cooled in an ice bath. Gradually, and with stirring, 7.78 ml (0.18 mol) of 4-methylpyridine diluted in 100 ml of p-xylene was added. As soon as the amine was added, a blue precipitate started to form. After 20 min of stirring, the resulting precipitate was separated by filtration and washed twice with 100 ml of cold heptane, and then dried in desiccator. Other clathrates and metal complexes were formed in this manner with the substitution and omission of appropriate compounds.

484

Table 1. Pervaporation data for separation of xylene isomers

Membrane	Feed o:m:p(w/w)	Temp. C	Permeation rate g/m^{2}hr(1mil)	Separation factor p/m[a]	p/o[b]	m/o[c]	Ref.
High density PE annealed in p-xylene at 100 C	30:65: 0.5	30	---	1.26	2.00	1.59	1
	90: 5: 5	30	---	1.24	1.99	1.60	1
PVdF with Ni(SCN)$_2$(a-MeBzAm)$_4$o-xylene	1: 1: 1	60	2.15[d]	1.26			2
Hydroxypropylmethyl cellulose including a-cyclodextrin	1: 1: 1	72	2	1.5	1.8-2.03		3

a:p/m= separation factor between para and meta isomers.
b:p/o= separation factor between para and ortho isomers.
c:m/o= separation factor between meta and ortho isomers.
d:Since it is reported that the compositon difference of p-xylene between the product and feed was 8 %, separation factor for para/meta was obtained on the assumption that the difference of 8 % was equally distributed between the increment of p-xylene and the decrement equally ascribed to o- and m-xylene.

Grafting Process

As grafting method we used the simultaneous irradiation technique in vacuum with ^{60}Co γ-ray. Freshly distilled 10ml 4-vinylpyridine monomer (58.0-60.0°C/11 mmHg) was poured into test tubes containing PTFE membranes (90 mm in diameter). The containers were degassed by alternately freezing and thawing in a vacuum line. The test tubes were placed into hot cape of radiation source (^{60}Co 3000 Ci) and subjected to a series of irradiation doses. The irradiation were carried out at room temperature at a dose rate of 0.182 Mrad/hr for the times required to achieve the desired irradiation dose.

The grafted membrane were then extracted with methanol at room temperature for more than 48 hr with stirring until the poly(vinylpyridine) homopolymer was completely extracted. Grafted membranes were stored dry in desiccator prior to successive chelation and clathration. Percent grafting was determined gravimetrically by the following expression.

The grafting yield (%) $= \dfrac{W_2 - W_1}{W_1}$ x 100 where W_2 and W_1 are the final and initial weights of the membrane, respectively.

Immobilization of Metal Complex and Clathrate Compound

The immobilization of clathrate compound into membrane was carried out in the following two ways.

1. The graft membrane was reacted with metal salt solution and then reacted with the solution of organic ligand finally to be combined with metal complex. When the graft membrane was dipped in the mixed solution of organic ligand and isomeric xylene in the latter process, the graft membrane reacted with clathrate compound would be obtained.
2. The graft membrane was reacted directly with the clathrate solution to obtain the clathrate bound to membrane.

The effects of reaction time, concentration, continuous or intermittent procedure, grafting percent and composition of ligand was investigated in the above-mentioned processes to search an optimum condition.

The uptake of chelate and clathrate compound was expressed as shown in the following expression.

The uptake percent of membrane (%) $= \dfrac{w_r - w_g}{w_g}$ x 100 where w_g and w_r are the weights of graft membrane and the reacted graft membrane respectively.

Pervaporation Measurement

The pervaporation experiments were carried out in an isostatic liquid permeation apparatus whose diagram is shown in Figure 1. Figure 2 is a cross section of liquid permeation cell made of corrosion resistant stainless steel with the exception of the joint parts connected to a dropping funnel and a capillary tube. The liquid mixture was supplied to the upper compartment(4) equipped with a Teflon stirrer(4) from the universal ground glass joint(3) using a dropping funnel. A Teflon gasket and the sample membrane were inserted between the upper compartment(4) and the lower compartment(8). The flanges were bolted together by means of volts and nuts(15) and inlet(10) was connected to circulation unit of the isostatic liquid permeation apparatus. The lower compartment is fitted with a porous sintered metal plate(4.5cm in effective diameter) to support the membrane.

486

The temperature was regulated by circulating preadjusted water through the inlet(13) and outlet(13) of the jacket(14).

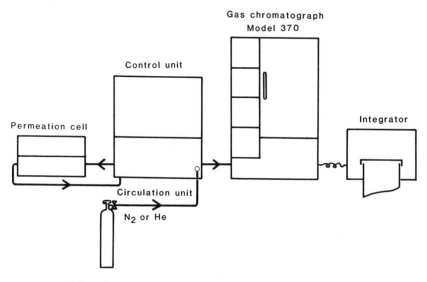

Fig. 1. Automatic liquid permeation apparatus.

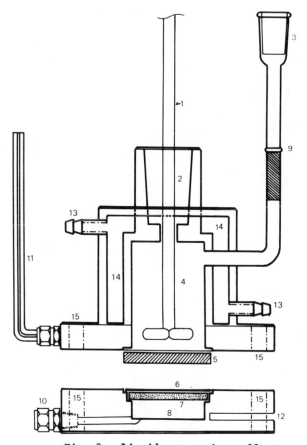

Fig. 2. Liquid permeation cell.

The isotatic liquid permeation apparatus consisted of (1) permeation cell, (2) control unit, (3) circulation unit, (4) gas chromatograph and (5) integrator as given in Figure 1. The operation of the apparatus is as follows. The permeated vapor through the membrane was circulated continuously over the membrane by nitrogen gas until the amount of permeated came to be detectable. Then the vapor was introduced into gas chromatograph by control unit to measure the composition of the permeate. The data was processed by integrator.

Knowing the composition of the mixture on the upstream and downstream sides of the membrane, it is possible to express the membrane selectivity in a ternary system (o-, m- and p-xylene) in terms of a separation factor (α), defined as the concentration ratio Y_p/Y_m in the pervaporate divided by concentration ratio X_p/X_m in the original liquid:

$$\alpha_{p/m} = \frac{Y_p/Y_m}{X_p/X_m}$$

Electron Microscopy Characterization

The surface and micropore of the membrane specimens were investigated by scanning electron microscopy (Akashi Manuf. Co. Mini SEM 6) after being evaporated with gold in vacuum.

RESULTS AND DISCUSSION

Graft Membranes

In order to add metal complex to the micropore of PTFE membrane through the exchange of the pyridyl group of graft chain and the pyridyl group of ligand in metal complex, 4-vinyl pyridine was grafted onto PTFE membrane by radiation. In Figure 3 is shown graft percent of 4-vinyl-pyridine onto porous PTFE membrane versus radiation dose of [60] Coγ-ray, as one run of several experiments. It can be seen that graft percent increased monotonously with an increase of radiation dose. Graft membrane swelled in proportion to graft percent and highly grafted membrane turned hard and brittle when it was dried. The graft membrane with graft percent of 30-80% was employed for the experiment of membrane separation. When

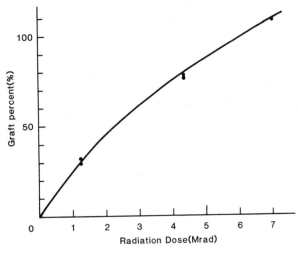

Fig. 3. Relation between graft% and radiation dose.

PTFE membrane was irradiated for grafting up to 27 Mrad PTFE membrane became completely ragged.

Time Dependence of Chelate and Clathrate Uptake

In order to attach clathrate compound to graft membrane more efficiently, two approaches were taken as shown in Figure 4. In the first method graft membrane was reacted directly with the clathrate solution to obtain clathrate membrane, which has clathrate compound along the polymer chain as a pendant (See Figure 4).

In the second method clathrate membrane was prepared through chelated membrane. Reaction of graft membrane with metal salt solution produced chelate membrane, which was then allowed to react with organic ligand to form clathrate membrane.

To optimize the condition of chelation and clathration of grafted PTFE membrane with 4-vinylpyridine, it seems appropriate to make clear the time dependence of chelate and clathrate uptake of graft membrane. Figure 5 shows the relation between chelate uptake and reaction time, when the graft membrane with 75% of grafting was reacted at room temperature with the $Co(SCN)_2$ solution comprising 1 g $Co(SCN)_2$, 10 ml H_2O and 40 ml CH_3OH. The uptake curve in Figure 5 was found to level off in nearly 10 hours.

Shown in Figure 6 is clathrate uptake versus reaction time, when the graft membrane with 92.3% of grafting was reacted at room temperature with the clathrate solution of 1 g $Ni(SCN)_2(4-vinylpyridine)_4$ p-xylene and 50 ml CH_3OH. Comparing with the result of Figure 6 it took as long as 100 hrs for the uptake curve to be saturated.

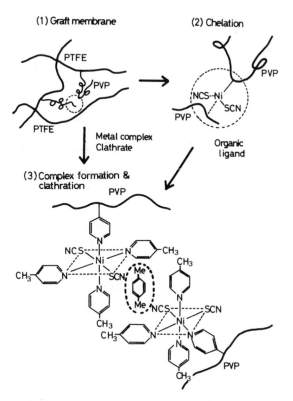

Fig. 4.   Two approaches to obtain clathrate membrane.

489

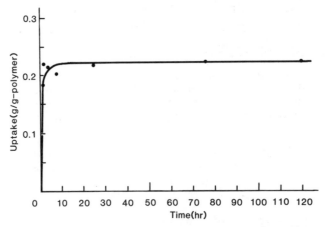

Fig. 5. Time dependence of chelate uptake of graft membrane reacted with metal salt solution. Graft membrane: 4.37 Mrad irradiated, 75 graft %. Metal solution: $CO(SCN)_2$ 1g, H_2O 10 ml, CH_3OH 40 ml. Temperature: room temperature.

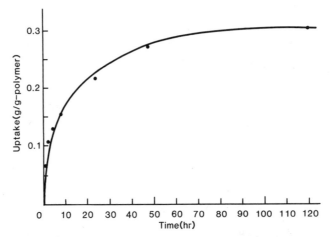

Fig. 6. Time dependence of clathrate uptake of graft membrane reacted with clathrate solution. Graft membrane: 4.37 Mrad irradiated, 92.3 graft %. Clathrate solution: $Ni(SCN)_2(4-ViPy)_4 \cdot$p-xylene 1g, CH_3OH 50 ml. Temperature: room temperature.

Concentration Dependence of Chelate and Clathrate Uptake

Figure 7 is the concentration dependence of chelate uptake of graft membrane with 118.1 graft %, when the graft membrane was immersed in chelate solution of various concentration. The uptake curve shown in Figure 7 appears to level off at the concentration of 0.2 g $Co(SCN)_2$ in 10 ml of mixed solvent of 15 ml H_2O and 75 ml CH_3OH. Similarly Figure 8 illustrates clathrate uptake of graft membrane versus concentration of clathrate solution. In the clathration of graft membrane the uptake curve increases with an increase of concentration of clathrate solution and the curve passes through a maximum value at the concentration of about 0.3g/10 ml. This can be explained in the following way. In the lower range of concentration the exchange of ligands occurs more favorably to form clathrate with increasing concentration, and after passing the maximum region the exchange of ligand works to decompose the clathrate compound.

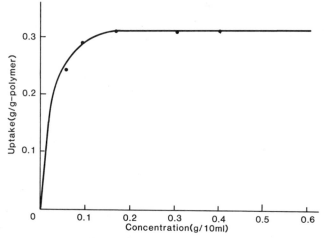

Fig. 7. Concentration dependence of chelate uptake of graft membrane reacted with metal salt solution. Graft membrane: 7 Mrad irradiated, 118.1 graft %. Metal solution: $Co(SCN)_2$ was dissolved in 10 ml mixed solvent of 15 ml H_2O and 75 ml CH_3OH. Reaction time: 48 hr. Temperature: room temperature.

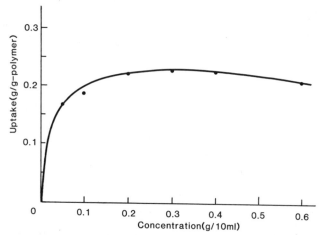

Fig. 8. Concentration dependence of clathrate uptake of graft membrane reacted with clathrate solution. Graft membrane: 43.7 Mrad irradiated, 102.2 graft %. Clathrate solution: $Ni(SCN)_2(4-ViPy)_4$. p-xylene was dissolved in 10 ml of CH_3OH. Reaction time: 48 hr. Temperature: room temperature.

Continuous and Intermittent Method

Two series of clathration with graft membranes were conducted by varying the clathration procedure. In one series continuous clathration was used as shown in solid circle plots of Figure 9, i.e. the graft membrane was clathrated being immersed in the clathrate solution all the time. In the second series clathration step was performed in cyclic fashion. The graft membranes were once immersed in clathrate solution. They were then removed from the clathrate solution and were allowed to stand in air for a certain period. Then they were allowed to re-equilibrate with the same clathrate solution. This procedure was repeated cyclically. Figure 9 shows clathrate uptake done by continuous and intermittent method versus

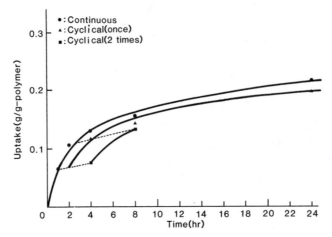

Fig. 9. Chelate uptake by various methods as a function of reaction time. Graft membrane: 4.37 Mrad irradiated, 92.3 graft %. Clathrate solution: $Ni(SCN)_2(4-ViPy)_4$p-xylene 1 g, CH_3OH 50 ml. Temperature: room temperature.

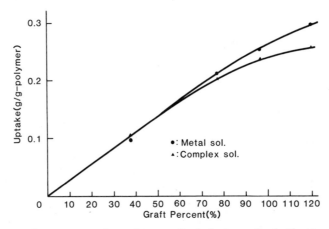

Fig. 10. Graft percent dependence of chelate and clathrate uptake of graft membrane reacted with metal salt solution and metal complex solution. Metal solution: $CoCl_2 6H_2O$ 3g, KSCN 2.4g, H_2O 20 ml, CH_3OH 80 ml. Metal complex solution: $Co(SCN)_2(4-MePy)_4$ 1g, CH_3OH 100 ml. Reaction time: 48 hr. Temperature: room temperature.

reaction time. However, the results reveal that there are substantially no difference in uptake of clathrate between the two series of clathration compared with each other on the basis of the same reaction time, although clathration slightly proceeded during the drying period.

Graft Percent Dependence of Chelate and Complex Uptake

As represented in Figure 10, chelate and clathrate uptake of graft membrane increased linearly in the lower range of graft percent and then the uptake curve diverged into two at the graft percent of 50-60. The chelate uptake was slightly higher than the metal complex uptake in the range of graft percent of more than 60.

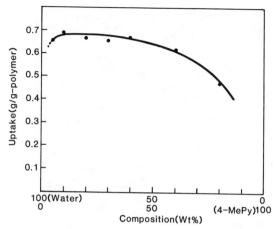

Fig. 11. Relation between clathrate uptake and the concentration of 4-methylpyridine solution. Membrane: the graft membrane (graft %: 78%) reacted with $Co(SCN)_2$ solution (uptake: 21.94%). Reaction time: Stopped when pink color appeared. Temperature: room temperature.

Clathrate Uptake Through Chelate Membrane

In order to clarify how much clathrate uptake occurs when chelate membrane is clathrated in the indirect manner as shown in Figure 4, Figure 11 illustrates the relationship between the complex uptake of the graft membrane combined with $Co(SCN)_2$, and the concentration of 4-methylpyridine as organic ligand. It can be seen that there is an appropriate region of concentration for clathrating the chelate membrane, whereas the chelate compound bound to the chelate membrane dissolved more easily than clathrate with organic ligand at both the extremely lower and higher concentration of aqueous 4-methylpyridine solution.

There has been no direct evidence that the crystalline structure of metal complex was maintained even in the graft membrane. However, it was confirmed in some cases by the color change of the metal complex linked with the graft membrane. The color of some metal complexes such as $Co(4-methylpyridine)_4(SCN)_2$ is very sensitive to temperature and solvation. The pink color of $Co(4-MePy)_4(SCN)_2$ did not change when the metal complex was bonded to the graft membrane.

Calculating from the weight ratio in the metal complex of $CO(4-MePy)_4$ $(SCN)_2$ bonded to the graft membrane, 3.85 was obtained for the coordination number of 4-methylpyridine to the central atom of cobalt. The coordination number should be 3 and 4 respectively when the metal complexes are completely linked with the graft membrane, and are entirely free from the graft chain. These data suggest that the metal complex was partially pendant to the graft chain of 4-vinylpyridine.

Characterization by Scanning Electron Microscopy

The following findings were made clear by observing the change in surface and micropore of the PTFE membrane in micrographs given in Figure 12.

In Figure 12-(1) fine fiber-like structure seems to disappear in the microvoids, whereas it could be seen before grafting. This can be attributed to the fact that grafting filled up the void partially. There is no

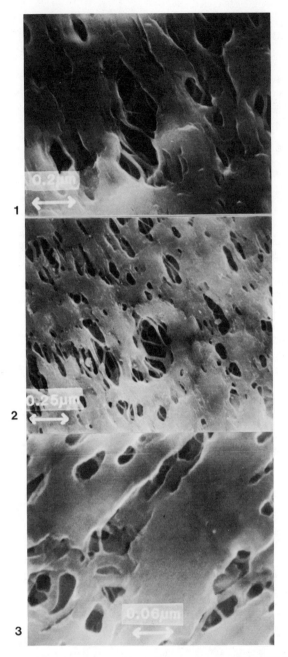

Fig. 12. SEM micrographs of surfaces of various PTFE membranes.
(1) Grafted PTFE; (2) grafted and chelate reacted PTFE;
(3) grafted and complex formated PTFE.

distinct difference between the chelate and complex membranes comparing the
two micrographs of (2) and (3). As can be seen from Figure 12-(3) the
porosity is not entirely covered even in complex membrane by grafting of
4-vinylpyridine and additional complex forming. However, on the whole the
size and percentage of the void in the membrane are clearly decreasing in
the order of grafting, chelating and complex forming of PTFE membrane.

Pervaporation

Several experiments of pervaporation were conducted by liquid permeation system with the downstream side being maintained under vacuum. The membranes bound to $Ni(SCN)_2(4-MePy)_4p$-xylene and $Co(SCN)_2$ $(4-MePy)_4p$-xylene could stand for initial scores of minutes but later collapsed (feed liquid emerged abruptly). In these cases the permeation rates were less than $1 \ g/m^2hr$. However the definite value of permeation rate could not be determined because the membrane was broken scores of minutes after starting.

In the isostatic pervaporation system several runs were done using the clathrate membranes reacted with $Ni(SCN)_2(4-MePy)_4$ p-xylene and $Co(SCN)_2$ $(4-MePy)_4$ p-xylene respectively. Finally the clathrate membrane was coated with thin top layer of PVA by dipping the clathrated membrane in 1% aqueous PVA solution to endow stability. When equal weight percent mixture of o,m and p-xylene was separated with the PVA coated clathrate membrane of $Ni(SCN)_2(4-MePy)_4$ p-xylene in the isostatic pervaporation system, the separation factor was 1.26 for p-xylene to m-xylene although the permeation rate could not be determined.

The transport and separation of xylene isomers can be interpreted in the following manner. Since the inclusion compound has an ability to encapture and release the specific xylene isomer (p-xylene in this case) reversibly, p-xylene species selectively diffuse jumping among the domain of inclusion compound aligned inside the microvoid of porous PTFE membrane.

At this moment the problems to be solved are as follows:

(1) effect of thin PVA coating on separation,
(2) effect of partial vapor pressure in the evaporation process of downstream side,
(3) determination of permeation rate,
(4) increasing the permeability and selectivity for p-xylene.

To clarify these problems more experiments and further consideration will be needed.

CONCLUSIONS

The main conclusion of this research effort was that it was found possible to attach metal complex to the microvoid of porous membrane by the exchange of the pyridyl group of metal complex and the pyridyl group of graft polymer, and thus provide a p-xylene selectivity to porous membrane, although few questions remained to be solved.

Furthermore, the effects of various factors such as reaction time, concentration, continuous or intermittent procedure, grafting percent and composition of ligand on immobilization of metal complex into membranes were elucidated to optimize the experimental condition of immobilization.

To extend the scope of future investigations, there seem to be scores of possible highly selective carrier materials to be examined. In particular, there are many different types of clathrates and metal complexes which seem to be worthy of investigation. Another important point in future study is development of better method for immobilizing metal complex in membrane so that the immobilized carrier may exhibit high selectivity.

REFERENCES

1. A. S. Michaels, R. F. Baddour, H. J. Bixler, and C. Y. Choo, Ind.Eng. Chem.Process Des.Dev., 1:14 (1962).
2. J. G. Sikonia and F. P. McCandless, Separation of isomeric xylenes by permeation through modified plastic films, J.Memb.Sci., 4:229 (1978).
3. C. H. Lee, Synthetic membranes containing schardinger cyclodextrin additives, J.Appl.Polym.Sci., 26:489 (1981).
4. W. D. Schaeffer, W. S. Dorsey, D. A. Skinner, and C. G. Christian, Separation of xylenes, cymenes, methylnaphthalenes and other isomers by clathration with inorganic complexes, J.Am.Chem.Soc., 79:5870 (1957).
5. P. de Radzitzky and J. Hanotier, Clathration of hard-to-separate aromatic mixtures with new werner complexes, Ind.Eng.Chem.Process Des.Dev., 1:10 (1962).

MATHEMATICAL MODELING OF COLLOIDAL ULTRAFILTRATION: A METHOD FOR ESTIMATION

ULTRAFILTRATION FLUX USING EXPERIMENTAL DATA AND COMPUTER SIMULATIONS

R. Lundqvist and R. von Schalien

Heat Engineering Laboratory, Department of Chemical
Engineering, Åbo Akademi, Biskopsgatan 8
SF-20500 Åbo 50, Finland

ABSTRACT

Ultrafiltration of a colloidal suspension, an oil-water emulsion, was
investigated in a pilot-plant apparatus with tubular membrane modules. A
model presented in the literature for describing the permeation rate based
on the Hagen-Poiseuille equation was used and modified. A method was
developed for estimating fluxes utilizing a small amount of experimental
data and using a parameter estimation procedure based on simulation.

INTRODUCTION

The increasing use of the ultrafiltration technique has created a
need for accurate mathematical models for predicting flux and retention
in ultrafiltration systems for design, process control and cost estimation
purposes. This has naturally led to several different approaches for
solving these problems being presented in the literature.

For macromolecular solutions the mass balance approach in the region
near the membrane surface resulting in the concentration polarization model
has generally been accepted. This model shows that the water flux J_w
through the membrane can be described by

$$J_w = k \ln\left(\frac{c_w - c_p}{c_b - c_p}\right)$$
(1)

Factor k is the mass transfer coefficient, the magnitude of which can
be expressed as a function of Reynolds and Schmidt numbers and the given
membrane module geometry.

In most ultrafiltration systems of interest the surface concentration
c_w often exceeds a certain concentration limit, resulting in a gel layer
with an assumed constant concentration c_g at the membrane surface. When
this occurs, c_w in Equation (1) is replaced with c_g.

When, however, Equation (1) or the modified "gel-polarized" form of
Equation (1) is used for estimating fluxes in ultrafiltration of colloidal

suspensions, the calculated values have often been shown to be one to two orders of magnitude lower than experimentally measured values. These discrepancies, which are partly due to low diffusivities of colloids, indicate that other transport phenomena than the convective transport of the solute towards the membrane surface and the backdiffusion due to the concentration gradient exist and substantially augment the permeation rate.

Several hypotheses with different mechanisms have been prepared for explaining the observed differences. Porter[1] has suggested that the radial migration of colloidal particles from the membrane surface, the so-called tubular pinch effect, increases the mass transfer in addition to that due to the concentration gradient. For ultrafiltration in the turbulent stream region Friedrich and Tietze[2] have proposed that turbulence balls affect the gel layer and augment the mass transfer. Madsen and Nielsen[3] have suggested that stresses in the membrane arise, causing some kind of oscillations in the membrane and the gel layer, and that these result in an increase in flux. Blatt et al.[4] have approached the problem by assuming that the fluid shear in the vicinity of the membrane surface causes the whole gel layer, or part of it, to flow in an axial direction in a membrane module, i.e. parallel to the membrane surface.

The first three mechanisms have all proved to exist in membrane ultrafiltration but not to such an extent as to quantitatively describe fluxes in colloidal systems.

Gernedel[5] has used the last hypothesis of the axial flow of the gel layer to determine fluxes in ultrafiltration of milk and whey. This approach is based on the Hagen-Poiseuille equation describing the flux by

$$J_w = \frac{\Delta p}{32} \frac{1}{\xi_m + \xi_g} \tag{2}$$

The resistance ξ_m is obtainable from pure water flux and thus has a constant numerical value, while the gel-layer resistance ξ_g is a function of the process variables, the physical properties of the bulk fluid and the membrane module geometry. The mathematical relationship for the gel-layer resistance obtained from experimental data and by theoretical calculation from the radial and axial transport mechanisms is given in Gernedels's approach by

$$\xi_g = \beta_1 \frac{1 + \beta_2 \Delta p}{1 + \beta_2 p_1} \frac{v_b c_b}{\tau_w} (\frac{L}{\beta_3})^{(\beta_4 + c_b)} \tag{3}$$

In this paper a similar approach is described to that of Gernedel in mathematical modelling of colloidal ultrafiltration. This includes a method for obtaining models in an easy way by making use of a small amount of experimental data and computer simulations.

EXPERIMENTAL

The experiments needed for the modelling were carried out in a pilot-plant apparatus sketched in Figure 1. The construction of this apparatus made it possible to perform experimental runs in a batchmode and in a continuous mode.

The modules used were two Abcor HFM-100 tubular, polyamide membrane modules with a nominal cut-off value of 10.000 Dalton connected by means of a U-bend. Two different set-ups of modules were employed, 2.76 m and

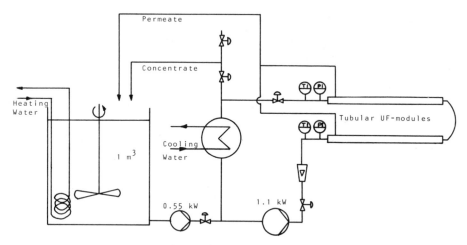

Fig. 1.   Pilot-plant apparatus.

1.355 m in membrane length, both with an inner diameter of 25.4 mm.   The
modules were in no way modified in the entrance and outlet regions, thus
assuring that the experiments were as realistic as possible.

A completely emulsified oil-water emulsion was chosen to serve as a
working fluid.   The oil, Shell Dromus BS, contained an emulgator to ensure
that the oil droplets in the solution remained in a stable form.   The water
used was softened and ultrafiltered, giving a numerical value of $2 \times 10^{11} \, m^{-1}$
for the membrane resistance ξ_m.   Given these circumstances, the working
fluid could be regarded as semi-realistic.

A well-chosen experimental plan was designed in order to minimize the
number of experiments.   The process variables were varied in the following
ranges; temperature 20-50°C, pressure 1.7-3.6 bars, volumetric flow rate
1.6-2.5 dm^3/s, oil concentration 1-35% (v/v).

During the experiments the temperature in the system and the pressure
before and after the modules were registered with visual temperature and
pressure indicators.   The volumetric flow rate in the system was registered
with a Fischer & Porter flowmeter and the flux was measured volumetrically.
The oil concentration in the permeate was measured with a Perkin-Elmer
UV-spectrophotometer at a wavelength of 300 nm.   After each experiment a
washing technique similar to that described in Madsen[6] was used.

In each experiment a rapid flux decline was observed during the first
five minutes of the experiments.   During the period following this a steady
state was obtained, which lasted throughout the experimental run time, i.e.
10 to 12 hours.   In every experiment the oil concentration in the permeate
was under 10 ppm.   After each experiment the system was flushed with water
restoring the flux to 70 - 90% of pure water flux.   This indicated that the
gel layer was of a porous structure, which is a condition when using the
Hagen-Poiseuille equation.

RESULTS AND DISCUSSION

As it was neither possible, desirable nor expedient for practical
reasons to determine the gel-layer resistance model by using only experimental data, a parameter estimation method based on simulation was
developed.

This approach in modelling comprises two separate parts, i.e. the hardware and the software of the model. The hardware of a model consists of mass or energy balance, which must always be valid in the model. The software of a model consists of expressions which usually describe transport phenomena specifying the redistribution rates of the balance elements.

The hardware in this case was the energy balance, which after some simplifications meant that the pressure loss Δp_{loss} due to friction in the membrane module system could be written:

$$\Delta p_{loss} = \frac{\rho_b w^2}{2} \left(\frac{L}{d}\zeta_d + \zeta_{el}\right) \tag{4}$$

As the total pressure loss in the module system could be measured, the factor ζ_d was easily obtained by means of Equation (4). This made it possible to determine the transmembrane pressure at every position in the modules.

The Hagen-Poiseuille Equation (2), including the gel-layer resistance ξ_g with an unknown structure, represented the software. The main point was to find the most suitable structure for this resistance, or in other words, to adapt the model to reality. A computer program for simulating the process, including an optimization procedure for obtaining a model of the gel-layer resistance, was therefore constructed.

With this program the process was simulated, utilizing the process variables from each experiment. The modules were divided into several sections, in which the local transmembrane pressure and the gel-layer and the membrane resistance determined the local flux. The simulated overall flux J_{ov}, which can be defined by

$$J_{ov} = \frac{\int_A J_w dA}{A} = \frac{\int_L J_w dL}{L} \tag{5}$$

was compared with experimental data and an algorithm, a Quasi-Newton optimization routine, was used to change the numerical values of the parameters in the gel-layer resistance model in order to minimize the least-squares of simulated and measured data. The integral in Equation (5) was solved with the aid of Euler's method and an integration length of 0.55 m for the 2.76 m membranes was observed to be sufficient to obtain the desired accuracy. The flowsheet for the whole procedure is presented in Figure 2.

Using the presented method, the following structure for the gel-layer resistance model was obtained for the 2.76 membrane modules

$$\xi_g = \beta_1 \cdot (\Delta p)^{\beta_2} \cdot \frac{v_b c_b}{\tau_w} \cdot \left(\frac{L \cdot \beta_3}{\tau_w}\right)^{(c_b + \beta_4)} \tag{6}$$

with the optimal numerical values β_i

$\beta_1 = 0.88 \ 10^{19} \ kg/m^4 s$

$\beta_2 = 2.80$

$\beta_3 = 78.0 \ kg/m^2 s^2$

$\beta_4 = 0.812$

The similarity of the expressions (3) and (5) is remarkable but with the latter expression the least-square values were reduced to half of those

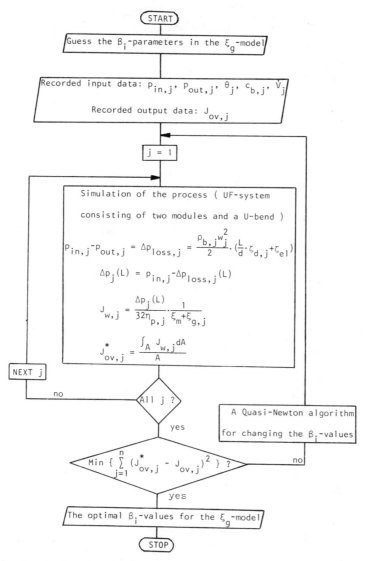

Fig. 2.   Flowsheet for the parameter adoption procedure. J_{ov}^* = simulated overall flux, J_{ov} = recorded overall flux.

obtained with expression (3). The term $\beta_1(\Delta p)^{\beta_2}$ derived from conventional filtration was physically interpreted with the compaction of the gel-layer occurring in ultrafiltration and making the flux almost invariant of the transmembrane pressure. The term $(L \beta_3/\tau_w)^{(c_b+\beta_4)}$ was explained with the gradual build-up of the gel-layer along the membrane channel due to the transport mechanisms involved in the vicinity of the membrane surface. This term indicated that a considerably higher flux was obtained in the entrance region of the modules (up to 1 m) than downstream from the modules. In our approach heavier emphasis was laid on the shear stress in the length term owing to the experimental device design, which was assumed to cause an enhanced mass transfer in the entrance region.

The influence of the process variables on flux behavior and the good agreement between simulated fluxes and experimental data are illustrated in Figures 3, 4, 5 and 6.

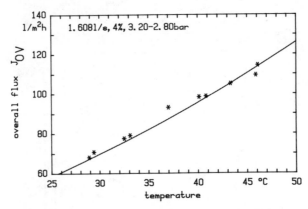

Fig. 3.   Ultrafiltration flux as a function of average transmembrane
pressure in the UF-system (* = experimental data).

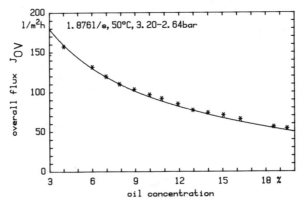

Fig. 4.   Ultrafiltration flux as a function of volumetric flow rate in the
UF-system (* = experimental data).

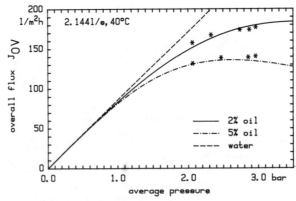

Fig. 5.   Ultrafiltration flux as a function of temperature (* =
experimental data).

    The obtained gel-layer resistance model (6) as well as the model
derived theoretically with the assumed transport phenomena predicted a
10 to 30%-increase in flux when using membrane modules of half the length

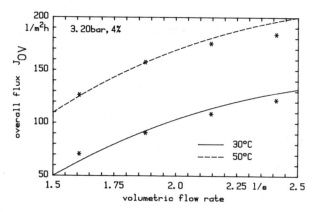

Fig. 6.   Ultrafiltration flux as a function of bulk concentration (* = experimental data).

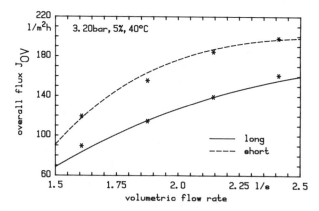

Fig. 7.   Ultrafiltration flux versus temperature for long and short membrane modules in the UF-system (* = experimental data).

of the previously used 2.76 m ones. To confirm the validity of this prediction an experimental series with 1.355 m modules was carried out under the same conditions as for the 2.76 m modules. In every experiment carried out an expected high flux value was observed for the shorter membranes. This is illustrated for some example in Figures 7 and 8. Thus, the conclusion was that the gel-layer was gradually built up in the membrane channel and that the assumed axial movement of the layer really occurred.

Slightly different numerical values for the parameters in the model (6) were obtained when the system with shorter membrane was simulated. This was explained by the sensitivity in measuring and determining the pressure drop in these shorter modules. The optimal parameters for the long and the short modules and the combination of both are presented in Table 1.

CONCLUSIONS

The described method could be used as an excellent tool to estimate fluxes for different kinds of modules and applications. The benefit of the procedure is that only a small amount of experimental data is needed, which

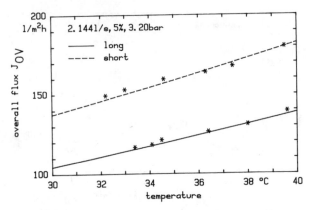

Fig. 8. Ultrafiltration flux versus volumetric flow rate for long and short modules in the UF-system (* = experimental data).

Table 1. The optimal β-parameters.

	$\beta_1 / (kg/m^4 s)$	β_2	$\beta_3 / (kg/m^2 s^2)$	β_4
short modules	$0.88 \cdot 10^{19}$	2.80	78.0	0.812
long modules	$0.70 \cdot 10^{19}$	2.71	104.0	0.910
combination	$0.77 \cdot 10^{19}$	2.74	97.5	0.844

is desirable in chemical engineering. An approach like this also implies that only the input variables - pressure, volumetric flow rate, temperature, bulk-stream concentration - and the output variable - flux - have to be measured, while the hardware and the software of the model explain what happens in the membrane modules. In other words, this means that no interference in the module structure itself is necessary to obtain the influence of membrane length on flux.

The working fluid used in these experiments could be regarded as semi-realistic. In real systems, however, with runs lasting for days or weeks fouling and incomplete retention of the solute occurs. Emphasis has to be laid on these problems in further developments.

The influence of membrane length on ultrafiltration flux implies that shorter membranes are more advantageous in full-scale ultrafiltration plants, as the permeation rate capacity is 10 to 30% higher compared with that of longer membranes. This has an economic consequence as the capital costs are in this way reduced.

Acknowledgements

The authors wish to thank Svenska Tekniska Vetenskapsakademien i Finland, Svenska Vetenskapliga Centralrådet and Stiftelsen för Teknikens Främjande for the financial support.

NOMENCLATURE

A	– membrane area (m^2)	r	– radius
c_b	– solute concentration in the bulk fluid (%)	$\dot{V}$	– volumetric flow rate (dm^3/s)
c_g	– solute concentration in the gel-layer (%)	w	– average velocity of the fluid (m/s)
c_p	– solute concentration in the permeate (%)	β_i	– proportionality constants (–)
c_w	– solute concentration at the membrane surface (%)	ζ_d	– friction factor in the UF-module (–)
d	– diameter (m)	ζ_{el}	– friction factor for the U-bend (–)
J_{ov}	– overall flux ($1/m^2h$)	η_p	– dynamic viscosity of the permeate (kg/ms)
J_w	– flux ($1/m^2h$)	Θ	– temperature (°C)
k	– mass transfer coefficient (m/s)	ν	– kinematic viscosity of the bulk fluid (m^2/s)
L	– membrane channel length (m)	ξ_g	– gel-layer resistance (m^{-1})
P_{in}	– inlet pressure in the UF-system (bar)	ξ_m	– membrane resistance (m^{-1})
P_{out}	– outlet pressure in the UF-system (bar)	ρ_b	– density of the bulk fluid (kg/m^3)
Δp	– transmembrane pressure (bar)	τ_w	– shear stress at the membrane surface (kg/ms^2)
Δp_1	– reference pressure (bar)		
Δp_{loss}	– pressure loss in the UF-system due to friction (bar)		

REFERENCES

1. M. C. Porter, Concentration polarization with membrane ultra-filtration, Ind.Eng.Chem.Prod.Res.Develop., 11:234 (1972).
2. F. Friedrich and R. Tietze, Beeinflussung und Vorausbestimmung der Filtrationsleistung bei der Ultrafiltration, Chem.Techn., 26:688 (1974).
3. R. F. Madsen and W. K. Nielsen, Thin-channel ultrafiltration, theoretical and experimental approaches, in: "Ultrafiltration Membranes and Applications," A. R. Cooper, ed., Plenum Press, New York (1980).
4. W. F. Blatt, A. Dravid, A. S. Michaels, and L. Nelsen, Solute polarization and cake formation in membrane ultrafiltration: Causes, consequences and control techniques, in: "Membrane Science and Technology," J. E. Flinn, ed., Plenum Press, New York (1970).
5. C. Gernedel, "Uber die Ultrafiltration von Milch und die den Widerstand der Ablagerungsschicht beeinflussenden Faktoren," Dissertation, Die Technische Universität, München (1980).
6. R. F. Madsen, "Hyperfiltration and Ultrafiltration in Plate-and-Frame Systems," Elsevier, Amsterdam (1977).

PERMEABILITY AND STRUCTURE OF PMMA STEREOCOMPLEX

HOLLOW FIBER MEMBRANE FOR HEMODIALYSIS

T. Kobayashi*, M. Todoki**, M. Kimura*, Y. Fujii*
T. Takeyama* and H. Tanzawa**

*Toray Industries, Inc. Otsu Shiga 520
Japan
**Toray Research Center, Inc. Otsu, Shiga 520
Japan

INTRODUCTION

The characteristics of the structure and the properties of polymethyl methacrylate (PMMA) hollow fiber are its uniformity of structure, hydrophobic property and wider performance compared with cellulosic membranes[1,2]. For these characteristics its structural analysis will be simplified.

As the transport model describing the flow of solute for a liquid-liquid system, the solution-diffusion theory for a reverse osmosis membrane[3] and a capillary model theory for ultrafiltration membranes[4] have hitherto been accepted. For hemodialysis membranes, the pore size of which is between that of the previous two, there are several arguments which are appropriate. We will discuss the properties of PMMA membranes using a capillary model theory in this study.

Recently Todoki and Ishikiriyama[5] found for PMMA hydrogel membrane a relation between capillary radius and the lowering of freezing point detected with DSC analysis. We assume that the capillary condensation effect holds for hydrogel membranes. The membrane structure derived from the DSC analysis showed relatively good coincidence with the structure assumed by a tortuous capillary model theory. As a result of this we can get discrete pore radii for several membranes used in hemodialysis.

BACKGROUND

When we have studied the freezing-melting behavior of PMMA hydrogel membranes we have found that some of the water exerts an affect on the freezing point. According to our studies the hydrogel membrane may contain water having a continuous freezing point distribution determined by pore radius. This is not bound water but capillary water which shows a capillary condensation effect based upon a radius distribution. DSC analysis has been utilized for the estimation of pore radius of inorganic porous materials like silica gels and porous glasses[6]. L.G. Homshaw is the only researcher known to us who has applied the method to a hydrogel polymer[7].

The relationship between pore radius and lowering of freezing temperature is described as follows[8]:

Using Kelvin's equation;

$$RT \ln P_r/P_o = (2\sigma V_m \cos\theta)/_r \tag{1}$$

where P_r is the vapor pressure in the capillary of radius r, P_o is the vapor pressure of free water; σ, V_m and θ are surface tension, molar volume and contact angle of the water, respectively, and relating this to Gibbs-Duhem's equation, one can obtain

$$\Delta T = (2\sigma_{iw} V_m T_o \cos\theta)/_{r\Delta Hm} \tag{2}$$

where ΔT is lowering of freezing point of capillary water, σ_{iw} is interfacial tension between ice and water, ΔH_m is transition enthalpy, T_o is 273.1K°. The relation between capillary radius and freezing point calculated according to the Equation (2) on the assumption that σ_{iw} is 0.01 N/m, V_m is 18.02 cm³/mol, θ is 0°, is shown in Figure 1.

A heating DSC curve was analyzed to obtain melting peak temperatures (T_p °C; freezing point) and Peak area (ΔH cal/g). The number of pores is assumed to be proportional to $\Delta H T_p^2$, because T_p is reciprocal to pore radius, while ΔH is proportional to total water content, that is, total pore volume, hence

$$\text{Number of pores} = \frac{\text{Total pore volume}}{\text{Volume of a pore } (\pi r^2 \ell)}$$

$$\propto \frac{\Delta H}{(1/T_p)^2} = \Delta H \cdot T_p^2 \tag{3}$$

The capillary model theory using tortuous factor was adopted to estimate the porous structure of PMMA membranes from the permeability data[9]. We assume that the membrane structure is uniform through the depth of membrane and that the pore radius is sufficiently large compared with that of a water molecule. In this case we can use Hagen-Poiseuille's law to show water flux J_ν as follows:

$$J_\nu = \frac{n R_p^4}{8\ell\eta} \tag{4}$$

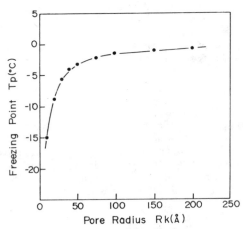

Fig. 1. Correlation between the freezing point of water in capillary and its radius based on a capillary condensation effect.

508

where R_p and n is pore radius and pore density, ℓ is the mean length of a pore, η is viscosity of water. Expressing the volume fraction of pores (porosity) in the membrane as ε

$$\varepsilon = \eta R_p^{\,2} \, \ell n/\lambda \tag{5}$$

where λ is wall thickness of membrane. t_w $(=\ell/\lambda)$ is defined as the tortuous factor. Giving a practical value to t_w, pore radius R_p and Pore density n (pore number/unit membrane area) can be estimated from the experimental values of permeability J_v and the membrane structural factors ε and λ. In this study, we give $t_w = 1.5$. This value was obtained from earlier studies about the permeability of water and urea for PMMA membranes[2].

EXPERIMENTAL

Polymers polymerized anionically and radically were used as isotactic and syndiotactic elements, respectively. The mixing ratio of isotactic to syndiotactic elements was varied from 1:5 to 1:2. Two polymer types were used to produce the membrane studied. One is Homo-PMMA (H-PMMA) which is composed of pure MMA monomer unit in either the isotactic or syndiotactic. The other is C_o-PMMA which is composed of pure isotactic PMMA and syndiotactic PMMA copolymerized with Sodium-p-styrenesulfonate. DMSO was used as a solvent. The polymer concentration in solution was varied between 20 and 35 wt%. These mixtures were heated and stirred above the gelation temperature of the solution to make the spinning solution. This solution was extruded through annular spinnerets into a cooling air zone, then in a coagulant bath to gel the structure and extract the solvent.

The water permeability coefficient L_p is defined by

$$J_V = L_p \, \Delta P$$

The expression of water permeability used is $\lambda L_p^{\,1}$.

A heating DSC measurement at a heating rate of 2.5°C/min was adopted to avoid any supercooling effect. Temperature range was from -45°C to +15°C. Heating curve for a typical sample is shown in Figure 2. In the Equation (1) temperature of Peak II was used as T_p (°C) and peak area as ΔH.

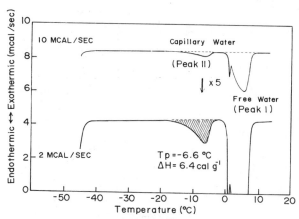

Fig. 2. A typical heating DSC curve for a PMMA hydrogel membrane.

RESULTS AND DISCUSSION

The dependence of water permeability (λL_p) on porosity (ε) is shown in Figure 3. Membranes can be formed from PMMA which have a wide range of performance permeability by varying the membrane forming conditions. Thus we can produce membranes for ordinary hemodialysis, high flux hemodialysis, hemofiltration and protein permeable hemofiltration. It is clear from Figure 3 that λL_p is directly related to porosity. The permeability λL_p of H-PMMA is higher than that of C_o-PMMA at the same porosity, which suggest that H-PMMA gives a membrane having a larger pore size and smaller pore density than that of C_o-PMMA.

In order to elucidate this observed difference, the pore structure analysis with DSC method and structural estimation from hydraulic data using tortuous capillary model were performed. The capillary radius (R_p) and pore density (n) were calculated according to the tortuous capillary theory from obtained porosity and permeabilities. Their relation to λL_p are shown in Figure 4. It is recognized from this figure that R_p increases and n decreases with an increase of λL_p. As to the polymer types, H-PMMA gives a larger pore radius and smaller pore number than C_o-PMMA at every

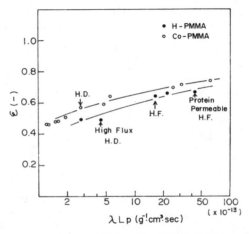

Fig. 3. Porosity vs. permeability for PMMA hollow fiber membranes.

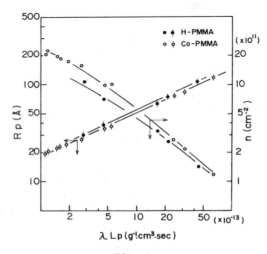

Fig. 4.

permeability. We suppose, therefore that H-PMMA is suitable as a high flux membrane for hemodialysis because of its good permeability of middle molecular weight substances for its large pore sizes. Figure 5 shows the relation between melting peak temperature and water permeability. The membrane having lower permeability shows a greater lowering of freezing temperature. As to the polymer types, H-PMMA gives a smaller lowering of melting peak temperature than Co-PMMA at every permeability. By using the relation between T_p and R_k shown in Figure 1 and the relation between T_p and λL_p, we may obtain the relation between R_k and λL_p shown in Figure 6. By calculation from Equation (3), we may derive the relationship between λL_p and $\Delta H T_p^2$. These are also shown in Figure 6. We find it remarkable that the results shown in Figure 6 agree so well with that obtained by use of that tortuous capillary model. The correlation of pore radius (R_p) from tortuous capillary model and R_k derived from DSC analysis is shown in Figure 7.

Since the agreement of these values is so close, we propose that the melting peak temperature can be used as measure of pore radius for PMMA hydrogel membranes. In many previous studies of water in hydrogels anal-

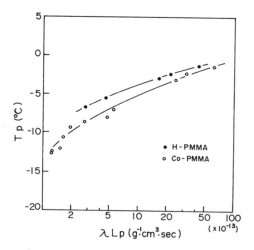

Fig. 5. Melting peak temperature on heating DSC curve vs. permeability for PMMA membranes.

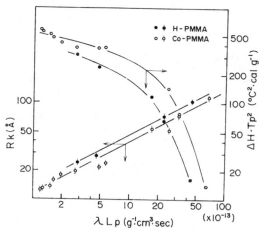

Fig. 6. R_k and $\Delta H \cdot T_p^2$ from DSC analysis vs. permeability for PMMA membranes.

yzed by DSC, lowering of freezing point is attributed to bound water[10]. In our studies of PMMA we attribute this lowering of freezing point to capillary water. There exists a good correlation between R_p and R_k in the region of 50Å to 100Å but R_p is greater than R_k in the region below 50Å. We may suppose that our assumption of a tortuousity factor t_w=1.5 is valid in the higher pore radius region but should be decreased in the lower region. By taking a value of t_w=1.0, R_k and R_p in the lower region coincide well with each other.

Tortuousity factor has been well investigated in the detailed studies of cake filtration mechanism[11]. According to Kozeny-Carman's equation t_w is predicted to be $\sqrt{2}$ [11]. The concepts underlying Kozeny-Carman's equation have been applied to membrane permeability investigation. It is very interesting that our t_w agrees well with that obtained from macroscopic filtration studies. The values of R_k for various PMMA membranes and the pore radii for other membranes obtained by Klein[4] are shown in Table 1.

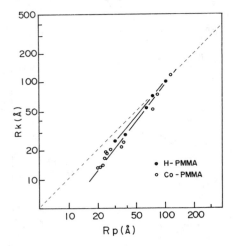

Fig. 7.   Correlation of R_k (Å) and R_p (Å).

Table 1. R_k for Various PMMA Membranes and the Pore Radii for Other Polymers Obtained by Klein[4].

PMMA membranes	R_k	Pore radius obtained by Klein[4]	
Ordinary Hemodialysis (B-2)	20Å	Cellulose (Cuprophan)	17.2Å
		Polycarbonate	22.4Å
High Flux Hemodialysis (B-1)	29Å	Cellulose acetate	25.2Å
Hemofiltration (B1-L)	53Å	Polyacrylonitrile	54.8Å
Protein Peameable Hemofiltration TK-401, BK	100Å		——

CONCLUSION

Membranes which have a wide range of permeability performances such as membranes for ordinary hemodialysis, high flux hemodialysis, hemofiltration and protein permeable hemofiltration, can be formed from PMMA stereocomplex polymers by varying the membrane forming condition, the ratio of isotactic to syndiotactic polymer, the concentration of polymer in solution and the cooling condition of the hollow fiber.

The pore structure analysis with DSC method and structural estimation from hydraulic data using tortuous capillary model were performed.

DSC method gives effective information about the structure of PMMA hydrogel membranes in the pore radius range of 15Å to 100Å. We can estimate pore radius from measurement of melting peak temperature of a heating DSC curve and it coincides well with that from the estimation using tortuous capillary model.

Pore radii of membranes for ordinary hemodialysis, high flux hemodialysis, hemofiltration and protein permeable hemofiltration were estimated as 20Å, 29Å, 53Å, 100Å respectively.

REFERENCES

1. Y. Sakai, S. Hosaka, and H. Tanzawa, J.Appl.Polym.Sci., 22:1805 (1978).
2. Y. Sakai, S. Hosaka, H. Tanzawa, and M. Itoga, Kobunshi-ronbunshu, 34:801 (1977).
3. S. Kimura and S. Sourirajan, AIChEJ., 13:497 (1967).
4. E. Klein, F. F. Holland, and K. Eberle, J.Memb.Sci., 5:173 (1979).
5. M. Todoki and K. Ishikiriyama, unpublished.
6. R. Kondo, "Takozairyo," Gihodo, Japan (1963).
7. L. G. Homshaw, J.Colloid Interface Sci., 9:2363 (1965).
8. S. Deodhar and P. Luner, "Water in Polymer," ACS Symp. Ser. (1980).
9. M. Nakagaki, Membrane, 6(3):1 (1981).
10. K. Nakamura, T. Hatanaka, and H. Hatakeyama, Text.Res.J., 51:607 (1981).
11. P. C. Carman, J.Soc.Chem.Ind., 58:1 (1939).

AN IMPROVED MODEL FOR CAPILLARY MEMBRANE FIXED ENZYME REACTORS

J. E. Prenosil and T. Hediger

Technisch-chemisches Laboratorium ETH
CH-8092 Zurich
Switzerland

ABSTRACT

A realistic model for the diffusional type of an asymmetric hollow fiber enzyme reactor with non-linear reaction kinetics is presented. An efficient collocation method was used for the steady-state solution resulting in a substantial decrease of the computing time. The results are presented in a concise graphical form.

A membrane reactor based on a hollow fiber module Amicon H1P3-20 was constructed for continuous hydrolysis of 5% lactose. Enzyme β-galactosidase was immobilized in the backflush mode into the porous structure of the membrane and high specific activity was achieved. A significant increase in the enzyme stability was observed after the cross-linking with glutaraldehyde. In this way an enzyme membrane reactor with performance comparing well with that of a fixed bed reactor of similar dimensions was developed.

The experimental results of the continuous lactose hydrolysis agreed very well with the values predicted by the model simulations. It could be shown the operation conditions of this type of enzyme membrane reactors are largely such that both kinetic and mass transfer effects are important. This means that the simplified asymptotical models, which have been frequently suggested in the literature, should not be used without great caution.

INTRODUCTION

Modern membrane systems provide a suitable means of retaining enzymes and cells within a biological reactor. Generally, two basic types with different modes of operation are possible:

a) Convection, or Ultrafiltration (UF) Type. The retention of enzymes or whole cells and their contact with substrate is achieved on one side of the membrane, usually in a stirred tank. One mode of operation would allow only the substrate and product to pass through the membrane, while the enzyme or cells are rejected and recycled at a high rate back through the reactor. Substrate is added continuously, and the net result is a continuous stirred tank reactor with essentially constant enzyme or cell concentration (Figure 1A). Such a reactor was successfully used for coenzyme depending reaction systems[1] and recently commercialized for 1-amino acid production by Degussa, West Germany. The main drawback of this type of reactor is that the enzyme retention is not absolute and even membranes with a very low nominal cut-off will let a certain, however small, portion of enzyme through. The enzyme losses become significant as the operation time of the reactor increases and an additional enzyme amount must be supplied to the reactor. Also, enzyme, present in solution, is usually less stable than in its immobilized form.

b) Diffusion Type. The biocatalyst is retained on one side of the membrane and the substrate flows tangentially on the other side. The reaction process requires the diffusion of substrate and products through the membrane. With modern asymmetric membranes, diffusion paths can dramatically be reduced by immobilizing the catalyst in the pores close to the inner membrane, which is the actual separation membrane. Both, flat and hollow fiber membrane systems can be operated in this fashion. The later system is shown in Figure 1B; the biocatalyst is retained on the shell side and the reacting substrate flows through the tubes. This type of membrane reactor has been increasingly studied[2-12].

The diffusional type of membrane reactor was chosen to be investigated here, because of its number of distinct advantages over the ultrafiltration type.

THEORY

The membrane may be applied not only as a barrier between two liquid phases, but also as a matrix for cell or enzyme immobilization. Let us consider such a reactor in which the biocatalyst is immobilized in the walls of a tubular membrane with the substrate flowing through its interior called lumen. The substrate diffuses from the solution into the porous support structure of the membrane capillary, undergoes an enzymatic reaction and diffuses back in the bulk solution. Such a system is shown schematically in Figure 2.

At steady-state, a mass balance on elements of fluid in the lumen (region 1), membrane skin (region 2) and porous structure (region 3) leads to a system of differential equations. Written in cylindrical coordinates:

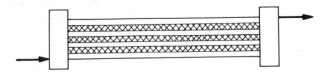

B) Continuous hollow fiber UF membrane reactor

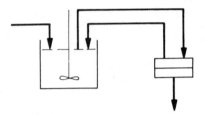

A) Tank reactor with UF membrane on exit stream

Fig. 1.

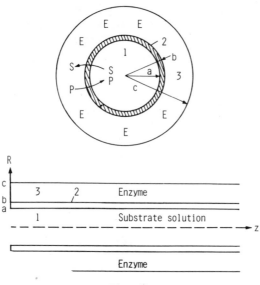

Fig. 2.

$$D_1 \left(\frac{\partial^2 c_1}{\partial r^2} + \frac{1}{r}\frac{\partial c_1}{\partial r}\right) = u_z \frac{\partial c_1}{\partial z} \qquad \text{Region 1} \qquad (1)$$

$$D_2 \left(\frac{\partial^2 c_2}{\partial r^2} + \frac{1}{r}\frac{\partial c_2}{\partial r}\right) = 0 \qquad \text{Region 2} \qquad (2)$$

$$D_3 \left(\frac{\partial^2 c_3}{\partial r^2} + \frac{1}{r}\frac{\partial c_3}{\partial r}\right) = v \qquad \text{Region 3} \qquad (3)$$

The terms on the left-hand sides represent the radial diffusion of substrate. The right-hand side of Equation (1) is the convection term, where for laminar flow:

$$u_z = u_o \left(1 - \frac{r^2}{a^2}\right) \qquad (4)$$

On the right-hand side of Equation (3) is a kinetic expression for the considered reaction.

Equation (1) is a parabolic partial differential equation, which is known as the Graetz Problem. An analytical solution is possible, depending on boundary conditions. Equation (2) has analytical solution, and Equation (3) can be solved analytically only in case of a linear kinetic expression.

The solution for a simplified linearized problem was found by Waterland et al.[4] in a numerical-analytical form. Lewis and Middleman solved the problem also in a simplified manner for low Thiele moduli ($\phi^2 < 1$). Mashelkar and Ramachandran[13] postulated an axial diffusion model for the linear case which could be solved analytically. Another simplified solution for the case of first-order kinetics was presented by Kim and Cooney[14]. However, in all these studies first-order kinetics were used which do not apply to enzymatic lactose hydrolysis.

Differential balance equations similar to Equations (1-3) can be written also for the reaction products. Whether they must be used depends on the kinetic model. In this case a simple Michaelis-Menten kinetics with product inhibition was used:

$$v = \frac{v_{max} \, c_3}{c_3 + K_m \{1 + (c_o - c_3/K_i)\}} \tag{5}$$

and the product concentrations could be calculated by stoichiometry. Boundary conditions in the lumen (region 1) are:

$$\frac{\partial c_1}{\partial r} = 0 \qquad\qquad \text{for } r = 0 \tag{6}$$

$$D_1 \frac{\partial c_1}{\partial r} = D_2 \frac{\partial c_2}{\partial r} \qquad\qquad \text{for } r = a \tag{7}$$

with an initial condition:

$$c_1(r,z) = c_o \qquad\qquad \text{for } z \leqq 0 \tag{8}$$

In the UF membrane (region 2):

$$D_2 \frac{\partial c_2}{\partial r} = D_3 \frac{\partial c_3}{\partial r} \qquad\qquad \text{for } r = b \tag{9}$$

and in the porous fiber structure (region 3):

$$\frac{\partial c_3}{\partial r} = 0 \qquad\qquad \text{for } r = c \tag{10}$$

Equation (6) provides for axial symmetry and Equations (7) and (9) reflect the continuity of the mass transfer across the region boundaries. The reflecting boundary at $r = c$ (no substrate can leave the fiber) is satisfied by Equation (10).

The equilibrium concentrations at the region boundaries are given by the partition coefficient γ, assumed to be the same for both boundaries:

$$\gamma c_1(a,z) = c_2(a,z) \tag{11}$$

$$c_2(b,z) = \gamma c_3(b,z) \tag{12}$$

Solutions to Equations (1-12) must be obtained by numerical methods; for this problem orthogonal collocation[15,16] seems to be the most suitable method and is used here.

Converting the model equations into dimensionless form suitable for orthogonal collocation leads to:

Dimensionless radial variables:

$$R = r/a \tag{13}$$

$$\varepsilon = (r-b)/(c-b) \tag{14}$$

so that $R, \varepsilon \in (0,1)$. Further:

$$\beta = b/(c-b) \tag{15}$$

is a dimensionless geometry parameter of the fiber. Several dimensionless groups were also used:

$$Z = L/(Pe\ a) \qquad\qquad \text{Graetz number} \tag{16}$$

$$Pe = u_o a/D_1 \qquad\qquad \text{Peclet number} \tag{17}$$

The Biot numbers resulted from the analytical solution of Equation (2) for region 2, combined with the boundary conditions, Equations (7) and (9):

$$Bi_1 = \gamma(D_2/D_1)\ \{1/(2\ \ln(b/a))\} \qquad \text{Biot number (region 1)} \tag{18}$$

$$Bi_3 = \gamma(D_2/D_3)\ \{1/(2\ \beta\ \ln(b/a))\} \qquad \text{Biot number (region 3)} \tag{19}$$

and Thiele modulus:

$$\phi^2 = \frac{v_{max}\ (c-b)^2}{K_m\ D_3} \tag{20}$$

Dimensionless concentrations and kinetic parameters can be written:

$$C_i = c_i/c_o \tag{21}$$

$$\xi = c_o/K_i \tag{22}$$

$$\theta = K_m/c_o \tag{23}$$

Collocation model:

First let us consider region 3. The above transformations applied to Equation (3) yield:

$$\frac{1}{\beta+\varepsilon}\ \frac{\partial}{\partial\varepsilon}\ \{(\varepsilon+\beta)\ \frac{\partial C_3}{\partial\varepsilon}\} = \frac{\phi^2\ \theta\ C_3}{C_3\ (1-\theta\xi) + \theta\ (1+\xi)} = v \tag{24}$$

It is convenient to introduce:

$$\eta = (1-\varepsilon)^2 \tag{25}$$

so that the boundary condition (Equation 10) will be automatically fulfilled. Equation (24) becomes then:

$$\frac{4\eta\ \partial^2\ C_3}{\partial\ \eta^2} + 2\ \frac{1 + \beta - 2\eta^{1/2}}{1 + \beta - \eta^{1/2}}\ \frac{\partial C_3}{\partial\eta} = \frac{\phi^2\ \theta\ C_3}{C_3(1-\theta\xi) + \theta(1+\xi)} \tag{26}$$

together with the remaining boundary conditions reduced into:

$$\frac{\partial c_3}{\partial\eta} = -\ Bi_3\ (C_3-C_1) \qquad\qquad \text{for } \eta = 1 \tag{27}$$

Equation (2) for region 2 was solved analytically, the result being implicitly expressed in terms of Biot numbers.

The collocation procedure requires a set of basic functions to describe the concentration profiles. In this solution the Lagrange interpolation polynom was used:

$$y_N(z) = \sum_{j=0}^{N+1} y_j \, L_j(z) \tag{28}$$

Its derivatives at z_i are given by:

$$\left. \frac{dy_N}{dz} \right|_{z=z_i} = \sum_{j=0}^{N+1} \left. \frac{dL_j}{dz} \right|_{z=z_i} y_j(z_j) \tag{29}$$

$$= \sum_{j=0}^{N+1} A_{ij} \, y_j(z_j)$$

and:

$$\left. \frac{d^2 y_N}{dz^2} \right|_{z=z_i} = \sum_{j=0}^{N+1} \left. \frac{d^2 L_j}{dz^2} \right|_{z=z_i} y_j(z_j) \tag{30}$$

$$= \sum_{j=0}^{N+1} B_{ij} \, y_j(z_j)$$

It had to be adjusted to satisfy Equation (27) exactly and Equation (26) at a set of N interior points. This procedure applied to Equation (26) gives for the derivatives:

$$\left. \frac{\partial^2 C_3}{\partial \eta^2} \right|_{\eta_i} = \sum_{j=1}^{N+1} B_{ij} \, C_{3j} \tag{31}$$

$$\left. \frac{\partial C_3}{\partial \eta} \right|_{\eta_i} = \sum_{j=1}^{N+1} A_{ij} \, C_{3j} \tag{32}$$

The boundary condition (Equation 27) becomes:

$$\sum_{j=1}^{N} A_{N+1,j} \, C_{3,j} + A_{N+1,N+1} \, C_{3,N+1} = -Bi_3 \, (C_{3,N+1} - C_{1,M+1}) \tag{33}$$

Solving for the concentration C_3 at the boundary (point N+1) yields:

$$C_{3,N+1} = \frac{-\sum\limits_{j=1}^{N} A_{N+1,j} \, C_{3,j} - Bi_3 \, C_{1,M+1}}{A_{N+1,N+1} + Bi_3} \tag{34}$$

Substitution of Equations (31–34) into Equation (26) with subsequent algebraic manipulation leads finally to:

520

$$\sum_{j=1}^{N} D_{ij}\, C_{3,j} -v(C_i) + Bi^* \, C_{1,M+1} = 0 \qquad\qquad i = 1,N \qquad (35)$$

where:

$$D_{ij} = 4n_i \sum_{j=1}^{N} \left(B_{ij} - \frac{B_{i,N+1}\, A_{N+1,j}}{A_{N+1,N+1} + Bi_3} \right) \qquad (36)$$

$$+ 2\, \frac{1 + \beta - 2n_i^{1/2}}{1 + \beta - n_i^{1/2}} \sum_{j=1}^{N} \left(A_{ij} - \frac{A_{i,N+1}\, A_{N+1,j}}{A_{N+1,N+1} + Bi_3} \right)$$

and:

$$Bi^* = 4n_i\, \frac{Bi_3\, B_{i,N+1}}{A_{N+1,N+1} + Bi_3} + 2\, \frac{1 + \beta - 2n_i^{1/2}}{1 + \beta - n_i^{1/2}}\, \frac{Bi_3\, A_{i,N+1}}{A_{N+1,N+1} + Bi_3} \qquad (37)$$

A similar procedure can be applied to region 1. Equation (2) after the appropriate changes can be written in dimensionless form as:

$$\frac{1}{R}\frac{\partial}{\partial R}\, R\, \frac{\partial C_1}{\partial R} = (1-R^2)\, \frac{\partial C_1}{\partial Z} \qquad (38)$$

Once more a transformation:

$$\rho = R^2 \qquad (39)$$

conveniently takes care of the boundary condition required by Equation (6) and Equation (38) can be rewritten:

$$4\rho\, \frac{\partial^2 C_1}{\partial \rho^2} + 4\, \frac{\partial C_1}{\partial \rho} = (1-\rho)\, \frac{\partial C_1}{\partial Z} \qquad (40)$$

together with the remaining boundary condition:

$$\left. \frac{\partial C_1}{\partial \rho} \right|_{\rho=1} = -Bi_1\, (C_1 - C_3) \qquad (41)$$

The same collocation procedure as for region 3 is applied for Equation (40), with the partial derivatives becoming:

$$\left. \frac{\partial^2 C_1}{\partial \rho^2} \right|_{\rho_i} = \sum_{j=1}^{M+1} \bar{B}_{ij}\, C_{1,j} \qquad (42)$$

and:

$$\left. \frac{\partial C_1}{\partial \rho} \right|_{\rho_i} = \sum_{j=1}^{M+1} \bar{A}_{ij}\, C_{1,j} \qquad (43)$$

Equation (41) for the boundary condition translates into:

$$\sum_{j=1}^{M} \bar{A}_{M+1,j} \, C_{1,j} + \bar{A}_{M+1,M+1} \, C_{1,M+1} = -Bi_1 \, (C_{1,M+1} - C_{3,N+1}) \tag{44}$$

from where it follows for the concentration C_1 at the region boundary:

$$C_{1,M+1} = - \frac{\sum_{j=1}^{M} \bar{A}_{M+1,j} \, C_{1,j} - Bi_1 \, C_{3,N+1}}{\bar{A}_{M+1,M+1} + Bi_1} \tag{45}$$

Equation (34) must be used to calculate the concentration $C_{3,N+1}$ needed in Equation (45). Equation (40) can be rewritten by means of Equations (42) and (43) into its final form:

$$\frac{dC_1}{dZ} = \sum_{j=1}^{M} \bar{D}_{ij} \, C_{1,j} + \bar{D}_{i,M+1} \, C_{1,M+1} \qquad i = 1,M \tag{46}$$

The boundary concentration $C_{1,M+1}$ comes from Equation (45) and $\bar{D}_{ij}$ is given by:

$$\bar{D}_{ij} = \frac{4\rho_i}{1-\rho_i} \bar{B}_{ij} + \frac{4}{1-\rho_i} \bar{A}_{ij} \tag{47}$$

The initial condition for Equation (46) is now:

$$C_{1,i}(o) = 1 \qquad i = 1,M+1 \tag{48}$$

The distribution of the collocation points is shown schematically in Figure 3.

Computational Procedure

1) The numbers of internal collocation points, M for region 1, N for region 3, must be chosen.
2) The M and N collocation points and the derivatives A_{ij}, B_{ij} (region 1) and $\bar{A}_{ij}$, $\bar{B}_{ij}$ (region 3) are calculated.

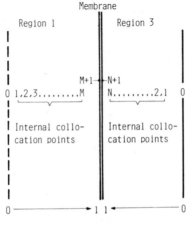

Fig. 3.

3) D_{ij} and $\bar{D}_{ij}$ are calculated from Equations (47) and (36); Bi* is given by Equation (37).
4) $C_{3,j}$ (j=1,N) is calculated numerically (secant method) using Equations (45), (35) and (34). For this purpose, the concentrations $C_{1,j}$ (j=1,M) from the previous integration step must be used.
5) $C_{1,M+1}$ is calculated using Equations (45) and (34).
6) Equation (46) will be integrated numerically one step ΔZ using the IMSL sub-routine DGEAR (Adam or Gear method).
7) Return to step 4, until the end of integration is reached.

The mixing cup concentration in the fiber lumen was calculated by means of the Gauss integral:

$$C_B(Z) = 2 \int_0^1 C_1(Z,\rho) \ (1-\rho) \ d\rho = \sum_{i=1}^{M+1} W_i \ C_{1,i}(\rho,i) \qquad (49)$$

The conversion $X = 1 - C_B(Z)$ is now a function of 7 dimensionless parameters:

$$X = f \ (Bi_1, \ Bi_3, \ \beta, \ \phi^2, \ \theta, \ \xi, \ Z) \qquad (50)$$

The advantage of this computation method can be seen in that no iteration is needed and a short computing time is required.

Given and Estimated Parameters

The dimensions of the reactor fibers are given as follows: a = 250 μm, b-a = 0.1 μm, c-b = 175 μm. The measured active length of the fiber was L = 0.165 m; the nominal length was found to be L = 0.203 m, and the reactor module contained 250 fibers. The kinetic parameters of β-galactosidase were taken from an unpublished work in our laboratory: K_m = 59.9 mM, K_i = 18 mM, v_{max} = 46.0 μmol/mg E min, at T = 40°C and pH = 4.5.

The value of the lactose diffusion coefficient in aqueous solution at 35°C was given by Robertson[18] as $D_1 = 5.6 \times 10^{-6}$ cm^2/s. The diffusion coefficient for restrictive diffusion of lactose in the membrane skin could not be calculated for the lack of knowledge of the parameters of the skin (porosity). According to the Literature[3-8,18], its value was assumed to be one order of magnitude smaller than that for aqueous solution given by Robertson et al.[18]: $D_2 = 5.6 \times 10^{-7}$ cm^2/s. Further, it was assumed that the macroporous support structure of the membrane did not influence the lactose permeation so that $D_3 = D_1$.

The value of the partition coefficient γ, was set to unity as, to our knowledge, the interaction of sugar molecules with the membrane material was negligible. The initial concentration of lactose was c_o = 146 mM (5 wt-%) corresponding roughly to the lactose concentration in milk or whey.

Using these values, the dimensionless parameters required for the solution were determinated. The solution of the governing differential equations was obtained in terms of the lactose concentration as a function of the radial position. An example of the radial concentration profiles at two longitudinal positions (Z = 0.1 and 0.3) and ϕ^2 = 10 is presented in Figure 4. With this value of ϕ^2, an expected significant effect of mass transfer on the reactor operation is confirmed by the relatively steep slopes of the radial profiles, especially in the fiber lumen. The concentration profile in the porous structure (region 3), where the reaction takes place, flattens out, suggesting that this part could possibly be described by a well mixed tank model (CSTR), which would greatly simplify the computation procedure. Also, at low values of ϕ^2, $\phi^2 \sim 0.1$, and at

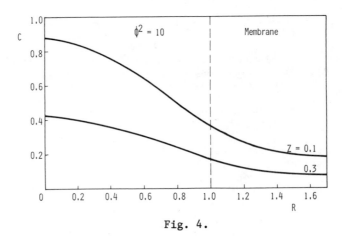

Fig. 4.

increased reactor length $Z \geq 5$, the profile becomes practically flat, independent of the radial position.

The parameters to work with are now ϕ^2 and X, and the solution is presented graphically. Figure 5 shows the dimensionless length required for a given X as a function of ϕ^2. For higher values of ϕ^2, $\phi^2 > 50$, log Z becomes assymptotically constant, i.e. independent of ϕ^2, indicating thus the reactor operations in fully diffusion controlled regime. It has the practical consequence in that under such conditions no additional amount of enzyme will further increase X. This insensitivity of X towards the enzyme concentration is also important with respect to measurement of enzyme stability. Mass transfer masks enzyme deactivation causing an increase in the apparent enzyme half-life. In the range of small values of ϕ^2, the reaction is kinetically controlled and log Z versus log ϕ^2 approaches a straight line.

The ratio of the diffusion coefficients D_1/D_2 and D_3/D_2 has always been assumed to be 10, as the value of D_2 was difficult to estimate from

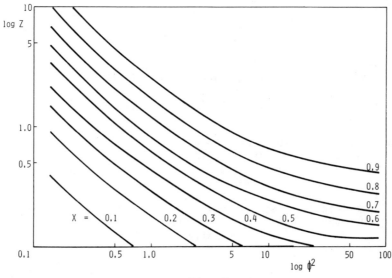

Fig. 5.

the membrane data. However, it was found that the solution was quite insensitive towards this ratio; its doubling to the value of 20 caused only negligible changes in X. This result is not surprising, considering how thin the actual membrane skin is (region 2).

The model equations were also solved using the finite element method and exactly the same results were obtained. However, the computer time required was much longer. For higher values of ϕ^2, the system becomes stiffer making the difference between these two methods even larger.

EXPERIMENTAL

Materials

- β–Galactosidase (E.C. 3.2.1.23) from <u>Aspergillus oryzae</u> (Lactase AO) was obtained from Miles Italiana S.p.A., I-20040 Davenago Brianza (Milano).
- D(+)-Lactose Monohydrate, Art.No. 7656, Merck GmbH, Darmstadt, West Germany.
- Glutaraldehyde, pract. 25%, Art.No. 49630, Fluka A.G., Buchs, Switzerland.
- Citric Acid Monohydrate, puriss., Art.No. 27490, Fluka A.G., Buchs, Switzerland.
- Phosphate buffer, pH = 4.5.

Analytical Instruments

- pH–Meter: Metrohm E-603, Metrohm AG, Herisau, Switzerland.
- Glucose analyzer: Beckman Glucose Analyzer 2, Beckman Instruments Inc., Morges, Switzerland.

Ultrafiltration Device

- Hollow-fiber membrane modules: H1P3-20 (cut-off value 3000), H1P10-43 (cut-off value 10'000), Amicon Corp., Denver, Mass., USA.

METHODS

Kinetics

The kinetic data for the lactase AO were measured by the initial rate method in batch experiments with the soluble enzyme, and the parameters of the kinetic model (Equation 5) were determined. Figure 6 shows the good agreement between measurements and simulation curves at different initial substrate concentrations.

Immobilization

The enzyme immobilization in the hollow fiber module was done in a backflush mode with the flow diagram shown in Figure 7. To avoid enzyme deactivation by a microbial infection, all the operations were done under sterile conditions. The enzyme was dissolved in 200 ml phosphate buffer with pH = 4.5 and pumped at ambient temperature through a sterile filter into the reactor shell. The enzyme solution was recycled for about 90 min.

Cross-linking of the Enzyme

In the first experiments, significant activity losses were observed within a short time, but it was found that this could be prevented by

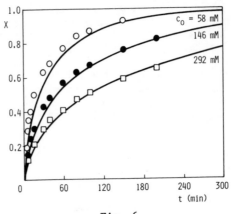

Fig. 6.

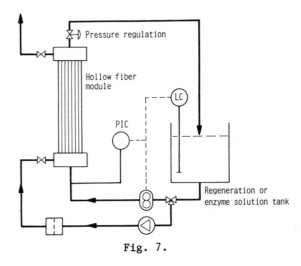

Fig. 7.

cross-linking the enzymes with glutaraldehyde. For this purpose, glutaraldehyde was added to the immobilization solution up to 1% and recycled for
6.5 hr. After this time, the reactor was drained and washed with the
buffer solution in the backflush mode to remove the rest of the glutaraldehyde. The enzyme loaded modules were stored at 3°C or directly used.

Hydrolysis Experiments

The experiments were carried out with 5% lactose substrate at a temperature of 40°C and a pH of 4.5 in an apparatus shown in Figure 8.
ϕ^2 and Z were varied by changing the flow rate and enzyme loading. The
samples were taken from the reactor effluent in steady-state and the
glucose content, measured by the Glucose Analyzer, was taken as a measure
of X. The only overall reaction assumed to proceed was the conversion of
lactose to an equimolar mixture of glucose and galactose. It was found
that not all of the fibers in the reactor are flown through. Assuming
Hagen-Poiseuille's law, the actual number of fibers (N) can be calculated
by knowing the pressure drop over the module:

$$N = \frac{8 \mu L F}{\pi a^4 \Delta P} \tag{51}$$

where μ (the viscosity) was taken to be 0.725 cp at 40°C. This gives a modified Z value:

$$Z = \frac{N \pi L D_1}{2 F}$$ (52)

RESULTS

The results of the hydrolysis experiments for various enzyme loadings are shown in terms of X as a function of Z in Figure 9 and 10. The solid lines represent the calculated X curves from the model. The agreement of the theoretical and experimental results seems to be excellent, especially in view of the complicated measurements and uncertainty of the kinetic model[17].

Effect of Cross-linking

The effect of cross-linking is presented in Figure 11. The conversion in a hollow fiber module without cross-linking falls rapidly within a few hours to about 60% of the theoretical value, whereas the conversion in the same module with cross-linking stays constant and near to the theoretical value represented in the drawing by the full straight line. A special cleaning procedure was developed for a complete stripping of the cross-linked enzymes and restoring the original permeability of the hollow fibers.

Effect of Enzyme Loading

It was observed that the amount of loaded enzymes had a small, nevertheless a systematic effect on the conversion. For the small loadings, the experimental values were always slightly lower than the theoretical expected values and for the high loadings vice-versa. This can be clearly seen in Figure 10.

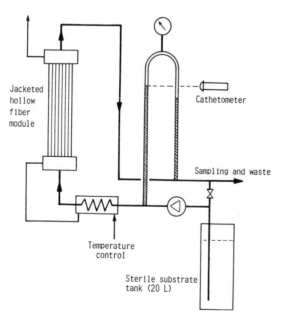

Fig. 8.

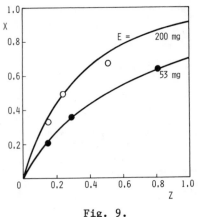

Fig. 9.

CONCLUSIONS

   In this report, it has been demonstrated how enzyme membrane reactors
of the diffusional type can be modelled by a kinetic model coupled with
external and internal mass transfer. The efficiency of orthogonal collocation, which is particularly suitable for problems of this type, has
been shown.

   Similar equations have been applied by Waterland[4] and Robertson[18]
using, however, different numerical methods for their solution. Checking
their results we have found significant deviations especially in the range
of low ϕ^2 (Figure 12). We have checked our computations using finite
element method in addition to orthogonal collocation, and both methods gave
consistently identical results throughout the whole range of ϕ^2.

   A comparison with extensive experimental data shows that the model is
sufficiently general to handle a wide range of reaction conditions. The
numerical techniques are efficient and can be easily adapted to different
situations.

   It should be pointed out that all the important parameters needed for
the computation were either known a-priori, e.g. the sugar diffusion coefficient (D_1), or were determined experimentally with a high degree of

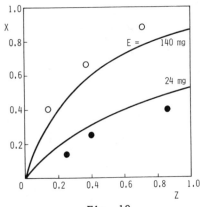

Fig. 10.

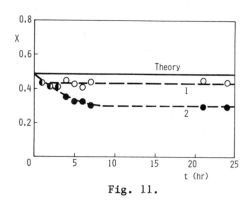

Fig. 11.

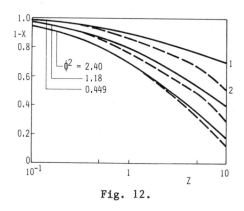

Fig. 12.

accuracy, e.g. the kinetic parameters. Luckily, the other parameters which had to be estimated, the diffusion coefficient D_2 in the membrane skin and the partition coefficient γ, or those which could not be determined accurately, like the thickness of membrane skin, do not seem to have a significant effect on the solution. The simulation runs with these parameters varying by factors exceeding the limits of expected estimation errors did not show any significant deviations of the final results.

Another uncertainty seems to be the model assumption that the enzymes are homogeneously distributed throughout the membrane porous structure, which may not be the case. The immobilization was done by pumping the enzyme solution in a backflush mode through the module and it is quite obvious that a gradient of enzyme concentration could be formed with its highest value towards the membrane skin. This would explain why in some experiments with high enzyme loadings the conversion was higher than that predicted by the theory and lower for the low enzyme loadings (Figure 10). The assumption of uniform enzyme distribution increases the effect of pore diffusion at high enzyme concentrations and vice-versa.

In the work of Waterland[4] such a gradient was probably not formed as the immobilization was done by a simple soaking of the hollow fibers with an enzyme solution. On the other hand, he observed a longitudinal enzyme concentration gradient formed during the reactor operation, which was not detectable in this work. All later experiments were done with crosslinking, eliminating thus the possibility of enzyme migration. The crosslinking also seems to improve the enzyme stability making it comparable with the enzyme immobilized on Duolite, which has been used in the fixed bed reactors of our pilot plant for whey hydrolysis.

A preliminary analysis shows that the volumetric performance of a diffusional membrane reactor will be comparable to that of a fixed bed reactor with β-galactosidase immobilized on Duolite. However, the membrane reactor could be used for the processing of whole whey or milk.

Acknowledgement

We would like to express our thanks to Professor H. Pedersen, Rutgers University, New Jersey, USA, for his generous assistance and help with the collocation procedure.

NOMENCLATURE

a	– Fiber inner radius (m)	u_z	– Axial velocity (m/s)
b–a	– Membrane thickness (m)		
Bi	– Biot number	v	– Reaction velocity (μM/min)
c	– Fiber outer radius (m)		
c	– Substrate concentration (mM)	v_{max}	– Maximum specific reaction velocity (μmole/mg E min)
C	– Dimensionless substrate concentration		
D	– Diffusion coefficient (m^2/s)		
E	– Amount of loaded enzymes	W(z)	– Weight function
F	– Flow rate (ml/min)	X	– Dimensionless conversion $(1 - c_B)$
K_i	– Inhibition constant (mM)		
K_m	– Michaelis constant (mM)	z	– Axial distance (m)
L	– Reactor length (m)	Z	– Dimensional axial length
$\vec{Ly}$	– Lagrange interpolation polynom	$\bar{A}_{ij}, A_{ij} = \dfrac{dL_j}{dz}\Big|_{z=z_i}$	in region 1 and region 3
M	– Inner collocation points (region 1)		
N	– Inner collocation points (region 3)	$\bar{B}_{ij}, B_{ij} = \dfrac{d^2L_j}{dz^2}\Big|_{z=z_i}$	in region 1 and region 3
ΔP	– Pressure drop (mm H_2O)		
Pe	– Péclet number (μL/D)	D_{ij}, D_{ij}, Bi^*	– Auxiliary variables defined by Equations (47), (36) and (37)
r	– Radius (m)		
R	– Dimensionless radius		
u_o	– Maximum velocity (m/s)		

Indices

o – Inlet

1 – Region 1

2 – Region 2

3– Region 3

B – Bulk

Greek Symbols

β – Dimensionless geometrical parameter, b/(c–b)

γ – Membrane partition coefficient

ε – Dimensionless geometrical parameter, (r–b)/(c–b)

η – $(1-\varepsilon)^2$

ρ – Fiber radius squared, R^2 (m^2)

θ – Dimensionless Michaelis constant, K_m/C_o

ξ – Dimensionless inhibition constant, C_o/K_i

ϕ^2 – Thiele modulus

μ – Viscosity (kg/m s)

REFERENCES

1. C. Wandrey, R. Wichmann, and A. S. Jandel, Multi-enzyme systems in membrane reactors, in: "Enzyme Engineering," I. Chibata, S. Fukui, and L. B. Wingard jr., eds., Plenum Press, New York, 6:61 (1982).
2. P. R. Rony, Multiphase catalysis. II: Hollow fiber catalysts, Biotechnol.Bioeng., 13:431 (1971).
3. P. R. Rony, Hollow fiber enzyme reactors, J.Am.Chem.Soc., 94:8247 (1972).
4. L. R. Waterland, A. S. Michaels, and C. R. Robertson, A theoretical model for enzymatic catalysis using asymetric hollow fiber membrane, AIChE.J., 20:50 (1974).
5. L. R. Waterland, A. S. Michaels, and C. R. Robertson, Enzymatic catalysis using asymmetric hollow fiber membranes, Chem.Eng.Commun., 2:37 (1975).
6. R. A. Korus and A. C. Olsow, Use of β-galactosidase, glucose isomerase and invertase in hollow fiber reactor, in: "Enzyme Engineering," E. K. Pye and H. H. Weetall, eds., Plenum Press, New York, 3: (1975).
7. J. C. Davis, Kinetics studies in a continuous steady state hollow fiber membrane enzyme reactor, Biotechnol.Bioeng., 16:1113 (1974).
8. W. Lewis and S. Middleman, Conversion in a hollow fiber membrane/enzyme reactor, AIChE.J., 20:1012 (1974).
9. B. R. Breslau and B. M. Kilcullen, Hollow fiber enzymatic reactors: An engineering approach, in: "Enzyme Engineering," E. K. Pye and H. H. Weetall, eds., Plenum Press, New York, 3: (1975).
10. J. M. Engasser, J. Caumon, and A. Marc, Hollow fiber enzyme reactors for maltose and starch hydrolysis, Chem.Eng.Sci., 35:99 (1980).
11. D. E. Kohlwey and M. Cheryan, Performance of a β-D-galactosidase hollow fiber reactor, Enzyme Microb.Technol., 3:64 (1981).
12. L. M. Huffman-Reichenbach and W. J. Harper, β-Galactosidase retention by hollow fiber membranes, J.Dairy Sci., 65:887 (1982).
13. R. A. Mashelkar and P. A. Ramachandran, A new model for a hollow fiber enzyme reactor, J.Appl.Chem.Biotechnol., 25:67 (1975).
14. S. S. Kim and D. O. Cooney, An improved theoretical model for hollow fiber enzyme reactors, Chem.Eng.Sci., 31:289 (1976).
15. J. Villadsen and H. Michelsen, "Solution of Differential Equation Models by Polynomial Approximations," Prentice Hall, Eglewood Cliffs, New Jersey (1978).
16. B. A. Finlayson, "Nonlinear Analysis in Chemical Reaction Engineering," McGraw Hill, New York (1980).
17. H. Betschart, Trägergebundene β-Galactosidase bei der Lactosehydrolyse in Molke unter Berücksichtigung der Oligosaccharide, Dissertation ETH No.7360, Zurich (1983).
18. C. R. Robertson, A. S. Michaels, and L. R. Waterland, Molecular separation barriers and their application to catalytic reactor design, Separation and Purification Methods, 5(2):301 (1976).

ETHANOL PRODUCTION BY COUPLED ENZYME FERMENTATION AND CONTINUOUS

SACCHARIFICATION OF CELLULOSE, USING MEMBRANE CELL-RECYCLING SYSTEMS

H. Hoffmann, M. Grabosh, and K. Schügrel

Institute of Chemical Engineering
University of Hanover

SUMMARY

Enzymes produced by Trichoderma reesei were used for the conversion of cellulose into ethanol in a three-stage continuous cascade. The fermentation process includes the enzyme production in the first stage, the cellulose decomposition in the second stage and the ethanol production by Saccharomyces cerevisiae in the third stage.

The whole system was set up as a recycling system by means of capillary cross-flow microfiltration and/or ultrafiltration membranes.

INTRODUCTION

The conversion of cellulose can be carried out in a single-stage process as proposed by Takagi et al.[1,2,3] or as a multistage process[4]. The advantages of the single-stage process are the reduced reactor volume and the elimination of the inhibition of glucose on the enzymatic decomposition of cellulose. The multistage process offers the advantages that each of the particular processes can be carried out at their optima, which differ considerably with regard to their temperature and pH values. Mainly because of the different temperature optima and the inhibitory effect of ethanol and glucose on the enzyme system[5], a multistage system was developed for this conversion of cellulose to ethanol. By using microfiltration modules for cell recycling it should be possible to increase the cell mass concentration in the first and third stages. The use of ultrafiltration modules should allow the recycling of the enzyme complex, especially β-glucosidase which is not adsorbed on the cellulose surface. At the same time, the inhibitory products are removed from the second and third stage of the systems. Each of these should allow an increase in the productivity of ethanol from cellulose. In the following, this multistage reactor system will be investigated.

MATERIALS AND METHODS

Organisms

Trichoderma Rut C30 was obtained from the Agricultural Research Service, Peoria, Illinois, USA. Inocula were prepared as described by

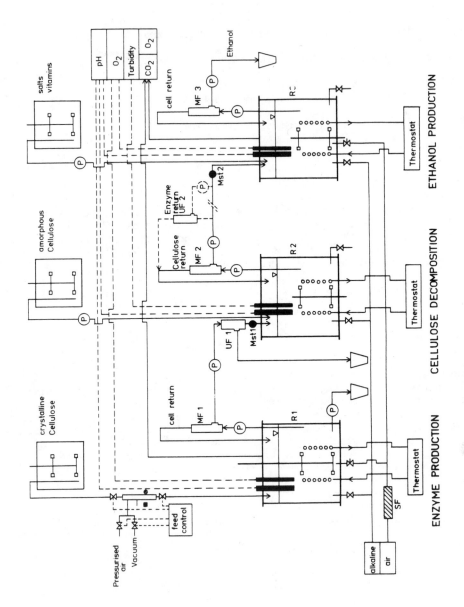

Fig. 1a. Illustration of the three fermentation stages.

Mandels and Reese[6]. Saccharomyces cerevisiae H 1022 was obtained from the Institute of Biotechnology ETH Zurich, Switzerland. The culture was maintained on a glucose-agar slant culture (4 g glucose, 10 g malt extract, 15 g agar and 4 g yeast extract in 1 liter of water). Both cell cultures were grown aerobically in shake flasks set in a rotary shaker at 27°C for 20 h.

Medium

A synthetic medium D 3% was used in cell experiments with yeast. It consisted of 240 g/l glucose with special nutrient salts. The medium was sterilized through a 0.2 μ micro-filter (Sartorius high pressure filtration module SM 16260). The pH of the medium was adjusted to pH 4. For the experiments involving Trichoderma reesei, a medium with 10 g/l (Serva Avicell 14205 SF prac. grad.) was used. The medium was sterilized by autoclaving for 1 h at 121°C. For the enzymatic hydrolysis of cellulose a medium with 30 g/l amorphous cellulose (Sigmacell Type 100) was used. The pH of the medium was 4.7, adjusted by a citrate buffer.

Fermentation Conditions

The experiments for the production of enzymes were conducted in a 10 l continuous-stirred-tank reactor (Biostat S Braun/Melsungen). The conversion of cellulose and the production of ethanol were performed in two continuously operated stirred tank vessels with an effective volume of 2.7 l (Figure 1a/1b). The whole system was set up as a recycling system by coupling the vessels to microfiltration hollow-fiber units from ENKA AG/Wuppertal. These modules have an effective filtration area of 0.05 m^2 and consist of seven hollow fibers of polypropylene with a nominal pore size of 0.2–0.45 μm. For the concentration of enzymes and the separation of glucose a hollow-fiber ultrafiltration unit with a nominal cut-off of 10,000 (Dalton) was used (Amicon H1P 10-8).

Analytical Methods

Total dry weight is expressed as cellulose and biomass dry weight per unit volume (g/l). Biomass dry weight was obtained by removing the cells from cellulose by treatment with nitric acid/acetic acid at 100°C, for 30 min. Glucose concentrations were measured by the glucose dehydrogenase

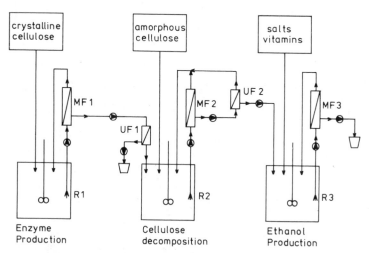

Fig. 1b. Schematic flow sheet diagram of the fermentation process (MF: microfiltration; UF: ultrafiltration).

method of Merck. Total reducing sugars were measured by the method of
Nelson[7]. The cellulose and β-D-glucosidase activity were estimated
with filter paper FPA[8], carboxy-methylcellulose CMC[8] and p-nitrophenyl-
β-D-glucopyranosid methods[9]. Ethanol was analyzed using a gas chromatograph (Perkin-Elmer Sigma 3B) employing a glass column with Chromosorb 101.
The column was operated thermically at 150°C with a flame ionisation
detector.

RESULTS AND DISCUSSION

Effect of Microfiltration

     Figure 2 shows typical data obtained from preliminary continuous tests
with Trichoderma reesei under identical conditions with and without feed-
back of cells. One fermenter was operated with a microfiltration unit.
The unit was started as soon as the fermenters worked continuously (after
50 h). Since we were able to keep the fermentation volume constant by
matching the permeate flux, the experiments were conducted at dilution
rates of 0.035, 0.054, 0.08 1/hr; recycling rate 0.62.

     Figure 2 shows the total dry weight for the different dilution rates.
It was found that the biomass concentration effects were appreciably
greater with cell recycling than without recycling.

     As is evident from the upper curve of Figure 2 the membrane filtration
leads to a maximum dry weight of approximately 12 g/1, corresponding to a
dilution rate of 0.08 1/h. For this dilution rate a total dry weight of
only 7 g/1 is obtained under standard conditions. Product concentrations
for these dilution rates are shown in Figure 3.

     These results demonstrate that enzyme activity of 10,000 - 18,000 IU/1
can be maintained for cell recycling experiments at a productivity which

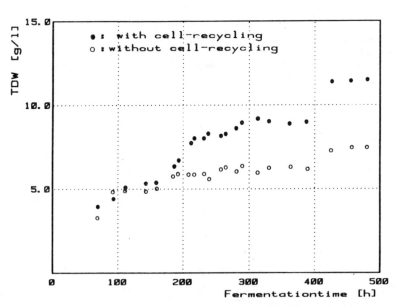

Fig. 2. Effect of cell-recycling on fermentation dry weight for several
         dilution rates (Trichoderma reesei Rut C30, feedback ratio
         a = 0.62, D = 0.035, 0.054, 0.08 (1/h) substrate concentration
         10 g/1).

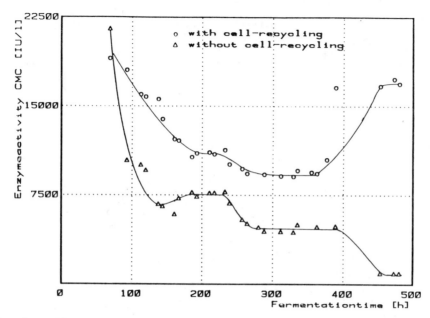

Fig. 3. Effect of cell-recycling on product concentrations for several
dilution rates (see Figure 2).

varies from 450 to 972 IU/l h. Product concentrations without feedback
represented by the lower curve lead to a maximum productivity of 250 IU/l h
with a significant decrease at high dilution rates.

Effect of Ultrafiltration

Figure 4 shows the effect of ultrafiltration on the time course of
saccharification with and without filtration. Due to low β-D-glucosidase
concentrations in vessel 2 significant levels of cellobiose accumulated at
the fermentation broth (Curve A). At the same time, the glucose concentration remained low. Since cellobiose is a stronger inhibitor for the
enzymatic decomposition of cellulose than glucose, the cellulose conversion
was also reduced.

By using ultrafiltration modules it is possible to keep the β-D-glucosidase in the second stage. This results in an increased β-D-glucosidase
concentration in vessel 2 leading to a faster saccharification. In this
case the cellobiose level is substantially reduced, as shown in curve B.
The corresponding glucose concentrations rise rapidly to a constant level
of 14 g/l (Curve C).

Biomass and Enzyme Activity

Data for strain Rut C30 growing on 10 g/l cellulose medium and strain
H1022 growing on glucose medium are shown in Figure 5 for dilution rates of
0.05 and 0.13 1/h, respectively. It is evident from the data that the
yeast biomass concentration declines rapidly after changing over from
discontinuous to continuous operation. After 50 h, we have a constant
biomass concentration of approximately 4 g/l.

These concentrations are very low for yeast fermentations under
recycling conditions, but it is possible to increase the cell concentrations by increasing the recirculation rates. The cell concentrations
of Tricoderma reesei show a course which is typical of fermentations under

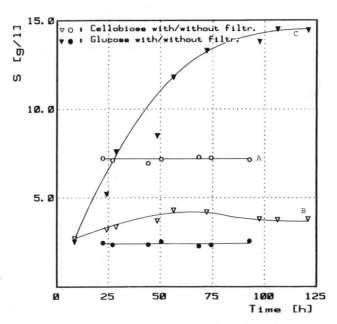

Fig. 4.   Effect of ultrafiltration on saccharification of cellulose.
(Temperature 50°C, substrate concentration 30 g/l).

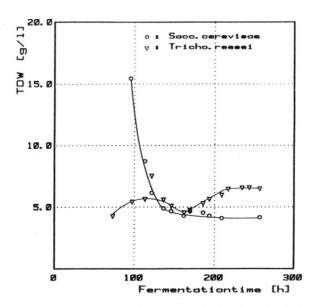

Fig. 5.   Total dry weight as a function of the fermentation time for
fermentations with yeast (o) and Trichoderma reesei (Δ).

recycling conditions. The total dry weight has a maximum of 6 g/l after
20 h, it declines to 4 g/l and then rises to a constant level of 7 g/l.

    In Figure 6, typical courses of cell (dry) mass and cellulose (dry)
mass concentrations which yield the total (dry) biomass are shown during a
batch cultivation. A maximum in the product concentration exists corresponding to the cell mass concentration.

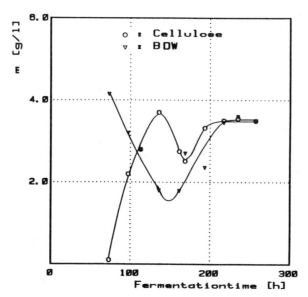

Fig. 6.   Cellulose and biomass contents of the total dry weight as functions of the fermentation time.

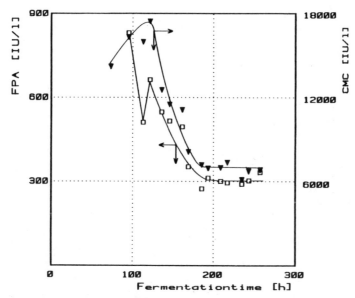

Fig. 7.   Enzyme activity for fermentation with Trichoderma reesei Rut C30 under cell-recycling conditions (D = 0.05 1/h, feedback ratio a = 0.6, temperature 30°C).

The enzyme activity with FPA and CMC tests at the steady state were 300 and 7000 IU/1, as shown in Figure 7. This activity is leading to a productivity of 350 IU/1·h (with CMC test) and 15 IU/1·h (with FPA test).

The fermentation broth was filtered through a microfiltration membrane and cellulose and cells of Trichoderma were recycled to the first vessel.

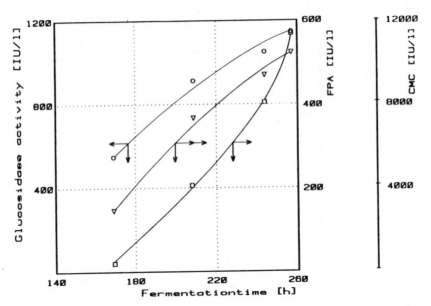

Fig. 8. Enzyme concentrations as functions of the fermentation time with the saccharification stage (□: FPA activity, ∇: CMC activity, o: Glucosidase activity).

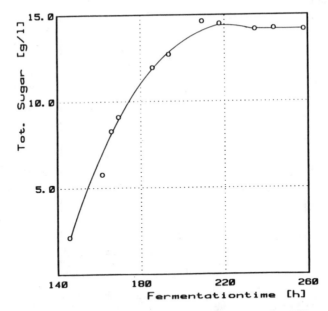

Fig. 9. Time course of sugars produced after saccharification stage as a function of the fermentation time.

The filtrate containing the enzyme was treated with an ultrafiltration membrane to increase the enzyme concentration for the saccharification in vessel 2 (Figure 8). The enzyme activity exhibits a considerable increase as a function of time.

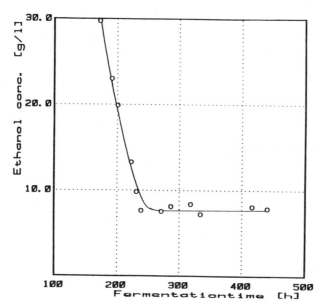

Fig. 10. Ethanol concentration at steady state as a function of the fermentation time (S_o = 14 g/l, temperature 30°C).

It was found, however, that the concentration of the glucose produced approached a constant final value after 210 h of fermentation time (Figure 9) while the enzyme concentrations were still increasing. Thus it is obvious that the enzyme activity was inhibited by the glucose in reactor 2.

The course of the ethanol produced is shown in Figure 10. During the 200 h startup phase of the process the ethanol was produced in the third vessel by using medium feed with 240 g/l glucose inlet concentration. After that the glucose containing medium from the second stage was used which was produced by enzymatic hydrolysis of cellulose in this stage. After the steady state was established a constant ethanol concentration of approximately 8 g/l was attained. By optimization of the system this concentration can certainly be further increased.

CONCLUSIONS

The results obtained from these experiments with coupled enzyme production cellulose hydrolysis and ethanol fermentation in a three-stage continuous process demonstrate that ethanol concentrations up to 8 g/l can be attained. The current system contains several parameters which have to be optimized. But the results show that the use of membrane filtration can improve the performance of these bioreactors considerably. In our system the modules have to be backflushed every 400-500 h in order to maintain these productivity levels. By optimizing the fermentation and filtration parameters, it seems to be possible to develop a process for the conversion of biological waste celluloses with practical applicability.

Acknowledgements

We gratefully acknowledge the financial support of BMFT, Bonn, and ENKA AG/Wuppertal for supplying the membrane modules.

REFERENCES

1. M. Takagi, S. Abe, S. Suzuki, G. H. Emert, and N. Yata, Proc.Bioconv. Symp;Delhi, 551-571 (1977).
2. A. W. Khan and W. D. Murray, Biotechn.Lett., 4(3):177-180 (1982).
3. P. Gosh, N. B. Pamment, and W. R. B. Martin, Enzyme Microb.Techn., 4:425-430 (1982).
4. S. Blanco, A. Camarra, L. Cuevas, and G. Ellenrieder, Biotechn.Lett., 4(10):661-666 (1982).
5. T. K. Ghose and K. Das, Adv.Biochem.Eng., 1:55 (1971).
6. M. Mandels and E. T. Reese, J.Bacteriol., 269-278 (1957).
7. N. Nelson, Journ.Biol.Chem., 153:375-380 (1944).
8. M. Mandels and J. Weber, Adv.Chem.Ser., 98:391-414 (1969).
9. T. M. Wood, Biochem.J., 109:217-227 (1968).

ENZYMATIC DETERMINATION OF CHOLIC ACIDS IN HUMAN BILE

BY AMPEROMETRIC OXYGEN ELECTRODE AS DETECTOR

L. Campanella, F. Bartoli, R. Morabito, and M. Tomassetti

Department of Chemistry,
University of Rome "La Sapienza"
Rome, Italy

SUMMARY

The enzymatic determination of cholic acids in human bile, by ampero-metric Clark oxygen electrode as detector, is described. The optimum experimental conditions are established by operating with both the employed enzymes in homogeneous phase.

INTRODUCTION

Faster and simpler methods of quantitative determination of cholic acids in biological and pharmaceutical solutions are to date of increasing importance. In previous papers we described the optimization of some known enzymatic[1] and spectrophotometric[2] methods, and the development of new methods, such as the potentiometric[3,4] and calorimetric ones[2,5]. Potentiometric methods, employing ion selective liquid membrane electrodes, afford good results in the analysis of aqueous solutions of products like drugs[4], but they are less suitable for the determination of compounds in biological fluids where laborious pretreatment operations are generally needed[6]. With enzymatic methods, the pretreatment generally is not required and if so it is quite simple and rapid. Generally in these methods the detection is performed by spectrophotometry[1,7] or fluori-metry[8]. However these methods are generally expensive and specific. In the last years, the application of the Clark oxygen electrode has been developed as amperometric detector in enzymatic methods to determine organic species in biological fluids[9,12].

Recently Cheng and Cristian showed the feasibility of the utilization of a NAD^+ dependent dehydrogenase, in presence of the peroxidase, for the determination of substrates such as ethanol[13] and lactate[14]. In the present paper we describe the application of this method to the determin-ation of cholic acids. The reaction involved are:

$$7\alpha- \text{ (or } 3\alpha-) \text{ hydroxysteroid} + NAD^+ \xrightleftharpoons{7\alpha- \text{ (or } 3\alpha-) HSD} 7 - \text{ (or } 3-) \text{ chetosteroid} + H^+ + NADH \qquad (1)$$

$$NADH + H^+ + 1/2\ O_2 \xrightarrow{\text{peroxidase}} NAD^+ + H_2O \qquad (2)$$

The resulting oxygen consumption is measured by a Clark electrode.

EXPERIMENTAL

Materials

The 7α- and 3α- hydroxysteroid dehydrogenase (7α-(or 3α-) HSD) were
supplied by Sigma (activity, 6.2 and 2.9 U/mg solid respectively); peroxidase was from Miles Laboratories (activity 230.9 U/mg solid); NAD$^+$ (grade
1) was from Boheringer. Analytical grade sodium cholate, potassium dihydrogenphosphate, sodium monohydrogenphosphate, manganese (II)-chloride
tetrahydrate, hydrazine monohydrate and all the other reagents were from
Merck.

We are indebted to the 2nd Medical clinic of Rome University for the
human bile samples obtained by duodenal aspiration from patients suffering
from gallstones.

Apparatus

The determinations were carried out in a jacketed reaction cell made
of glass (inner diameter, 3 cm; height, 14 cm) equipped with a magnetic
stirrer, an oxygen probe and a constant temperature water circulator
(ultra-thermostat Colora).

The monitoring system consisted of an oxygen probe (amperometric Clark
electrode, Orion model 97-08), a potentiometer (Orion research micro-
processor ionlayzer 901) and a Leeds & Northrup recorder.

Method

The reaction cell was filled with 10 ml of a buffer solution containing 0.2 mg/ml 7α-(or 3α-) HSD, 0.6 mg/ml peroxidase, 6.8 mg/ml NAD$^+$. To
the reaction mixture 300-500 μl of the cholate solution were added. From
the decreasing of the oxygen concentration, monitored by the oxygen probe,
the amount of cholate was calculated.

RESULTS AND DISCUSSION

Several experiments were carried out by varying the temperature in the
20-31°C range and the pH in the 4.5-9.0 range. With both 7α-HSD and 3α-
HSD, the temperature resulted to be a critical factor. In fact, the
temperature must be higher than 27°C, otherwise, even with the optimum pH
and ionic strength values, the experimental data are not reliable, as no
meaning answer can be recorded. Therefore we operated at 30°C, a value in
good agreement with those utilized by others[13] in similar enzymatic
systems.

As regards to the choice of the working pH, the following considerations were taken into account: (i) the optimum pH for the peroxidase is
near the neutrality and for the 3α-or 7α-HSD is at about 9.0; (ii) the
Reaction (1) is shifted to the right at alkaline pH values as also shown by
Cheng and Christian[13] for similar reactions. Based on these considerations, we tested our system, with both the 3α-HSD and the 7α-HSD, at pH
9.0, 8.0, 7.0. No substantial oxygen consumption was observed at the above
pH, nor at pH 4.5. Really, at pH values lower than 5.0 it occurs the
denaturation of the enzymes and the precipitation of cholic acids. However
when the pH was kept in the 5.5 ÷ 6.0 range with the 7 α-HSD, the addition
of known amounts of cholate to the buffer solution resulted in consistent
variations of the oxygen concentration, the latter being of the same order
of magnitude of the theoretical values. However, with 3α-HSD and at the
same pH values, the recorded oxygen consumption never exceeds 15% of the

theoretical value. This suggests that the system is controlled by the Reaction 2. This is in agreement with literature data showing that the NADH oxydation is favored by pH values in the range we found[16-19].

The buffer system for the experiments, in the pH range 5.5 ÷ 6.0 was sodium monohydrogenphosphate, potassium dihydrogenphosphate, 0.25 mol/1. This buffer concentration ensures a stable pH value, within 0.2 units, mainly when acid reagents such as NAD^+ are added. A more concentrated buffer, for instance 0.5 mol/1, is able to ensure a stable pH value within 0.1 pH units, but, in comparison with the concentration of 0.25 mol/1, it gave a longer response time and a lower oxygen composition.

In conclusion, the optimum composition of the reaction mixture for the amperometric determination of the cholic acids was found to be as it follows: phosphate buffer, 0.25 mol/1; pH, 5.8; temperature, 30°C; 7α-HSD and peroxidase as enzymatic catalysts. The recorded variations of the oxygen concentration in the reaction mixture, following each addition of successive aliquots of 500 µl of 10^{-2} mol/1 cholate solution, are reported in Table 1. It was observed that the initial response time decreases by passing from the first to the second addition, and increases at the third one. The total response time remains practically unchanged for the first two additions, and increases markedly for the third one. The shape of the peaks is changed too (Figure 1): from the first to the third addition lower and wider peaks are noted. Since Mn^{2+} and hydrazine are known to catalyze the oxidation of NADH by peroxidase[20] and shift to the right the reaction (1)[21] respectively, we tested also these compounds in our assay.

However it was found that hydrazine (500 µl of a 0.1 mol/1 solution in 10 ml of the reaction mixture), by itself, takes up an amount of oxygen equivalent to an addition of 500 µl of 10^{-3} mol/1 sodium cholate solution. Nevertheless some other considerations must be made on the hydrazine. It could favour the reaction, by reacting with the cholestenone produced forming the hydrazone[22], but it also inhibits the 7α-HDS and could complex Mn^{2+}[23] avoiding their catalyzing action. Finally a competitive action of hydrazine with NADH in the reduction of oxygen, on the basis of the standard redox potential values, can be foreseen:

$$\frac{1}{2} O_2 + 2 H^+ + 2 e \rightleftharpoons H_2O \qquad E_o = + 1.23 \text{ V} \qquad [24]$$

$$N_2 + 5 H^+ + 4 e \rightleftharpoons NH_2-NH^+_3 \qquad E_o = - 0.2 \text{ V} \qquad [22]$$

$$NAD^+ + H^+ + 2 e \rightleftharpoons NADH \qquad E_o = - 0.32 \text{ V} \qquad [25]$$

In fact in presence of hydrazine, no improvements of the system response were recorded.

In order to verify the effect of Mn^{2+}, 0.02 - 0.1 mg/ml $MnCl_2$ $4H_2O$ were added to the above described reaction mixture. The addition of Mn^{2+} to the reaction mixture (see Table 1) positively affect the system response. The total and initial response time become constant during the successive additions of cholate and also the consumption of oxygen is increased. Also with 3α-HSD, the addition of Mn^{2+} results in an improvement of the system response, but the experimental values of the oxygen consumption never exceeds 19% of the stoichiometric value.

Some preliminary tests were performed on human biliar samples with the reaction mixture containing also Mn^{2+}. The observed values of oxygen composition and response times were very similar to those reported in Table 1. Nevertheless a careful examination of the recorded peaks shows a more complex answer of the biliar sample, in comparison with pure cholate, characterized by a series of successive peaks rather than an only one.

Table 1. Values of the oxygen consumption, of the initial and total response times, detected by the oxygen electrode by adding constant volumes of a cholate solution to 10.0 ml of a 0.25 mol/1 phosphate buffer solution (pH 5.8), containing 7α-HSD, peroxidase, NAD$^+$, in the optimum reported conditions.

Constant addition of 500 µl, 10^{-2} mol/1, of sodium cholate solution	Variation of oxygen concentration (ppm ± 0.05)	% Consumptions of oxygen, toward theoretical values (found % value ± 0.5)	Initial response time (sec ± 2)	Total response time (min ± 0.5)
1st add.	5.00	62.5	35	8.0
2nd add.	2.76	34.5	20	8.0
3rd add.	1.71	21.4	48	13.0

Values obtained by adding 0.25 mmol/1 of Mn^{2+}

1st add.	6.96	87.0	5	8.0
2nd add.	4.95	61.9	5	8.0
3rd add.	3.80	47.5	5	8.0

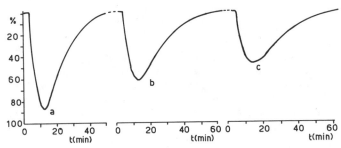

Fig. 1. Consumption of oxygen as % theoretical values by addition of successive (a→b→c) constant volumes of cholate solution.

CONCLUSION

In the present work it has been demonstrated that the determination of cholic acids can be performed utilizing an amperometric Clark electrode as detector and a polyenzymatic system based on the 7α-HSD and the peroxidase in the presence of Mn^{2+}. This method requires an expensive enzyme (7α-HSD), however, it was demonstrated that the same reaction mixture could be utilized for more than one sample. Nevertheless, further work must be done to maintain constant the response of the system during the successive additions of cholic acid.

As regards to the determination of biliar samples, it is necessary to verify if the complex system of peaks observed is due to the different biliar acids present in the sample or to adsorption phenomena on the membrane of the electrode.

Work is now in progress on the immobilization of 7α-HSD and peroxidase on nylon nets since the immobilized enzymatic system could be reutilized and the nylon net could protect the electrode.

Acknowledgement

This work was financially supported by Italian CNR.

REFERENCES

1. U. Biader Ceipidor, R. Curini, G. D'Ascenzo, M. Tomassetti, A. Alessandrini, and C. Montesani, Determination of lithogenic index in human bile: procedure of calculation and evaluation of some analytical method for phosphate and biliar acids, G.It.Chim.Clin., 5:127-151 (1983).

2. U. Biader Ceipidor, R. Curini, G. D'Ascenzo, and M. Tomassetti, Analytical comparison among calorimetric, enzymatic and chemical methods for quantitative determination of cholic acids, Thermochim.Acta, 46:269-278 (1981).

3. L. Campanella, L. Sorrentino, and M. Tomassetti, Determination of cholic acids by ion selective liquid membrane electrode in pharmaceutical products, Anal.Letters, 108:1515-1522 (1982).

4. L. Campanella, L. Sorrentino, and M. Tomassetti, Cholate liquid membrane ion-selective electrode for drug analysis, Analyst, 108:1490-1494 (1983).

5. U. Biader Ceipidor, R. Curini, G. D'Ascenzo, and M. Tomassetti, Cholic acids determination by melting heats, using differential scanning calorimetry. Application to pharmaceutical products, Termochim.Acta, 46:279-287 (1981).

6. L. Campanella, M. Tomassetti, and M. Cordatore, Determination of cholic acids in human bile and in drugs by cholate ion-selective liquid membrane electrode. VIII Congress of Italian Society of Clinical Biochemistry, Florence, Italy, 26-10-1983.

7. O. Fausa and B. A. Skalhegg, Quantitative determination of bile acids and their conjugates using thin-layer chromatography and a purified 3α-hydroxysteroid dehydrogenase, Scand.J.Gastroent., 9:249-254 (1974).

8. F. Mashige, K. Imai, and T. Osuga, A simple and sensitive assay of total serum bile acids, Clin.Chim.Acta, 70:79-86 (1976).

9. G. G. Guilbault, Ion-selective electrodes applied to enzyme systems, Ion-Sel.Elec.Rev., 4:187-231 (1982).

10. R. L. Solsky, Ion-selective electrodes in biomedical analysis, Crit. Rev.Anal.Chem., (CRC Press), 14:1-52 (1982).

11. "The Determination of Oxygen: Scientific and Technical Aspects," Seminary of Dept. of Chemistry of University "La Sapienza", Rome 7-6-1983.

12. M. Mascini, M. Tomassetti, and M. Iannello, Determination of free and total cholesterol in human bile samples using enzyme electrode, Clin.Chim.Acta, 132:7-15 (1983).

13. F. S. Cheng and G. D. Christian, Enzymatic determination of blood ethanol, with amperometric measurement of rate of oxygen depletion, Clin.Chem., 24:621-626 (1978).

14. F. S. Cheng and G. D. Christian, A coupled enzymatic method to measure blood lactate by amperometric monitoring of the rate of oxygen depletion with a Clark oxygen electrode, Clin.Chim.Acta, 91:295-301 (1979).

15. D. M. Small, "The Bile Acids," Vol.1, Padmanabhan P. Nair, ed., Plenum Press (1971).

16. T. Akazawa and E. E. Conn, The oxidation of reduced pyridine nucleotides by peroxidase, J.Biol.Chem., 232:403-415 (1958).

17. S. J. Klebanoff, An effect of thyroxine on the oxidation of reduced pyridine nucleotides by the peroxidase system, J.Biol.Chem., 234:2480-2485 (1959).

18. V. P. Hollander and M. L. Stephens, Studies on phenol-activated oxidation of reduced nucleotides by rat uterus, J.Biol.Chem., 234:1901-1906 (1959).

19. J. B. Mudd and R. H. Burris, Participation of metals in peroxidase-catalyzed oxidations, J.Biol.Chem., 234:2774-2777 (1959).

20. F. S. Cheng and G. Christian, Amperometric measurement of enzyme reactions with an oxygen electrode using air oxidation of reducer nicotinamide adenine dinucleotide, Anal.Chem., 49:1785-1788 (1977).

21. P. Talalay, Enzymatic analysis of steroid hormones, Methods Biochem. Anal., 8:119-143 (1960).

22. B. A. Skalhegg and O. Fausa, Enzymatic determination of bile acids. The NADP-specific 7α-hydroxysteroid dehydrogenase from P. Testosteroni (ATCC 11966), Scand.J.Gastroent., 12:433-439 (1977).

23. G. Charlot, "Analisi Chimica Qualitativa, Equilibri in Soluzione," p.402, Piccin, ed., Padova (1977).

24. "CRC Handbook of Chemistry and Physics," R. C. Weast, ed., CRC Press, INC 59th Ed., (1978-79).

25. A. L. Lenninger, "Biochimica," S. P. A. Zanichelli, ed., Bologna, Italy (1975).

SYNTHESIS OF POLYURETHANE MEMBRANES FOR

THE PERVAPORATION OF ETHANOL-WATER MIXTURES

L. T. G. Pessôa, R. Nobrega, and A. C. Habert

Coppe/Ufrj
Federal University of Rio de Janeiro
Brazil

ABSTRACT

The application of polyurethane membranes to the fractionation of ethanol-water mixtures by pervaporation was investigated in this study. Sorption tests with two polyurethanes selected, one based on polyether, the other on polyester, showed large differences of swelling in water and ethanol. For the first polyurethane, permation for a 94.6% ethanol mixture, at 40°C, resulted in a flux of 1.02 ℓ/h m^2 and a selectivity factor of 3.8 towards water. Membranes were synthesized with the polyurethane based on polyester by phase inversion and total solvent evaporation techniques. Their microstructures were characterized by scanning electron microscopy and their permeability and selectivity were evaluated. The study revealed that membranes synthesized by phase inversion were asymmetric while they have a dense structure if prepared by total solvent evaporation. The porosity of the membranes increased when both solution polymer concentration and solvent evaporation time were reduced. Permeation fluxes obtained at 25°C ranged from 0.1 to 3.5 ℓ/h m^2 with selectivity factors varying from 5.5 to 1.4, respectively. The influence of the composition on the pervaporation of ethanol-water mixtures was also studied, indicating that ethanol dehydration using polyurethane membranes is feasible.

INTRODUCTION

The use of membrane processes for the fractionation of ethanol-water mixtures is being considered as an interesting alternative to conventional alcohol production technology. Solubility and diffusivity differences between the components of a mixture in a polymeric membrane constitute some of the principal factors for the choice of the suitable polymer for a required separation. The reasonable difference observed between water and ethanol in polyurethane suggests this class of polymer as candidate. Hopfenberg et al.[1] and Huang and Autian[2] have already studied the permeation and diffusion behavior of a variety of monohydric alcohols and water, attempting to explain the transport mechanism. On the other hand, Koenhen et al.[3] have investigated the asymmetric polyurethane membrane formation and its application in desalination by reverse osmosis.

549

The physical and chemical membrane properties are influenced by the polymer nature and the membrane manufacture conditions. These factors will determine the structure, and consequently the permeability characteristics of the membranes obtained. The elucidation of membrane structure and morphology is extremely important for a best knowledge of the permeation performance. The classical work of Riley et al.[4] demonstrating the asymmetric structure of Loeb-Sourirajan[5] cellulose acetate membranes by electron microscopy constitute an important contribution in this aspect.

In this work, the technical feasibility of the use of polyurethane membranes for the fractionation of ethanol-water mixtures by pervaporation was investigated. The membranes were synthesized with varied structure, dense and asymmetric. The evaluation of the asymmetric membranes in pervaporation was stimulated by the relatively low permeation fluxes obtained with dense membranes. The membranes were characterized by electron microscopy to interpret their permeability and selectivity results and also to evaluate the synthesis variables influence. The influence of ethanol-water feed composition was also studied, checking for the best range of pervaporation application, and in particular, for the breaking up of the azeotrope[6].

FUNDAMENTALS

The performance of the membranes in the pervaporation process is described by the permeation flux (J) and by the selectivity factor (α) defined for a binary mixture by:

$$\alpha_{i/j} = \frac{Y_i/Y_j}{X_i/X_j} \tag{1}$$

where X and Y are the weight fractions of the components i and j in the liquid phase (feed) and in the permeate, respectively. If the transport is exclusively diffusive, it can be expected the vality of a phenomenological relation based on Fick's law:

$$J_D = \frac{P}{L} = \frac{D \cdot S}{L} \tag{2}$$

where P, D and S are, respectively, the permeability, diffusivity and solubility coefficients of the mixture in the membrane, and L is its thickness. Since D and S are usually concentration - dependent for liquids, these coefficients were here assumed as average values.

When a convective flow through pores is to be considered, the flux can be expressed by a model based on Hagen-Poiseuille's law:

$$J_c = \frac{\varepsilon r_1^2 \, \Delta P}{8 \eta \, L} \tag{3}$$

where ε is the membrane porosity, r_1 is the pore average radius, η is the mixture viscosity and ΔP is the pressure gradient.

The permeation behavior is also evaluated by the permeation ratio (θ), introduced by Huang and Lin[7] as a measure of deviation of the ideal behavior of the mixture. It is obtained by a relation between the real permeation flux (J) and the calculated if the system were ideal (J^o):

$$\theta = \frac{J}{J^o} \tag{4}$$

The permeation ratio of each component can also be determined knowing the permeate mixture composition. Thus, in a non ideal permeation, the permeation ratio can be higher or lower than unity, representing the positive or negative deviation from ideality, respectively, due to effects caused by one component on the permeation of the others. Plasticization and clusters formation are examples of such effects.

EXPERIMENTAL

Membrane Synthesis

Two polyurethanes were selected to be evaluated as membrane materials for this study. With the first polyurethane, the membranes were synthesised by "in situ" polymerization from a liquid prepolymer (VIBRATHANE B-601) based on polyether (polytetramethyleneetherglycol) and toluene diisocyanate (TDI), using 4,4' methylene bis (2 chloro-aniline) (MOCA) as chain extensor, both supplied by UNIROYAL (São Paulo). A suitable casting solution was prepared with 2,5 g of prepolymer and 0.45 g of MOCA to each millilitre of methylethylketone. All cast films were cured at 50°C for 20 h to complete conversion of the reactants. The residual solvent was extracted in water baths periodically renewed. The membranes obtained, MPE series, were coded MPE-X, where X is an order number.

The other polyurethane used was based on polyesther (polyethyladipate) and diphenyl methane - 4,4' diisocianate (MDI), and was suggested by the work of Koenhen et al.[3]. It was supplied by COFADE (São Paulo) as a 38% wt solution of dimethylformamide and toluene (Elastolan V445). With this thermoplastic polyurethane the membranes were synthesized by total evaporation of the solvent and by phase inversion techniques to obtain dense and asymmetric membranes, respectively. Two synthesis parameters were varied: the concentration of the polymer solution, diluting the polymer solution with dimethylformamide, and the partial evaporation time before the phase inversion of the liquid films in water-methanol (1:1) baths. The evaporation temperature was 55°C. These membranes, M series, were coded M-X-Y, where X is the polymer solution concentration (% wt) and Y is the partial evaporation time of the solvent (min) for the asymmetric membranes. Dense membranes were obtained at evaporation times of 180 min and were coded M-X-D. After synthesised, the membranes were immersed in water baths periodically renewed and subsequently in 95% ethanol baths for 48h. Before being used in the pervaporation tests, the membranes were soaked in the ethanol-water mixtures.

Membrane Characterization and Evaluation

The influence of the composition of the ethanol-water mixtures in the sorption and pervaporation behavior was evaluated for dense membranes of the two polyurethane types (MPE and M). The M membranes (dense and asymmetric) were also characterized by scanning electron microscopy and by pervaporation of a 95% ethanol mixture. A gravimetric technique was employed in the sorption tests, after equilibrating membrane samples (minimum of 3 for each test) in the ethanol-water mixtures. The evaporation in weighing was compensated by extrapolation to zero time of the weight vs. time curve. The swelling extent of the mixtures was obtained by the expression

$$S = \frac{W_1 - W_0}{W_0} \tag{5}$$

where W_0 and W_1 denote the weight of dry and swollen membranes, respectively.

The pervaporation system is showed in Figure 1. The permeation area of the cells used in the MPE and M membranes tests were 28.3 and 63.6 cm^2, respectively. The permeate downstream pressure was maintained at about 1 mm Hg and the liquid feed temperature was kept at 25± 0.1°C. No influence on membrane permeabilities was found for operating downstream pressure up to 10 mmHg. The permeation fluxes were measured volumetrically and the feed and permeate samples were analyzed by gas chromathography.

RESULTS AND DISCUSSION

MPE - Membranes

The sorption behavior of MPE-membranes (Table 1) shows the swelling increase with ethanol concentration, indicating the better compatibility of ethanol with polyurethane. The results are in good agreement with Hung and Autian's data[2] who worked with a similar polyurethane, although they did not report on synthesis conditions.

The higher swelling extent of the polyurethane membranes by the ethanol can be attributed to its larger tendency to bind to a polyurethane

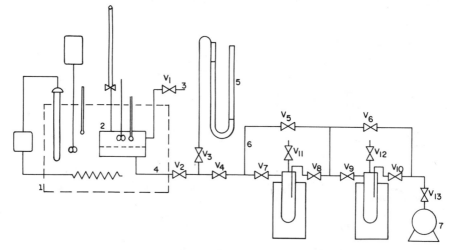

Fig. 1. Pervaporation system: (1) Thermostatic bath, (2) pervaporation cell, (3) non-permeate line, (4) permeate line, (5) U. gauge, (6) permeate collect system, (7) vacuum pump.

Table 1. Polyurethane MPE - Membrane Swelling in Ethanol-Water Mixtures at 25°C.

MIXTURE COMPOSITION (% wt ETHANOL)	SWELLING EXTENT (g MIXTURE/g DRY MEMBRANE)	
	PRESENT WORK	HUNG AND AUTIAN'S[2] DATA
0	1,75	1,9
50	14,3	14
99,8	37,8	28

site in comparison with water. This hypothesis is based in the early work
of Hung et al.[2] who studied the clusters formation in the ethanol-water-
polyurethane system through the Zimm-Lundberg function.

The fractionation study of ethanol-water by pervaporation with the MPE
membranes also indicates the higher swelling extent due to the ethanol.
The permeation fluxes increased from 0.17 to 1.02 ℓ/hr m^2 when feed ethanol
concentration changed from 9,4 to 94.6% ethanol. The separation factor α
water/ethanol = 3.8 obtained for the 94.6% ethanol feed confirmed the
potential of this polymer. In order to obtain a larger control on membrane
structure and explore the properties of asymmetric membranes further work
proceeded with the thermoplastic polyurethane.

M - Membranes

Figures 2-4 show the results of characterization of the M-membranes by
scanning electron microscopy. As it can be seen in Figure 2, the membranes
synthesized by phase inversion are clearly asymmetric contrasting with
those obtained by total solvent evaporation, which are dense, as visualized
in Figure 3d. Figure 2 also shows the effects of the increase of polymer
solution concentration and the partial evaporation time in the phase
inversion technique. As observed, the result was a pore density reduction
that can be related to the higher difficulty of solvent diffusion[8,9].
This effect was more pronounced with the evaporation time increase. Figure
3 also illustrates this fact showing the cross section of membranes synthesized from a same casting solution (31% wt). The increase of dense
regions can be best seen in this figure. The extreme case of absence of
pores (at the level of microscope resolution obtained) in the membranes
synthesized by total evaporation (Figure 3d) allowed their classification
as dense membranes.

Structural irregularities in the membranes top surface, in the form of
cellular configuration, can still be noticed in Figure 2. This "orange-
peel effect[8]" has been attributed to local heterogenities of the casting
solutions which would cause vortex formation and deformed surfaces. Top
and lower faces of the membranes have different aspect (Figure 4) although
macroscopically the difference tended to disappear as roughness in both
surfaces increased with polymer concentration and evaporation time. It is
also interesting to observe that membranes with relatively higher surface
pore densities exhibited more bulk dense regions.

Primary evaluation of the M-membranes transport properties was made by
pervaporation of a 95% wt ethanol solution. As indicated in Table 2, the
higher permeation fluxes were obtained with the more porous membranes
reaching a maximum of 3.54 ℓ/h m^2. The increase in both polymer concentration and evaporation time caused a reduction in the permeation flux of
the membranes, suggesting the effect of pore density reduction. The lower
permeation fluxes were obtained with the dense membranes. The effect of
the evaporation time in the the permeation flux appeared to be more drastic
when polymer concentration was increased. Therefore, the permeation fluxes
of the M-26-Y membranes were less modified (approximately 15%) than the
permeation fluxes of the M-35-Y membranes (approximately 370%). It is also
interesting to notice the difference in thickness of dense and asymmetric
membranes. As polymer concentration and evaporation time increased, the
variation of thickness tend to be smaller, probably due to pore density
reduction.

The permeability coefficients (P) of the M-membranes were calculated
from the permeation fluxes (J) and the membrane thicknesses. However, the
obtained values with the asymmetric membranes should not be considered
"true" permeability coefficients. For dense membranes, the permeability

553

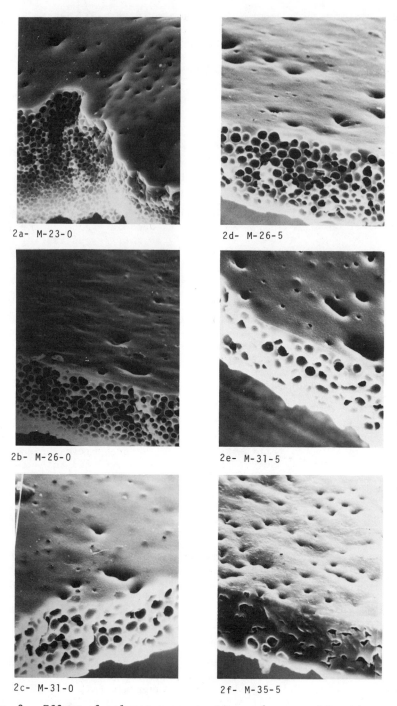

Fig. 2. Effect of polymer concentration and evaporation time on PU
         membrane structure – top surface and cross section
         (magnification: 1000X)

coefficients were approximately constant with polymer concentration as it
should be expected. As can be seen in Figure 5, the plot of the permeation
fluxes of the dense membranes vs. their thicknesses suggests the diffusion

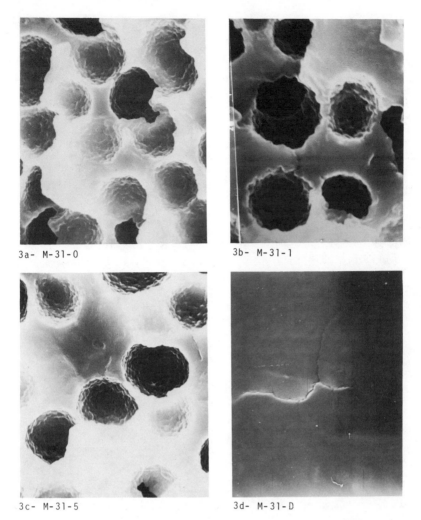

3a- M-31-0

3b- M-31-1

3c- M-31-5

3d- M-31-D

Fig. 3. Effect of the evaporation time of the solvent on the porosity of
the PU membranes - cross section (magnification: 5000X).

transport of the ethanol-water solution through the membranes. This
behavior is typical of dense membranes[8,10,11,12]. In this case the
calculated permeability coefficient from the slope of Figure 5 is 1.30 x
10^{-8} cm^2/s for the 95% wt ethanol solution at 25°C.

The separation factors higher than 1, showed in Table 2, indicate the
higher selectivity of these polyurethane membranes with respect to water.
The separation factors increased with the evaporation time and were more
sensitive to polymer concentration variations. The dense membranes showed
the higher values with an average separation factor of 5.5.

The permeation fluxes and selectivities of the asymmetric membranes
were correlated in Figure 6. It can be observed that an increase in flux,
related with an increase in the membrane porosity, is obtained at the
expenses of selectivity. This is in accordance with effects of membrane
synthesis variables upon the final structure as previously discussed. The
higher permeation fluxes of the more porous membranes may have been caused
by a convective component of the permeation flux due to a possible inter-

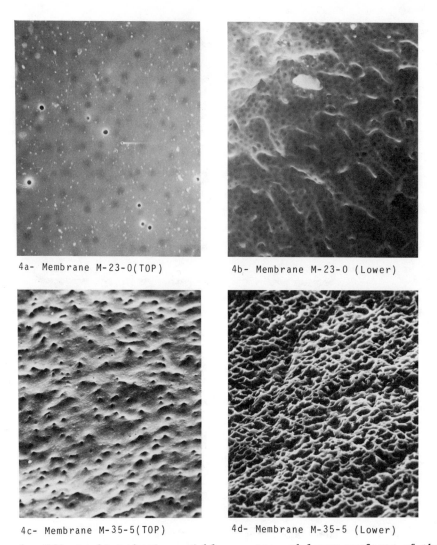

4a- Membrane M-23-0(TOP)

4b- Membrane M-23-0 (Lower)

4c- Membrane M-35-5(TOP)

4d- Membrane M-35-5 (Lower)

Fig. 4. Effect of synthesis variables on top and lower surfaces of the
PU membranes (magnification: 500X).

connection between the surface and bulk pores of the membranes. Therefore,
this convective transport would add to the diffusive one occurring in the
dense regions. Membrane selectivity would accordingly be higher when
diffusive transport predominates, or is absolute, as in the dense membranes.

Despite of the higher separations obtained with the dense membranes,
the maximum permeation flux through them was 0.31 ℓ/h m^2 (M-23-D). Thus,
efforts must be employed to obtain diffusive skin type membranes so that
the selectivity factor of 5.5 already reached can be maintained with higher
permeation fluxes.

Influence of Feed Composition

Figure 7 presents the sorption behavior of a M-26-D (20µ thick)
membrane in ethanol-water mixtures. It is interesting to observe the
almost linear increase of the swelling extent up to approximately 75%

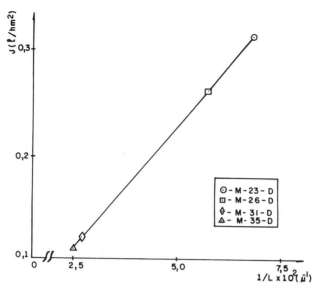

Fig. 5. Effect membrane thickness on the permeation flux of the dense PU
- membranes - pervaporation tests at 25°C - feed: 95% ethanol.

Table 2. Influence of Synthesis Variables on Permeability and Selectivity
of the Polyurethane M-membranes - Pervaporation Tests at 25°C -
Feed: Ethanol 95% wt.

POLYMER SOLUTION COMPOSITION C_p (%)	EVAPORATION TIME θ_V (min)	MEMBRANE THICKNESS $L(\mu)$	PERMEATION FLUX $J(\ell/h\ m^2)$	PERMEABILITY COEFFICIENT $P \times 10^8 (cm^2/s)$	SEPARATION FACTOR α water/ethanol
23	180	15	0,310	1,29*	5,1
26	0	41	3,05	34,7	1,5
	1	59	3,54	58,0	1,4
	5	49	2,61	35,5	1,7
	180	18	0,258	1,29*	5,4
31	0	49	1,97	26,8	1,7
	1	57	1,05	16,7	2,6
	5	48	0,41	5,49	3,3
	180	37	0,122	1,25*	5,6
35	0	51	0,96	13,6	2,6
	1	52	0,46	6,59	3,4
	5	44	0,20	2,48	4,6
	180	40	0,112	1,25*	5,7

*Diffusive

ethanol. Above this composition, the swelling extent is relatively constant and equal to 12.8 x 10⁻² g mixture/g dry membrane, which is 9 times
larger than the pure water value.

The pervaporation results are presented in Figure 8-10. Figure 8
shows the mixture permeation flux variation with the ethanol concentration
in the feed. One can observe the non ideal behavior of the mixture, with a
maximum permeability being reached at 75% ethanol mixture. In connection
with sorption results (Figure 7), the decrease in membrane permeability at
concentrations higher than 75% ethanol may be associated to a reduction in
the mixture diffusion coefficient, since it also directly influences the
permeability.

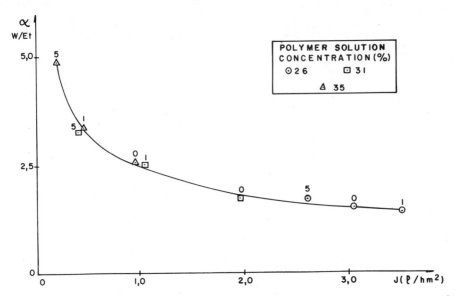

Fig. 6. Pervaporation behavior of asymmetric PU – membranes – T = 25°C –
feed: 95% ethanol. The 0,1 and 5 indices refers to the partial
evaporation times (min).

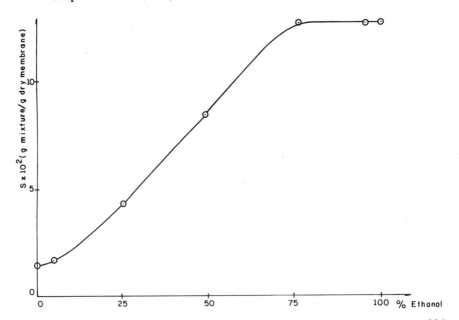

Fig. 7. Influence of ethanol-water mixture composition on the swelling
of PU – membranes at 25°C.

The non ideality of mixture permeation can be better visualized in
Figure 9 where the calculated permeation ratios (θ) of the mixture and each
component are plotted in function of ethanol concentration in the feed.
One can also observe the non ideal ethanol and water behavior in all the
concentration range, probably due to interactions between the mixture
components, and among them and the polymeric membrane.

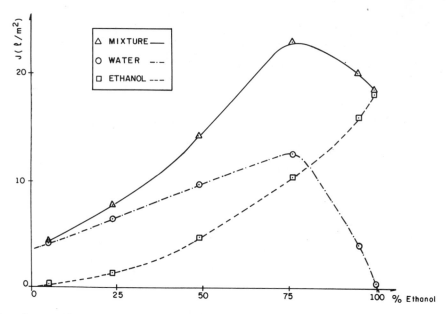

Fig. 8.   Influence of ethanol-water mixture composition on the permeation
          flux.   Membrane M-26-D (L = 20 μm). T = 25°C.

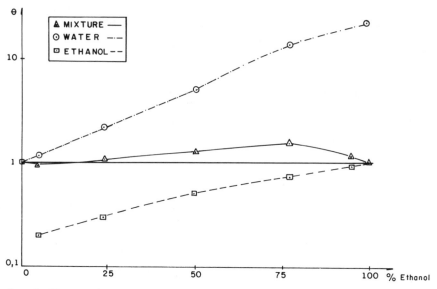

Fig. 9.   Influence of ethanol-water mixture composition on the permeation
          ratio - membrane M-26-D (L = 20 μm).   T = 25°C.

         According to Huang and Lin[7], the permeation ratio of a pure compon-
ent can be interpreted as a measure of the effect of the other component in
its permeation flux.   For example, plasticization of the membrane associated to the greater solubility component can cause an increase in the
permeability of the lower solubility component.   Another effect than can be
also present in competition to plasticization is the clustering tendency
within the mixture and/or among polymer/mixture components molecules, which
can both cause a reduction in the permeability.   Several systems already

studied try to explain their permeation behaviors taking into account these two effects[1,13-17].

In the present study a qualitative interpretation of the observed results is presented here as a preliminary hypothesis. It is also based in the competition between plasticization, due to the swelling difference of the polyurethane membrane in water and ethanol, and clusters formation, due to the polar character of the system.

Figure 9 shows the increase of water permeability with ethanol concentration in the feed in all concentration range and the decrease of ethanol permeability in relation to its ideal behavior. The water facilitated transport may be associated with the plasticization of the membrane by the ethanol and the decrease in ethanol permeation with clusters formation like ethanol-ethanol, ethanol-water or ethanol-polyurethane by attractive bonds like Van der Walls type. According to Hung et al.[2], in the ethanol-water-polyurethane system, water does not present any tendency either to cluster with an identical molecule nor to bind to active sites of the polymer. Water would only be present in the form of water/ethanol species bound to the polymer chain. The authors also reported that the polymer interactions only occur at concentrations higher than 53% ethanol.

The increase in the mixture permeability up to 75% ethanol, as can be seen in Figure 8 is probably due to the stronger ethanol plasticization effect. At higher concentrations, the clustering effect seems to predominate over the plasticization effect, causing a decrease in the mixture permeability because of the hindered transport of greater molecular permeant species like ethanol-ethanol type through the polymer diffusional vacancies. The water permeability also decreased in this concentration range but is still greater than its ideal behavior (Figure 9). This positive deviation can be attributed to the swelling of the membrane by the ethanol.

Selectivity Data. Figure 10 shows the influence of mixture composition in the selectivity of the polyurethane membrane. The poor separ-

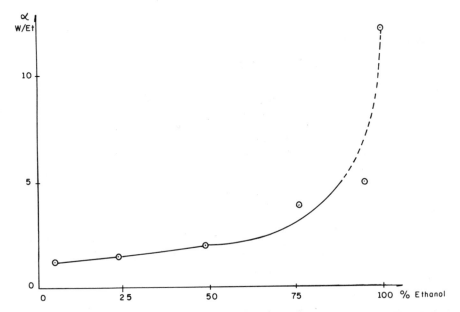

Fig. 10. Influence of ethanol-water mixture composition on selectivity — membrane M-26-D (L = 20 μm). T = 25°C.

ation factors obtained at lower ethanol concentrations can also be associated with the ethanol plasticization effect. When the ethanol concentration increases, the water separation becomes more pronunciated. As analyzed above, despite of the increase of membrane swelling, clusters formation begins to compete with the plasticization effect. The preferential water permeation may be attributed to the faster transport of the smaller ethanol-water clusters than the ethanol-ethanol type. Therefore a relatively high separation factor can be achieved when the mixture becomes very poor in water.

It is worth to point out that results of this work clearly illustrate the case where pure component sorption measurements are not a sufficient criteria for predicting selectivity of a membrane towards component of a mixture. Whereas it may occur, interactions within the membrane may reverse sorption behavior and analysis of composition of the sorbed mixture must be undertaken.

An azeotropic ethanol-water dehydration test was also done in this study with the same membrane. The pervaporation cell was fed with 125 ml of a 95% ethanol mixture. After a 24 h permeation at 25°C, the 95 ml of non permeate mixture had a 97.5% ethanol composition. Therefore, the pervaporation process using polyurethane membranes of this type can be considered as a promising alternative process to be further explored.

REFERENCES

1. H. B. Hopfenberg, N. S. Schneider, and F. Votta, J.Macromol.Sci.,Phys, B3(4):751 (1969).
2. G. W. Hung and J. Autian, J.Pharm.Sci., 61:1094 (1972).
3. D. M. Koenhen, M. H. V. Mulder, and C. A. Smolder, J.Appl.Pol.Sci., 21:199 (1977).
4. R. J. Riley, U. Merten, and J. O. Gardner, Science, 143:801 (1964).
5. S. Loeb and S. Sourirajan, Advan.Chem.Ser., 38:177 (1962).
6. L. T. G. Pessôa, M.Sc. Thesis, COPPE/UFRJ (1983).
7. R. V. M. Huang and V. J. C. Lin, J.Appl.Pol.Sci., 12:2165 (1968).
8. R. E. Kesting, "Synthetic Polymeric Membranes," McGraw-Hill Book Company, USA, 1971.
9. H. Strathman, Desalination, 35:39 (1980).
10. W. Push and A. Walch, J.Membr.Sci., 10:323 (1982).
11. R. C. Binning, R. J. Lee, J. F. Jennings, and E. C. Martin, Ind.Eng. Chem., 53(1):49 (1961).
12. A. S. Michaels, R. F. Baddour, H. J. Bixler, and C. Y. Choo, Ind.Eng. Chem., 1(1):14 (1962).
13. A. E. F. Campello and R. Nobrega, Revista Brasileira de Engenharia, Cadernos de Engenharia Quimica, 1(2):49 (1983).
14. W. Featherstone and T. Cox, Brit.Chem.Eng., 16(9):817 (1971).
15. A. C. Habert, M.Sc. Thesis, COPPE/UFRJ (1971).
16. B. T. Pierzynsky, R. Nobrega, and M. L. Santos, Revista Brasileira de Tecnologia, 5:25 (1974).
17. M. L. Santos and D. M. Leitão, Boletim Técnico da Petrobrás 16(4):275 (1973).

SEPARATION OF BENZENE-METHANOL AND BENZENE-CYCLOHEXANE

MIXTURES BY PERVAPORATION PROCESS

E. Nagy, J. Stelmaszek* and A. Ujhidy

Research Institut for Technical Chemistry of the Hungarian
Academy of Science, P.O.B. 125, Veszprem, Hungary
*Technical University of Wroclaw, Poland

SUMMARY

Pervaporation of binary components mixtures of benzene-methanol and
benzene-cyclohexane was investigated by using polyethylene and differently
orientated and non-orientated polypropylene membranes. The permeability
coefficient was calculated from the experimental results by means of a
phenomenological equation. The value of this coefficient depends very
strongly on the composition of the feed but it remains practically unchanged by temperature and by pressure on the down-stream side of the
membrane.

INTRODUCTION

It is well known that the pervaporation process can be applied
successfully for the separation of liquid mixtures of two components.
These mixtures, such as azeotropic mixtures and close-boiling mixtures
can not be separated or can only be separated with great effort by conventional methods i.e. distillation, fractional crystallization and
extraction.

According to the literature summarizing the pervaporation methods
it can be seen that in the last 10-20 years, a great number of liquid
mixtures have been investigated. Thus, separation of the mixture of
benzene-methanol and benzene-cyclohexane by different polymer membranes
is discussed in many publications[1-8]. Azeotropic mixture of benzene-
methanol (x_B = 0.386 mol/mol) was investigated by applying the membranes
of polyethylene[1,2] and polytetrafluorethylene-polyvinylpirrolidone[3].
The value of the selectivity coefficients (α) is 2-10, which slightly
decreases with the temperature. The pervaporation rate ranges between
1×10^{-4} and 20×10^{-4} kg/m²s. Pervaporation of the mixture of benzene-
cyclohexane was investigated in great detail in the literature. Using
the polyethylene membrane[4,5,6], the pervaporation rate of (1-10) x
10^{-4} kg/m²s and α = 1 - 2.5 was found. The pervaporation rate depends
greatly on the composition of the liquid mixture; by increasing of the
benzene concentration, the maximum rate will be between x_B = 0.5 and 0.7,
while the selectivity decreases monotonously with the concentration[6].
Much higher selectivity (α = 2 - 15) was obtained by McCandless[7] by
means of the vinilydene-fluoride membrane softened with 3-methyl-sulfolene

but the pervaporation rate was lower by about one order of magnitude as in the case of polyethylene membrane. Favorable selectivity ($\alpha = 5$) was also achieved by a modified cellulose-ester[8] for the mixture of benzene-cyclohexane.

Two characteristics of the efficiency of the pervaporation process i.e. the pervaporation rate and the selectivity depend on many factors. Thus, these characteristics depend considerably on the so-called operational properties, which are independent of the properties of membrane and liquid mixtures, i.e. temperature, pressure. The effect of these characteristics can be seen firstly on the evaporation of the liquid transported via the membrane from the surface. The component transfer via the membrane is a very complicated process. Accurate description of this process has still to be presented. Mathematical model was developed by Greenlaw et al.[9], assuming diffusion component transport depending on the concentration. This model was used by Rautenbach and Albrecht[6] for the description of the pervaporation of benzene-cyclohexane. A very simple phenomenological equation was used by Hoover and Hwang[10], in which for example, the pervaporation rate of component B through the membrane can be calculated as follows:

$$J_B = Q_B \ (P_B^o x_B - P_2 Y_B / \gamma_B) / \delta \tag{1}$$

From the Equation (1), the value of the permeability coefficient Q_B can be determined but this is only characteristic for the component transport via the membrane. The effect of the external factors (temperature, pressure) is taken into consideration in the second term of the Equation (1). Naturally, the value of Q is more or less, indirectly affected by the temperature and pressure. The Equation (1) is valid if there is equilibrium between membrane and permeate and if the diffusion through the membrane is the slowest, the so-called rate-determining step. These assumptions are probably not fulfilled when the downstream pressure is too high - close to the saturation vapor pressure the downstream desorption will be the rate-determining step - or when it is too low. Probably at low pressure the equilibrium can not be established. Despite this fact for most part of pressure range the value of Q is typical of the mass transport through the membrane and permits a better comparison of the experimental results derived under differing conditions.

In our work, the experimental results obtained by means of the different polymer membranes (polyethylene, polypropylene) are compared based on the values of Q calculated from Equation (1). The effect of the composition in the feed and temperature and pressure on the downstream side of the membrane is investigated.

EXPERIMENTAL

The apparatus for the separation consisted of a plexiglass cell, 0.154 m in diameter and 0.115 m high, with a stirrer. The supported membrane is placed between two stainless steel flanges. The support is made from stainless steel with 650 holes of 2.5×10^{-3} m I.D. In the cell a heating spiral casing, a thermometer and a cooler on the top of cell can be found. The vapor chamber is connected to vacuum via a condensing system. This contains an annular glass condenser cooled by water which is followed by a trap cooled by aceton-dry ice mixture to -60°C. The schematic diagram of the pervaporation apparatus is shown in earlier work[11].

The "stage cut" i.e. the fraction of feed permeated was 0.01-0.03. The product was analyzed either by a Perkin-Elmer series 900 gaschromatograph or by refractometer (Carl Zeiss, Jena, GDR). The results are the mean values of 3-8 pervaporation experiments. The liquids used in the experiments were benzene-methanol and benzene-cyclohexane mixture of different compositions. Commercially available polyethylene membranes having a thickness of 5.3×10^{-5}m (made in Poland) were used. The polypropylene membranes are also commercial membranes produced in Italy. Three types of polypropylene membranes were applied: oriented in one direction, 3.9×10^{-5}m thick, oriented in two directions, 4×10^{-5}m thick and an unoriented one, 4.2×10^{-5}m thick.

Some sorption experiments were carried out by soaking the membranes for 24 hr at a constant temperature in the two binary mixtures mentioned above. The weight of the membranes was considered as the sorbed mass.

RESULTS AND DISCUSSION

The capacity of the different polymer membranes as a function of the liquid composition is shown in Figure 1. In the case of the benzene-methanol, the amount of the absorbed material slightly increases as the benzene content of the liquid increases up to the values of $x_B = 0.8 - 0.9$. The expansion is about doubled but the highest value of the absorbed amount is rather low, less than 0.1 g/g. In the case of the benzene-cyclohexane, a greater change was observed. By increasing the concentration of the benzene a decrease of the absorbed liquid was achieved. It was found that the polypropylene orientated in two directions can absorb much more liquid than the polypropylene oriented in one direction. This great difference is somewhat surprising since it is largely unexpected that the sorption capacity of the polypropylene orientated in two directions and, therefore well-ordered, is higher. The effect of this can also be seen in the pervaporation characteristic of the membrane as it is shown in the Figures.

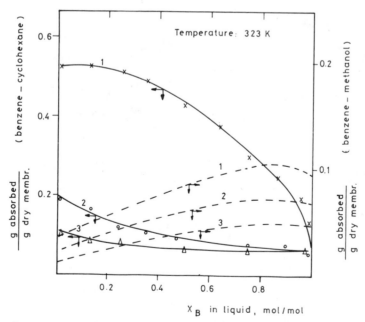

Fig. 1. Sorption capacity of different membranes in benzene-cyclohexane and benzene-methanol mixtures. (1) Polypropylene oriented in two directions, (2) polypropylene oriented in one direction, (3) polyethylene.

In the case of benzene-methanol mixtures, the changes of the pervaporation flux and the selectivity as a function of the benzene concentration is shown in Figure 2A. Three different membranes were used. The pervaporation rate changes almost two order of magnitudes with the changes of liquid composition. Despite its low sorption capacity, polyethylene has a very high pervaporation rate. The selectivity of the polyethylene and the polypropylene is completely different, the polyethylene has a much high selectivity. In the case of all three membranes, the selectivity goes through a maximum when x_B is progressively increased, but this maximum is different: $x_B \cong 0.2$ for polypropylene and $x_B \cong 0.6$ for polyethylene. The values of the permeability calculated with Equation (1) and based on the data of Figure 2A are shown in Figure 2B. The value of activity coefficient (γ) was calculated by the Wilson-equation[12]. In the case of both membranes, firstly the permeability of benzene increases more quickly, while in the range of $x_B > 0.5$, Q_M changes faster. The increasing Q_B value means a higher selectivity and the increase of Q_M results in lower selectivity. The very strong effect of the liquids on the permeability of components is proved by the fact that the permeability of methanol practically does not or slightly decrease (in the case of polyethylene membrane it is only temporary) as the concentration of benzene increases. According to that, the pervaporation rate of methanol, (J_M), also increases up to x_B = 0.08 and above this value it decreased slowly.

Experimental results derived with benzene-cyclohexane mixture are shown in Figure 3A. Polyethylene and polypropylene orientated in one direction were used. The pervaporation properties of polyethylene are more favourable but the selectivity of both membranes changes similarly. Practically, the selectivity coefficient monotonously decreases with the increase of the benzene concentration. The permeability calculated from Equation (1) is shown in Figure 3B. Activity coefficient and vapor pressure of the benzene-cyclohexane mixture were derived from the book of Kogan

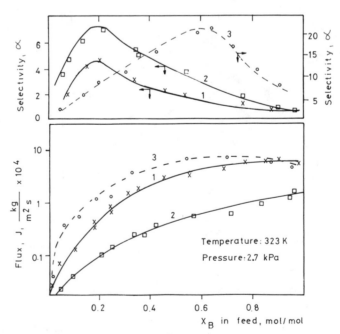

Fig. 2A. Pervaporation of binary benzene-methanol mixture. (1) Polypropylene oriented in two directions, (2) polypropylene oriented in one direction, (3) polyethylene.

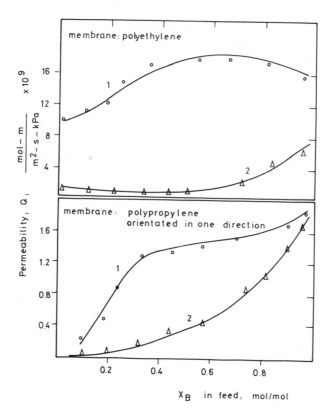

Fig. 2B.   Permeability of benzene and methanol at 323 K.   Pressure:
2.7 kPa.   (1) Benzene, (2) methanol.

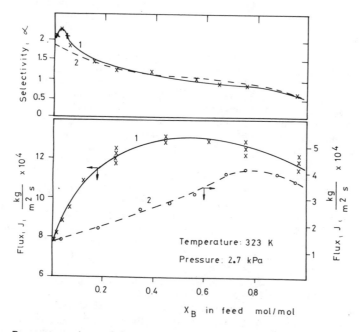

Fig. 3A.   Pervaporation of binary benzene-cyclohexane. (1) Polyethylene,
(2) polypropylene oriented in one direction.

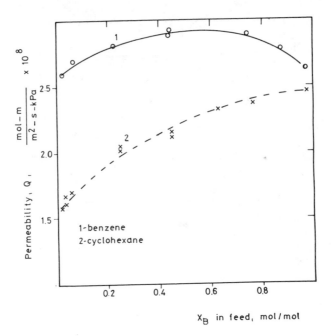

Fig. 3B. Permeability of benzene and cyclohexane in polyethylene membrane
at 323 K. Pressure: 2.7 kPa.

and Fridman[13]. The permeability of the cyclohexane changes to a higher
extent with concentration while, at the beginning, the permeability of the
benzene increases and, after it passes the maximum, decreases again. The
decrease of the selectivity with the concentration is caused by the strong
increase of Q_C. In the case of the polypropylene membrane, the value of Q
similarly changes. The effective pervaporation rate, J_B increases with the
increase of the benzene concentration, while that of the cyclohexane
decreases with the higher rate.

Furthermore, effect of the pressure on the downstream side of membrane
and the temperature on the permeability will be investigated shortly.
Pervaporation rate and selectivity of the azeotropic mixture of benzene-
methanol is shown in Figure 4A using polypropylene membranes. The values
of J and α decrease with increasing pressure and this is in close agreement
with the literature data of other mixtures[6,11]. The permeability calculated from these experimental results is shown in Figure 4B. The permeability of compounds does not essentially change in the pressure range
investigated. It is proved that the value of J is decreased due to the
decrease of the rate of the downstream desorption from the membrane surface
to the vapor phase. Similar results were obtained for the effect of
temperature. The effect of the temperature for a given composition of
benzene-cyclohexane mixture is shown in Figure 5A, using three membranes.
The pervaporation rate exponentially increases with the temperature while
the permeability of the components is almost stable. It can be concluded
from the above that the second term of Equation (1) increases to nearly the
same extent as the increase of the pervaporation rate. This term is the
'driving force' increase derived from the increasing vapor pressure at the
component transfer between the membrane and the vapor phase. The difference between the pervaporation properties of oriented and unoriented
polypropylene can also be seen in this Figure. A much higher pervaporation
rate was obtained by the non-oriented polypropylene but its selectivity was
much lower, as well.

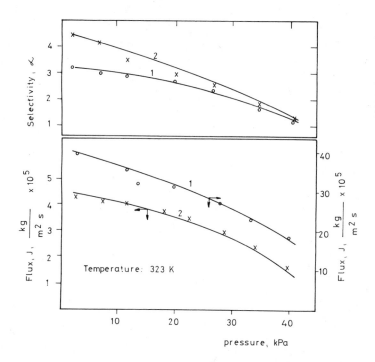

Fig. 4A. Pervaporation of benzene–methanol mixture in function of pressure (x_B = 0.386 mol/mol). (1) Polypropylene oriented in two directions, (2) polypropylene oriented in one direction.

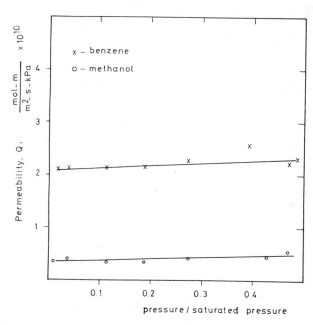

Fig. 4B. Permeability of benzene and methanol in polypropylene membrane oriented in one direction in function of the pressure at 323 K.

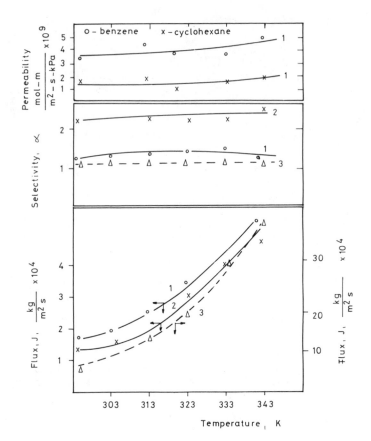

Fig. 5. Pervaporation of benzene-cyclohexane mixture ($x_B = 0.75$) in function of the temperature. Pressure: 6.5 kPa. (1) Polypropylene oriented in two directions, (2) polypropylene oriented in one direction, (3) unoriented polypropylene.

NOMENCLATURE

J – pervaporation rate, kg/m²s
Q_o – permeability, mol m/m²s kPa
P^o – saturated vapor pressure, kPa
x – mole fraction in liquid
P_2 – pressure on the vapor side of membrane, kPa
y – mole fraction in vapor phase
γ – activity coefficient
δ – membrane thickness, m

Subscripts

B – benzene
M – methanol
C – cyclohexane

$$\alpha = \frac{y_B/(1-y_B)}{x_B/(1-x_B)}, \text{ selectivity coefficient}$$

REFERENCES

1. J. W. Carter and B. Jagannadhaswamy, Separation of organic liquid by selective permeation through polymeric films, Brit.Chem.Eng., 9:523 (1964).
2. J. I. Dytnerskij, Membranprozesse zur Trennung flüssiger Gemische, VEB Deutscher Verlag für Grundstoffindustrie, Leipzig (1977).

3. P. Aptel, N. Challard, J. Cuny, and J. Neel, Application of the pervaporation to separate azeotropic mixtures, J.Membrane Sci., 1:271 (1976).

4. A. S. Michaels and H. J. Bixler, Membrane permeation: theory and practice, in: "Progress in Separation and Purification," E. S. Perry, ed., Wiley, New York (1968).

5. R. Y. Huang and V. J. C. Lin, Separation of liquid mixtures by using polymer membranes. I. Separation of binary organic liquid mixtures through polyethylene, J.Appl.Polym.Sci., 12:2615 (1968).

6. R. Rautenbach and R. Albrecht, Separation of organic mixtures by pervaporation, J.Membrane Sci., 7:203 (1980).

7. F. P. McCandless, Separation of aromatics and naphthalenes by permeation through modified vinylidene-fluoride films, Ind.Chem. Proc.Des.Dev., 12:354 (1973).

8. E. C. Martin, R. C. Binning, L. M. Adams, and R. J. Lee, Diffusive Separation, U.S. Patent 3,140,256, July 7th (1964).

9. F. W. Greenlaw, R. A. Shelden, and E. V. Thompson, Dependence of diffusive permeation rates on upstream and downstream pressures. II. Two-component permeant, J.Membrane Sci., 2:333 (1977).

10. K. C. Hoover and S-T. Hwang, Pervaporation by a continuous membrane column, J.Membrane Sci., 10:253 (1982).

11. E. Nagy, O. Borlai, and J. Stelmaszek, Pervaporation of alcohol-water mixtures on cellulose hydrate membranes, J.Membrane Sci., 16:79 (1983).

12. G. M. Wilson, Vapour-liquid equilibrium. XI. A new expression for the excess free energy of mixing, J.Am.Chem.Soc., 86:127 (1964).

13. W. B. Kogan and W. M. Fridman, Handbuch der Dampf-Flüssigkeit-Gleichgewichte, Deutscher Verlag der Wissenschaften, Berlin (1961).

INDUSTRIAL SEPARATION OF AZEOTROPIC MIXTURES BY PERVAPORATION

E. Mokhtari-Nejad and W. Schneider

GFT Ingenieurbüro für Industrieanlagenplanung
Gerberstr 48
6650 Homburg/Saar, FRG

Although pervaporation has, in the past few years, become of increasing interest for numerous industrial applications, relatively little is
known about its fundamentals, its limits and possibilities. We will,
therefore, attempt in this paper to summarize some empirically established
relationships and will propose criteria and methods, enabling characterization of membranes in such a way that measured data can be easily used to
establish reliable information for both comparison and engineering purposes. With regard to theoretical aspects, we would like to refer to the
publications of Hwang and Kammermeyer, as well as of Binning and Neel.

The important difference between pervaporation or liquid permeation
(as it is sometimes called) and other membrane processes, is the phase-
changing of the component permeating through the membrane. The difference
between the latter and conventional distillation is that the boiling
temperatures of the various components in the feed mixture are practically
of no importance and the separation of a certain component from the mixture
is only dependent on the 'selectivity' of the membrane. The driving force
in the pervaporation process is the difference between the 'chemical
potential' of the substance on the liquid feed side and the vaporous
permeate side. This means that condensation on the permeate side, causing
the chemical potential difference to tend to zero, and subsequently the
separation activity as well, has to be avoided by all means. The easiest
way to do this is to sweep the permeate away by a carrier gas or to 'pump'
it away by a vacuum system. The characteristics of a membrane depend on
the composition of the feed mixture, as well as on the membrane material
itself. Whilst for CA or CTA membranes the flux varies strongly with the
thickness of the active layer, this is not the case with other polymers.
This means that no single mathematical model can be applied for all types
of pervaporation membranes. This shows clearly that different data will
be applied in order to enable the design of a pervaporation system for
every membrane type and every feed mixture. Using as a common example
the separation of water-ethanol mixtures by means of recently developed
membranes - now commercially available - we will try to propose a common
characteristical basis for chemists and engineers.

Figure 1 shows the behavior of a PVA membrane in ethanol/water feed
at various concentrations. Curve pv shows the compositions of the pervaporation permeate as parts by weight. By comparing these to those of
the normal vapor equilibria (curve eq) it becomes quite obvious that

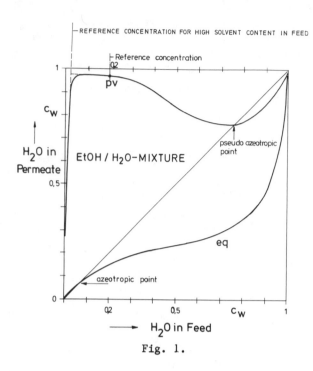

Fig. 1.

pervaporation performs best where distillation is quite inefficient or impossible (see azeotropic point on curve eq). Curve pv shows further that optimal performance of this membrane for the specified feed mixture is for an EtOH contents of approx 65% b.w. or higher. At a feed concentration of 98% b.w. the water content is greatly increased. As curve pv is well away from the 45 degree line, this means for the membrane that there is no azeotropic point in this concentration range. Curve pv crosses the 45 degree line at approx. 25% b.w. EtOH concentration. Therefore, this membrane is capable of removing EtOH from fermentation broths for example. Curve pv goes through 0 as does the vapor equilibrium curve. This means that theoretically almost any EtOH concentration can be reached (in fact 99.9% b.w. is possible).

Figure 1 also shows clearly that, if a concentration over almost the whole range is required, a combination of distillation and pervaporation offers a more effective separation than each process alone. As can be concluded from Figure 1 as well, it is difficult to define one separation factor valid for the whole concentration range. If such is formulated by:

$$S = \frac{(water/ethanol)\ permeate}{(water/ethanol)\ feed} \tag{1}$$

which is analogous to the "relative volatibility" in distillation, S can reach values from minus three to some magnitudes of ten. The widely used separate factors 'α' or 'β' do not permit the design of a plant, i.e. exact determination of permeate composition. In a strict sense, therefore, the selectivity of a pervaporation membrane has to be given by the complete separation curve (curve pv in Figure 1).

For the practical use of this membrane type it has been established that it is sufficient to characterize the separation properties in the range in which the membrane will be used, as using membranes at an EtOH concentration of below 65% b.w. does not make sense.

For the membrane on which curve pv on Figure 1 is based, as well as for similar membranes in the range of approx. 65% b.w. to approx. 98%, there is a relatively constant area for the permeate composition, little affected by the feed composition. We specify the selectivity of our membranes by measuring at 80% b.w. constant EtOH concentration and at a feed temperature of 80°C (isotherm. conditions). By specifying the point at which the permeate water content starts to decrease strongly, similar membranes can be compared and the approximate composition of the total permeate over the whole working range can be determined. Typical H_2O concentrations in the permeate at 80% b.w. EtOH in feed and 80°C are, for our membranes, 85, 90 and 95% b.w. This gives overall EtOH concentrations for the practical working range of 10, 20 and 30 % b.w. As in most cases in membrane technology, a higher separation factor can be attained, at the expense of the flux or vice versa. In many cases it is useful to select different membranes for various feed concentration ranges.

Similar to the selectivities, the fluxes of pervaporation membranes must be given with the exact conditions under which they have been measured or for which conditions they are valid. Figure 2 shows the total flux ϕ_t

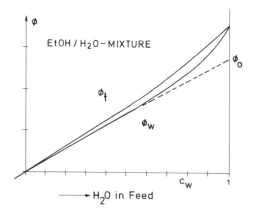

Fig. 2.

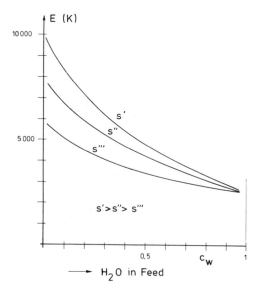

Fig. 3.

and the partial water flux ϕ_w as a function of the feed composition. For the type of membrane manufactured by us, both the total flux and the partial water flux go through 0. The partial water flux is a linear function within the practical working range. Only at EtOH concentrations of less than 50% b.w. in the feed does the partial water flux increase. Therefore, we have introduced the "pure water flux ϕ_o".

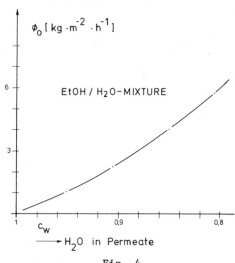

Fig. 4.

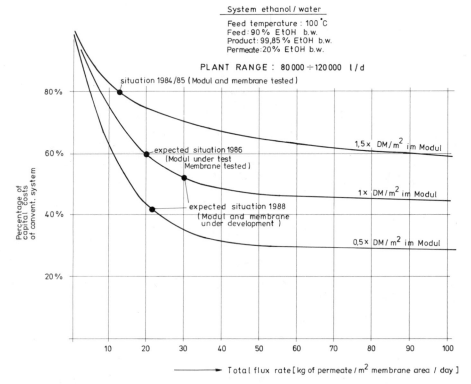

Fig. 5. Influence of flux on costs for pervaporation systems.

$$\phi_o = \phi_w / C_w \tag{2}$$

where C_w stands for the water content of the feed measured at the same point as O_w. The standard O_o is based on a temperature of 80°C. Typical values for O_o are 1.0, 2.5 and 4.0 kg of water removed per hour and m². Figure 4 shows the dependence of our membrane O_o on water concentration in the permeate. In contrast to separation, where only a minor temperature dependence can be observed, the temperature influence on the flux is significant. The temperature influence on the flux can be expressed by an Arrhenius relationship:

$$\phi = \phi_\gamma \cdot \exp- E(\frac{1}{T} - \frac{1}{T_\gamma})$$

The factor E depends on various factors and has to be established experimentally. T_r and ϕ_r are the reference temperature and reference flux respectively, for example, ϕ_o and $T_o = 80°C$. In Figure 3 the dependence of E on the H_2O content in the feed is shown for 3 membranes with different selectivities S.

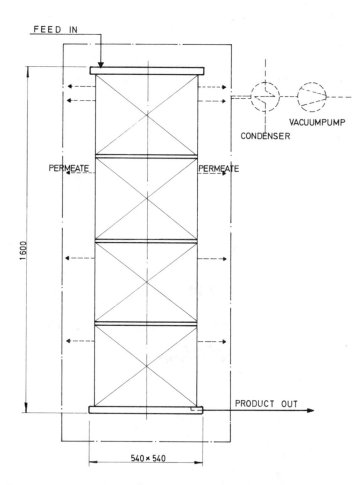

Fig. 6. Pervaporation standard-module. 1 module equipped with GFT-membrane 6/85 can dehydrate approx. 2000 1/D EtOH/H_2O mixture from 90% b.w. EtOH to 99.5% b.w. EtOH with an H_2O-content of 75% in the permeate.

Worthy of note is the increase of E at higher EtOH concentrations in the feed. The concentration dependent on increase of E indicates that at high EtOH contents the process is no longer controlled by the permeation solely, but the transfer of water to the membrane surface increasingly determines the flux rate. Measurements with our membranes show that the pure water flux ϕ_o can be doubled with every 10°C increase, i.e. for the same membranes as mentioned before the fluxes are increased at 90°C to 2.5 and 8 and at 100°C to 4.10 and 16 kg of water removed per hour and m^2. Temperature-dependence mentioned before does not apply to lower temperatures (<50°C), where the rate changes abruptly.

In summary we can say that we found it quite useful to characterize the membranes at the following conditions:

a) Fluxes are given as "pure water flux ϕ" measured at T_o = 80°C and 80% b.w. EtOH in the feed, as defined in Figure 2 and can be used for the pervaporation working range, i.e. EtOH concentrations in the feed of more than 65% b.w.

b) ϕ_r will be corrected in accordance with the stated Arrhenius relationship within the temperature limits of 60°C to 120°C

c) For the separation two values are given: the H_2O concentration in the permeate measured at ϕ_o as specified under a) and the concentration point, up to which the H_2O concentration is nearly constant.

The above-mentioned method of characterizing membranes allows the engineer to make predictions and optimize industrial applications. In Figure 5 we show the influence of module type and the flux ϕ_{eff} on costs. This figure confirms what is valid for most membrane systems; namely, reduction of costs for module and membranes is more important than drastic improvement in membrane properties. Therefore, the engineer must always

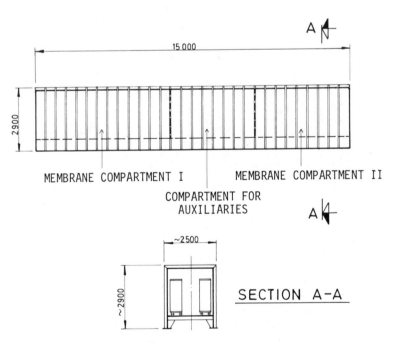

Fig. 7. Typical arrangement for a pervaporation plant for EtOH-dehydration. Plant size: 100 000 1/d (100% b.w. EtOH).
Feed: 91% b.w. EtOH. Product: 99.8% b.w. EtOH.

578

consider the membrane properties in conjunction with the price. Figure 5 shows that pervaporation systems – besides lower operation costs – now already offer investment costs, which are considerably lower than those of conventional systems. Figure 6 shows a standard module and Figure 7 the outlines of a 100.000 l/d plant for dehydration of an EtOH/water mixture from an EtOH concentration of 90% b.w. up to 99.85%. The last figures are also valid for similar mixtures, for example an 1PA/water mixture.

CONCLUSION

It can be said that sufficient experience and information is available for the introduction of pervaporation on an industrial scale. The first small plants have been built or are under construction, and at the beginning of next year a 6000 l/d plant, capable of dehydrating an organic solvent/water mixture from approx. 75% b.w. solvent content up to 99.7% b.w. will be started up. We expect not only a strong demand for industrial systems for separation of simple binary azeotropic mixture, but especially for complex multi-component mixtures, as here the technical and economical advantages are even larger.

REFERENCES

1. S. T. Hwang and K. Kammermeyer, "Membranes in Separations," J. Wiley & Sons, New York (1975).
2. R. C. Binning, R. J. Lee, J. F. Kennings, and E. C. Martin, Ind.Eng. Chem., 53:45 (1961).
3. J. Stelmaszek, Chemie Technik., 12 (1979).
4. J. Neel and P. Aptel, Entropia, 18(104):1-92 (1982).

ECONOMICS OF INDUSTRIAL PERVAPORATION PROCESSES

H. E. A. Brüschke and G. F. Tusel

GFT Ingenieurbüro für Industrieanlagenplanung
Gerberstr. 48
6650 Homburg/Saar, FRG

Separation of liquid mixtures by means of membrane, whereby a vaporous product is obtained on the permeate side of the membrane, has been known for about 80 years. Thirty years ago, the term "pervaporation" was introduced to describe the presence of both liquid and vapor phase in contact with the membrane. In their fundamental work, Binning and co-workers described pervaporation processes for the separation of a number of liquid mixtures. They emphasized the high efficiency and economical advantages of pervaporation processes compared to conventional techniques, especially for the separation and dehydration of azeotrope-forming aqueous-organic mixtures.

The first approach in the use of pervaporation membranes for the separation of azeotrope-forming mixtures is shown in Figure 1. A binary mixture, forming a minimum boiling point azeotrope, is first distilled in column I and separated into the pure component A and the azeotrope. The latter leaves the top of the column; in the membrane unit it is split into two streams, one enriched in component A (lower than azeotropic concentration) and the other enriched in component B (higher than azeotropic concentration). This stream is distilled in a second column II, producing the pure component B at its bottom and the azeotrope at the top. Membranes with modest selectivity only were required for such a scheme arrangement, as marginal shifting of the concentration below or above the azeotropic composition was sufficient.

Unfortunately, the liquid-vapor equilibrium curves of most azeotrope-forming mixtures of common organic solvents with water do not favor such a procedure. In many cases only minimal differences between the boiling point of the azeotrope and the boiling point of one pure component are observed, leaving only a very narrow gap between the vapor-liquid equilibrium curve and the 45°-line (Figure 2, isopropanol-water system). Distillation of mixtures with compositions above the azeotropic point would require columns with large numbers of theoretical plates and operation at high re-flux ratios, rendering such a process much less economical than conventional techniques. This is the reason why such schemes have never been applied in industry.

Recent developments have led to new membranes, which combine high selectivities over a broad range of feed compositions with high fluxes. The characteristics of these membranes have been described in another paper

581

during this conference, and I will, therefore, restrict this paper to the economical aspects of the industrial application of these membranes.

From Figure 2 it is evident that these membranes can not only be used to split an azeotrope, but also to dehydrate aqueous-organic mixtures to very high purities. Even if no azeotrope is found in the vapor-liquid equilibrium of a certain mixture, pervaporation processes may still be more economical for final dehydration than normal distillation.

The technical and economical advantages of a pervaporation process, employing these recently developed membranes for the purification and dehydration of organic solvents from their aqueous mixtures, can be summarized as follows:

- No entrainers are needed to break down azeotropic mixtures, thus reducing both costs and environmental hazards.
- At low water concentration of the feed mixture, pervaporation is more selective than distillation, reducing investment costs for distillation columns.

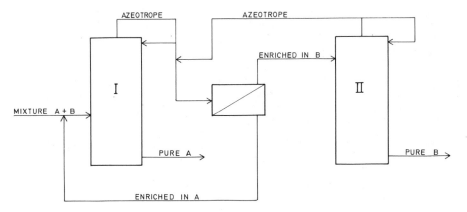

Fig. 1. Process schema for combination of distillation and pervaporation using membranes with low selectivities.

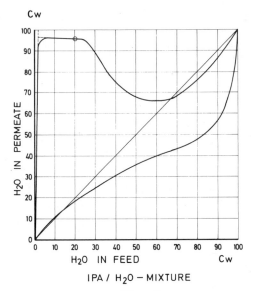

Fig. 2.

- Starting concentrations for pervaporation processes can be selected from a broad composition range.
- Energy requirements for separation are drastically reduced, as only the permeate stream has to be evaporated, which, in any case, mainly comprises the water to be removed.
- Water can be easily removed from multi-component organic mixtures.
- Distillation and pervaporation can be combined in a hybrid process, offering a minimum of both investment and operation costs.
- Membrane systems are easy to operate, they run continuously without requiring regeneration.
- Pervaporation processes can be operated within a broad range of temperature if necessary, as well as above or below the boiling point of all components.

A few examples are given to explain the technical and economical advantages of pervaporation processes. As evident from its separation characteristics, the greatest benefits from the use of this membrane can be expected in the separation of azeotropic mixtures. Figure 3 shows a schematic flow diagram for the dehydration of an ethanol-water mixture.

In the first heat exchanger the feed mixture is preheated to 100°C and pumped to the membrane system. As the mixture passes over the membrane and permeate evaporates, the mixture is cooled down. In order to supply the necessary heat for the evaporation of the permeate, the total membrane area is sub-divided into several equally-sized sections. Between every two sections the feed mixture is reheated, in order to operate the system at as high a temperature as possible. The final product leaving the last membrane section is cooled down, the sensitive heat is recovered by preheating the feed. The number of membrane sections depends on initial and final concentration of the mixture and on the membrane area necessary for the special separation.

EXAMPLE 1

A feed stream of 2000 kg/h of ethanol, close to azeotropic concentration (94% b.w.), is concentrated to a final product of 1867 kg/h at 99.85% b.w.. The permeate stream of 133 kg/h with 23.4% b.w. is condensed and further separated in a small column. As the permeate contains not more than 1.6% of the initial amount of alcohol, only a stripping column for water removal is needed; the remaining alcohol is recycled with the fuel mixtures. If 12 membrane sections, each of 50 m² are used, with a small degree of energy recovery, a specific steam consumption of only 0.1 kg/l of dehydrated alcohol is required. Investment costs for this system, including column, are only 72% of the costs of a conventional system. Operation costs, including savings due to non-necessity of an entrainer, amount to only 60% of those for the most modern conventional system.

EXAMPLE 2

100,000 l/d of dehydrated ethanol (99.85% b.w.) is produced from a feed mixture of 90% b.w. Including stripping column for permeate purification, specific investment costs are 135,- DM per kg of water removed per day. Savings in operation costs are even higher than in example 1, equally compared to the most modern conventional system.

EXAMPLE 3

A pre-distilled stream of isopropanol, at a concentration of 85% b.w. is dehydrated in a membrane system to a purity of 99.0%. For a plant with

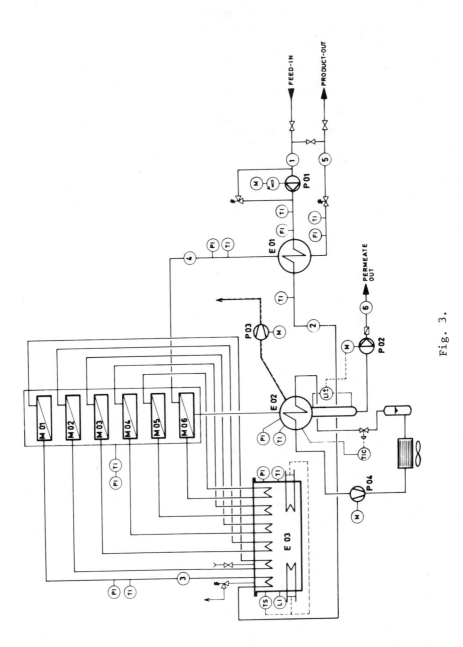

Fig. 3.

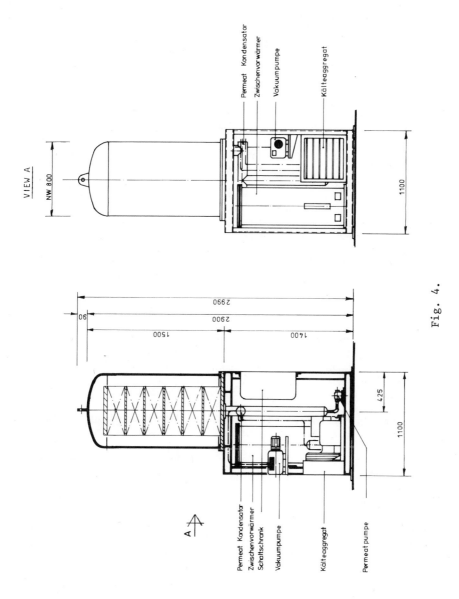

VIEW A

NW 800

Permeat Kondensator
Zwischenvorwärmer
Vakuumpumpe
Kälteaggregat

1100

2990
90
2900
1500
1400

425

1100

A

Permeat Kondensator
Zwischenvorwärmer
Schaltschrank
Vakuumpumpe
Kälteaggregat
Permeatpumpe

Fig. 4.

585

a capacity of 100 tons/d of IPA, investment costs are only 52% of those of azeotropic distillation, with savings in operation costs of the same magnitude.

EXAMPLE 4

In the previous examples only binary mixtures are mentioned. If multi-component mixtures are dehydrated by pervaporation processes, savings compared to conventional techniques are even much higher, as no distillation processes exist as an alternative. Out of a quaternary mixture of organic solvents, 6% b.w. of water has to be removed. Investment costs for a membrane system only amount to approximately 30% of those of a conventional molecular sieve system, savings in operation costs are as high as 65%, at a plant capacity of 300 t/d of feed mixture.

With the recently developed pervaporation membranes, dehydration of aqueous-organic mixtures by pervaporation processes has become an industrially viable technology. In particular for the separation of azeotrope-forming mixtures, pervaporation processes are highly competitive with existing technology, with significant savings in both investment and operation costs. Additionally, many other organic solvents, which do not form azeotropes with water, can be dehydrated economically by pervaporation processes, if combined with a distillation system. Energy savings may be increased if vapors from the top of the distillation column are used as a heating medium. Using vacuum and a condensing system on the permeate side of the membrane contributes to minimization of operation costs. Membranes now under development, which will not only enable water removal, but even solvent-solvent separation, will enlarge the applicability of pervaporation processes.

REFERENCES

1. R. C. Binning, R. J. Lee, J. F. Kennings, and E. C. Martin, Ind.Eng. Chem., 53:45 (1961).
2. J. Stelmaszek, Chemie Technik., 12 (1979).
3. A. H. Ballweg, H. E. A. Brüschke, W. H. Schneider, G. F. Tusel, K. W. Böddeker, and A. Wenzlaf, "Pervaporation Membranes – an Economical Method to Replace Conventional Dehydration and Rectification Columns in Ethanol Distilleries," paper presented at V. Int. Symposium on Alcohol Fuel Technology in Auckland, New Zealand (1982).
4. G. F. Tusel and A. Ballweg, "Method and Apparatus for Dehydrating Mixtures of Organic Liquids and Water," US Patent No. 4,405,409, Sept. 20 (1983).
5. Mulders, PhD Thesis (1983).

NONISOTHERMAL MASS TRANSPORT OF ORGANIC AQUEOUS

SOLUTION IN HYDROPHOBIC POROUS MEMBRANE

Z. Honda, H. Komada, K. Okamoto and M. Kai

Daicel Chemical Industries Ltd., Research Center
1239 Shinzaike, Aboshi-ku, Himeji
671-12 Japan

INTRODUCTION

Membrane separation by the use of temperature difference as a driving force has been studied by many investigators. It is divided into four categories by the type of membranes employed:

1) thermoosmosis by dense membranes[1],
2) thermodialysis by porous membranes filled with solution[2],
3) membrane distillation by hydrophobic porous membranes filled with vapor[3],
4) thermopervaporation by dense membrane[4].

We investigated nonisothermal transport of various organic aqueous solutions and pure water by membrane distillation. In this paper the mechanism of transport of organic aqueous solutions and that of water flux are discussed.

EXPERIMENTAL

The membrane used in this investigation was Polyflonpaper[TM] 5L (Daikin Ltd.) made of polytetrafluoroethylene (PTFE) fiber. The thickness is 0.45 mm and the maximum and the average pore diameter are 45 μm and several μm respectively. The scanning electron micrographs of the surface and the cross-sectional area of Polyflonpaper[TM] are shown in Figure 1.

The apparatus employed is shown schematically in Figure 2. It consists essentially of two cells separated by the membrane. The temperature of each cell was maintained constant by circulating thermostated fluid through the jacket. The solution was constantly agitated by a stirrer. The flux was measured by weighing the solution which overflowed out from the capillary attached to the top of the cold cell. The solution was supplied from a small tank to the warm cell through the other capillary.

When the permeate in the form of vapor through the porous membrane was collected directly, a metal plate with gutter was placed at the cold side of the membrane to condense the vapor.

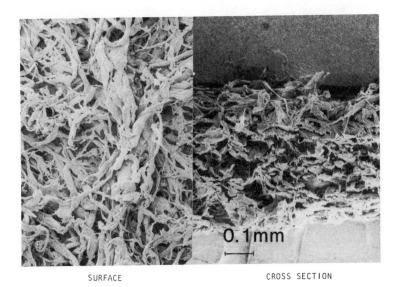

SURFACE   CROSS SECTION

Fig. 1.   Scanning electroscopic picture of the cross section of
          Polyflonpaper^TM.

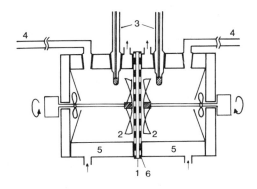

Fig. 2.   Schematic diagram of apparatus. (1) Membrane, (2) stirrer,
          (3) thermometer, (4) glass capillary, (5) jacket, (6) perforated
          supporting metal.

RESULTS AND DISCUSSION

Separation of Various Organic Aqueous Solutions

     Nonisothermal transport experiments of various organic solutions were
performed by applying the predetermined temperature difference across PTFE
porous membrane. The solution fluxes were always observed from the warm to
the cold cell. The concentration change of the cold solution was dependent
on the relative vapor pressure of solute and water. The results are
summarized in Table 1. In the case of the solute of which concentration in
the vapor phase is larger or smaller than that in the liquid phase, the
solution was concentrated or diluted in cold cell, respectively. And it
was confirmed that under the conditions indicated in Table 1 all the
solutions examined cannot penetrate the pore of the membrane due to low
surface tension of PTFE. However, in the case of 70 wt% ethanol aqueous
solution, which penetrated into the PTFE porous membrane, no change was
observed in both cells. These facts led us to the following idea. Vapor-

Table 1. Nonisothermal Separation of Various Organic Aqueous Solution*.

Organics concentrated in cold cell	Organics concentrated in warm cell
Methanol, Ethanol, Propanol, Butanol, Acetone, THF, 1,4 - Dioxane etc.	Acetic acid, Formic acid, Propionic acid, DMSO, DMF, 1,4 - Butanediol etc.

* 1 wt% solution, warm cell; 51 - 52°C, cold cell; 15 - 16°C.

Table 2. Permeate Concentration in the Nonisothermal Transport of Alcoholic and Acidic Aqueous Solutions.

Solute	Feed[1] [A]	Permeate[2] Calculated [B]	Observed	Flux kg/m^2·hr	Separation factor α
Water	–	–	–	7.4	–
Methanol	5.0	20.2	10.9	9.2	2.3*
Ethanol	4.9	15.2	12.2	8.8	2.7*
2-Propanol	4.5	35.7	13.0	10.2	3.2*
Formic acid	5.0	3.1	2.1	9.2	2.5#
Acetic acid	5.0	4.0	2.1	8.1	2.5#
Propionic acid	5.0	–	3.0	9.2	1.7#

[1] 55.0 - 56.5°C.   [2] 16.5 - 20.7°C.

* $\alpha = ((100 - [B])/[B]) / ((100 - [A])/[A])$.

$\alpha = ([B]/(100 - [B])) / ([A]/(100 - [A]))$.

liquid equilibrium exists at the interface of the solution and the membrane, and the vapor mixture produced at warm side diffuses through the membrane and condenses at the interface of cold side.

In order to test experimentally the idea described above, alcoholic and acidic aqueous solutions were chosen as model solutions. In the experiments, permeates were condensed on chilled surface adjacent to the membrane and sampled directly. The results are listed in Table 2 together with the values calculated on the basis of the idea. As expected, the permeates of alcoholic or acidic aqueous solutions were concentrated or diluted, respectively. The solute concentration of the permeate was always smaller than that calculated. This could be explained as follows: the vapor mixtures permeate by the Knudsen flow through PTFE porous membrane. So, the permeability of water vapor is always larger than that of solutes with higher molecular weight. Therefore, the concentration of solution obtained is less than the calculated one.

Aqueous solutions of nonvolatile solutes were concentrated more efficiently than those of volatile solutes, because only water vaporizes and permeates through the membrane. The concentration of nonvolatile sodium gluconate aqueous solution was carried out and the results are shown in Figure 3, together with the calculated values (dotted line). It was found that the observed concentration change was significantly larger than

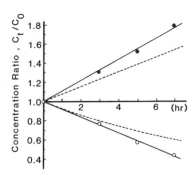

Fig. 3. Separation of sodium gluconate aqueous solution. Dotted line and solid line indicates the calculated and observed ratio, respectively. ●: 43.1°C, ○: 11.1°C.

that calculated. This unexpected observation might be explained by an assumption that sodium gluconate aqueous solution permeates in the liquid state from the cold cell to the warm cell. On this assumption the concentration of the solution in the cold cell (C_L) is given by Equation (1),

$$C_L(t) = C_0 - \frac{1}{V} J_a \, C_L(t)dt - \frac{1}{V} J_1 \, C_L(t)dt \tag{1}$$

where J_a is the observed water flux and J_1 is the backward solution flux into the warm cell. C_0 is the concentration of the initial as well as the additional solution, and V is the volume of each cell (270 ml). The net water flux, J_g, is then given as $J_a = J_g - J_1$. From Equation (1) $C_L(t)$ is given as,

$$\ln C_L(t) = - \frac{J_g}{V} t + const. \tag{2}$$

As shown in Figure 4, a linear relationship expressed by Equation (3) was obtained experimentally between $\ln C_L(t)$ and t.

$$\ln C_L(t) = - 8.19 \times 10^{-7} t + const. \tag{3}$$

From Equations (2) and (3), J_g was calculated to be 2.21×10^{-4} (g cm^{-2}s^{-1}). As J_a was 1.47×10^{-4} (g cm^{-2}s^{-1}), the value of 7.4×10^{-5} (g cm^{-2} s^{-1}) was obtained for J_L. The solution flux (J_L) amounts to one third of water vapor flux. When sodium chloride was used instead of sodium gluconate as a nonvolatile solute, the observed concentration change was in close agreement with that calculated on the assumption that only water vapor transports (in Equation (1), $J_1 = 0$).

The difference might be due to the difference in solute–membrane interaction so that sodium gluconate aqueous solution manages to penetrate into a portion of membrane pores, whereas sodium chloride aqueous solution does not. Still, the driving force that moves the liquid phase in the membrane backward remains unknown.

Pure Water Flux

Theory. By use of irreversible thermodynamics, the driving force of water flux was analyzed. In the present paper, pure water was chosen as the simplest liquid system. Three assumptions were made in order to analyze the pure water flux:

1) The temperature of water and that of water vapor are the same.

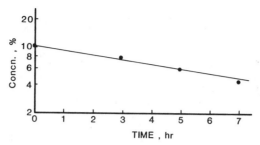

Fig. 4. Concentration change of sodium gluconate in cold cell (11.1°C).

2) The air flow in the membrane is negligible and the permeate consists of only water vapor.
3) All the vapor produced in the warm cell permeates through the membrane in gas phase without leaving condensed water in the membrane.

The last assumption is supported by the observation that no continuous liquid water phase across the membrane was formed during a seven-hour test run. In the case of transport of only water vapor, water flux is given as[5]

$$J_w = -D \text{ grad } C_w - \frac{L_{1q}}{T} \text{ grad } T \tag{4}$$

The diffusion coefficient (D) of water vapor is the function of temperature (at 760 mmHg) and it is reported[6] to be expressed as

$$D = D_s \left(\frac{T}{273.1}\right)^{1.75} \tag{5}$$

where D_s is the diffusion coefficient at standard conditions (273.1 K, 760 mmHg). The relationship between the partial pressure of water vapor and the temperature of water is given by the Antoine equation as

$$\log P = a - \frac{b}{c + (T - 273.1)} \tag{6}$$

where a, b and c are 8.2773, 1838.7 and 241.41 in the temperature range from 273 to 303 K, and 8.0790, 1733.9 and 233.67 from 304 to 333 K, respectively.

The molar concentration of water vapor (mole cm^{-3}) is

$$C_w = \left(\frac{1}{2.24 \times 10^4}\right) \left(\frac{273.1}{T}\right) \left(\frac{P}{760}\right) \tag{7}$$

When logarithmic average temperature is used, from Equation (4), (5), (6), (7) water flux is expressed as in Equation (8).

$$J = \frac{1}{2.24 \times 760 \times 10^4} \left(\frac{T_{av}}{273.1}\right)^{0.75} \frac{D_s}{\ell} \left(\frac{10^{P_1}}{T_1} - \frac{10^{P_2}}{T_2}\right) + \frac{L_{1q}}{\ell} \ln\frac{T_1}{T_2} \tag{8}$$

$$P_1 = \log P_1 = a_1 - \frac{b_1}{c_1 + T_1 - 273.1}$$

$$\underline{P}_2 = \log P_2 = a_2 - \frac{b_2}{c_2 + T_2 - 273.1}$$

where ℓ is the membrane thickness.

Temperature at the Interface of Water and Membrane. To analyze the water, it is necessary to measure the actual temperature at the interface of water and the membrane. However, the heat transported across the membrane makes temperature polarization at the interface, so, the actual temperature at the membrane surface cannot be measured directly. We applied the following method. Warm fluid from the predetermined temperature reservoir was circulated at a high velocity through warm cell jacket, and warm bulk water temperature was measured by changing the rotation speed. Cold cell temperature was also done by circulating cold fluid in a similar manner as above. The observed bulk water temperature (T_b) was plotted against the reciprocal of the number of rotation, as shown in Figure 5. A linear or a slightly curved relationship was observed. The actual temperature (T_i) will be obtained by extrapolating the number or rotation to infinity ($N_R^{-1} = 0$). In the case of 400 rpm, under these conditions, the relationship between T_i and T_b is expressed as

$$T_i = 0.99 \, T_b + 0.5 \quad \text{(warm cell)} \tag{9}$$

$$T_i = 0.94 \, T_b + 2.0 \quad \text{(cold cell)} \tag{10}$$

Diffusion Coefficient and Thermal Diffusion Coefficient. Water flux increases with the temperature difference and the average temperature as shown in Figure 6 and Figure 7. The optimum solution of D_s and L_{1q} were obtained by the Symplex method using the data of Figure 6 and Figure 7. The results are shown in Table 3. Diffusion coefficient, D_s by concentration gradient was from about 1.1 to 0.94 x 10^{-2} (cm^2 s^{-1}). Each run in Table 3 was carried out using nominally the same membrane which did not always show the same performance. The difference in D_s and L_{1q} of

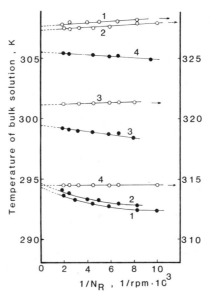

Fig. 5. Relationship between bulk water temperature and the reciprocal of the number of rotation. Number indicates the series of the experiments. ● : Warm cell, ○ : cold cell.

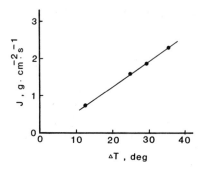

Fig. 6.   Effect of temperature difference on water flux at average
temperature 304.3 K.

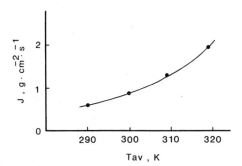

Fig. 7.   Effect of temperature on water flux at water temperature
difference 19.0 degree.

two series of the run might be due to the (individual) membrane performance.

In order to compare the contribution to water flux of concentration
diffusion with thermal diffusion, the data of Run 2 in Table 3 served as
a typical example. The water flux effected by concentration diffusion
and thermal diffusion were calculated, 8.34×10^{-6} and 2.07×10^{-6} (mol
$cm^{-2}s^{-1}$), respectively. The former is about 4 times larger than the
latter. As L_{1q} is reportedly in proportion to the concentration, thermal
diffusion coefficient, D^T, can be defined as $D^T = L_{1q}/C_w T$. When the
logarithmic average value is used for C_w and T, thermal diffusion coefficient is given as

$$D^T = \frac{L_{1q}}{\left(\dfrac{T_1 - T_2}{\ln (T_1/T_2)}\right)\left(\dfrac{C_1 - C_2}{\ln (C_1/C_2)}\right)} \tag{11}$$

In the case of $T_1 = 319.4$ (K), $T_2 = 291.0$ (K), $C_1 = 3.88 \times 10^{-6}$ (mol cm^{-3}),
$C_2 = 8.52 \times 10^{-7}$ (mol cm^{-3}), D^T is 1.680×10^{-3} (cm^2 s^{-1} K^{-1}). Thermal
diffusion phenomenon is characterized by the Soret coefficient, which is
defined as the ratio of thermal diffusion coefficient to concentration
diffusion coefficient. The Soret coefficient of water vapor through PTFE
porous membrane was $S = D^T/D = 1.55 \times 10^{-2}$ (K^{-1}).

We tried to evaluate the membrane tortuosity. The membrane thickness
(ℓ) used in Equation (8) was the straight thickness measured by a micro-
gauge. The actual passage for water vapor in the membrane is considered to

Table 3. Calculation Results of D_s and L_{lq}.

	Fig. 5				Fig. 6			
Run	1	2	3	4	5	6	7	8
D_s *	1.083×10^{-1}				9.430×10^{-2}			
L_{lq} **	1.023×10^{-6}				1.096×10^{-6}			
error %	-3.4	-1.0	-0.1	2.2	-0.6	-1.7	-0.6	1.8

* $cm^2 \cdot s^{-1}$ ** $mol \cdot cm^2 \cdot s^{-1}$

run 2: $T_1 = 319.4$ K, $T_2 = 291.0$ K.

be bent. The tortuosity of 2.03 was obtained for the PTFE porous membrane, by dividing the reported D_s (0.219 $cm^2 s^{-1}$)[6] measured in the absence of membrane by D_s (0.108 $cm^2 s^{-1}$) calculated from Equation[8].

CONCLUSION

The separation of volatile organic aqueous solution driven by temperature difference across a porous PTFE membrane is performed by the transport of vapor mixture in the direction from warm to cold side of the membrane unless the solution wets the membrane. The composition of the permeate is mainly determined by vapor-liquid equilibrium at the warm interface and slightly modified by transport resistance.

The separation of nonvolatile NaCl aqueous solution is done by water vapor transport and the concentration change with increase in permeate is predictable. The concentration change of nonvolatile sodium gluconate aqueous solution was found unexpectedly to be more than that calculated, suggesting the co-existence of liquid phase transport of the solution from cold to warm side of the membrane.

The driving force of membrane distillation was analyzed with pure water system and shown to consist of differences in vapor concentration and temperature. The flux effected by concentration difference was about 4 times that effected by temperature difference.

Acknowledgment

This work was performed under the management of the Research Association for Basic Polymer Technology for synthetic membranes for new separation technology as a part of a project on Basic Technology for Future Industries sponsored by the Agency of Industrial Science and Technology, and the Ministry of International Trade and Industry.

REFERENCES

1. R. P. Rastogi and K. Singh, Trans Faraday Soc., 62:1764 (1966).
2. F. S. Gaeta and D. G. Mita, J.Membr.Sci., 3:191 (1978).
3. K. Y. Cheng and S. J. Wiersma, U. S. P., 4316772 (1982).
4. P. Aptel, N. Challard, J. Cunny and J. Neel, J.Membr.Sci., 1:271 (1976).
5. A. Katchalsky and P. F. Curran, "Nonequilibrium Thermodynamics in Biophysics," Harvard University Press, Cambridge (1965).
6. D. E. Gray, "American Institute of Physics Handbook," McGraw-Hill, New York, (1957).

OBSERVATIONS ON THE PERFORMANCES OF

PERVAPORATION UNDER VARIED CONDITIONS

R. Rautenbach and R. Albrecht

Institut für Verfahrenstechnik, RWTH Aachen (FRG)
D-5100 Aachen
Turmstraße 46

INTRODUCTION

The high selectivity of pervaporation has already been verified in
laboratory experiments for a number of systems including water/organic
mixtures as well as organic mixtures. So far, research has concentrated
on the development and optimization of membrane materials, only little is
known about module- and process design. For process- as well as module
design equations for the calculation of selectivity and permeate flux at
different operation conditions are necessary. Such design equations should
include the least possible number of material properties and these material
properties should be determinable from simple experiments.

This paper will concentrate on such a model for the description
of pervaporation-material-transport in the membrane. Furthermore, the
separation potential of pervaporation will be compared with the separation
potential of RO and the selectivity enhancement of certain additives forming a "dynamic-membrane" will be discussed.

The Separation Potential of Pervaporation

The description of the material transport in the membrane is based on
the sorption-diffusion-model, i.e. assumes a sequence of 3 steps: sorption,
diffusion and desorption. The application of Fick's law in its correct
form results in the following equation for the permeate-flux through a
membrane[1,2]:

$$J_i = - D_i \frac{d\rho_i}{dx} - D_i \frac{w_{iM}}{1-W_{iM}} \frac{d\rho_i}{dx} = - \frac{\rho_M \cdot D_i}{1-W_{iM}} \frac{dw_{iM}}{dx} \tag{1}$$

The diffusion coefficient cannot be assumed to be constant across the
membrane. Several molecular models have been proposed for the description
of the concentration dependency of the diffusion coefficient. The most
widely accepted model is the free-volume-theory of Fujita which has been
adapted to the permeation of binary mixtures through a membrane by Fels and
Hwang[3,4]. The disadvantage of this model is, that the resulting transport equations are rather complicated and - even more important - that the
parameters of the model (the material properties) can only be determined by
highly sophisticated experiments. Furthermore, the model is very sensitive

with respect to the accuracy of these measurements and it will fail completely in case of anomalies in diffusivity. For these reasons, a phenomenological approach seems to be of more practical value - at least at present. In this case, the necessary material properties can be determined from relatively simple, steady-state experiments.

Two parameter equations have been proposed by several authors but we found that even a one-parameter equation of the form

$$D = D_o \cdot w_{iM} \tag{2}$$

is sufficient for engineering calculations. Equation (2) is valid in the case of the pervaporation of a pure substance through a membrane. For binary mixtures, the coupling of fluxes has to be taken into account. Shelden and coworkers[5] proposed the following equations ($i \triangleq$ more permeative component):

$$D_{iM} = D_{io} \cdot (w_{iM} + \alpha w_{jM}) \tag{3.1}$$

$$D_{jM} = D_{jo} \cdot (w_{jM} + \beta w_{iM}) \tag{3.2}$$

According to these equations, the transport of each species of a binary mixture in the membrane is determined by the local concentration of each component of the mixture.

The boundary conditions for the numerical solution of Equation (1) are determined by assuming thermodynamic equilibrium at both sides of the membrane:

$$W_{iM1} = \phi_{i1} \gamma_i \, x_{i1} \quad \text{and} \quad W_{jM1} = \phi_{j1} \, \gamma_j \, x_{j1} \tag{4}$$

$$W_{iM2} = \phi_{i2} \frac{P_{i2}}{P_i^\circ} \quad \text{and} \quad W_{jM2} = \phi_{j2} \frac{P_{j2}}{P_j^\circ} \tag{5}$$

Equation (5) has been derived for moderate feed pressures which are typical for pervaporation. The transport equations have to be solved numerically[6]. An analytical solution can be obtained for small concentration differences $W_{iM1} - W_{iM2}$ and in cases where the coupling of the fluxes can be neglected. These assumptions are generally valid if the solubilities of the permeating components in the membrane are low. In this case the permeate flux can be calculated by

$$J_i = - \rho_M D_i \cdot \frac{dw_{iM}}{dx} \tag{6}$$

Assuming $\alpha = \beta = 1$, i.e. no coupling of fluxes, Equation (7 and 8) can be derived:

$$X_{i2}^2 - X_{i2} \left\{ 1 - \frac{P_j^\circ}{P_2} \frac{K_M \gamma_{i1} X_{i1} + \gamma_{j1} X_{j1}}{\dfrac{\phi_{i2}}{\phi_{j1}} - K_M \dfrac{P_i \phi_{i2}}{P_i^\circ \phi_{i1}}} \right\} - \frac{P_j^\circ}{P_2} \frac{K_M \gamma_{i1} X_{i1}}{\dfrac{\phi_{j2}}{\phi_{j1}} - K_M \dfrac{P_j \phi_{i2}}{P_i^\circ \phi_{i1}}} = 0 \tag{7}$$

$$J_i = A_i \left\{ \frac{1}{2} \left[1 + K_M \frac{\phi_{j1}}{\phi_{i1}} \cdot \frac{1 - X_{i2}}{X_{i2}} \right] \right\} \left[(\gamma_{i1} X_{i1})^2 - \left(\frac{\phi_{i2}}{\phi_{i1}} \frac{P_2}{P_i^\circ} X_{i2} \right)^2 \right]$$

$$\tag{8}$$

596

$$+ \frac{\phi_{j1}}{\phi_{i1}} \left[\gamma_{j1}X_{j1} - K_M \frac{1-X_{i2}}{X_{i2}} \gamma_{i1}X_{i1} \right] \left[\gamma_{i1}X_{i1} - \frac{\phi_{i2}}{\phi_{i1}} \frac{P_2}{P_i^o} X_{i2} \right] \Bigg\} \qquad (8)$$

The parameters

$$A_i = \frac{\rho_M \phi_{i1}^2 D_{i0}}{\delta_M} \quad , \quad A_j = \frac{\rho_M \phi_{j1}^2 D_{j0}}{\delta_M}$$

$$K_M = \frac{M_j D_{i0} \phi_{i1}}{M_i D_{j0} \phi_{j1}} = \frac{M_j}{M_i} \frac{\phi_{i1}}{\phi_{i1}} \frac{A_i}{A_j}$$

must be considered as the membrane-constants for the ideal pervaporation of a binary mixture. They are sufficient for the description of mass-transport in the membrane if

$$\phi_{i1} = \phi_{i2} = \phi_i \quad ; \quad \phi_{j1} = \phi_{j2} = \phi_j$$

A_i and A_j can be determined from <u>pervaporation-experiments with the pore</u>

components, K_M (resp. $\frac{\phi_j}{\phi_i}$) can be determined from <u>sorption measurements with</u>

the pure components or from pervaporation experiments with the binary mixture.

EXPERIMENTS

The accuracy of the transport equations has been checked experimentally with the system benzene/cyclohexane and with symmetric PE-membranes (ρ_M = 930 kg/m^2, δ = 36 μm. The module and the experimental technique have been described in detail elsewhere[7].

<u>Discussion of the Transport Parameters</u>

The general equations for the mass-transport in the membrane contain 6 parameters which have to be determined experimentally: the 2 sorption-coefficients ϕ_i and ϕ_j, the 2 diffusion-coefficients D_{io} and D_{jo} and the coupling-parameters α and β (Table 1).

Table 1. Transport Parameters

Parameter	Dependence	Determination
D_{io} Diffusion coefficient D_{jo}	temperature	Pervaporation of pure components
α Coupling of fluxes β	temperature	Pervaporation of binary mixture
ϕ_{i1} Sorption coefficient ϕ_{j1}	temperature feed composition	Sorption measurements

Sorption Coefficients

The solubility of benzene and cyclohexane in the membrane has been measured in the vapor-phase[6]. According to our experiments, the solubility of benzene can be described by

$$\phi_i = \frac{W_M}{a_L} = 0.065 \cdot e^{\frac{760}{T_o}(1 - \frac{T_o}{T})} \left[1 + (2.12 \cdot e^{\frac{645}{T_o}(1 - \frac{T_o}{T})} - 1)a_L^{1.8}\right] \qquad (9)$$

and the solubility of cyclohexane by

$$\phi_j = \frac{W_M}{a_L} = 0.0915 \cdot e^{\frac{1594}{T_o}(1 - \frac{T_o}{T})} \left[1 + (2.077 \cdot e^{\frac{865}{T_o}(1 - \frac{T_o}{T})} - 1)a_L^{2.4}\right] \qquad (10)$$

$T_o = 318, 6$ K is chosen as the reference-temperature, the validity of the functions has been verified in the temperature range

$$298 \text{ K} \leq T \leq 323 \text{ K}.$$

In our calculations of mass transport in the membrane, the sorption coefficients have been calculated for liquid binary mixtures as the surrounding phase ($a_L = y_i x_i + y_j x_j$) taking into account the consumption of the feed. The sorption coefficient has been assumed to be constant across the membrane, i.e.

$$\phi_{i1} = \phi_{i2} = \text{const}, \quad \phi_{j1} = \phi_{j2} = \text{const}.$$

Diffusion Coefficients

The diffusivities D_{io} and D_{jo} can be determined from pervaporation experiments with the pure components i and j. The temperature-dependence is of the Arrhenius-type.

$$D_{io} = D_0^* \exp\left[\frac{c}{T^*}\left(1 - \frac{T^*}{T}\right)\right] \qquad (11)$$

From our experiments, the data listed in Table 2 have been derived.

Parameters α, β

The transport-model favored in this paper accounts for the coupling of the fluxes of the components i and j by two empirical parameters α and β. These parameters, which are independent of pressure and concentration but temperature-dependent, have to be determined from pervaporation-experiments

Table 2. Temperature Dependence of Diffusion Coefficients

	D_0^* (10^{-9} m²/sec)	c (K)
Benzene	2,151	5215
Cyclohexane	0,7603	3597

Reference temperature: $T^* = 318,6$ K

598

with a binary mixture of any composition and any permeate pressure. Figure 1 is the result of more than 100 experiments at different operating conditions. According to Figure 1 the coupling of fluxes is negligible at low operating temperatures but becomes more important with increasing temperatures. The selectivity of the process is always lowered by the coupling of fluxes. It must be concluded that the increasing coupling of fluxes is a direct consequence of the increase in sorption. As long as the sorbed amount of components in the membrane is small, pervaporation can be considered to be ideal.

Calculation of Pervaporation Processes

The result of a numerical solution of the transport equations is shown in Figure 2, where the permeate-concentration is plotted against the feed-concentration. The separation characteristics is shifted to much more favorable figures compared to the thermodynamic equilibrium, the azeotropic point can be supressed. Parameter in the diagram is the ratio of total pressure at the permeate side to saturation pressure $p_i^o = \gamma_i x_i p_i^o + \gamma_j x_j p_j^o$. The permeate pressure p_2 controls the concentration w_{iM2} in the membrane and, as a consequence, the concentration difference across the membrane. An increasing concentration difference across the membrane increases the transport potential, difference in sorption ϕ_i, ϕ_j and the differences in diffusivity D_{io}, D_{io} become increasingly important. Figure 2 shows that the selectivity decreases with increasing pressure at the permeate side. For a ratio of $p/p_L^o = 1$ finally, the separation characteristic is identical to the thermodynamic equilibrium curve (for ν = const)-the separation potential of the membrane is zero.

According to Figures 3 and 4 the calculations based on the transport-equations are in good agreement with experiments within the total concentration- and pressure range within the experimentally covered temperature range $25°C < T < 50°C$.

In general, the selectivity $S_{ij} = \dfrac{x_{i2}/x_{j2}}{x_{i1}/x_{j1}}$ decreases with increasing concentration of the preferentially permeating component in the feed and with increasing permeate-pressure (Figure 3). This has to be accepted

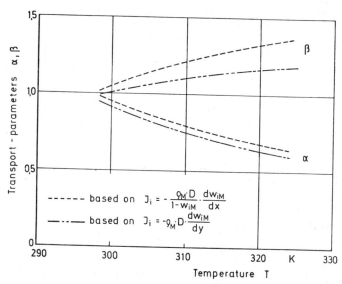

Fig. 1. Temperature dependence of transport parameters α, β.

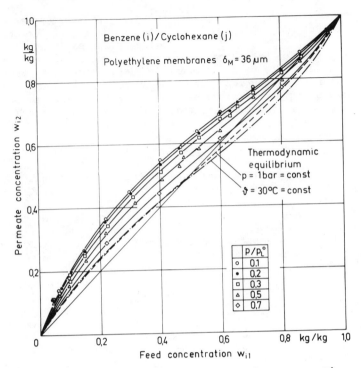

Fig. 2. Separation characteristics of pervaporation.

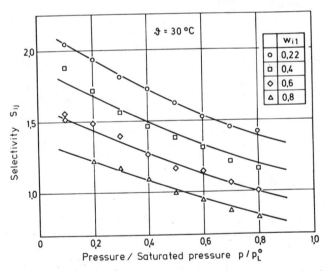

Fig. 3. Pressure dependence of selectivity (——— calculated).

according to thermodynamics for all membrane materials. The temperature-dependence of the selectivity (Figure 4) has to be viewed differently. The decrease of selectivity with increasing operating temperature is a consequence of the membrane swelling and the increasing rate of coupling between the fluxes. This will however be less pronounced for highly cross-linked membrane materials and for this reason the temperature dependence of the selectivity must be considered membrane-specific.

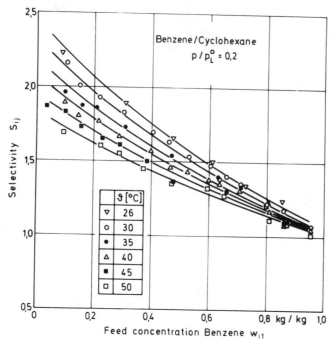

Fig. 4. Influence of feed composition and temperature on selectivity
(—— calculated).

Figure 5 compares experiments and theory with respect to fluxes. A
sensitivity analysis with respect to α and β reveals, that the increase of
the total flux in the concentration range $0 \leq w_i < 0.6$ is not caused by
coupling-effects. The flux is primarily determined by the dependence of
the sorption coefficients on feed-consumption. According to Figure 6, the
discrepancy between theory and experiment is somewhat larger with respect
to the pressure-dependence of the flux. However, the significance of this
result should not be overestimated. Since the selectivity of pervaporation
strongly decreases with increasing pressure, the operating conditions will
be limited to a relatively narrow range of p/p_L°. In the range of $p/p_L^\circ \leq$
$0,5$, difference between theory and experiment are 20% or less. Since the
predicted flux is always lower than the actual flux, the calculation will
be conservative, i.e. on the safe side.

ENHANCING THE SEPARATION POTENTIAL BY ADDITIVES

The selectivity of pervaporation can be improved by the addition of
such non- or heavy volatile components to the feed-solution which passes a
preferential solubility for the preferential permeating component of the
mixture. The additive is transported to the membrane surface by convection
and forms a concentration boundary layer because of its low volatility.
This concentration boundary layer must be regarded as an additional selective barrier for the less permeating component. Dextrane, sodium-chloride
and sodium-citrate for example improve the selectivity of pervaporation of
water-ethanol-mixtures markedly[8]. Figure 7 shows some results of our
experiments with cellulose-acetate membranes.

A NaCl-saturated feed solution has been used in these experiments.
According to Figure 7, the selectivity of the process is markedly enhanced
– independent of the ethanol-concentration of the feed! It should be
noticed that the water flux is independent of the water-ethanol concentration of the feed. According to the sorption-diffusion model, the

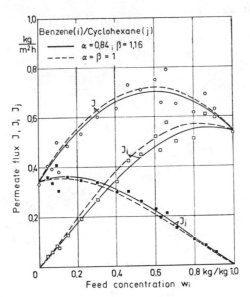

Fig. 5. Influence of feed composition on permeate flux ($p_2/p_L^o = 0.2$; $\nu = 30°C$) (=== calculated).

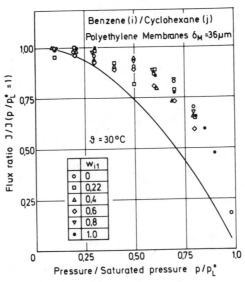

Fig. 6. Pressure dependence of permeate flux (—— calculated).

feed-side-membrane-concentration w_{iM1} must be independent of the feed-composition in this case which can be explained by an additive layer with a low ethanol-concentration. The enhancement of the selectivity of per-vaporation by low-volatile additives is not limited to water-organic feed solutions.

According to Figure 8, the selectivity for the system benzene/cyclo-hexane can be improved – though not as markedly – by adding small amounts of tetraethylene glycol (TEG) or polyethylene glycol (PEG). TEG is effec-tive in the total range of feed-consumption while PEG is limited to mix-tures with a high benzene-concentration. This is a consequence of the

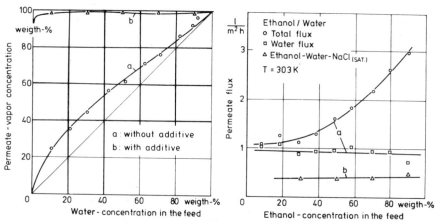

Fig. 7. Pervaporation of water/ethanol through cellulose acetate membranes
a) no additive, b) with additive (sodium chloride) saturated for
each feed composition.

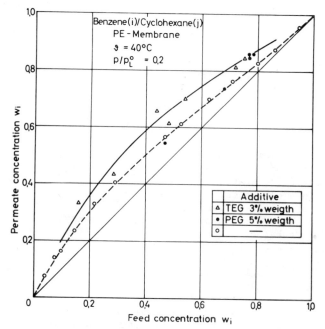

Fig. 8. Influence of additives on the separation of benzene/cyclohexane.

different solubilities of both additives in the system benzene/cyclohexane.
For benzene concentrations below 70%, PEG is practically insoluble and
consequently cannot be transported properly to the membrane surface.

In our experiments the membranes have been pretreated with pure PEG
resp. TEG for 1 hour at 45°C. Despite this pretreatment the selectivity
increased during the first 2 hours of pervaporation which demonstrates
clearly that the selective barrier in front of the membrane is formed as a
consequence of the lateral transport to the membrane surface. Correspondingly the (total) flux decreases for about 25% (Figure 9). At present the
question remains unanswered whether these effects can be utilized economically for large-scale pervaporation processes. In principle however such
'dynamic' membranes seem to be an interesting possibility for a 'high-
selectivity-high flux' process.

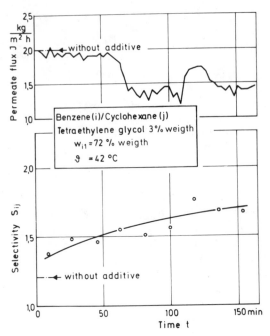

Fig. 9. Pervaporation characteristics of 'dynamic membranes'.

COMPARISON OF PERMEATION (PV) AND REVERSE OSMOSIS (RO)

Reverse osmosis and pervaporation are based on the same transport mechanism which can be described mathematically by Fick's law of diffusion. Assuming a linear concentration dependence of the diffusion coefficients, transport equations can be derived for reverse osmosis which are similar to these for pervaporation. In Table 3 the boundary conditions for both processes are compared. In RO-membranes the concentration w_{iM2} is primarily influenced by the feed pressure p_1 whereas in PV-membranes, operating at low pressure differences, the concentration gradient is created by lowering the permeate pressure. The maximum concentration difference across the membrane is achieved in pervaporation for $p_2 \to 0$ and in reverse osmosis for $\Delta p = p_1 - p_2 \to \infty$.

In Figure 10 reverse osmosis and pervaporation are compared on the basis of proper transport equations. The same assumptions have been made for both separation processes: linear concentration dependence of the diffusivities in the membrane, no coupling of the fluxes ($\alpha = \beta = 1$). The results are calculated for the permeation of benzene and cyclohexane through polyethylene membranes. The calculation indicates that the fluxes and the selectivities of pervaporation can be achieved by reverse osmosis – at least in theory. The necessary operating pressures must be high and, accordingly, the membrane quality must be extremely high with respect to compaction and to imperfections. Though selectivity and flux of PV and RO can be the same in theory, the separation potential of RO is markedly lower in reality.

Figure 11 shows the result of some of our experiments with pure cyclohexane and PE-membranes. In Figure 11, the product of flux and membrane thickness is plotted against the driving force. A relationship of this form follows directly from the transport equation for a pure substance: Assuming a linear concentration dependence of the diffusivity

Table 3. Boundary Conditions

Pervaporation	$w_{iM1} = \phi_i \gamma_i x_{i1}$	$w_{iM2} = \phi_i \cdot \dfrac{P_{i2}}{P_i^\circ}$ $w_{jM2} = \phi_j \dfrac{P_{j2}}{P_j^\circ}$
Hydraulic Permeation (Reverse Osmosis)	$w_{jM1} = \phi_j \gamma_j x_{j1}$	$w_{iM2} = \phi_i \gamma_{i2} x_{i2} \cdot \exp - \dfrac{\bar{V}_i}{RT}(p_1 - p_2)$ $w_{jM2} = \phi_j \gamma_{j2} x_{j2} \cdot \exp - \dfrac{\bar{V}_j}{RT}(p_1 - p_2)$

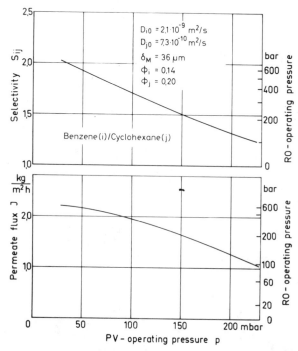

Fig. 10. Comparison of pervaporation and reverse osmosis. R.O. pressure needed to obtain the same separation as in pervaporation.

in the form $D = D_o w_M$, the integration of Equation (6) gives

$$J\delta_M = \frac{\rho_M D_o W_{M1}^2}{2}\left[1 - \left(\frac{W_{M2}}{W_{M1}}\right)^2\right] \qquad (12)$$

Independent of the nature of the membrane process (RO or PV), the concentration w_{M1} is determined by the sorption coefficient ϕ for $\gamma_i x_{i1} = 1$. The concentration w_{M2} depends on the process. For pervaporation, w_{M2} is determined by the ratio p_{i2}/p_i°, for RO by the pressure difference $(p_1 - p_2)$ (Table 3).

Some details of the experiments should be noted since they reveal the difference between RO and PV existing in reality: for pervaporation,

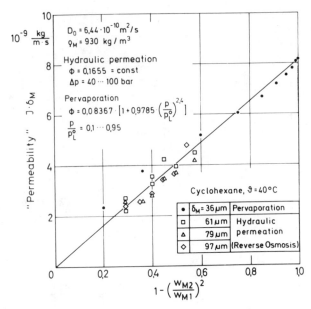

Fig. 11.   Permeate flux of pervaporation and hydraulic permeation (reverse osmosis).

membranes of a thickness of 36 μm have been employed. Since the trans-membrane pressure difference in pervaporation is small ($p_1-p_2 < 1$ bar), the convective transport through imperfections (micropores) is negligibly small compared to diffusive transport. For the RO-experiments, membranes of a thickness of 61-97 μm had to be installed in order to avoid convective flow through micropores.

The operating conditions have been varied between $0,1 < p/p° < 0.95$ for pervaporation and between 40 bar $< p <$ 100 bar for RO. Figure 10 confirms the sorption-diffusion model: permeate-flux depends on the concentration-difference in the membrane $w_{iM1}-w_{iM2}$ regardless whether this concentration-difference is caused by lowering the pressure at the permeate-side or by rising the pressure at the feed-side. The experiments indicate that in reality, the separation potential of pervaporation is markedly larger than that of RO since it is extremely difficult for RO to achieve figures of $1 - (\frac{w_2}{w_1})^2$ in the range 0.8 ÷ 0.9.

It should be noted, that, so far, RO and PV have been compared only with respect to the transport-mechanisms and on the basis of laboratory experiments focused on the membrane transport characteristics - in no way however with respect to process design and economics!

p	– pressure	bar
P_i	– partial pressure	bar
$p°$	– saturation pressure	bar
R	– gas constant	$J\ mol^{-1}\ K^{-1}$
S_{ij}	– selectivity	--
T	– temperature	K
$\bar{V}$	– molar volume	$m^3\ mol^{-1}$
w_i	– mass fraction for the ith component	--
x_i	– molar fraction for the ith component	--
x	– length	m
α	– coupling coefficient	--
β	– coupling coefficient	--
γ	– activity coefficient	--
δ_M	– membrane thickness	m
ν	– temperature	°C
ρ	– density	$kg\ m^{-3}$
$\phi_i \phi_j$	– sorption coefficient	--

Indices

i,j	component
l	liquid solution
M	membrane
0	reference state
1	feed side
2	permeate side

REFERENCES

1. D. R. Paul and O. E. Ebra-Lima, The mechanism of liquid transport through swollen polymer membranes, J.Appl.Polym.Sci., 15:2199–2210 (1971).

2. D. R. Paul, Further comments on the relation between hydraulic permeation and diffusion, J.Polymer Sci.Polymer Phys.Ed., 12:1221–1230 (1974).

3. M. Fels and R. Y. M. Huang, Theoretical interpretation of the effect of mixture composition on separation of liquids in polymers, J.Macromol.Sci., 5:89–110 (1971).

4. M. Fels, Permeation and separation behaviour of binary organic mixtures in polyethylene, AIChE-Symp.Ser., 68(120):49–57 (1972).

5. F. W. Greenlaw, R. A. Shelden, and E. V. Thompson, Dependence of diffusive permeation rates on upstream and downstream pressures. II. Two component permeant, J.Membrane Sci., 2:333–348 (1977).

6. R. Albrecht, Pervaporation: Beiträge zur Verfahrensentwicklung Dissertation, RWTH Aachen (1983).

7. R. Rautenbach and R. Albrecht, Separation of organic binary mixtures by pervaporation, J.Membrane Sci., 7:203–223 (1980).

8. E. G. Heisler, Solute and temperature effects in the pervaporation of aqueous alcoholic solutions, Science, 124:77–78 (1956).

CHEMICAL ENGINEERING ASPECTS OF MEMBRANE PROCESSES

AND ADAPTATION OF THE EQUIPMENT TO THE APPLICATION

J. Wagner

DDS RO-Division
P.O. Box 149, DK-4900 Nakskov
Denmark

INTRODUCTION

The number of mistakes are numerous when laboratory membranes and laboratory equipment are to be adapted to full scale production. Some of the difficulties are described in this paper.

CONSIDERATIONS

Though it may be strange to mention, the first step in the evaluation should be an economic one. Most RO-fans have the tendency to rush into technical details, which indeed can be extremely fascinating. But for those who live by their sale, the approximate net costs per unit treated product should be calculated as well as the typical investment in a plant. Since the variable costs are quite often in the range of 0.5 - 1.0 US$/ 1000 liters permeate (water removal), that will eliminate quite a few of the requests for waste water treatment which are pouring in continuously.

The investment is quite a bit more difficult to estimate due to rather big differences in flux rate. But since membrane filtration equipment is rather expensive, that will again eliminate a number of potential users who have too small volumes or far too big volumes. I consider it reasonably accurate to say that investments in production plants are rarely below 50,000 US$ and rarely above 500,000 US$, at least as long as the water desalination plants are excluded. This kind of money is not a problem for a big industrial group. But for a small company the investment may be prohibitively high.

In the literature you will find very many examples of papers which are technically very interesting. But economically the process never had any success.

The technical aspects can be divided into three groups:

The product which is to be treated
The equipment on which the product is to be treated
The operating conditions

Concerning the product, there are a number of points to bear in mind.

When testwork is done in the laboratory, it is rarely possible to get fresh liquid for the trials. It is in this respect unfortunately so that chemical reactions occur as time goes by and consequently the liquid sample, which is 24 or 48 hours old, may be entirely different from the fresh sample.

It is an annoying fact that the change may not be detectable by a chemical analysis. Polymerizing of lignosulfonates or change of crystal form for $CaHPO_4$ is not easily seen. But the changes may have as a result that only little fouling is experienced in the laboratory, whereas a commercial plant will have serious problems with fouling.

Ageing of a product strangely enough often gives an unrealistic high flux. Still, this statement is only true as long as we talk about chemical reactions only. When microorganisms are involved it may be quite a different story.

Quite a few of the products treated by UF and RO are good substrates for microbes, and that makes it tricky to avoid growth.

It is common to add chemicals to prevent microorganisms from multiplying. But it is also common that the chemicals added, change the behavior of the product. If testwork is done on such a preserved solution it is essential to evaluate whether the result can be considered representative and to consider to which degree the results have been influenced.

The only thing is to run trials at a factory and thus ensure a steady supply of fresh product.

Microbiological growth will for several reasons influence the UF/RO process in a negative way. Growth itself causes a change of composition of the product, since one or more solutes are consumed and converted into something else. This "something else" is very often polysaccharides and other slimy substances which foul membranes and cause severe and sometimes irreversible drop of flux. One should therefore be very much aware if growth takes place in a test solution.

If growth is excessive the increase of the protein content may be a limiting factor due to viscosity. We have, as a matter of fact, not seen this in lab. tests, but we have seen it at a pharmaceutical company who could not control growth in the liquid sent to a big RO plant.

Pretreatment prior to membrane filtration is another chapter where it is difficult to go from laboratory scale to production scale. There is a tendency to make pretreatment in the laboratory too good, maybe because the chemists are working well and according to the book, and maybe because there is so little liquid that even extensive pretreatment is not too much hard work. But it is very important to consider what the effect of the pretreatment in the lab. is and to consider how a good pretreatment can be achieved in commercial operation. Unless this is done, the pretreatment section may easily be sacrificed during a commercial struggle to get an order.

So far I have been discussing testwork done with a reasonably representative liquid on normal membranes. A lot of tests have also been made with solutions consisting of one solute in one solvent, e.g. 2000 ppm NaCl in water or trypsin in water. These investigations are necessary in order to characterize membranes. But it is almost never possible to be quite sure how a 'real' solution behaves, since it is a mixture of numerous solutes.

The inorganic salts rarely give surprises. They behave mostly as if they were alone in the solution and thus the results from tests with a pure solution can be used to predict how a mixture behaves.

If there are organic solutes as well as inorganic solutes it is time to beware, because they may interact. A classic example is calcium which is chemically bound to milk proteins. Therefore, the Ca-permeability seems to be much lower than it should be. The answer is of course that the permeability of free calcium is as it should be and the permeability of bound calcium is nil.

An organic solute may show a permeability radically different from the prediction based on molecular weight. An example is congo red, with an MW around 700. It does not go through a UF membrane with a cut off value of 20,000 because it forms a tight dynamic membrane which retains itself. Linear, hydrophilic, synthetic polymers may do the reverse. Poly-vinyliden pyrrolidon with MW 50,000 can give permeabilities >100%, even with a cut off 20,000 membrane. The reason is that the concentration polarization helps to get the solute through the membranes, in contrast to congo red. Experience does give some rules-of-thumb which can be quite good. But you can still be fooled, e.g. by misleading information from a potential customer. He may for example tell you that the MW is 5000. And he is right, as long as you have only one solute. We have an example where in the commercial solution we found the solute linked to so many other solutes that it was colloidal in nature. And that ruined the prediction that this particular solute would go through the membranes.

There is one point on which everybody not only tend to, but have to make unrealistic work, and that is the ratio between the amount of product and the membrane area in a test rig. Knowing the vast span you see in reality, it is an average figure if you say that $1 m^3$ of product is treated on $1 m^2$ of membrane area in laboratory testwork. In a commercial installation it is quite different, often between 500 and 1000 m^3 to 1 m^2 of membrane area during the lifetime of the membranes. And that means that if there is a tiny amount of solutes which can precipitate and which does not get removed during a normal cleaning cycle, then it is less than likely that laboratory trials will show that something will cause problems. But in the full scale plant the problems will arise for sure.

Water desalination plants are probably the plants receiving the best pretreated product. The water has been pH adjusted and filtered thoroughly. Antiscaling agents are quite often added and Fe and Mn salts are virtually removed. When all the pretreatment works well, the RO plant is likely to last for 20,000 hours and sometimes more than that. The RO plant is rarely cleaned during this period.

Process streams in the biochemical industry represent the opposite extreme. It is rarely possible to adjust pH, addition of chemicals is not allowed, removal of Fe and Mn is not even considered and filtering may or may not be done, consequently cleaning has to be performed frequently, i.e. up to twice a day. The membrane lifetime is therefore typically 8000-12000 hours, sometimes less, rarely more.

The physical size of laboratory equipment is quite often different from the commercial size equipment. But with a little experience it is not too difficult to make a realistic scale-up from the laboratory equipment. The relatively small laboratory equipment has another effect, and that is that testwork is made batch-wise. It is quite easy to operate batch-wise, but it will inevitably expose the product to far more pumping than it will receive in a commercial plant. It will of course also pass through the back pressure valve many times. The combined effect of pumping and throttling may be damaging to shear-sensitive products such as enzymes. The shear is often quite a lot more than necessary, since laboratory pumps tend to be oversized and valves not to be low shear type. So if a product can survive a laboratory unit it is virtually sure that it can survive in a commercial plant. If the product is damaged, it is time to consider whether a modification of the plant is called for. Egg white is probably the most shear-sensitive product treated by RO. And it only works well in a full size commercial plant where all equipment creating shear has been modified and built especially for this product. An RO plant for egg white is one of few RO plants without valves on the product side.

Good operators are of great value. The capability of the operator may be the difference between a successful laboratory trial and a complete failure. It is of course very difficult to measure the effect of the operator quantitatively. But most engineers are likely to know how the operators are and make allowances for the different understanding of the process.

When evaluating testwork, one of the best tools is a log book giving all relevant measurements. It is really surprising how often it is neglected to measure and record necessary data. When tests are made at a client, the RO companies have only themselves to blame, because then they have not given sufficiently good instructions.

Newcomers in RO are surprisingly ingenious when it comes to finding new ways to create unnecessary and useless data. The only way to avoid that is to supply a log sheet, tailormade for each RO user. That does call for a little thinking before the testwork is done. But it is a very good thing to make it quite clear in beforehand what can and must be measured and recorded to satisfy the RO supplier and what is of interest for the user. That may be quite different data.

The operating conditions are the easiest to keep stable and at representative values. There is, however, a tendency to use too high flow rates in laboratory equipment, simply because it is so easy to get and because energy consumption is of no concern. In scaling up it may not be possible or desirable to keep the same flow rate, e.g. because of energy costs. And that may result in a big plant with significantly poorer performance than expected, simply because the membranes are not flushed well enough. Pressure, temperature, pH etc. can without diffusivity be kept at realistic value.

FINAL REMARKS

From the above it may seem impossible that anyone can make a reasonably sure scale-up. But in reality it works quite well. The main variations arise from the product and only a few from the equipment. But reality shows that some good tests together with a lot of experience and a little intuition make the scale-up possible, without more than normal uncertainty.

CONTROL OF GAS TRANSFER IN POLYMERIC MEMBRANES

H. Ohno and K. Suzuoki

Technical Research Laboratory,
Asahi Chemical Industry Co. Ltd
2-1 Samejima, Fuji-shi, Shizuoka-ken, Japan

SUMMARY

Correlation between gas permeabilities and intermolecular terms of second moment of broad line NMR, $\langle \Delta H^2 \rangle_{inter}$, were measured for rigid polymer such as aromatic polyamides, and flexible polymers such as poly (2,6-dimethyl-1,4-oxyphenylene), poly(methyl methacrylate), and various polystyrene derivatives. A good correlation between gas permeability and $\langle \Delta H^2 \rangle_{inter}$ was obtained. The results indicated that the intermolecular packing of the polymers evaluated by $\langle \Delta H^2 \rangle_{inter}$ may control the gas permeability of the glassy polymers.

INTRODUCTION

Gas permeation phenomenon in the non-porous and homogeneous polymeric membrane is interpreted as the transfer of gases through the small openings between polymer molecules. In this report, special attention was paid to the openings of the polymer molecules through which gas permeates, and the relationship between gas permeation coefficients and the packing states of the molecules which constitute membranes were investigated.

The free volume theory was used so far for the theoretical treatment of the gas permeation through polymers. As the new indication of the packing state of polymers, we introduced the second moment of the broad line NMR in the solid state. As the intermolecular term of the second moment, $\langle \Delta H^2 \rangle_{inter}$, is the function of the distance between the protons belonging to different polymer molecules, it is more closely related to the actual path of gases.

The relation between the gas permeability and the $\langle \Delta H^2 \rangle_{inter}$ of various kinds of polymers were examined.

THEORY

NMR spectrum in the solid state becomes very broad because of the influence of the local magnetic field produced by the magnetic moment of the neighboring protons. Therefore the spectrum is called the broad line NMR.

When the molecular motions are completely frozen, second moment $\langle \Delta H^2 \rangle$ of the broad line NMR spectrum is given as a function of the distances between protons existing in the system, as shown in Equation (1)

$$\langle \Delta H^2 \rangle = \frac{\int_{-\infty}^{\infty} H^2 f(H)\,dH}{\int_{-\infty}^{\infty} f(H)\,dH} = 716.2 \times \frac{1}{N} \sum_{i>j} r_{ij}^{-6} \qquad (1)$$

where H is the deviation of magnetic field from the resonance center, $f(H)$ is NMR adsorption intensity, N is the number of protons in the system and r_{ij} is the proton-proton distance (A).

$\langle \Delta H^2 \rangle$ in Equation (1) consists of two terms, the intramolecular term ($\langle \Delta H^2 \rangle_{intra}$) and the intermolecular term ($\langle \Delta H^2 \rangle_{inter}$) as shown in Equation (2).

$$\langle \Delta H^2 \rangle_{total} = \langle \Delta H^2 \rangle_{intra} + \langle \Delta H^2 \rangle_{inter} \qquad (2)$$

$$\langle \Delta H^2 \rangle_{intra} = 716.2 \times \frac{1}{N} \sum_{\substack{intramolecular \\ protons}} r_{ij}^{-6} \qquad (3)$$

$$\langle \Delta H^2 \rangle_{inter} = 716.2 \times \frac{1}{N} \sum_{\substack{intermolecular \\ protons}} r_{ij}^{-6} \qquad (4)$$

$\langle \Delta H^2 \rangle_{total}$ is determined by the measured NMR spectrum and $\langle \Delta H^2 \rangle_{intra}$ is calculated from Equation (3) assuming appropriate geometry of the molecule. Then $\langle \Delta H^2 \rangle_{inter}$ can be determined by the difference of $\langle \Delta H^2 \rangle_{total}$ and $\langle \Delta H^2 \rangle_{intra}$ as shown in Equation (5).

$$\langle \Delta H^2 \rangle_{inter} = \langle \Delta H^2 \rangle_{total} - \langle \Delta H^2 \rangle_{intra} \qquad (5)$$

$\langle \Delta H^2 \rangle_{inter}$ thus obtained, which is the function of proton-proton distance (r_{ij}), can be regarded as a new indication of intermolecular packing state of molecules. Qualitatively smaller $\langle \Delta H^2 \rangle_{inter}$ means looser intermolecular packing state. When some part of molecule rotates in the solid state, $\langle \Delta H^2 \rangle$ becomes smaller than that in the frozen state by the factor of $1/4(1-3\cos^2\beta)^2$.

$$\langle \Delta H^2 \rangle_{moving} = 1/4(1-3\cos^2\beta)^2 \langle \Delta H^2 \rangle_{frozen} \qquad (6)$$

where β is the angle from vector between rotating protons and that of the rotating axis.

We assumed that both methyl and t-butyl groups of the molecules are freely rotating in the solid state.

The calculated result of Equation (6) shows that $\langle \Delta H^2 \rangle_{moving}$ is 1/4 of $\langle \Delta H^2 \rangle_{frozen}$ for methyl group and the $\langle \Delta H^2 \rangle_{moving}$ is 1/144 of $\langle \Delta H^2 \rangle_{frozen}$ for the t-butyl group.

EXPERIMENTAL

Measurement of Broad-Line NMR Spectrum

Polymer sample packed in a pyrex tube (18 mmϕ) was dried in vacuum for two days at the temperature 20°C lower than its glass transition temperature and then the sample tube was sealed.

The NMR spectrum was measured with a JEOL JNM PW60 spectrometer (60 MHz) under the condition of RF level 50 dB, modulation width 0.5 Gauss and scanning rate 6.25 Gauss/min.

Measurement of Gas Permeation Coefficient

Gas permeation coefficient was measured by using an apparatus shown in Figure 1. Permeated gas was analyzed by a gas chromatography with a thermal conductivity detector (TCD) and a photo ionization detector (PID).

RESULTS AND DISCUSSION

The basic idea of this report is that the intermolecular packing state of the polymer, which is described by $\langle \Delta H^2 \rangle_{inter}$, control the gas permeation property of the polymer. This $\langle \Delta H^2 \rangle^{inter}$ is calculated by the Equation (3) and (5). In calculating $\langle \Delta H^2 \rangle^{inter}_{intra}$ in Equation (3), it is necessary to know all proton-proton distances in the molecule. Therefore it is easy to deal with the polymers with rigid chain and fixed geometry. Several aromatic polyamides were chosen as this type of polymer and the relationship between their gas permeability and their intermolecular packing state, i.e. $\langle \Delta H^2 \rangle_{inter}$ was studied at various temperatures.

This idea was also applied to polymers whose main chains are more flexible, poly(2,6-dimethyl-1,4-oxyphenylene) (PPO), polystyrene (PSt) and poly(methylmethacrylate) (PMMA). As PPO and PSt have the highest gas permeabilities among the glassy polymers, it is interesting to know how the molecules are packed in these polymers.

Furthermore, the effect of ring substituents in polystyrene were studied. t-Butylated polystyrenes were investigated in detail as the polymer with a very bulky substituent.

Aromatic Polyamides

Plots of $\langle \Delta H^2 \rangle_{inter}$ of aromatic polyamides

$$\{NH\text{-}\bigcirc\text{-}NHCO\text{-}\bigcirc\text{-}CO\}_n \qquad (R=H,\ CH_3,\ OCH_3)$$

and their gas permeability of hydrogen and nitrogen are shown in Figure 2. A good linear relationship was observed between the logarithm of permeability coefficient of H_2 and N_2 and $\langle \Delta H^2 \rangle_{inter}$ for both gases. This suggests that the intermolecular packing state of the polymer, which is expressed in terms of $\langle \Delta H^2 \rangle_{inter}$, governs the gas permeability.

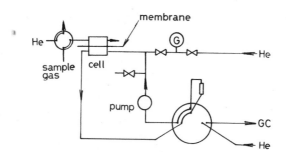

Fig. 1. Apparatus for measuring gas permeability.

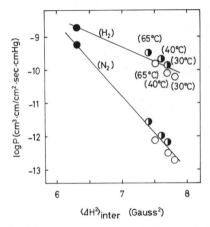

Fig. 2. Relationship between gas permeability coefficients (P) and intermolecular terms of second moment of broad line NMR ($<\Delta H^2>_{inter}$) of
aromatic polyamides at various temperatures.
o: R=H; ◑: R=CH$_3$; ●: R=OCH$_3$.

It is shown in Figure 2 that the same relationship between $<\Delta H^2>_{inter}$
and gas permeability is valid when the measuring temperature is changed.
This fact indicates that the loosening of intermolecular packing state
caused by the thermal movement of the polymer chains and that causued by
the substituents give almost the same effect on the gas permeability. This
observation is considered to give a useful instruction for the design of
gas separation membrane.

Thus it has been proved that the intermolecular packing state expressed by $<\Delta H^2>_{inter}$ controls the gas permeability of aromatic polyamides.

Flexible Polymers

Data of $<\Delta H^2>_{inter}$ against the gas permeability of PPO, PSt and PMMA
were plotted in Figure 3, together with the data of aromatic polyamides.
As shown in the figure, the linear relationship between gas permeability
and $<\Delta H^2>_{inter}$ was also obtained for PPO, PSt and PMMA. This indicates
that the intermolecular packing state, which is described by $<\Delta H^2>_{inter}$,
also controls the gas permeabilities of these flexible polymers.

However, the relation between the gas permeability and $<\Delta H^2>_{inter}$ in
the flexible polymers is different from that in the rigid aromatic polyamides. Aromatic polyamides give the larger $<\Delta H^2>_{inter}$'s than the flexible
polymers at the same gas permeability coefficients. In the partially
crystallized polymer such as the aromatic polyamides, permeability
decreases only proportionally with crystallinity while crystalline part, in
which the distance between protons are smaller than those in amorphous
part, gives far larger influence on $<\Delta H^2>$ because of its minus sixth power
dependency on the distance between protons.

Polystyrene Derivatives

Plots of the gas permeability against $<\Delta H^2>_{inter}$ for some polystyrene
derivatives were shown in Figure 4. When the substituent group is chlorine
or 0-methyl, the plots of the gas permeability against $<\Delta H^2>_{inter}$ are on
the same line as that of PPO, PSt and PMMA. This indicates that the
intermolecular packing state is properly evaluated by $<\Delta H^2>_{inter}$ for these
polymers.

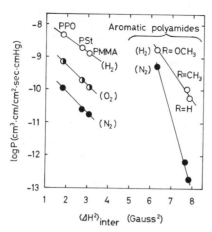

Fig. 3. Relationship between gas permeability coefficients (P)
and intermolecular terms of second moment of broad line NMR
($<\Delta H^2>_{inter}$) of various polymers at 30°C. o: H_2; ◑: O_2; ●: N_2.

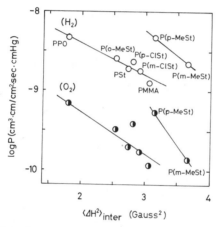

Fig. 4. Relationship between gas permeability coefficients (P) and intermolecular terms of second moment of broad line NMR ($<\Delta H^2>_{inter}$) of
polystyrene derivatives. o: H_2; ◑: O_2.

However in the case of p-methyl and m-methyl substituted polystyrenes,
their plots deviated from the line described above. Particularly poly-
(p-methylstyrene) had $<\Delta H^2>_{inter}$ 1.4 Gauss^2 larger than that of PPO,
whereas the gas permeabilities of these two polymers were about the same.
While these two polymers have about the same vacancy between polymer chains
as an average, distance between p-methyl groups in poly(p-methylstyrene) is
shorter because they are attached far off the center of the chain, causing
greater $<\Delta H^2>_{inter}$.

t-Butylated Polystyrenes

t-Butylated polystyrenes with various degree of t-butylation were
synthesized by reacting polystyrene with t-buthlbromide. Plots of gas
permeability of hydrogen, oxygen and nitrogen against the degree of t-
butylation are shown in Figure 5. Gas permeation increases as the degree
of the substitution increases, and the maximum gas permeation was observed
at 70 mol % substitution for all the gases. Further t-butylation adversely

decreases the permeation. As far as we know, this is the first case that the copolymer showed the maximum permeation at a certain composition. The openings between polymer molecules in the 70 mol % t-butylated polystyrene may be optimum in size and number for the permeation of gases.

Figure 6 shows the broad line NMR spectra of the samples with various degree of t-butylation measured at room temperature. The spectrum became sharper as the content of t-butyl group increased, indicating that the protons in t-butyl group are freely rotating. $\langle \Delta H^2 \rangle_{intra}$'s listed on Table 1 were calculated assuming that t-butyl group as a whole is freely rotating. The value of 7.12 Gauss^2 was obtained for 97 mol % t-butylated polymer assuming that t-butyl group is frozen and only the methyl groups therein are rotating. This value is larger than $\langle \Delta H^2 \rangle_{total}$ (6.26 Gauss^2), indicating that t-butyl group is freely rotating at room temperature.

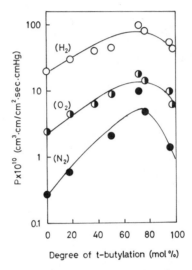

Fig. 5. Gas permeability coefficients of t-butylated polystyrenes.

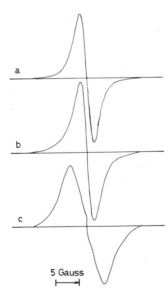

Fig. 6. Broad line NMR spectra of t-butylated polystyrene. The degree of t-butylation is a) 97 mol%; b) 50 mol%; c) 0 mol% (polystyrene).

$\langle \Delta H^2 \rangle_{inter}$, was plotted against the degree of t-butylation in Figure 7. $\langle \Delta H^2 \rangle_{inter}$ exhibited minimum value at 70 mol % t-butylation. This shows that t-butylated polystyrene at this composition has the loosest packing state. Thus, $\langle \Delta H^2 \rangle_{inter}$ was again shown to be an excellent parameter for gas permeation also in this system.

The densities and the glass transition temperatures of t-butylated polystyrenes with various compositions are shown in Figure 8 and 9, respectively. These results do not correspond with those of gas permeability measurement shown in Figure 5. Thus it is shown that $\langle \Delta H^2 \rangle_{inter}$ is the best indication of intermolecular packing state which is closely associated with the gas permeation behavior.

CONCLUSION

The intermolecular term of the second moment of the broad line NMR ($\langle \Delta H^2 \rangle_{inter}$) of various kind of polymers had a good correlation with

Table 1. Gas Permeability Coefficients and Second Moments of Broad Line NMR of t-butylated Polystyrenes

t-Bu [a] (mol%)	$\langle \Delta H^2 \rangle$ (Gauss^2)			$Px10^{10}(cm^3 \cdot cm/cm^2 \cdot sec \cdot cmHg)$ [b]		
	total[c]	intra	inter	H_2	O_2	N_2
0	10.66	7.69	2.97	20.1	2.45	0.27
18	9.58	6.97	2.61	30.6	4.53	0.59
37	9.03	6.21	2.82	40.5	6.38	–
50	8.26	5.70	2.56	45.7	9.08	2.17
71	7.08	4.86	2.22	97.2	17.9	10.1
76	6.87	4.66	2.21	81.1	14.2	4.8
95	6.36	3.90	2.46	54.9	9.79	1.39
97	6.26	3.82	2.44	42.1	6.18	–

a) Degree of t-butylation
b) Gas permeability coefficient (30°)
c) 30°C

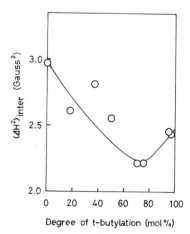

Fig. 7. Intermolecular terms of second moment of broad line NMR ($\langle \Delta H^2 \rangle_{inter}$) of t-butylated polystyrenes.

619

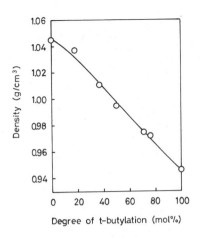

Fig. 8. Density of t-butylated polystyrenes measured by density-gradient
tube method using isopropanol/ethylene glycol system at 25°C.

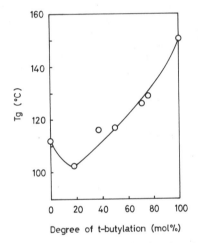

Fig. 9. Glass transition temperatures (Tg) of t-butylated polystyrenes
measured with DSC.

their gas permeability coefficients. $\langle \Delta H^2 \rangle_{inter}$ was found to be a new
quantitative parameter for the packing state of polymers which is supposed
to control gas permeabilities.

Acknowledgements

We express our gratitude to Prof. Dr. R. Kitamaru and Dr. F. Horii
for their valuable advice on measuring broad line NMR. Grateful acknowledgement is made to Dr. T. Tanabe for his constructive comments and
helpful advice on completing this report. This work was performed under
the management of the Research Association for Basic Polymer Technology
for synthetic membranes for new separation technology as a part of a
project on Basic Technology for Future Industries sponsored by the Agency
of Industrial Science and Technology, Ministry of International Trade and
Industry.

DEVELOPMENT OF HYDROGEN SEPARATION MEMBRANES

FOR "C_1 CHEMISTRY" IN JAPAN

K. Haraya, Y. Sindo, N. Ito, K. Obata,
T. Hakuta and H. Yoshitome

National Chemical Laboratory for Industry
Tsukuba, Ibaraki, 305 Japan

ABSTRACT

The pure gas permeation rates of H_2 and CO through the asymmetric membranes prepared from polyimide and polysulfone and through the plasma polymerized membranes were measured. These membranes are currently under development in C_1 chemistry R&D Project. Some of these were tested as separation membranes using binary mixtures of gases. Depending on membrane preparation conditions, ideal separation factors varying from 20 to 130 and permeation rates varying from 1×10^{-6} to 1×10^{-4} $cm^3(STP)/cm^2 \cdot s \cdot cmHg$ were observed. Actual separation efficiencies of polyimide and plasma polymerized membrane agreed well with the ones predicted on the basis of the pure gas permeabilities. In addition a three stage separator to produce the syngas with various ratios of H_2/CO was investigated by computer simulation.

INTRODUCTION

In Japan, C_1 chemistry R&D Project is carried out by National Chemical Laboratory for Industry and some private companies. The project will be completed at F.Y. 1987. "C_1 chemistry" denotes chemical processes that form compounds with carbon-carbon bonds from synthesis gas (syngas), which is obtainable from gasification of coal, LNG, residual heavy oil, biomass and so on. The syngas with various mole ratio of H_2/CO would be required to produce commodity chemicals. The R&D Project plans to use membrane separators to adjust each suitable H_2/CO ratio for ethylene glycol, ethanol, acetic acid and olefins.

Asymmetric membranes of polyimides and polysulfones, and composite membranes with plasma polymers on the surface of porous PTFE are being developed as a high selectivity membrane. Porous glass membranes and porous polyimide membranes are also being developed as a high flux membrane. This paper presents the interim results of an investigation of high selectivity membranes in the separation of hydrogen and carbon monoxide, a system which has considerable importance in C_1 chemistry.

PREPARATION OF MEMBRANE

Development target of membrane configuration is hollow fiber type. The companies in cooperation with NCLI are shown in Table 1. Membrane preparation and their transport properties were tested initially using flat sheet membrane to screen available material. Subsequently, hollow fiber membranes were prepared from promising polymers.

Suitable polymers for membranes were selected from several types of polyimides and polysulfones, synthesized from various monomers. Basic properties of typical polymers are shown in Table 1. The ratio of permeability coefficients of the polyimide to H_2 and CO was above 80 and the ratio of the polysulfone to which amino group was added was 85. Asymmetric membranes were prepared from these polymer solutions which were cast into flat sheets or spun into hollow fibers, partially evaporated solvent, gelled in a water bath and finally dried.

Plasma polymerized layers were prepared on the surface of the porous PTFE which were coated with a silicon resin. Namely, the membrane consists of three layers: ultrathin plasma polymer, silicon resin and porous substrate of PTFE. Active plasma layers were prepared from a variety of organic monomers through utilization of a rf(13.56 MHz) glow discharge. The power used in preparation ranged from 5 to 150 W. The pressures of monomer were between 0.05 and 0.6 torr, and the flow rates of monomer ranged from 0.5 to 6 cm^3/min.

PURE GAS PERMEATION TEST

A number of membranes were tested using pure H_2 and CO at various temperatures and the pure gas permeation rates were measured. The area of the tested membranes of a flat sheet was about 10 cm^2 and of a hollow fiber was about 200 cm^2. The permeation rates for H_2 at 30°C and their ratios to

Table 1. Basic Properties of Membrane Materials

membrane	monomers	cooperative company

polyimide
  Tg=262°C, P_{H2}=1.6–1.8x10^{-10}(35°C), P_{H2}/P_{CO}=80–108
		UBE Industries
		  Ltd.

3,3',4,4'-Benzophenontetracarboxylic Dianhydride, 4,4'-Diaminodiphenyl Ether		
polysulfone Tg=250°C, P_{H2}=4.5x10^{-10}(25°C), P_{H2}/P_{CO}=85		TOYOBO Co.,Ltd.
Bis(4-(4-aminophenoxy)phenyl)sulfone, Bis(4-hydroxyphenyl)sulfone, Isophthaloyldichloride		
plasma polymerization		SUMITOMO Electric Industries, Ltd.
hydrocarbon, silane compounds	(substrate: porous PTFE)	

P_{H2}: permeability coefficient for hydrogen $\left[cm^3(STP).cm/cm^2.s.cmHg\right]$

622

CO, which give an ideal separation factor for H_2/CO binary mixture, are shown in Figure 1 and Figure 2. The permselectivities of asymmetric membranes depended on concentration of polymer in dope solution, period of solvent evaporation, composition of aqueous solution for coagulation bath and temperature for drying membrane. Higher permeation rate for H_2 than $1x10^{-5}$ $cm^3(STP)/cm^2.s.cmHg$ was attained successfully in the hollow fiber membrane of polyimide which maintained ideal separation factor over 80. In the polysulfone membrane, the permeability of hollow fiber was inferior to that of flat sheet at the same selectivity level. It is important to make the active layer thinner in preparing hollow fiber membrane of polysulfone.

The selectivity of plasma polymerization membranes was strongly dependent on the kinds of monomers which formed ultra-thin polymer layers. Promising monomers in this study were silane compounds which contained unsaturated bond like a vinyl group. It is necessary at present to coat silicon resin more than 10 μm thick prior to plasma polymerization in order

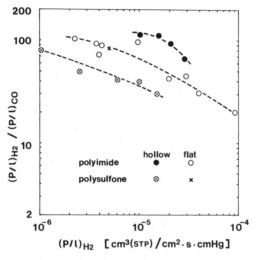

Fig. 1.   Permeation rates and selectivities of asymmetric membranes prepared from polyimide and polysulfone to pure H_2 and CO at 30°C.

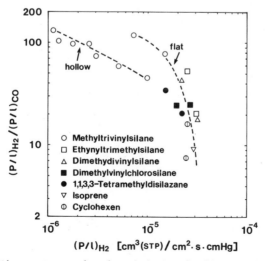

Fig. 2.   Permeation rates and selectivites of plasma polymerized membranes prepared from various monomers to pure H_2 and CO at 30°C.

to withstand large pressure differences. It is very important for the
hollow type plasma membrane to make thinner silicon resin layer and
increase the permeability.

Dependency of the permeation rates in the three kinds of membranes on
temperature between 25 and 100°C is shown in Figure 3. In both polyimide
and polysulfone, permeability increased exponentially and selectivity
decreased when the temperature was raised. But the ideal separation factor
of the polyimide was 68 and of the polysulfone was 40, namely these membranes still maintained high selectivity at 100°C. In the plasma polymerized membrane, both permeability and selectivity increased as the
temperature raised. The ideal separation factor increased from 65 at 25°C
to 255 at 95°C as illustrated in Figure 3. Similar phenomena were observed
in many membranes prepared from plasma polymers.

SEPARATION OF H_2/CO BINARY MIXTURES

The degree of separation was measured for H_2/CO binary mixtures at
various feed compositions and various pressures. In all of the binary
tests, the ratios of permeation rate to feed rate (i.e. CUT) were maintained below 0.01. As a result, the measured concentration at the permeate
side and at the reject side were sure to be very similar to point composition on both sides of the membrane. The results are shown in Figure 4,
Figure 5 and Table 2.

In binary separation of H_2 and CO, McCandless[1] reported that
observed separation factors in Kapton film were less than those predicted
from the pure gas permeability data. Therefore we studied differences
between predicted and observed values, by use of the membrane efficiency
which was defined by Benedict and Pigford[2]. In general, analysis is

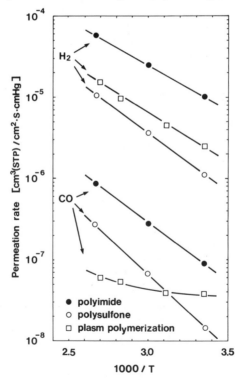

Fig. 3. Dependence of permeability on temperature.

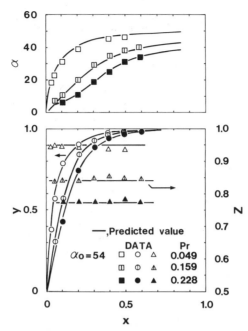

Fig. 4. Effects of composition and pressure ratio on the separation degree of H_2/CO in flat sheet polyimide membrane at 100°C and p_1 = 1.03 kg/cm².

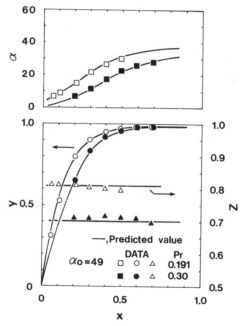

Fig. 5. Effects of composition and pressure ratio on the separation degree of H_2/CO in flat sheet plasma polymerized membrane at 30°C and p_1 = 1.03 kg/cm².

carried out assuming that the individual components do not interact with each other and the permeability coefficients for these components are independent of pressure. The permeation flux of the components A and B of a binary gas mixture are described by Equations (1) and (2).

memb.	y/x	Z	1-Pr	P_1/P_h	α_o	P'_{H2}	Temp
PI	0.905/0.172	0.905	0.897	4.2/40.6	90.8	2.5	60
	0.868/0.188	0.872	0.865	5.5/40.5	68.2	6.0	100
PS	0.806/0.173	0.889	0.946	2.7/50.4	40	1.0	100
	0.937/0.483	0.687	0.782	11.0/50.2	39	0.75	100
PLS	0.987/0.641	0.632	0.636	3.0/8.4	100	0.26	30
	0.985/0.632	0.579	0.576	3.6/8.5	100	0.26	30

PI: polyimide membrane , PS: polysulfone membrane ,

PLS: plasma polymerized membrane

P'_{H2}: permeation rate for hydrogen $\left[cm^3(STP)/cm^2.s.cmHg\right] \times 10^5$

$P_h, P_1 \left[kg/cm^2\right]$, Temp$\left[°C\right]$

$$N_A = \frac{P_A \ P_h}{1}(x - Pr\ y) \tag{1}$$

$$N_B = \frac{P_B \ P_h}{1}\ (1 - x - Pr(1-y)) \tag{2}$$

The membrane separation efficiency is defined[2] as Equation (3)

$$Z = \frac{y-x}{y-x*} \tag{3}$$

Where x* is described by Equation (4) which is derived from Equations
(1) and (2) when Pr is equal to zero.

$$x* = \frac{y}{y + \alpha_o(1-y)} \tag{4}$$

$$\alpha_o = P_A/P_B \tag{5}$$

Equation (6) is obtained by substituting Equation (4) into Equation (3)

$$Z = \frac{(y-x)(y+\alpha_o(1-y))}{(\alpha_o - 1)y(1-y)} \tag{6}$$

Combining Equations (1), (2) and (6), one obtains

$$Z = 1 - Pr \tag{7}$$

Relation between the separation factor and the separation efficiency is
described as

$$\alpha = \frac{y}{1-y} \Big/ \frac{x}{1-x}$$

626

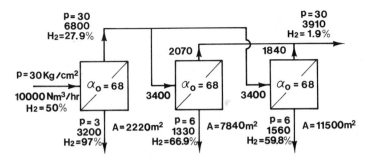

Fig. 6. Estimated performance of three stage separator with countercurrent
flow modules, calculated base on permeability data of polyimide
(100°C) in Table 2.

$$= 1 + \frac{y(\alpha_o - 1)\ Z}{x\ (y + \alpha_o(1-y))} \tag{8}$$

If measured compositions of the reject stream and of the permeate
stream are approximately equal to point compositions on both sides of the
membrane, the separation efficiency is able to be calculated from Equation
(6). Consequently, one can find deviation from ideality in comparing
actual values from Equation (6) with Equation (7).

As shown in Figure 4, Figure 5 and Table 2, actual separation efficiencies agree well with the predicted values from Equation (7) in the
polyimide and in the plasma polymer. Therefore, the separator using these
membranes can be designed based on Equation (1) and (2). In the polysulfone, actual separation efficiencies are less than the predicted values.
It seems that there is some interaction between components of the binary
mixture during permeation. However, exact analysis requires more data.

ROLE OF MEMBRANE SEPARATOR IN C_1 CHEMISTRY

The role of membrane separator in C_1 chemistry process is briefly
discussed. The syngas with various ratios of H_2/CO would be demanded to
produce chemicals in current and future C_1 chemistry processes. For
example, the stoichiometric ratio for ethylen glycol and acetaldehyde is
3:2, for ethanol and ethylen 2:1 and for acetic acid (by methanol carbony-
lation) 0:1. To obtain these three kinds of syngas, a three stage
separator was studied. Computer simulation was carried out for a 10000
Nm^3/hr feed stream of equimolar mixture of H_2 and CO at 100°C and 30
kg/cm^2, using permeability data of the polyimide in Table 2. Counter-
current plug flow was assumed to be and the solving method was similar to
Blaisdell and Kammermeyers'[3]. The results of computation are shown in
Figure 6. A three stage separator can produce pure hydrogen, 2:1 syngas,
3:2 syngas and pure carbon monoxide.

CONCLUSION

This paper is an interim report of performance of membranes which are
currently under development in C_1 chemistry R&D Project. Hereafter,
membrane performances will certainly be improved. Some of the membranes
examined in this study are sure to be useful for adjusting the ratio of
H_2/CO in syngas, and, of course, for the recovery of hydrogen from purge
gases in current industrial plants.

Acknowledgement

The authors deeply appreciate contribution from Ube Industries Ltd, Toyobo Co., Ltd and Sumitomo Electric Industries Ltd.

NOMENCLATURE

1 – thickness of membrane, cm
N – permeation flux, cm/s
P – permeation coefficient, $cm^3(STP)-cm/cm^2-s-cmHg$
Pr – ratio of total pressures on the two sides of the membrane (p_l/p_h)
x – point composition of faster permeating gas in high pressure side
x^* – composition defined by Equation (4)
y – point composition of faster permeating gas in low pressure side
Z – membrane separation efficiency defined by Equation (3)
α – separation factor
α_o – ideal separation factor defined by Equation (5)

REFERENCES

1. F. P. McCandless, Ind. Eng. Process Des. Develop. 11:470 (1972).
2. M. Benedict and T. H. Pigford, "Nuclear Chemical Engineering," McGraw-Hill, New York (1957).
3. C. T. Blaisdell and K. Kammermeyer, Chem.Eng.Sci., 28:1249 (1973).

APPLICATION OF GAS-LIQUID PERMPOROMETRY

TO CHARACTERIZATION OF INORGANIC ULTRAFILTERS

C. Eyraud

Laboratoire de Chimie Appliquée et Génie Chimique
CNRS, ERA No. 300, Université Claude Bernard
Lyon, France

High permeability inorganic membranes for ultrafiltration are made
of two or many porous layers, either superposed (Figure 1a) or overlapping
(Figures 1b). In order to characterize the texture of the separative
layer, supposed to be uniform, of a ceramic membrane of type a, it is
possible to sample scraps from the layer, and to submit them to one of the
following methods: mercury porometry[1], BJH porometry[2,3], thermoporometry[4,5,6]. In the case of structure (b), these three methods are
unworkable, for the three following reasons:

1) the mesoporous stuffings are not of uniformed texture;
2) they are difficult to sample;
3) their relative volume is too small.

One is then led to resort to non-destructive dynamic techniques
allowing to characterize a membrane as such. These techniques are biliquid
permporometry[7,8] and gas-liquid permporometry[9].

The first is based upon the Laplace relation ruling the mechanical
equilibrium at interface:

$$\Delta p = \frac{2\sigma}{r} \tag{1}$$

The experiment consists in measuring the water flow rate through a
membrane impregnated with isobutanol as a function of the pressure difference Δp. The second is based upon the Kelvin relation ruling the liquid-
vapor thermodynamic equilibrium of a capillary condensate:

$$\ln \frac{P}{Po} = - \frac{2\sigma}{r} \frac{v}{RT} \tag{2}$$

In this case the experiment consists in measuring the flow rate of
helium containing a minute amount of a condensable vapor such as ethanol,
as a function of the relative pressure of this vapor. The gas flow regime
is of Knudsen type in all open pores. Both methods allow to express the
membrane permeability K as a function of the pore radius, either as a
cumulative value $K = f(r)$ or as a differential one $\frac{\Delta K}{\Delta r} = g(r)$.

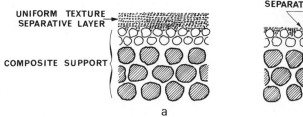

Fig. 1. Structure of an asymmetrical inorganic ultra-filter (a) with
superposed porous layer, (b) with overlapping separative layer.

We have especially been more interested in the second method. It has
the advantage of not accumulating at the level of the separative layer any
pollutant which would induce changes in the liquid-liquid interfacial
energy and in the three phases contact angle. The realizations were those
of Eyraud[9] and of Katz[10]. They corresponded to a stationary flow
regime. However, they suffered from many drawbacks:

1) measurement of small gas flows is difficult, especially in a wide
variation interval;
2) it is not easy to realize a feed at variable flow of a gaseous mixture
containing a vapor at a relative pressure very precisely known at the
vicinity of $\frac{P}{P_0} = 1$;
3) at high relative pressures, the intervention of a capillary flow of
condensate induces a relatively important decrease of condensable gas
concentration on the inlet side.

This is why we chose a pseudo-stationary technique depicted by
Figure 2[11,12,13]. The membrane separates two closed compartments, of
respective volumes V_1 and V_2, about 1 liter each. Volume V_1 is connected
to bellows made of corrugated metal, with a backward and forward motion of
volumic amplitude of about 1 cm^3. The membrane is previously brought to
equilibrium with ethanol vapor by helium circulation between an evaporator
at temperature T and the membrane holder at temperature T_0. If T_A is the
ambient temperature, condensation in the tubing is avoided by maintaining
$T < T$. When the four valves are closed, the phase difference ϕ between
the variation of volume V_1 and the pressure difference between the two
compartments is expressed by the relation:

$$\omega \tan \phi = A \, K \, RT_0 \left(\frac{1}{V_1} + \frac{1}{V_2}\right) \qquad (3)$$

A, ω, V_1 and V_2 are known. K is deduced from the measurement of ϕ. Thus,
the variations of $\omega \tan \phi$ versus p/p_0 can be represented by the curve of
Figure 3. The distribution $K_k = f(r_k)$ can be deduced by using relations
(2) and (3). To improve the method, it is necessary to take into account
the thickness of the layer of molecules adsorbed at the surface of open
pores. The actual pore radius r_p is related to KELVIN radius r_k by:

$$r_p = r_k + e \, n \qquad (4)$$

where e is the unimolecular layer thickness (e = 0.44 nm for ethanol) and
n is the layer number given by the Halsey relation:

$$N = \left(\frac{B}{\ln p/p_0}\right)^{1/x} \qquad (5)$$

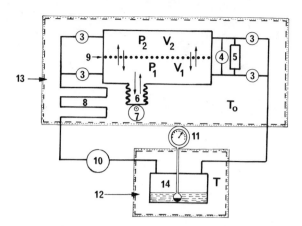

Fig. 2. Permeameter with alternate flow of a non condensable gas mixed
with a condensable vapor. (1) Closed compartment 1, (2) closed
compartment 2, (3) electric valves, (4) by-pass, (5) pressure
differential sensor, (6) metallic bellows, (7) cam, (8) heat
exchanger, (9) ultra-filter, (10) membrane pump, (11) absolute
pressure manometer, (12) thermostat enclosure at temperature T,
(13) enclosure at temperature T_o, (14) ethanol evaporator.

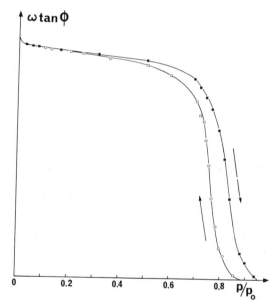

Fig. 3. Phase difference between the volume and the pressure difference
variations, vs the vapor relative pressure.

B and x result from experimental determinations carried out for the
adsorbate-adsorbent couple, outside the domain of capillary condensation.
This is the way in which the distribution curve of Figure 4 is obtained.
The permporometry radius at 50% of membrane permeability is $r_{pp}50 = 5.5$ mm,

the permporometry radius at maximum of distribution curve $\frac{\Delta K_p}{\Delta rp} = g(r_p)$ is
$r_{ppmax} = 5.2$mm.

In order to check this method, we used a cylindrical composite alumina
ultrafilter, of the type shown in Figure 1a. The separative layer is

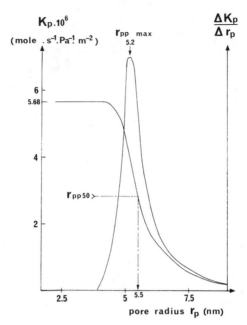

Fig. 4. Distribution of helium permeability vs pore radius.

considered to be of uniform texture and its thickness is known exactly.
The support permeability is at least one hundred times larger than that of
the fine layer. By assuming for the latter a model of cylindrical pores
perpendicular to the membrane sides, it is possible to pass from the

permeability distribution $\frac{\Delta K_p}{\Delta r_p} = g(r_p)$ to the porosity distribution $\frac{\Delta \epsilon}{\Delta r_p} = h(r_p)$

We write that the class of open pores of Kelvin radius r_k takes part to
permeability for the value:

$$d\ K_k = s\ \frac{dN}{\ell}\ \sqrt{\frac{2\Pi}{MRTo}}\ r_k^{\ 3} \qquad\qquad (6)$$

where ℓ is the thickness of the separative layer, N the number of pores per
surface unit. It is assumed that the gliding coefficient at the wall s is
equal to 1.

Figure 3 represents the pore size distribution obtained by the three
following methods:

1) thermoporometry, effected on scraps of the separation layer;
2) BJH porometry. In this case, the isotherm of nitrogen desorption at
   liquid nitrogen temperature is obtained by thermogravimetry;
3) helium-ethanol permporometry, effected on a complete membrane.

It is observed that in this case the porosity ϵ (Figure 5b) determined
by permporometry is about three times smaller than that determined by the
other two methods. In fact, the dynamic technique reach an 'effective'
porosity, it does not take into account either the 'one-eyed' pores or
those which are parallel to the membrane sides.

In conclusion, we think we have shown the efficiency of this new
method 'gas-liquid permporometry' for the analysis of the permeability
distribution of a composite rigid ultrafilter as a function of the pore
radius of the separative layer.

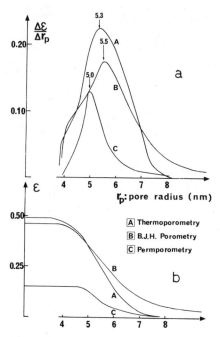

Fig. 5.   Distribution of porosity vs pore radius.

Acknowledgement

We gratefully acknowledge the part taken by the Commissariat à l'Energie Atomique in the funding of the experiments, and in the tuning up of the two new techniques aforementioned: thermoporometry and gas-liquid permporometry.

NOMENCLATURE

A   – area of the ultrafilter
K   – permeability $(mole.m^{-2}.s^{-1}.Pa^{-1})$
p   – vapor pressure
P_o – saturated vapor pressure at temperature T
Δp – pressure difference between the two sides of a membrane
r   – average curvature radius of an interface

r_k – pore KELVIN radius
r_p – actual pore radius
T   – absolute temperature
v   – molal volume of the capillary condensate
σ – interfacial energy or tension
ϕ – phase difference (radian)
ω – pulsation

REFERENCES

1. B. Rasneur, G. Schnedecker, and J. Charpin, "Silicates Industriels," 1301–1305 (1972).
2. E. P. Barrett, L. G. Joyner, and P. H. Halenda, JACS, 73:373 (1951).
3. C. Pierce, J.Phys.Chem., 57:149 (1953).
4. G. Boutillon, C. Eyraud, and M. Prettre, C.R.Acad.Sci., Paris, 240:756 (1955).
5. M. Brun, P. Eyraud, L. Eyraud, M. Richard, and C. Eyraud, C.R.Acad. Sci., Paris, B-272:565 (1971).
6. M. Brun, A. Lallemand, J. F. Quinson, and C. Eyraud, Thermochimica Acta, 21:59 (1977).

7. P. Grabar and S. Nikitine, Chim.Phys., 33:721 (1936).

8. G. Capanneli, F. Vigo, and S. Munari, J.Membr.Sci., 15:289 (1983).

9. C. Eyraud, J. Bricout, and G. Grillet, C.R.Acad.Sci., Paris, 257:2460 (1963).

10. M. G. Katz, World Filtration Congress III, Downingtown, Pa., p. 508, Sept. 13-17 (1982).

11. S. Bienfait, Thèse de Doctorat d'Etat ès Sciences (mention Sciences Appliquées), Lyon (1967).

12. C. Eyraud, S. Bienfait, and D. Massignon, Chimie et Industrie - Génie Chimique, 102:652 (1969).

13. C. Eyraud, M. Betemps, J. F. Quinson, F. Chatelut, M. Brun, and B. Rasneur, Bull.Soc.Chim.Fr., (in press).

AN EXPERIMENTAL STUDY AND MATHEMATICAL SIMULATION

OF MEMBRANE GAS SEPARATION PROCESS

E. Drioli, G. Donsi', M. El-Sawi, U. Fedele,
M. Federico and F. Intrieri

Dipartimento di Chimica – Università della Calabria
87030 Arcavacata & Dipartimento di Ingegneria Chimica
Università di Napoli

ABSTRACT

An experimental investigation of the behavior of different membranes in the separation of gaseous mixtures of industrial interest has been carried out. Polybutadiene, polyurethane and cellulose acetate flat sheet membranes as well as cellulose acetate spiral wound moduli have been used for the permeation of pure O_2, CO_2, C_3H_6, C_3H_8, CH_4, and then of mixtures of these components. A mathematical model simulating the separation process has been formulated and tested for adequacy.

Although the fractionation, purification and concentration of gaseous mixtures by permeation through membranes has been a well known technique since the end of the last century, only recently, as a result of improvements in polymeric membranes science and technology, the large scale application of this technique is gaining more interest.

This work aims at developing techniques for the characterization of membranes as applied to the separation of gaseous mixtures. Thereby, a mathematical model for the simulation of a single stage gaseous permeator has been formulated and a computer simulation program has been developed and tested for adequacy. Both co-current and counter-current flow patterns (Figures 1-3) have been considered with and without purge on the permeate side.

MATHEMATICAL MODEL

The mathematical model of a gaseous process consists, in general, of the following equations.

1. Permeation Rate Equation

Different mechanisms are suggested for mass transfer through membranes, e.g. molecular diffusion, Knudsen flow, surface flow, etc. resulting in a variety of rate equations. Whatever the permeation mechanism is, it can be regarded as consisting of the following steps:

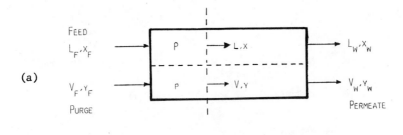

(a)

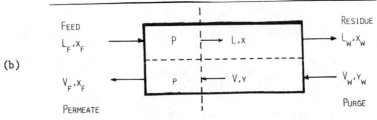

(b)

Fig. 1. (a) Co-current permeator, (b) counter-current permeator.

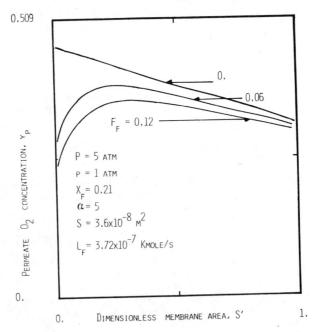

Fig. 2. Oxygen enrichment of air on cellulose acetate spiral wound modulus in co-current flow pattern, at different purge levels.

1) diffusion from the feed-side to the membrane surface;
2) sorption into the membrane;
3) "transfer" through the membrane;
4) desorption out of the membrane;
5) diffusion on the permeate-side from the membrane surface.

In analogy to the overall interphase mass transfer coefficient $\underline{K}$, a permeation coefficient (Q_i (permeability of the component i)/1) is defined as follows:

$$J_{is} = (Q_i/1) \ (P_i - p_i) \tag{1}$$

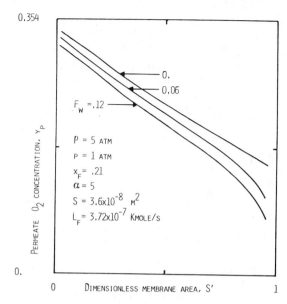

Fig. 3. Oxygen enrichment of air on cellulose acetate spiral wound modulus
         in counter-current flow pattern, at different purge levels.

where

      P_i is the partial pressure of component i on feed-side
      p_i is the partial pressure of component i on permeate-side
      l is the membrane thickness.

      Due to its semiemperical nature, this film-theory based equation can
be applied to describe the permeation flux relative to average molar
velocity in the s-direction independently of the transfer mechanism inside
the membrane.

2. Mass Balance Equation

      Concentration variations in the permeation equipment are described by
means of a differential mass balance for each component on both sides of
the membrane together with the initial and boundary conditions. Thus n
differential component mass balances on either side of membrane, together
with n algebric macroscopic component mass balance between a general point
in the permeation equipment and either the inlet or outlet, are required to
describe the concentration variations in an n-component permeation system.

      The general differential mass balance of a component i in a homogene-
ous phase can be formulated as follows:

$$\delta \rho_i / \delta t = - (\nabla \underline{n}_i) + r_i \qquad (2)$$

where

      ρ_i mass concentration of component i
      $\underline{n}_i$ mass flux of component i relative to fixed coordinates
      r_i rate of mass generation of cpt. i by chemical reaction.

      In the absence of chemical reaction, considering the bulk motion as
the only mechanism of mass transfer within the phase, and assuming:

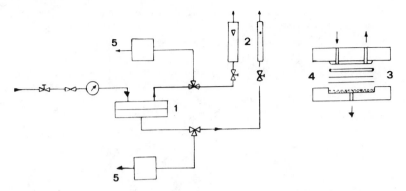

Fig. 4. FSP - Flow sheet of experimental plant for flat sheet membrane.
(1) Steel modulus, (2) flowmeter, (3) permeation cell details,
(4) membrane 12.5 cm², (5) gas chromatograph.

- concentration varies only along the z-direction axis (direction of
flow);
- the mass average velocity v has a single component in the z direction.

Equation (2) at steady-state condition is reduced to

$$d (c_i v_z) / dz = S (Q_i/1) (P_i - P_i) \qquad (3)$$

where S is the permeation area for each unit volume, parallel to the flow
direction with the boundary condition

$$c_i (0) v_z (0) = c_i^* v_z^*$$

where the superscript indicates the inlet conditions.

The algebraic component mass balance between the inlet and an arbitrary point along the permeate is expressed as:

$$L_f x_f + V_f y_f = Lx + Vy \qquad (3.1)$$

3. Pressure Drop Equation

Due to the negligible pressure drop in the small permeators used in
this study, the assumption of constant pressure has been made. However,
different formula for calculating pressure drops which could be substituted
in differential momentum balance equation, are available in the literature[2,4].

4. Enthalpy Balance

Due to practically absent heat effects during gas permeation, the
enthalpy balance has not been considered in the model.

EXPERIMENTAL RESULTS AND DISCUSSION

Two experimental apparatus have been realized for characterization of
lat sheet membranes (Figure 4) and spiral wound moduli (Figure 5). Chemi-
l analysis has been carried out by gas chromatography.

Permeability flow rates have been evaluated measuring in the flat
ator the permeate flows of different pure gases, at constant tempera-

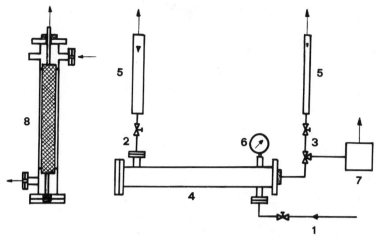

Fig. 5. CASWM – Flow sheet of experimental plant for spiral wound moduli.
(1) Feed, (2) residue, (3) permeate, (4) permeation cell,
(5) flowmeter, (6) manometer, (7) gas chromatograph, (8) membrane
modulus details.

Table 1. Permeability Flow Rate for Flat Membrane
$(cc/(cm^2 \ s \ cmHg))x10^7$.

Gas	Polybutadiene	Polyurethane	Cellulose acetate
N_2	1.228	0.41	22.8
O_2	3.719	0.85	96.5
CO_2	18.175	8.48	967.7 – 1050
CH_4			27.9
C_3H_6			12.2
C_3H_8			2.3

ture and pressure by closing the residue outlet (Table 1). This procedure
has been repeated using cellulose acetate spiral wound modules in the
spiral permeator (Figure 6).

Based on these measurements the separation factors of gas pairs
α_t ($\alpha_t = Q_a/Q_b$, where Q_a and Q_b are the permeability flow rates as
reported in Table 1, at 298°K and 1 bar) have been calculated. The
results are summarized in Table 2.

Flat Sheet Separation

Oxygen enrichment of air using cellulose acetate flat sheet membrane
(FSM) has been carried out by flat permeator. As shown in Figure 7 where
experimental permeate flow-rate and corresponding oxygen concentration are
presented, a 52.5% O_2 concentration has been reached for a Δp = 11 bar.

Spiral Wound Module Separation

Using spiral permeator (CASWM), the following separations have been
carried out:

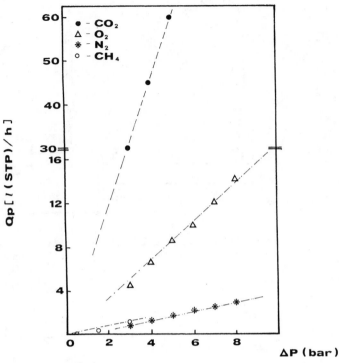

Fig. 6. Permeation flow-rates v.s. Δp - CASWM (1") by Separex.

Table 2. Separation Factors.

	Polybutadiene	Polyurethane	CA Flat membrane	CA Spiral membrane
O_2/N_2	3	2.1	4.23 – 5.6	4.86
CO_2/N_2	14.8	20.7	39. – 46	32.28
CO_2/O_2	4.9	9.97	7.6 – 10.9	6.65
CO_2/CH_4			34.6 – 37.5	28.46
CO_2/C_3H_6			79.4 – 86.18	
CO_2/C_3H_8			425.4 – 457.3	
CH_4/C_3H_6			2.3	
C_3H_6/C_3H_8			5.3	
CH_4/N_2			1.19	1.25

Oxygen Enrichment of Air

Oxygen enrichment of air has also been conducted in the spiral permeator using a cellulose acetate spiral wound modulus by Separex ($S = 400cm^2$). Experimental results are shown in Figure 8, oxygen enrichment up to 42% has been obtained already for a $\Delta p = 5$ bar.

Oxygen Enrichment of a Typical Industrial Mixture

An O_2/N_2 mixture containing 10.7% O_2 has been used as feed current and, as shown in Figure 9, oxygen enrichment up to 25% has been achieved for a $\Delta p = 6$ bar.

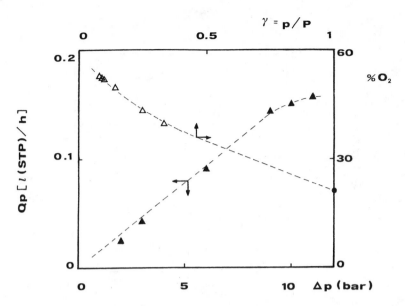

Fig. 7. Oxygen enrichment of air and corresponding flow rate on FSP.
P and p are feed and permeate pressure, respectively.

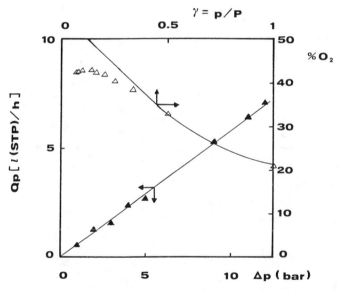

Fig. 8. Oxygen enrichment of air and corresponding flow rate - CASWM.
Continuous line represents simulated behavior.

CO_2 Concentration Processes

Two CO_2 concentration processes using CO_2/CH_4 mixture containing 30%
CO_2 and $CO_2-N_2-O_2$ mixture containing 29.9% CO_2 and 15% N_2 have been conducted. In the former a CO_2 enrichment up to 82% has been reached (Figure
10), while in the latter (Figure 11) a 70% CO_2 concentration could be
reached.

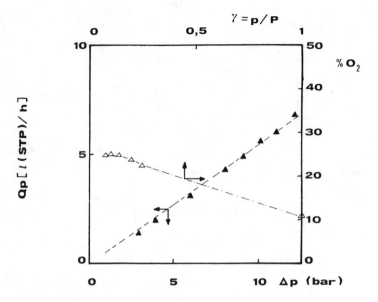

Fig. 9. O_2 enrichment of O_2/N_2 mixture (10.7% O_2) and corresponding flow
rate - CASWM.

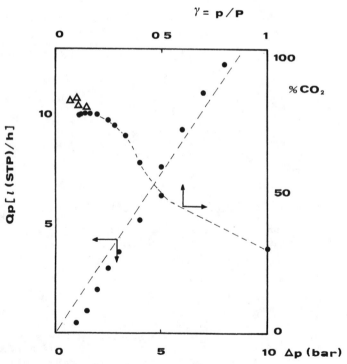

Fig. 10. CO_2 enrichment of CO_2/CH_4 mixture (30% CO_2)-Δ is obtained by
vacuum in permeate flow - CASWM.

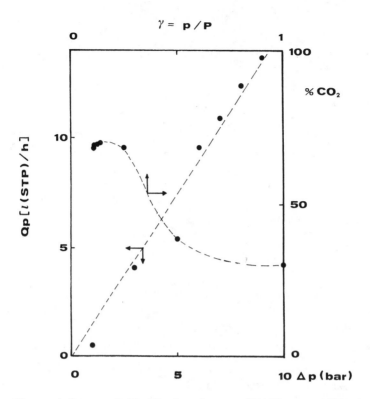

Fig. 11. CO_2 enrichment of CO_2-N_2-O_2 mixture (29.9% CO_2, 15% N_2) - CASWM.

Table 3. Composition of Gas Streams in the Purification of a
Cloro-methane Process Feed Current - CASWM Single.

Current Gas	F % Vol	P % Vol	R
CH_4	86.90	91.07	
O_2	0.18	0.28	
N_2	5.60	5.44	
C_2H_6	5.20	2.50	
C_3H_8	1.50	0.49	
iso-C_4H_{10}	0.23	0.07	
n-C_4H_{10}	0.31	0.11	
iso-C_5H_{12}	0.04	0.02	
n-C_5H_{12}	0.03	0.02	
P (atm)	2.0	2.0	
Q (l/h)			10

As expected theoretically, a linear relationship between flow rate and
ΔP (P feed - p permeate) has been obtained in all experimental tests.

Some experiments have then been performed for reducing the concentration of hydrocarbon components heavier then CH_4 in a cloro-methane

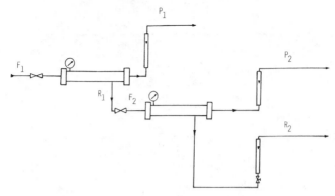

Fig. 12.   Cascade configuration of methane current concentration. F, P and R are feed, permeate and residue stream, respectively of the various stages.

process feed current at different pressure drop. A single spiral permeator (Table 3) as well as a cascade configuration (Figure 12, Table 4-6) were used.

A considerable reduction of heavy components has been obtained, in a current where also a slight concentration reduction is economically valuable.

The simulation of oxygen enrichment of air for a single stage spiral wound modulus has then been performed, solving numerically Equation (3) for

Table 4.   Composition of Gas Streams in the Purification of a Cloro-methane Process Feed Current – CASWM in Cascade.

Current Gas	F_1 % Vol	P_1 % Vol	P_2 % Vol	R_2 % Vol
CH_4	99.90	99.93	99.93	99.89
C_2H_6	0.067	0.05	0.05	0.0691
C_3H_8	0.0162	0.01	0.015	0.0244
iso-C_4H_{10}	0.0081	0.005	0.0061	0.0091
n-C_4H_{10}	0.001	–	–	0.003
P (atm)	1.5	1.0	1.0	1.47
Q (l/h)		0.066	0.138	40.0

Table 5.   Composition of Gas Streams in the Purification of a Cloro-methane Process Feed Current – CASWM in Cascade.

Current Gas	F_1 % Vol	P_1 % Vol	P_2 % Vol	R_2
CH_4	99.90	99.95	99.93	
C_2H_6	0.067	0.0381	0.0469	
C_3H_8	0.0162	0.0109	0.0153	
iso-C_4H_{10}	0.0081	0.0033	0.0051	
n-C_4H_{10}	0.001	–	0.002	
P (atm)	2.5	1.0	1.0	2.46
Q (l/h)		0.257	0.45	40.0

Table 6. Composition of Gas Streams in the Purification of a Cloro-methane Process Feed Current - CASWM in Cascade.

Current Gas	F_1 % Vol	P_2 % Vol	R_2	P_1
CH_4	99.90	99.93		
C_2H_6	0.067	0.0474		
C_3H_8	0.0162	0.0161		
iso-C_4H_{10}	0.0081	0.006		
n-C_4H_{10}	0.001	0.001		
P (atm)	4.0	1.0	3.96	
Q (1/h)		0.872	40.0	0.358

counter-current flow pattern and no purge on permeate side (Figure 6). In Figure 8 the experimental as well as the calculated values of air enrichment are reported (continuous lines). A fair argument between experimental and calculated oxygen enrichment can be noted especially at high values of γ. To improve the fitting between calculated and experimental results an optimization technique could be used.

CONCLUSIONS

On the basis of the work performed so far, the following conclusions can be drawn:

- Membrane permeation as a separation technique for all cases considered appears very promising. In particular, oxygen enrichment up to 52.5% in flat membrane and up to 43% in spiral wound moduli has been reached, while 82% has been obtained in the case of CO_2 process and, on the same modulus, the purification of a methane current has been carried out. The mathematical model tested seems to be adequate. The comparison reported in Figure 5 between experimental and predicted values for air separation with spiral moduli, results satisfactory, especially at high values of γ.

The work is still in progress, and applications to other processes are being studied.

REFERENCES

1. S. L. Matson, J. Lopez, and J. Quinu, Chem.Eng.Comm., (1983)
2. C. Y. Pan, AICHE Journal, 29:4, 545 (1963).
3. C. Y. Pan, and H. W. Habgood, Ind.Eng.Fund., 13:4 (1974).
4. A. S. Bernard, Laminar flow in channels with porous walls, J.Appl.Phys., 24:1232 (1953).

Acetic acid, carrier-facilitated transport and extraction through ion-exchange membranes, 309-317

Acetic anhydride, precipitation of microporous membranes, 128, 129

Acetone, precipitation of microporous membranes, 128, 129

Activating species, 56

Active transport of ions, 387-403

Alkaline water electrolysis
asbestos diaphragm, 319
experimental set-up, 324
performances, 323-324
polyantimonic acid based membranes, 319-325

Amicon hollow fiber membrane
experimental parameters, 248
hydraulic permeability, 246
modules, 242

Aminoacid separation by charged ultrafiltration membranes, 191-198

p-aminohippurate transport, 2-4
transepithelial, 6

Aminomannose(galactose)-cholesterol, as liposome marker, 33

Ammonia, carrier-facilitated transport and extraction through ion-exchange membranes, 309-317

Amperometric oxygen electrode, as detector in enzymatic determination of cholic acids in human bile, 543-546

Amphoteric ion exchange membranes, 85-92
cation fluxes, 88
conductances, 88
diffusional, 88, 90
electrolyte activity, 87
and conductance matrix elements, 90
experiments, 86
ion flux, 89

Amphoteric ion exchange membranes, (continued)
membrane conductance, 87
electroosmotic effect, 91
membrane potential, 87, 88
water flux, 88, 89

Angiotensin, electrolytic deionization, 307

Anhydro-4-epitetracycline, inducing Fanconi syndrome, 23-30

Anion-selective membranes, 301

Antibody
as liposome marker, 33
liposomal targeting with, 35

Antifoams
characterization, 223-224
'cloud point', 230-231
effects of flux decay, 227
effects on cross-flow filtration of microbial suspensions, 223-232
flux and differential pressure, 228
hydrodynamic conditions, 225-228
influence of
concentration, 228, 229
membrane pore size, 228
influence on retention of enzymes, 225, 226
physico-chemical parameters, 228-231
redilution experiment, 229
retention values and critical solubility, 230, 231
Selection, 223-224
temperature effect, 230-231
variation in membrane time behavior, 227

Antimicrobial therapy, liposome applications in, 36

Antiporter, 405

Aromatic polyamides, gas permeability, 615-616

Arthritis, liposome applications in, 36

Asialofetuin
    as liposome marker, 33
    pronase treated, *see* Pronase-
        treated asialofetuin
Asialoorosomucoid, as liposome
    marker, 33
Asymmetric membranes (*see also*
        Polymeric membranes),
        163-178
    facilitated transport of ions,
        77-84
Azeotropic mixtures, pervaporation
    of, 573-599

'Back-diffusion', 302
Basolateral membranes
    p-aminohippurate uptake, 2-4
    transport mechanisms of organic
        ions in, 1-8
    tetraethylammonium, 4-6
Benzalkonium chloride, 216
    interaction with membrane polymer,
        221
Benzene
    permeability of, 567, 568, 569
    pervaporation through NBR
        membrane, 338-339
Benzene-cyclohexane mixture
    separation by pervaporation,
        563-570
    influence of additives, 602-603
    sorption capacity of different
        membranes, 565
Benzene-methanol mixture
    separation by pervaporation,
        563-570
    sorption capacity of different
        membranes, 565
Bilayer membranes, planar, *see*
        Planar bilayer membranes
Bile, human, enzymatic determination
    of cholic acids in, 543-546
Biochemical industry, process
    streams in, 611
Black lipid membranes
    elution profile in complement
        experiment, 13
    fusion of complement-containing
        vesicles with, 12-14
    fusion of hemocyanin-containing
        vesicles with, 11-12
    membrane current time course
        in complement experiment, 13
        in hemocyanin experiment, 12
    preparation, 11
Boric acid, carrier-facilitated
        transport and extraction
        through ion-exchange
        membranes, 309-317
Bovine serum albumin, electrolytic
        deionization of, 306-307

Brevibacterium spec. suspension,
        effect of antifoams in
        filtration, 224, 225
Brush border membranes
    p-aminohippurate uptake, 2-4
    chromatofocusing of protein
        extracts from, 412
    coupled transport disorder, 23-30
    effect of EPI-TC, 25
    D-glucose uptake under voltage-
        clamped condition, 26, 27
    L-glucose efflux, 28
    isolation of, 1-2
    Na^+ gradient dependence of uptake
        of glycine, 411
    permeability, 26
    phospholipids, 27-28
    specific activities of marker
        enzymes, 2
    tetraethylammonium transport, 4-6
    transport mechanisms, 1-8
Bubble effect, 286-288, 294, 296,
        297
Butadiene, pervaporation through NBR
        membrane, 339-340

C_1 chemistry
    definition of, 621
    hydrogen separation membranes for,
        621-627
    role of membrane separator in, 627
Cancer chemotherapy, liposome appli-
        cations in, 36
Capillary distillation, 350
    theory of, 354-356
Capillary membrane fixed enzyme
        reactors
    analytical instruments, 525
    computational procedure, 522-523
    convection (ultrafiltration) type,
        515-516
    cross-linking of enzyme, 525-526,
        527
    diffusion type, 516
    effect of enzyme loading, 527
    given and estimated parameters,
        523-525
    hydrolysis experiments, 526-527
    immobilization, 525
    improved model for, 515-530
    kinetics, 525
    materials, 525
    theory, 516-525
    ultrafiltration device, 525
Capillary model, mass flow equation,
        328
Carbon dioxide enrichment, 641, 643
Carboxylic membranes, S18 SAFT, *see*
        S18 SAFT membranes
Cardiolipin, 21
Carman-Kozeny equation, 199, 201

Carrier-mediated process, criteria
      for, 3
Carrying species, definition of, 57
Casting solution layer, 129
Cathode polarization curves, 290
Cell break down voltage, 293-296
'Cell-pair', electrodialytic, 302
Cellulose
   conversion to ethanol, using
      membrane cell-recycling
      systems, 533-541
Cellulose acetate (see also Membrane
      polymers)
   coefficients for KCl, 48
   comparison of anionic surfactant
      solutions, 219
   coupling coefficients for KCl, 50
   differential conductance
      coefficients, 46-49
   electrical measurements, 44-45
   hyperfiltration studies, 40-52
   in DMAc, properties, 123
   osmotic flow rates, 45-46
   separation characteristics,
      209-214
   transport behavior, influence of
      surfactant/electrolyte
      system, 215
Cerebrosides, as liposome marker,
      33
Charge density, membrane, asymmetry
      in facilitated ion
      transport, 81-82
Charge ultrafiltration membranes
   aminoacid rejection, 192-195, 197
      effect of pH, 194, 195
      mixed acids, 196
      osmotic coefficients, 192, 194
      permeabilities, 192, 193, 194
      reflection coefficients, 192,
         193, 195
   characterization, 192
   in separation of aminoacids,
      191-198
Cheesemaking
   electrodialysis in whey-
      deionization, 306
   ultrafiltration in, 256, 257
   viscosity of liquid precheese, 257
Chelate membrane, clathrate uptake
      through, 493
Chelate uptake
   and reaction time, 492
   concentration dependence of,
      490-491
   graft percent dependence, 492
   time dependence of, 489-490
Chelation therapy, liposome appli-
      cations in, 36
Chemical engineering aspects of
      membrane processes, 609-612

Chemical potential gradient as
      driving force for mass flux,
      127
Chemicals, separation of, electro-
      dialysis in, 299-307
Chitin, deacetylation scheme, 138
Chitosan hollow fibers, 137-142
   burst pressure tests, 139, 141-142
   dialytic properties, 137
   diffusion coefficients, 142
   diffusion measurements, 140-141
   diffusion methods, 138
   hydraulic permeability, 142
   hydrophilic nature of, 140
   solvent permeability measurements,
      139, 141
   structure, 141
Cholesterol derivative, in liposomal
      targeting, 35
Cholic acids in human bile,
      enzymatic determination by
      amperometric oxygen
      electrode as detector,
      543-546
Clark oxygen electrode, see
      Amperometric oxygen
      electrode
Clathrate-forming metal complexes,
      483-495
   characterization by scanning
      electron microscopy, 488,
      493-494
   continuous and intermittent
      methods, 491-492
   grafting process, 486
   graft membranes, 488-489
   immobilization, 486
   pervaporation, 495
      measurement of, 486-488
   preparation, 484
   time dependence of chelate and
      clathrate uptake, 489-490
Clathrate membranes, approaches to
      obtaining, 489
Clathrate uptake, 493
   concentration dependence of,
      490-491
   graft percent dependence, 492
   through chelate membrane, 493
   time dependence of, 489-490
'Clock pump', 64
Cloromethane process feed current,
      645
   purification of, 644
Coagulation conditions, 200
Coenzyme Q
   flip-flop mechanism in bilayer
      membranes, 20
   in photochemically activated
      electron transport, 17-22
Cohesive energy balance, 479, 480

Colloids, ultrafiltration of, *see*
    Ultrafiltration, colloidal
Complement, in membrane reconstitution, 10-11, 12-14
Composite membranes, thin film, *see*
    Thin film composite
    membranes
Concentration polarization, 285-286,
    304, 447, 497
Concentration profiles, 130, 131,
    132
Concentration waves, oscillating,
    see Oscillating concentration waves
Conductance matrix, in amphoteric
    ion exchange membrane
    transport, 85-92
Converting enzyme inhibitor,
    electrolytic deionization,
    307
Cotransport
    conductance change, 407-408
    electrophysiological properties,
        406-410
    flow coupling in translocation
        step, 409-410
    mechanism of charge transfer, 413
    $Na^+/K^+/Cl^-$, 413-416
    putative models, 410-412
    stereospecific binding of
        substrate, 408-409
Cotransport-associated depolarization
    saturable nature of amplitude, 406
    voltage dependence, 406-407
Cotransporters, 405
    molecular properties, 410-412
Cross-flow filtration, effects of
    antifoams, 223-232
Current density scale-up
    bubble effect, 294
    versus
        SPE cell technologies, 281-298
        V/A characteristics, 292
        zero gap membrane cell, 281-298
Cyclohexane, permeability of, 568

Dairy industries, ultrafiltration
    in, 255-262
    evolution of membrane areas, 256
Deissler equation, 211, 212, 213-214
Depolarization, cotransport-
    associated, *see* Cotransport-
    associated depolarization
Desalination
    by electrodialysis, 305
    by hyperfiltration, 39-40
    by membrane distillation, 347
    plants, 611
Despić and Hills equation, 85, 91

Dextran
    calibration curves, 185, 186
    elution curves, 185, 186
    in ultrafiltration experiments,
        180
    retention versus molecular weight,
        188
Diabetes, liposome applications in,
    36
Dialysis, chitosan in, 137
DIDS, effect on p-aminohippurate
    uptake, 2-4
Diffusion coefficient
    and Fick's law, 127
    binary system with limited
        miscibility, 126-127
    derived from phase change, 71
Diffusivities, 'LONG-model'
    expressions, 335-336
Dipalmitoyl-PC/Chol/stearylamine
    liposomes, 32-33
Disinfection, electrodialytic, *see*
    Electrodialytic disinfection
Donnan effect, 50, 110, 309
Donnan equilibrium, in facilitated
    ion transport, 79
DPPC membranes, 455-459
    differential thermal analysis, 457
    solute permeability, 457-458
    specific permeabilities, 458, 459
Driving species, 56
DSC analysis, 507, 508, 511, 513
'Dusty-gas-model', 329
Dynamics, membrane, response to
    oscillating concentration
    waves, 69-75

Electrodes, catalytic activity of,
    288-291
Electrodialysis
    apparatus, 302-304
    applications, 305-307
    applied cell voltage, 267
    'back-diffusion', 302
    cell-pair, 302
    commercial use, 299, 300, 301
    concentration polarization,
        285-286, 304, 447, 497
    continuous operation, 299-300
    disinfection by, *see* Electro-
        dialytic disinfection
    in deionization of saline water,
        305
    in seawater concentration, 305
    in separation of chemicals,
        299-307
    intermembrane turbulance-promoting
        gasket, 303
    laboratory test unit, 265
    limiting voltage, 304-305
    modern membranes, 300-302

Electrodialysis (continued)
  multi-compartment process, 300
  of dilute strontium cations,
      263-272
  polarization effects, 305-306
  symmetrically reversing type
      plants, 305
Electrodialytic disinfection
  germicidal efficiency, 424, 425,
      426
  schematic diagram, 422
  viability cells and current
      density, 427
Electrolysis, alkaline water, see
      Alkaline water electrolysis
Electrolytes
  transport behavior through
      synthetic membranes,
      influence of surfactants,
      215-222
  uni-univalent system, facilitated
      transport equations, 79-80
Electroneutrality condition, 80
Electron transport
  cadiolipin in, 21
  photochemically activated, 17-22
Energy dispersive analysis of
      X-rays, 320
  polyantimonic acid based
      membranes, 321-322
Energy-efficient membrane distil-
      lation process, 343-348
Enthalpy balance, 638
Enzyme entrapment, liposome
      applications in, 36
Enzyme reactors, capillary membrane,
      see Capillary membrane fixed
      enzyme reactors
Enzyme replacement therapy, liposome
      applications in, 36
EPI-TC, inducing Fanconi syndrome,
      see Fanconi syndrome,
      experimental
Epithelial cell membranes, Na^+-
      coupled cotransport in,
      405-416
Escherichia coli suspension, effect
      of antifoams in filtration,
      224
Ethanol
  aqueous
      pervaporation of, 549-562
      separation by thin film
          composite membranes, 381-382
  production from cellulose, using
      membrane cell-recycling
      systems, 533-541
Ethylendimethacrylate, 234
Euglobins, precipitation from plasma
      by electrodialysis, 300

Evaporation, effect on membrane
      properties, 134-135
Exchange fluxes, S18 SAFT membrane,
      105, 106
Extraction, carrier-facilitated,
      through ion-exchange
      membranes, 309-317

Facilitated transport, see under
      Transport
Fanconi syndrome, experimental
  coupled transport disorder in,
      23-30
  distribution space of D-glucose,
      25
  enzymatic characterization of
      vesicle preparation, 24-25
  materials and methods, 23-24
  permeability of BBMV, 26
  phospholipid analysis, 24
  time course of D-glucose uptake,
      26
  uptake measurements, 23-24
Fermentation processes, antifoam
      use, 223-224
Fick's law of diffusion, 69, 127,
      141 550, 595
Film-theory model, 180
  equation, 211, 213-214
Filtration properties, phase
      inversion membranes, 123,
      124
Final-porous model, 181
Finger-structured membranes, 133,
      134, 156
  PDC polymer, 148
  polyantinomic acid based
      membranes, 319
  polyvinylidene fluoride, 165
Flat sheet separation, 639
Flavinmononucleotide, 17-18
Flip-flop mechanism, in bilayer
      membranes, 20
Flocculating colloids, in unstirred
      ultrafiltration, 199-208
Fluid polymer layer, 129
Flux-time behavior, flocculating
      colloids in unstirred
      ultrafiltration, 199-208
Formic acid, in precipitation of
      microporous membranes, 128,
      129
Friction coefficients
  binary molar, 49
  in membrane polymers, 39
Fujita free-volume theory, 595
Fungal diseases, liposome appli-
      cations in, 37

Galactose-PE, as liposome marker, 33
Ganglioside, as liposome marker, 33

Gas-liquid permporometry
  application to characterization of
        inorganic ultrafilters,
        629-633
  helium permeability versus pore
        radius, 632
  pore size distribution, 631-632
Gas separation technology, porous
        membranes in, 327-333
Gas sticking phenomenon, *see* Bubble
        effect
Gas transfer, control in polymeric
        membranes, 613-620
Gaucher's disease, liposome appli-
        cations in, 37
Gel permeation chromatography, 210
Glass membranes
  microporous asymmetrically
        structured, 115, 116
  thermogelation in preparation of,
        117
Glycolipids, as liposome markers, 33
Glycoproteins, as liposome markers,
        33
Goldmann's approximation, 95
Goldmann's hypothesis, 93, 94
Graetz problem, 517
Guanidine hydrochloride, recovery by
        electrodialysis, 300

Hagen-Poiseuille's law, 508, 526,
        550
  equation, 497, 498, 500
Hamaker constant, 203
Hemocyanin
  cation selectivity, 11
  in membrane reconstitution, 10-12
Hemodialysis, PMMA hollow fiber
        membrane for, 507-513
Henderson's equation, 109
Henry's law, 335, 336
n-Heptane, pervaporation through NBR
        membrane, 338-339
Hittorf's transport numbers, 267
Hollow fiber reactors with yeast
        cells, 241-254
  batch hydrolysis, 242-243
  continuous hydrolysis, 244,
        246-247
  experimental apparatus, 245
  experimental parameters, 248
  glucose production versus
        hydrolysis time, 245
  hydrolytic yield, 249
  Lineweaver-Burk plots, 244
  Michaelis-Menten plots, 243
  plug-flow scheme, 249
  simplified model, 247-251
Hollow fibers, outer-skinned,
        151-161
  air gap in spinning, 156, 159

Hollow fibers, outer-skinned,
        (continued)
  characterization, 153
  diameter estimates, 158
  diffusion phenomena, 158
  flow rate in spinning, 154-155,
        156, 158
Hydraulic permeability, 154, 155,
        157, 158, 159, 160
  coefficient, 153
  influence of additives, 153-154
  nature of core fluid, 153-154
  polymer materials, 152
  principles of preparation, 152
  spinning line, 152-153
  spinning variables, 154-159
  wall thickness, 160
  wet-dry spinning technique, 151
Hydraulic permeation, *see* Reverse
        osmosis
Hydrocarbons
  permeation through liquid
        membranes, influence of
        surfactants, 475-480
  separation of mixtures by
        pervaporation, 335-341
Hydrogels, polymeric, with entrapped
        yeast cells, 233-240
  flux decay rate, 235-236
  invertase activity, 238, 239
  kinetic characterization, 236-239
  membrane performances, 235-236
  permeate flux, 237
    versus applied pressure, 237
    versus process time, 236
Hydrogen separation membranes
  asymmetric, permeation rates and
        selectivities, 623
  basic properties of materials, 622
  composition and pressure ratio,
        625
  for C_1 chemistry, 621-627
  permeability and temperature, 624
  preparation, 622
  pure gas permeation test, 622-624
  separation of H_2/CO binary
        mixtures, 624-627
  three-stage separator, 627
Hydroxymethylmethacrylate, (*see also*
        Hydrogels, polymeric), 234
Hyperfiltration
  cellulose acetate, 45
  desalination by, 39-40
  membrane polymers, 40-52
  non-equilibrium thermodynamics,
        40, 41-44
Hypothermia, local, in liposomal
        targeting, 35

Ideal tracer, law of, 103
Immunopotentiation, liposome
        applications in, 36

Infrared spectroscopy, in study of
    water structure in membrane
    models, 361-370
Interferon eluate, reduction of salt
    concentration by electro-
    dialysis, 300
Intra-articular administration,
    liposome applications in, 36
Ion-exchange membranes (*see also*
    Ion-selective membranes)
  carrier facilitated transport and
      extraction, 309-317
  chloride ion concentration, 278
  chloride ion mobility, 279
  critical conditions, 314, 316
  critical fluxes, 317
  electric conductivity, 278
  electric resistance measurements,
      274-276
  electric resistivity, 278
  electrodialytic disinfection, *see*
      Electrodialytic disinfection
  experimental fluxes, 315
  experimental systems, 314-317
  flux and permeant concentration,
      311, 312
  hydrogen ion concentration, 278
  hydrogen ion mobility, 279
  inorganic, 463-467
    classification of, 462
    heterogeneous, 462, 463
    homogeneous, 463
  inorganic-organic
    classification of, 463
    heterogeneous, 469-470
    homogeneous, 467-469
  measuring cell, 274
  new inorganic and inorganic-
      organic, 461-473
  obtained by crystals, 464-465
  passive fluxes, 317
  polarization phenomena, 314
  potential distribution
      measurement, 275-276
  proton permselectivity, 465, 466
  single crystal, 463-464
  water dissociation effect in
      disinfection of *E.coli*,
      421-417
  with amorphous network, 462,
      466-467
  with crystalline network, 462
Ionic channels, reconstitution into
    planar lipid membranes, 9-14
Ionic concentration, effect on IR
    and NMR parameters of
    reverse micelles, 365
Ionic flux simulation, in
    facilitated ion transport,
    82-84

Ionic permeability
  passive membranes, 94-95, 97
  S18 SAFT membrane, 101-112
Ions (*see also* Ion-exchange
    membranes; Ion-selective
    membranes)
  active transport, 387-403
  facilitated transport in
    asymmetric membranes, 77-84
    general equation, 77-79
    membrane charge density, 81
    partition coefficient, 81
    simulation of ionic flux, 82-84
    uni-univalent electrolyte
      system, 79-80
  friction coefficients in membrane
    polymers, 39
  organic, *see* Organic ions
  transports through amphoteric ion
    exchange membrane, 85-92
Ion-selective membranes, (*see also*
    Ion-exchange membranes)
  for redox-flow battery, 273-280
  industrial applications, 301
Isobutene, pervaporation through NBR
    membrane, 339-340

Knudsen-diffusion, 327, 329, 332,
    589
  thermodialysis, 355
Kozeny-Carman's equation, 512
Kynar (polyvinylidene fluoride)
    membranes, 469
  proton permselectivity, 471

Lactose-PE, as liposome marker, 33
Lipid membranes
  black, *see* Black lipid membranes
  planar, reconstitution of ionic
    channels into, 9-14
Lipid quinones, in photochemically
    activated electron
    transport, 17-22
Liposomes
  medical application of, 31-37
    clearance from blood
      circulation, 32
    criteria for, 31-32
    encapsulated invertase
      distribution, 35
    glycolipids as markers, 34
    in vivo, 32
    lactocerebroside as marker, 34
    local hypothermia, 35
    marker molecules, 33
    pronase-treated asialofetuin
      injection, 33
    stability in blood circulation,
      32
    targeting, 32-37
    with antibody, 35

Liposomes (continued)
  unilamellar
    electron transport across, 17–22
    uptake of picrate anion
      due to redox reaction,
      401–402
Liquid membranes, influence of
      surfactants on permeation of
      hydrocarbons through,
      457–480
Liquid permeation apparatus,
      automatic, 487
Loeb-type CA membranes, 215
'LONG-model' expressions for
      diffusivities, 335–336
Lysosomal storage disease, liposome
      applications in, 36

Mass balance equation, 637–638
Mass transfer coefficient, 180, 181
Mass transport
  non-isothermal, see Non-isothermal
      mass transport
  zetameter, 203
Membrane bonded electrode, (see also
      SPE cell technology), 281
Membrane cell-recycling systems
  analytical methods, 535–536
  biomass and enzyme activity,
      537–541
  conversion of cellulose to
      ethanol, 533–541
  effect of microfiltration, 536–537
  effect of ultrafiltration, 537
  fermentation conditions, 535
  fermentation stages, 534
Membrane charge density, asymmetry
      in facilitated ion
      transport, 81–82
Membrane distillation (see also
      Pervaporation)
  applications, 347–348
  conductance distillate versus
      concentrate solution, 347
  energy-efficient process, 343–348
  influence parameters, 344–345
  membrane requirements, 343–344
  realization of process, 345–347
  versus temperature solution, 345
  with heat recovery, 346
Membrane gas separation process,
      635–645
  carbon dioxide concentration
      processes, 641
  co-current permeator, 636
  counter-current permeator, 636
  enthalpy balance, 638
  experimental results, 638–645
  flat sheet separation, 639
  mass balance equation, 637–638
  oxygen enrichment, 636, 637, 640,
      644–645

Membrane gas separation process,
      (continued)
  permeation rate equation, 635–637
  pressure drop equation, 638
  separation factors, 640
  spiral wound module separation,
      639
Membrane models, water structure in,
      361–370
Membrane polymers
  cellulose acetate, see Cellulose
      acetate
  conduction coefficients, 41–43
  coupling coefficients for KCl, 50
  flux coupling, 49–51
  friction coefficients of water and
      ions, 39–52
  hyperfiltration studies, 40–52
  osmotic flow rates, 45–46
  specific properties, 40–41
Membrane reactors
  hydrogel films in (see also
      Hydrogels, polymeric),
      233–240
  in dairy industries, 260–262
  performance, 235–236
  preparation of membrane, 235
  yeast cells in, 233–240, 241–254
Membrane reconstitution, 9–14
  buffers, 10
  materials and methods, 10–11
  planar bilayer experiments, 11
  reagents, 10
Membranes (see also individual types
      of membrane)
  asymmetric, see Asymmetric
      membranes
  passive, see Passive membranes
Metal membranes, thermogelation in
      preparation of, 117
Metal storage disease, liposome
      applications in, 36
Methane current concentration, 644
Methanol
  permeability of, 567, 569
  precipitation of microporous
      membranes, 128, 129
Micelles, reverse, see Reverse
      micelles
Michaelis constant, 233
Michaelis–Menten Kinetics, 410, 415,
      518
  rate equation, 238
Microbial suspensions, cross-flow
      filtration, effects of
      antifoams, 223–232
Microcapsules, polyactide, see
      polylactide microcapsules
  effect of pH on degradation,
      372–374

Microfiltration membranes
  addition of nonsolvent to
      homogeneous polymer
      solution, 118-119
  recycling systems, 536-537
  structure, 115, 116, 120, 121
  symmetrical, 115, 116
Microporous membranes, (see also
      Phase inversion membranes)
  addition of nonsolvent to
      homogeneous polymer
      solution, 118-119
  asymmetric, structure, 121
  characterization of structure, 120
  evaporation of volatile solvent
      from three-component polymer
      solution, 117-118
  experimental procedures, 119-120
  experimental results, 120-135
  miscibility gap, 117, 118, 125
  phase inversion, see Phase
      inversion membranes
  structural variation, 119
  symmetric, structure, 120, 121
  thermodynamics of formation, 117,
      118, 119
  thermogelation of two-component
      mixture, 116-117
  transport properties, 120
Milk
  electrodialysis in whey-
      deionization, 306
  organoleptic defects, 259
  pseudo-plastic rheological
      behavior, 256
  ultrafiltration treatment, 255,
      259, 260
Millipore membranes, 93, 455-459
  differential thermal analysis, 457
  solute permeability, 457-458
  specific permeabilities, 458, 459
Miscibility gap, 117, 118, 125
M.M.V. cheesemaking processes,
      ultrafiltration, 256, 258
Molecular sieve model, 484
'Moon-Light Project', 273
Multimembrane composite systems,
      transport-reaction in, 65

Nafion membrane
  critical and passive fluxes, 317
  facilitated extraction of ammonia,
      315
NASICON membranes, 470
  proton permselectivity, 471
Nelson colorimetric method, 234
Nernst-Einstein relation, 85, 91
Nernstian responses, deviations
      from, in ion-exchange
      membranes, 109
Nernst-Planck equation, 93, 310

Neutrality disturbance phenomenon,
      423, 426
NMR see Nuclear magnetic resonance
Non-isothermal mass transport
  bulk water temperature and
      rotation number, 592
  diffusion coefficient, 592-594
  in hydrophobic porous membranes,
      587-594
  membrane tortuosity, 593-594
  of acidic aqueous solutions,
      permeate concentration, 589
  of alcohol, permeate concen-
      tration, 589
  pure water flux, 590-594
  temperature at water/membrane
      interface, 592
  temperature difference and water
      flux, 593
Non-solvent precipitation, in
      preparation of phase
      inversion membranes, 119-120
Nuclear magnetic resonance
  in study of water structure in
      membrane models, 361-370
  measurement of broad-line
      spectrum, 614-615
  quadrupole splitting measurements,
      368
Nuclear waste solution, electro-
      dialysis of, 263-272
Nucleopore membranes, 93, 96, 441
  pore distribution, 441
  water flow distribution, 442

Onsager reciprocity principle, 49
Orange-peel effect, 553
Organic ions, transport mechanisms,
      1-8
Oscillating concentration waves
  experimental studies, 73-74
  mathematical models, 70-73
  response of membrane systems,
      69-75
Oscillating reaction pump, 59
Outer-skinned hollow fibres, see
      hollow fibres, outer-skinned
Oxygen electrode, amperometric, see
      Amperometric oxygen
      electrode
Oxygen enrichment, 640, 641, 642,
      644-645

Palmitoyl homocysteine, in liposomal
      targeting, 35-36
Partition coefficient, asymmetry in
      facilitated ion transport,
      81
Passive membranes
  ionic permeability, 94-95, 97
  ion transport, 93-99

Passive membranes (continued)
  porosity, 93-99, 96, 98
  selectivity, 93-99
  system permeability, 94-95
  transport numbers, 96, 97
PDC polymer, 143-150
  configuration of membrane, 148
  effects of additive reagents, 145,
      146
  effects of concentration, 145, 146
  evaporation time, 147
  gelation temperature, 147
  hydrolytic stability, 144
  hydrophilic property, 144
  Loeb-Sourirajan technique, 144
  pore structure, 148
  properties, 144
  relative humidity, 147, 148
  thermal stability, 144
Pellicular zirconium phosphate, 465
Permeameter, 631
Permeation apparatus, automatic, 487
Permeation rate equation, 635-637
Permporometry, gas-liquid, see
      Gas-liquid permporometry
Pervaporation, 483, 495
  and capillary distillation, 350
  azeotropic mixtures, 581, 586
  boundary conditions, 605
  calculation of processes, 599-601
  combined with distillation,
      process schema, 582
  compared with reverse osmosis,
      598, 604-606
  'dynamic membrane' character-
      istics, 604
  economic advantages, 582-583
  economics of industrial processes,
      581-586
  efficiency of, 564
  fluxes, 575-579
  industrial separation of
      azeotropic mixtures by,
      573-579
  in separation of benzene-methanol
      and benzene-cyclohexane
      mixtures, 563-570
  measurement of, 486-488
  of ethanol-water mixtures,
      polyurethane membranes for,
      549-562
  of hydrocarbon mixtures through
      rubbers, 335-341
  performances under varied
      conditions, 595-607
  permeate flux, 606
    influence of feed composition,
      602
    pressure dependence of, 602
  plant for EtOH-dehydration, 578
  selectivity, 336, 574-575

Pervaporation (continued)
  selectivity (continued)
    influence of feed composition
      and temperature, 601
  pressure dependence of, 600
  selectivity factor, 550
  separation characteristics, 600
  separation potential, 595-597
    enhanced by additives, 601-604
  sorption coefficients, 598
  standard-module, 577
  systems, influence of flux on
      costs, 576
  technical advantages, 582-583
  transport parameters, 597, 599
  water/ethanol through cellulose,
      603
Phase inversion membranes (see also
      Microporous membranes)
  characterization of, 429-446
  experimental results, 120-135
  filtration properties, 123
  finger-like structures, see
      Finger-structured membranes
  polymer concentration, 122
  polymer types, 122
  porosities, 123
  precipitation process, 129-133
  precipitation rate, 120, 122-124
  preparation by non-solvent
      precipitation, 119-120
  selection of polymer-solvent-
      precipitant systems, 133-134
  structures, 122, 129-133
    related to polymer concentration
      in casting solution, 134-135
  sponge-like, 133
Phase-inversion process
  control by kinetic phenomena, 174
  experimental procedures, 119-120
  in preparation of microporous
      membranes, 115-135
  kinetics, 126-129
  phenomenology of membrane
      preparation, 116-119
  polyvinylidene membranes, 163, 164
  thermodynamics, 124-126, 129-133
Phosphatidyl serine, 21
Phospholipid-containing membranes,
      27-28, 455-459
  differential thermal analysis, 456
  solute permeability, 457-458
  specific permeabilities, 458, 459
Phospholipid polar heads, in reverse
      micelles, 363, 367
Photoredox reaction
  and ion transport, 395-400
  anion transport in liquid
      membrane, 399
  as function of light intensity,
      400

Planar bilayer membranes
  lipid quinone transport across,
    17-22
  reconstitution experiments, 11
Plasma proteins, electrolytic
    deionization, 307
Polarization
  effects in electrodialysis,
    305-306
  ion exchange membranes, 314
  of concentration, 285-286, 304,
    447, 497
Polyamide(s)
  aromatic, see Aromatic polyamides
  in DMAc, properties, 124
  in DMSO, properties, 123
Polyamide membranes, 432-438, 439,
    440
  asymmetric structure, 435
  cast at 30°C, 434, 435
  chemical structure of pore-former,
    436
  composition of casting solution,
    437
  control of pore formation, 432-438
  difficulty of dissolving
    polyamide, 432
  precipitation rate, 436
  temperature of bath, 437
Polyantimonic acid based membranes
  characterization, 320-321
  experimental results, 321-325
  in alkaline water electrolysis,
    319-325
  preparation, 320
  resistance, 322-323, 324
Polyarylsulfone membranes see also
    PDC polymer), 143-150
Polybuffer exchanger, 410
Polybutadiene acronitrile rubber,
    pervaporation through,
    338-340
Polyethylene, microporous, 117
Polyethyleneglycol, 234
  as antifoam, 224
  calibration curves, 185, 186
  elution curves, 185, 186
  for selectivity improvement in
    pervaporation, 602-603
  in ultrafiltration experiments,
    180, 181, 183, 185-187
  retention versus molecular weight,
    188
Polyhydroxymethylmethacrylate (see
    also Hydrogels, polymeric),
    234
Polylactide microcapsules, 371-378
  degradation, 371-372
    effect of pH on, 372-374
    effect of temperature on, 374
    effect of urea, neutral salts
      and enzyme on, 375-377

Polylactide microcapsules
    (continued)
  degradation (continued)
    estimation of, 371
    gel permeation chromatogram, 374
    lactic acid generated from, 375,
      376
    preparation of, 371
Polymeric hydrogels, see Hydrogels,
    polymeric
Polymeric membranes (see also
    Polyamide membranes;
    polysulfone membranes)
  asymmetric, see Asymmetric
    polymeric membranes
  control of gas transfer, 613-620
  gas flow equations, 440
  intermolecular packing state, 615
  measurement of broad-line NMR
    spectrum, 614-615
  measurement of gas permeation
    coefficient, 615
  phase inversion, see Phase
    inversion membranes
  pore distribution, 438-445
  water distribution, 443, 444, 445
Polymer relaxation, effect on
    membrane properties, 134-135
Polymers
  concentration in casting solution,
    and membrane structure,
    134-135
  concentration profiles through
    precipitating membrane, 130
  flexible, gas permeability of, 616
  membrane, see Membrane polymers
Polymethylmethacrylate hollow-fiber
    membranes
  freezing-melting behavior, 507
  heating DSC curve, 509
  melting peak temperature, 511
  permeability, 511-513
  pore radius, 508, 512
  porosity versus permeability, 510
  structure, 511-513
  tortuosity factor, 511, 512
Polyoxyethylene-polyoxypropylene,
    216
Polypeptide solutions, deionization
    by electrodialysis, 300, 307
Polypropylene glycol, as antifoam,
    223-224, 225-228
Polystyrene derivatives, gas
    permeability of, 616-617
Polystyrenes, t-butylated
  broad-line NMR spectra, 618-619
  density and degree of
    t-butylation, 620
  gas permeability, 617-619
  glass transition temperatures, 620
Polysulfone, in DMF properties, 123

Polysulfone membranes, 429–432
cast at 30°C, 430
cast at 50°C, 432, 433
casting bath temperature and flow
rate, 433
cast in water, 430
control of pore development,
429–432
preparation, 320
resistance, 322, 323
separation characteristics,
209–214
temperature of bath, 437
Polytetrafluoroethylene membranes,
484
in non-isothermal mass transport
study, 587–594
in thermal separation of liquid
mixtures, 349–357
Polyurethane membranes
azeotropic ethanol-water
dehydration test, 561
characterization and evaluation,
551–552
effect of synthesis variables,
556, 557
evaporation time
and membrane structure, 554
and porosity, 555
influence of feed composition,
556–560
membrane thickness and permeation
flux, 557
M type, 553–561
MPE type, 552–553
permiation flux and ethanol
concentration, 557, 559
polymer concentration and membrane
structure, 554
selectivity data, 560–561
sorption results, 557, 558
synthesis for pervaporation of
ethanol-water mixtures,
549–562
Polyvinylidene fluoride based
membranes (see also
Polymeric membranes),
163–178
burst pressure evaluation, 164
effect of additives
in casting solution, 165–169
in coagulation bath, 172–173,
175
effect of coagulation bath
temperature, 176
effect of exposure condition
on burst pressure, 173
on flux, 169–172
on rejection, 170, 172
effect of solvent, 164–165
effect of temperature of
coagulation medium, 173–174

Polyvinylidene fluoride based
membranes
evaporation conditions of casting
solution, 170–171
finger-like structure, 165
morphology studies, 164
performance, 165–166
preparation, 164
rejection reduction, 168
ultrafiltration performance, 164
Polyvinylidene fluoride (Kynar), 469
proton permselectivity of
membranes, 471
Polyvinylpyrrolidon, in ultrafiltration experiments, 180
Polyvinylpyrrolidone, 234
Pompe's disease, liposome
applications in, 37
Pore densities, 94
Pore flow theory, for ultrafiltration membranes,
448–449
Porosity
of passive membranes, 96, 98
of phase inversion membranes, 123,
124
Porous membranes
cascade, 330
'dusty-gas-model', 328, 329
in gas separation technology,
327–333
in module arrangements, 330–332
mass flow through, 328
pressure level and pressure ratio,
332
pressure ratio and cascade size,
331
process-parameters affecting
separation factor, 328–329
'pseudo-capillary-model', 328
rectification, 330
separation factor and cascade
size, 331
separation processes, influence of
numbers of stages, 331
Potassium, ion transport numbers
compared with sodium, 110
Precipitant, concentration profiles
through precipitating
membrane, 130
Pressure drop equation, 638
Pronase-treated asialofetuin as
liposome marker, 33
subcellular distribution, 33
Protein(s)
deionization by electrodialysis,
300
salting-out by electrodialysis,
300, 307
Proteoliposomes, in reconstitution
of ionic channels, 9–14

Pseudo-capillary-model, 328
Pumps
  C*-C, 59, 63
  D*-D, 59, 63-64
  D-sink, 62
  D-source, 62
  oscillating reaction, 59
  primary, 65
  S-sink, 64
  S-source, 64
  secondary, 62, 65
  time-oscillation, 64
  uphill transport, 58, 60-62, 65

Radioimmunodetection, liposome
      applications in, 36, 37
Radionuclides
  effect of Sr complexing agents,
      270-271
  electrodialysis of, 263-272
  laboratory test unit, 265
  pH variation, 266
  transport numbers, 267, 269
RAI membranes, 264
Redox couple, 63
Redox-flow battery, ion-selective
      membranes, 273-280
Redox reaction (see also photoredox
      reaction), 388-403
  action spectra, 398
  and transport in liquid membrane,
      388-396
  as function of reducing agent, 401
  rate
    compared with K^+ transport rate,
      395
    dependence on various anions,
      392
    under varying perchlorate
      concentration, 394
  transmembrane, see transmembrane
      redox reaction
  uptake of picrate anion in
      liposomes, 401
Relaxation time, longitudinal, 365,
      367
Renal brush border, see Brush border
Reverse micelles
  aminoacid effects, 364
  effect of water-soluble chemical
      species on, 364-369
  electrolyte effects, 365
  electrostatic screening, 367
  ion binding, 367
  peptide effects, 364
  phospholipid polar heads, 363, 367
  water in, 362-363
Reverse osmosis
  boundary conditions, 605
  compared with pervaporation,
      604-606

Reverse osmosis (continued)
  permeate flux, 606
Reverse osmosis membranes (see also
      Thin film composite
      membranes)
  addition of nonsolvent to
      homogeneous polymer
      solution, 118-119
  asymmetrically structured, 115,
      116
    correlation with dense
      membrane, 452-453
  dense, nature of, 451-452
  evaporation of volatile solvent
      from three-component polymer
      solution, 117-118
  exclusion term, 454
  frictional interpretation, 453-454
  kinetic term, 454
  structure, 121
  transport equations, 447-448
  transport mechanism, 450-454
  transport parameter estimation,
      450
Rubbers, hydrocarbon separation by
      pervaporation through,
      335-341

S18 SAFT membrane
  biionic membrane potentials, 104,
      105
  cationic character, 104
  cationic transport numbers and
      electrolytic concentration,
      107
  dissociation versus pH, 111
  exchange flues, 105, 106
  experimental details, 102-103
  ionic diffusion coefficients, 110
  ionic permeability, 101-112
  ionic self diffusion, 112
  ionic transmembrane fluxes,
      102-103
  potentiometric responses, 109
  transport numbers, 101, 105-108
  water structure inside, 109
  zero current membrane potential,
      102, 103-105
Salt water, desalination of, see
      Desalination
'Salting-out' effect, 129
Scale-up of membrane processes,
      609-612
Scanning electron microscopy, 320
  polyantimonic acid based
      membranes, 321-322
Scatchard-Hildebrand theory of
      solubility, 475
Seawater, electrodialytic
      concentration, 305

SeCDAR-analysis
  in classification of transport-
      reaction systems, 55-67
  notations and conventions, 56,
      60-61
  SeCDAR-1 systems, 62
  SeCDAR-2 systems, 62-63
Separation factor (*see also*
      Pervaporation selectivity),
      336
  porous membranes, 328-329
Silicon oil emulsion, as antifoam,
      224
Skin-type membranes
  addition of non-solvent to
      homogeneous polymer
      solution, 118-119
  concentration profiles of
      precipitant, 132
  finger-like structure, 133
  sponge-like structure, 133
Sodium, ion transport numbers
      compared with potassium, 110
Sodium chloride
  and amphoteric ion exchange, 85-92
  surfactant rejection, 219, 220
Sodium dodecylbenzenesulfonate, 216,
      217, 218
Sodium gluconate, concentration
      change in cold cell, 591
Sodium 1-octanesulfonate, 216
Solid polymer electrolyte, cell
      technology, *see* SPE cell
      technology
Solid polymer layer, 130-133
Solubility, Scatchard-Hildebrand
      theory, 475
'Solution-diffusion' process, 39-40,
      335
Solvent concentration profiles
      through precipitating
      membrane, 130
Solvent AR-16, influence of membrane
      thickness on selectivity
      coefficient, 479
Sorption-diffusion model, 595
SPE cell technologies
  and current density scale-up,
      281-298
  and zero gap membrane, 281-298
  bubble effect, 286-288, 292, 294,
      296, 297
  catalytic activity of electrodes,
      288-291
  cathode potential versus current
      density, 291
  cell break-down voltage, 289,
      293-296
  cell configuration, 281
  electric field, 284-285
  electrolyte ohmic drops, 298

SPE cell technologies (continued)
  nickel cathodic current lead, 283
  permselective bilayer membrane,
      283
  schematic view of system, 284
  titanium anodic current lead, 283
  V/A characteristics, 292, 297
    versus caustic strength, 288
    versus operation time, 289
    voltage gap, 292
Specific permeability, 456
Spectroscopic studies, water
      structure in membrane
      models, 361-370
Spinning line, for outer skinned
      hollow fibres, 152-153
Spiral wound module separation, 639
Sponge-structured membranes, 133
Strontium cations (*see also*
      Radionuclides)
  electrodialysis of, 263-272
Strontium complexing agents, 270
Substrate, stereospecific binding
      of, 408-409
Sucrose, and amphoteric ion
      exchange, 85-92
Sulfatide, as liposome marker, 33
Sulfonated polysulfone
  membrane, in separation of
      aminoacids, 191-198
Surface tension, effect on membrane
      properties, 134-135
Surfactants
  anionic, influence on transport
      behavior, 216-219
  cationic, influence on transport
      behavior, 219
  characteristics, 477
  effect on
    permeation of hydrocarbons
      through liquid membranes,
      475-480
    transport of electrolytes
      through synthetic membranes,
      215-222
  nonionic, influence on transport
      behavior, 219
  rejection, 219, 220
Symmetric microfiltration membranes,
      addition of nonsolvent to
      homogeneous polymer
      solution, 118-119
Synthetic membranes, influence of
      surfactants on transport
      behavior of electrolytes
      through, 215-222

Tafel's lines, 290
Tautomers, 63
Teflon cell, 101, 102
Teorell's 'half time', 94

Tetraethylammonium, transport of, 4-6
  gradient, 5
  intravesicular acidification, 5
  membrane potential, 5
  temperature dependence, 4
Tetraethylammonium perfluoroctano-sulfonate, 216
Tetraethylene glycol, for selectivity improvement in pervaporation, 602-603
Tetraphenylphosphonium cation, 397
Thermodialysis
  experimental apparatus, 350
  heat transfer coefficients, 356
  hydrophobic membranes, 349-357
  Knudsen diffusion, 355
  theoretical predictions and experimental data, 356-357
Thermogelation, in preparation of microporous membranes, 116-117
Thin film back transport model, 199
Thin film composite membranes
  acetic acid-NaCl-water systems, 384, 385
  acetic acid-water systems, 384
  anionic charge, 385
  carriers, 393-394
  electron microscopic study, 379-385
  ethanol/alkali metal salts/water system, 382, 383
  ethanol-KCL-water systems, 384
  ethanol-NaHCO3-water systems, 384
  fabrication, 380
  mediators, 393-394
  picrate transport rate, 391
  separation of aqueous actic acid solution, 383
  separation of aqueous ethanol, 381-382
  solute transport, 379-385
  structure, 380-381
Three-component polymer solution evaporation of volatile solvent from, 117-118
Toluene-heptane system
  experimental data, 478
  selectivity coefficient, 478
Transepithelial transport
  p-aminohippurate, 6
  tetraethylammonium, 6
Translocation step, flow coupling in, 409-410
Transmembrane distillation, see Membrane distillation
Transmembrane and redox reaction
  anion transport driven by, 388-390
  driven by diffusion of ions, 390-393

Transport
  amphoteric ion exchange membrane, 85-92
  electrons, see Electron transport
  equations, 447-448
    concentration polarization, 447
    inside membrane, 447-448
  facilitated
    through asymmetric membranes, 77-84
    through ion-exchange membranes, 309-317
  ions, see under Ions
  non-isothermal, see non-isothermal mass transport
  concentration polarization, 447
  inside membrane, 447-448
Transport mechanism
  reverse osmosis membranes, 450-454
  ultrafiltration membranes, 448-449
Transport numbers
  passive membranes, 96, 97
  S18 SAFT membrane, 105-108
Transport-reaction systems
  activating species, 56
  carrying species, 56
  driving species, 56
  facilitations by S+C<=>S* equilibria, 58, 59-60
  multimembrane systems, 65
  oscillating reaction pump, 59
  passive, 58, 59
  pumps see Pumps
  representations, 58
  SeCDAR-analysis, see SeCDAR-analysis
  sinks, 58, 59, 62
  sources, 58, 59, 62
  uphill transports, 58, 60-62
  valves, 56
Triturus kidneys, 413-416
Tropical diseases, liposome applications in, 36
'Tubular pinch' effect, 199
Two-component mixture, thermogelation of, 116-117

Ultrafilters, inorganic, gas-liquid permporometry in character-ization of, 629-633
Ultrafiltration
  chitosan in, 137
  colloidal
    computer model, 201-202
    experimental details, 203-206
    final flux, 206
    floc sizes, 203
    flux-time behavior, 199-208
    mathematical modelling of, 497-505
    optimal beta-parameters, 504

Ultrafiltration (continued)
  colloidal (continued)
    parameter adoption procedure,
        501
    permeability of retentate layer,
        201
    pilot-plant apparatus, 499
    specific resistance, 206, 207
    T/V versus V curves, 205-206
    zeta potential, 203, 204, 205,
        207
  flux
    estimation by computer
        simulation, 497
    versus bulk concentration, 503
    versus temperature, 502, 503
    versus transmembrane pressure,
        502
    versus volumetric flow rate,
        502, 504
  in dairy industries, 255-262
  in membrane cell-recycling
        systems, 537
  polyarylsulfone membranes, see
        Polyarylsulfone membranes
Ultrafiltration membranes
  addition of nonsolvent to
        homogeneous polymer
        solution, 118-119
  bulk and permeate comparison, 185,
        186
  charged, see Charged ultra-
        filtration membranes
  circulation velocity, 213, 214
  concentration polarization, 212
  Deissler equation, 211, 212,
        213-214
  diffusion coefficients, 182
  film model equation, 211, 213-214
  final-porous model, 181
  finely porous model, 179, 189
  gel permeation chromatography, 210
  hydrodynamic radius, 212, 213
  light scattering measurements, 210
  mass transfer coefficient, 180,
        181
  mole weight distribution, 211, 212
  pore flow theory, 448-449
  retention, 211, 212, 213, 214
    and permeate flux, 183, 184
    data, 185, 187, 188, 189
  separation characteristics, 179,
        209-214
  solute radius and molecular
        weight, 182
  structure, 121
  transport equations, 447-448
  transport mechanism, 448, 449
Ultrafiltration-thermization at farm
        level, 259
  equipment, 260

Uniporter, 405
Uni-univalent electrolyte system,
        79-80
Uphill transports, 58, 60-62

Vapor phase precipitation, 131
Vasopressin, electrolytic
        deionization, 307
Vitamin K-1
  flip-flop mechanism in bilayer
        membranes, 20
  in photochemically activated
        electron transport, 17-22
Voltage measurement cell, 275, 276
  ion exchange membrane, 275, 276
    effect of current density, 277
    effect of HCl concentration, 277

Water
  dissociation effect, 421-427
  electrical transport of, 301
  precipitation of microporous
        membranes, 128, 129
  structure, in S18 SAFT membrane,
        109
  ultrapure, produced by membrane
        distillation, 347
Whey
  electrodialytic production, 306
  treatment by ultrafiltration, 255

X-rays, energy dispersive analysis,
        320
Xylene
  influence of surfactant kind on
        selectivity coefficient, 479
  isometric, separation by membranes
        containing clathrate-forming
        metal complexes, 483-495

Yeast cell entrapment
  in polymeric hydrogels, see under
        Hydrogels
  in membrane reactors, see under
        Membrane reactors

Zero gap membrane
  and current density scale-up,
        281-298
  and SPE cell technologies, 281-298
  bubble effect, 286-288, 292, 294,
        295
  catalytic activity of electrodes,
        288-291
  cathode potential versus current
        density, 291
  cell, 283
    break down voltage, 289, 293-296
  electric field, 284-285
  systems, 282-283
  V/A characteristics, 292

Zero gap membrane (continued)
  V/A characteristics (continued)
    versus caustic strength, 288
    versus operation time, 289
  voltage gap, 292
Zetameter, 203
Zeta potential, 203, 204, 205
Zirconium bis(benzenephosphonate),
    467

Zirconium bis(carboxymethane
    phosphonate), 467, 468
Zirconium phosphate, 465, 467, 469
  alpha-layered, 467, 468, 470, 471
  pellicular, 465